Handbook of Sensory Physiology

Volume IX

Development of Sensory Systems

By

C. M. Bate V. McM. Carr P. P. C. Graziadei H. V. B. Hirsch
A. Hughes D. Ingle A. G. Leventhal G. A. Monti Graziadei
E. W. Rubel R. Saxod A. B. Scheibel M. E. Scheibel J. Silver

Edited by

Marcus Jacobson

With 231 Figures

Springer-Verlag Berlin Heidelberg New York 1978

Marcus Jacobson
Department of Anatomy, University of Utah, College of Medicine, Medical Center, Salt Lake City, Utah 84132, USA

ISBN-13:978-3-642-66882-1 e-ISBN-13:978-3-642-66880-7
DOI: 10.1007/978-3-642-66880-7

Library of Congress Cataloging in Publication Data Main entry under title: Development of sensory system. (Handbook of sensory physiology; v. 9) Bibliography: p. Includes indexes. 1. Senses and sensation. 2. Developmental neurology. I. Bate, Christopher Michael, 1943— II. Jacobson, Marcus, 1930— III. Series. [DNLM: 1. Sense organs—Growth and development. WL700 H236 v. 9] QP351.H34 vol. 9 [QP431] 591.1'82'08s [596'.01'8] 78—5975

Softcover reprint of the hardcover 1st edition 1978

2122/3120-543210

Preface

This preface is addressed to the reader who wishes to inquire into the prevailing concepts, hypotheses and theories about development of sensory systems and wants to know how they are exemplified in the following chapters. I believe that science is hypothesis and theory and that the growth and evolution of any branch of science can be measured by the degree to which its theories have been reified. By that standard, one must conclude that developmental neurobiologie is in its infancy. The rapid accumulation of observations which has occurred in this branch of science in the past century leads to progress only to the extent that the facts validate or falsify hypotheses. The following chapters show that we have a plethora of facts but a dearth of hypotheses.

Another index of the maturity of any branch of science is its level of historical self-awareness. Because the history of any branch of science is essentially the history of ideas and of the rise and fall of theories, the level of historical awareness is related to the extent to which reification of its hypothetical constructs has advanced. It is largely because few theories of development of sensory systems, or indeed, of developmental neurobiology, have progressed far in the process of reification that the history of developmental neurobiology remains unwritten. The subject of this volume is hardly mentioned in the many books devoted to the history of related disciplines. Histories of sensory perception or of experimental psychology (BORING, 1942, 1950; PASTORE, 1971) touch on development of sensory systems only briefly in relation to the conflict between nativism and empiricism. Histories of neurophysiology (BRAZIER, 1959; LIDDELL, 1960), or of neuroanatomy (CLARKE and O'MALLEY, 1972), impinge on the subject of this volume very indirectly. The lack of any historical treatise on the subject of development of sensory systems is not due to lack of facts. Although this branch of science is in its infancy, if the development of its theories is taken as the index of its maturity, nevertheless it is armed with many techniques with which data have rapidly been amassed. Perhaps this accumulation has occurred too rapidly to allow time for much of the data to be related to hypotheses.

Discovery has often not required new facts, only a new synthesis of old data. Copernicus, Harvey, Darwin, and Einstein have shown the crucial importance of breaking away from accepted interpretations and creating new interpretations of the existing data. Such a new synthesis remains to be achieved in the field of development of sensory systems. It seems very likely, however, that the facts are present, only the new synthesis is lacking. Progress in this field is more likely to come from rearranging the existing data and breaking out of traditional moulds of thinking than from further massive accumulation of facts. There is an instructive example in this field: Both Golgi and Ramón y Cajal were in possession of the same facts, but their interpretations were different; the microscope did not reveal to Golgi the concept of the neuron or of its functional polarization, but it did so to Cajal. It is instructive, in this regard, to look at drawings of the neuron published during the period of reification of the neuron theory, from 1880, shortly after the Golgi technique came into use, to 1906, when Golgi and Cajal shared the Nobel Prize. Neither the improvements in the microscope nor advances in histological techniques can account for the radical changes in depicting neurons and in showing how they are connected. One can say that the microscope did not reveal things until the theory made it possible to observe them. This dictum can be supported by many other examples. Even such obvious structures as the cerebral convolutions were depicted like coils of intestines, and accurate drawings appeared only near the end of the eighteenth century (CLARKE and DEWHURST, 1972). It seems that the unaided eye, before it was guided by the inductive construct of localization of function in the nervous system, specifically of cerebral localization of function, was blind to the pattern of cerebral convolutions that now appears self-evident. The significance of these examples is not diminished, but rather increased by other examples in which false hypothetical constructs led to false observations, such as the construct of preformation which led observers to see a homunculus in the spermatozoon. The point is that the same observations can be made to support many constructs.

The conditions in the field of development of sensory systems, in which there is now a wealth of facts but a dearth of hypothetical constructs, seems to be especially advantageous for those who want to erect new constructs on the existing facts. Another advantage is that there are already many such theoretical constructs concerning the relationship between sensory functions and the fully developed structure of the nervous system (BORING, 1942; PASTORE, 1971). However, there seems to have been a reluctance on the part of those who study development of sensory systems to borrow hypotheses and theories from those who investigate fully developed sensory systems. This is one reason why studies of development tend to lag behind studies of the fully developed system. To take a well-known example from the seventeenth century, DESCARTES was not inhibited from postulating a mechanism of sensory perception, replete with structure-function correlation, erroneous as we now know, but he did not stop to reflect on how such a system could have developed before birth (SOUQUES, 1938). His successors during the following two centuries gave little more attention to this question, which is the principal theme of this volume. It seems as if there is an irrational reluctance to investigate the ontogeny of any system whose final structure and function are largely unknown. Yet there are many cases where observations on the developing nervous system have been the means of understanding the fully developed structure (CLARKE and O'MALLEY, 1968).

The science of developmental neurobiology, of which the subject of this volume forms an important part, is in its infancy, and it has had to borrow concepts and theories from the older neurosciences and from developmental biology (JACOBSON, 1978, pp. 345). The principle ideas, concepts, hypotheses and theories that are disclosed here all stem directly from the older scientific disciplines. It is thus remarkable to find little interest in the relationship between development of the nervous system and its evolution. The subject is not even considered in a recent book on the evolution of the nervous system (SARNAT and NETSKY, 1974). Nevertheless, as more efforts are made to understand how ontogeny of the nervous system relates to its phylogeny, new support has been provided for the theory of recapitulation, with modifications to accommodate the theory to new facts (REMANE, 1960; SPATZ, 1966; JACOBSON, 1974, 1975, 1978, pp. 302–307). However, new facts are not sufficient, and may not be necessary to enlarge the theory of brain evolution. For example, no new facts were required to advance the hypothesis that anomalies and variations of the nervous system that are functionally neutral under normal conditions may become advantageous under different conditions (JACOBSON, 1974, 1975). The observed structural polymorphism, especially in the local circuit neurons (RAKIC, 1977), is well known, but before this hypothesis was advanced, such variations were not seen as a possible means of evolution of the nervous system.

It should be noted that this hypothesis as well as most others about evolution of the nervous system belong to a class of hypotheses that are separated from the phenomena by many intervening variables, which are not easily verified or falsified by experiments. By contrast, the majority of hypotheses that relate to the subject matter of this volume are of a second class in which the hypotheses are simply and directly related to the phenomena. They are easily tested by relating them to additional observations and are thus easy to verify or falsify directly by experiment.

Hypotheses of development of sensory systems that belong to the second class are the various hypotheses concerning the mechanisms of assembly of neuronal circuits. For example, there are hypotheses of the importance of spatio-temporal regularities during development, and hypotheses of neuronal interactions and neuronal recognition. How and why these cellular events and interactions occur leads to additional hypothetical constructs regarding biochemical and molecular mechanisms, and these in turn lead to general theoretical constructs regarding the relations between changing structures and changing functions during development. To understand these relations one has to go beyond the questions "what" to ask "how" and "why". One can answer the question "what" by identifying and describing the components of the developing nervous system and their changing relations in space and time during development. To answer the question "how", one has to understand those changes in terms of biochemical and molecular mechanisms. To answer the question "why", one has to elucidate the genetic mechanisms underlying those developmental programs and determine their evolutionary history.

The reader will be disappointed if he expects to find answers to all those questions in the following chapters, or even if he expects to find hypothetical constructs, freely used in the attempt to obtain answers. The hypotheses are frequently implied rather than brought directly to the readers attention. Experimental observations are more often related to the research tools and techniques by which the facts were obtained than to the hypotheses which initiated the experiments. There should be little dissent from the statement that scientific techniques and the facts that they produce have little significance unless placed in a more general theoretical context, yet the reader will note that the contributors to this volume differ greatly in the breadth of context in which they prefer to place their facts. They all seem to agree, however, on maintaining the smallest possible distance between their facts and their hypotheses. It will be left to the reader, then, to judge whether that strategy is justified by the results and to ask again with BORING (1953) "should theory be kept simple and near description..., or should we generate hypothetical constructs and intervening variables freely for use in conceptual structures, in the hope that they may in due course harden into the stable realities of science and thus promote more progress than timorous description alone would yield?"

Salt Lake City, Utah, November, 1977 MARCUS JACOBSON

References

Boring, E.G.: Sensation and perception in the history of experimental psychology. New York: Appleton-Century-Crofts 1942.

Boring, E.G.: A history of experimental psychology, 2nd ed. New York: Appleton-Century-Crofts 1950.

Boring, E.G.: The role of theory in experimental psychology. Amer. J. Psychol. **66**, 169–184 (1953).

Brazier, M.A.B.: The historical development of neurophysiology, pp. 1–58. In: Handbook of physiology, I. Washington, D.C.: American Physiological Society 1959.

Clarke, E., Dewhurst, K.: An illustrated history of Brain function. Berkeley: University California Press 1972.

Clarke, E., Dewhurst, K., O'Malley, C.D.: The human Brain and spinal cord. A historical study illustrated by writings from antiquity to the twentieth century. Berkeley: University California Press 1968.

Jacobson, M.: A plenitude of neurons. In: Studies on the development of behavior and the nervous system, Vol. 2, pp. 151–166, ed. G. Gottlieb. New York: Academic Press 1974.

Jacobson, M.: Development and evolution of type II neurons: conjectures a century after Golgi. In: Golgi centennial symposium: perspectives in neurology, pp. 147–151, ed. M. Santini. New York: Raven Press 1975.

Jacobson, M.: Developmental neurobiology, 2nd ed. New York: Plenum 1978.

Liddell, E.G.T.: The discovery of reflexes. Oxford: Clarendon Press 1960.

Pastore, N.: Selective history of theories of visual perception 1650–1950. London: Oxford University Press 1971.

Rakic, P.: Local circuit neurons. Cambridge, Mass.: M.I.T. Press 1976.

Remane, A.: Die Beziehung zwischen Phylogenie und Ontogenie. Zool. Anz. **164**, 306–337 (1960).

Sarnat, H.B., Netsky, M.C.: Evolution of the nervous system. New York: Oxford University Press 1974.

Souques, A.: Descartes et l'anatomo-physiologie du système nerveux. Rev. Neurol. **70**, 221–245 (1938).

Spatz, H.: Gehirnentwicklung (Introversion-Promination) und Endocranialausguß. In: Evolution of the forebrain, pp. 136–152, eds. R. Hassler, H. Stephan. Stuttgart: Georg Thieme Verlag 1966.

Contents

List of Contributors

BATE, C. M., Department of Neurobiology, Research School of Biological Sciences, The Australian National University, Canberra City, A.C.T., Australia

CARR, V. McM., Department of Biological Sciences, Northwestern University, Evanston, Illinois 60201, USA

GRAZIADEI, P. P. C., Department of Biological Sciences, Florida State University, Tallahassee, Florida 32306, USA

HIRSCH, H.V.B., Center for Neurobiology, The University at Albany, Albany, New York 12222, USA

HUGHES, A.F.*, Department of Anatomy, Case-Western Reserve University, Cleveland, Ohio 44106, USA

INGLE, D., Brandeis University, Waltham, Massachusetts 02154, USA

LEVENTHAL, A.G., Center for Neurobiology, The University at Albany, Albany, New York 12222, USA

MONTI GRAZIADEI, G.A., Department of Biological Sciences, Florida State University, Tallahassee, Florida 32306, USA

RUBEL, E.W, Department of Otolaryngology, University of Virginia, Medical Center, Box 430, Charlottesville, Virginia 22901, USA

SAXOD, R., Laboratoire de Zoologie et Biologie Animale Université Scientifique et Médicale, BP 53-38041 Grenoble, France

SCHEIBEL, A.B., Department of Anatomy, University of California, School of Medicine, Los Angeles, California 90024, USA

SCHEIBEL, M. E.*, Department of Anatomy, University of California, School of Medicine, Los Angeles, California 90024, USA

SILVER, J., Department of Neuroscience, The Children's Hospital Medical Center, 300 Longwood Avenue, Boston, Massachusetts 02115, USA

* deceased

Chapter 1

Development of Sensory Systems in Arthropods

C. M. BATE, Canberra City (Australia)

With 16 Figures

Contents

A. Introduction

The arthropods are equipped with a wide variety of receptors, each of which is produced by modified epidermal cells. Groups of these cells cooperate in the construction of a specialized sensory structure such as a lens, a bristle, or a strand, which is linked with the dendrites of a peripherally placed sensory neuron. The first receptors to differentiate appear during embryogenesis, new ones are added in subsequent larval life and in the imaginal discs of the holometabolous insects, and at each moult the cuticular parts of the existing receptors are replaced. Because the sense organs develop quite separately from the central nervous system, the type of receptors, their number, and their distribution are determined by mechanisms that operate at the surface and not by the central connections of the sensory neurons. On the contrary, the connections made by the ingrowing axon of an arthropod sense cell are dictated by the kind of receptor that it innervates and where it lies at the surface, so that new input elements are continually being added to the nervous system by an independent set of morphogenetic mechanisms.

A simple theory for the stereotyped pattern of connections that develops between the sensory cells and their target interneurons suggests that each neuron is assigned a tag by its developmental history and that junctions are formed selectively between the axons of cells with congruent labels. In this view, the assembly of the sensory system is a process of sorting out, which is explained if the mechanics of labelling and recognition are understood. Normal development is a highly ordered process, however, which impresses a characteristic organization on the completed system by regular sequences and patterns of cell proliferation and differentiation. The most telling feature of the way the sensory system develops in arthropods is that it is progressive. At each level the framework of later innervation is laid down in preceding stages, so that the axons of differentiating sense cells take part in a refined game of follow-my-leader, in which they conform to existing routes and provide a pathway for the next generation of nerves. Although each cell may have a type of connection written into its phenotype, where it forms this connection, in what orientation and in what position relative to neighbouring axons depends on the organized environment in which it grows and to which it contributes.

This review is a summary of what is known of this process. It deals in turn with the development of spatially organized systems of receptors and diverse populations of peripheral neurons, the coordinate differentiation of sensory and second-order neurons, the growth of peripheral nerves, and lastly with the key question of how the identity to which each neuron is assigned is translated into characteristic connections with appropriate target cells.

The material is treated as a sequence of separate but interdependent growth processes, different kinds of receptors being grouped together despite the necessity of crossing taxonomic boundaries for this end. The results are sometimes uncomfortable, but for the moment this should be taken as a sign of relative ignorance rather than of a fundamental difference between the construction of a crustacean ommatidium and an insect bristle. We know most about these processes in insects, so with some exceptions the review is concerned with the development of the insect sensory system. No attempt has been made to cover the differentiation of receptive structures except insofar as this relates to the differentiation of the sensory neurons themselves.

The growth of the arthropod nervous system is constrained by the normal development of the whole animal, in particular by the division of the life cycle into a series of instars that are interrupted by the synthesis of a new cuticle and the shedding of the old, and in the holometabolous insects by the sequestering of prospective adult cells in undifferentiated groups known as imaginal discs. Readers who are unfamiliar with the vocabulary and details of insect development should consult the reviews by WIGGLESWORTH (1972) and COUNCE and WADDINGTON (1973). Several aspects of the development of arthropod nervous systems have previously been reviewed (EDWARDS, 1969; NÜESCH, 1968; YOUNG, 1973), including a comprehensive description and thoughtful analysis of the development of the compound eye and optic lobe in insects (MEINERTZHAGEN, 1973). The structure of arthropod sense organs and nervous systems in general has been fully described in a valuable review by HORRIDGE (1965).

B. The Peripheral Origin of the Sensory Neurons

Early descriptions of the construction of arthropod sense organs and the development of their connections with the central nervous system were muddled by confusion over the nature of the sensory cells. It was variously stated that the neurons were primary units with axons that grew inwards to the central nervous system (VOM RATH, 1888; HILTON, 1902), secondary units innervated by the centrifugal axons of central neurons (VOGEL, 1923) or glandular cells (BERLESE, 1909). As to their origins, they were either taken to be modified epidermal cells (VOM RATH, 1888) derived from the division of pre-existing sense cells (SOROKINA AGAFONOWA, 1924), or produced by the division of peripherally placed neuroblasts which had migrated from the central nervous system (FRANZL, 1941). Although the histological methods of workers such as VOGEL (1911) and ZAWARZIN (1912) showed that the sensory neuron was united by a fibre with the central nervous system, they did not distinguish the origin of this process as an outgrowth of the sense cell. The extensive Golgi studies by SANCHEZ (1919a and b) on the inward growth of axons in the eye of *Pieris* seem to have been ignored in this respect, so that at the time of Snodgrass's review (1926), in which he summarizes the inconsistencies of the alternative theories, the best evidence that the developing sensory neurons send their axons inwards to the central nervous system was the negative observation that the cell bodies of the sensory neurons could not be found in the central ganglia. For this reason much of the early work is coloured by the view that as in the vertebrates the sensory axons grow outwards, and the differentiation of the receptors depends on this innervation.

This uncertainty was decisively resolved by developmental observations which showed: a) that the sensory neuron is one of a group of cells underwriting the construction of a sensillum, derived from a common epidermal mother cell, b) that the first sign of the differentiation of the sensory neuron is the outgrowth of a centrally directed process (HENKE and RÖNSCH, 1951; KRUMINS, 1952: WIGGLESWORTH, 1953), c) that sensory axons grow inwards from the newly differentiated limbs of lepidopteran pupae from which the central ganglia have been removed (NÜESCH, 1952).

To ZAWARZIN (1912) we owe the division of the peripheral sense cells into type-1 and type-2 neurons. The type-1 cells are bipolar and unequivocally derived from the epidermis. The origins of the multipolar type-2 cells that ramify beneath the epidermis are unknown. Recently the idea that sensory cells are derived from the central nervous system (HAFFER, 1921) has been revived by RICE (1974) who suggests that the developing type-2 cells migrate outwards from the central ganglia. In crustacea the neurons of decapod myochordotonal organs (ALEXANDROWICZ and WHITEAR, 1957), nerve cord stretch receptors (HUGHES and WIERSMA, 1960; KENNEDY et al., 1966) and multipolar cutaneous mechanoreceptors (PABST and KENNEDY, 1967) are found in the central nervous system. Further analyses will probably turn up more of these central sensory neurons, but the overwhelming majority have a peripheral origin, so that the pattern of sensory innervation is determined by processes that regulate the differentiation of epidermal cells to form sensilla.

C. The Distribution of Receptors

The epidermal receptors form the elements of simple spatial arrays that are the object of intensive study as a general model for the formation of patterns (LAWRENCE, 1973a; WOLPERT, 1971). For this reason the characteristics of the process that determines the distribution of receptors are well known. The arrangement may be precise (e.g. *Drosophila*, STERN, 1954), or relatively so (crayfish, LETOURNEAU, 1976a), but many receptors are dispersed at a uniform density rather than in exact positions, and others are concentrated in well-defined groups such as the hair plates at joints and hinges in the exoskeleton. The mechanisms that produce these patterns do so by organizing the spatial distribution of prospective sensilla—regulating their density and their position and defining the boundaries of fields within which cells are competent to form structures such as ommatidia.

I. Dispersed Receptors

Receptor density is regulated by a homeostatic process which commits additional cells to form receptors in the spaces between existing sensilla as the number of intervening cells increases (WIGGLESWORTH, 1940). Experiments that disturb the normal sequence of cell proliferation and moulting show (in the bug *Rhodnius*) that the density-dependent signal that tells epidermal cells to form mechanoreceptive bristles precedes cell proliferation in any instar, so that an increase in bristle number occurs in the instar that follows an increase in cell number. Possible the factor that regulates the density of bristles is a substance produced by the epidermal cells, which commits them at some threshold value to divide a form a receptor cell group. Existing bristles take up the substance so that maxima occur in the intervening spaces and when a new receptor is formed it inhibits further determinations in its immediate surroundings. Alternatively a cooperative interaction may be necessary between a number of epidermal cells to determine the formation of a bristle at the centre of the group (LAWRENCE, 1970).

The density of any kind of sense organ varies in different parts of the same animal, at different times in the life cycle, and between the sexes. This characteristic pattern depends on regional variations in a heritable property of the epidermal cells. Cells that are induced to divide and migrate by wounding (WIGGLESWORTH, 1940) express their own receptor density and not that of the region to which they are displaced. The autonomous expression of this density at the margins of transplanted patches of cuticle suggests that the inherited variable is sensitivity to a uniform concentration of bristle substance, which is produced by all epidermal cells (LAWRENCE, 1969, 1973a). In this way relatively unspecific signals can produce precise changes in the pattern of receptors by triggering a change in the threshold sensitivity of limited groups of epidermal cells. In the cockroach *Leucophaea* the density of the thin-walled olfactory sensilla on the distal segments of the antenna increases by 70% at the imaginal moult (SCHAFER, 1973). The specific increase in the density of these specialist olfactory receptors coincides with the onset of sensitivity to pheromones produced by the opposite sex (ROTH and BARTH, 1967) and in *Periplaneta*, where only the female produces a pheromone, the increase in the olfactory sensilla is confined to the adult male (SCHAFER and SANCHEZ, 1973). In both cases the change in the density determining system is triggered by the fall in the level of juvenile hormone at the final moult. The normal increase fails in supernumerary instars and when hormone is applied to the larval antenna, but occurs precociously in premature adults induced in early instars by removing the source of the juvenile hormone, the *corpora allata* (SCHAFER and SANCHEZ, 1974).

II. Receptor Patterns

Where receptors occur precisely arranged in patterns or fields they indicate that information generated in the epidermis commits cells to alternative developmental fates according to their position. In the cricket *Teleogryllus*, for example, the ability to form tympanal cuticle at the adult moult is restricted to a few cells at the proximal end of the prothoracic tibia (BALL, unpubl.). Transplantations show that commitments of this kind are progressive and that the prospective fate of the cells in any particular position becomes more and more restricted as development proceeds. Nonetheless, the information that endows a field of cells with positional cues remains, so that if competent cells are present an appropriate structure will be formed. If the anlage of the eye in *Galleria* is removed even during a late larval instar, the wound tissue will still produce a normal eye at pupation if it receives an adequate stimulus to differentiate (WACHMANN, 1965).

A precise pattern of receptors could be produced by an interaction between competent cells (competence here implying having the appropriate genotype for the pattern concerned) and a *prepattern* peculiar to the structure involved (STERN, 1968). Experimental evidence suggests, however, that information generated in fields of epidermal cells is not specific to any particular structure but consists of a set of general positional cues to which cells respond according to their developmental history. In such a system receptor patterns are produced in a field of cells by the interaction of competent cells (peculiar to the field) with *positional information* (common to all fields) (WOLPERT, 1969). Cells assess their position and then differentiate according to their previous history of determination (which commits them progressively to form leg cells, for example, then prothoracic leg cells, and then prothoracic tibial cells). Mosaics of homoeotic mutants are an informative test of this theory, since they switch patches of tissue between appendages, so assaying each appendage for the generality of the cues that allow cells to assess their location within it. In *Drosophila* homoeotic prothoracic leg

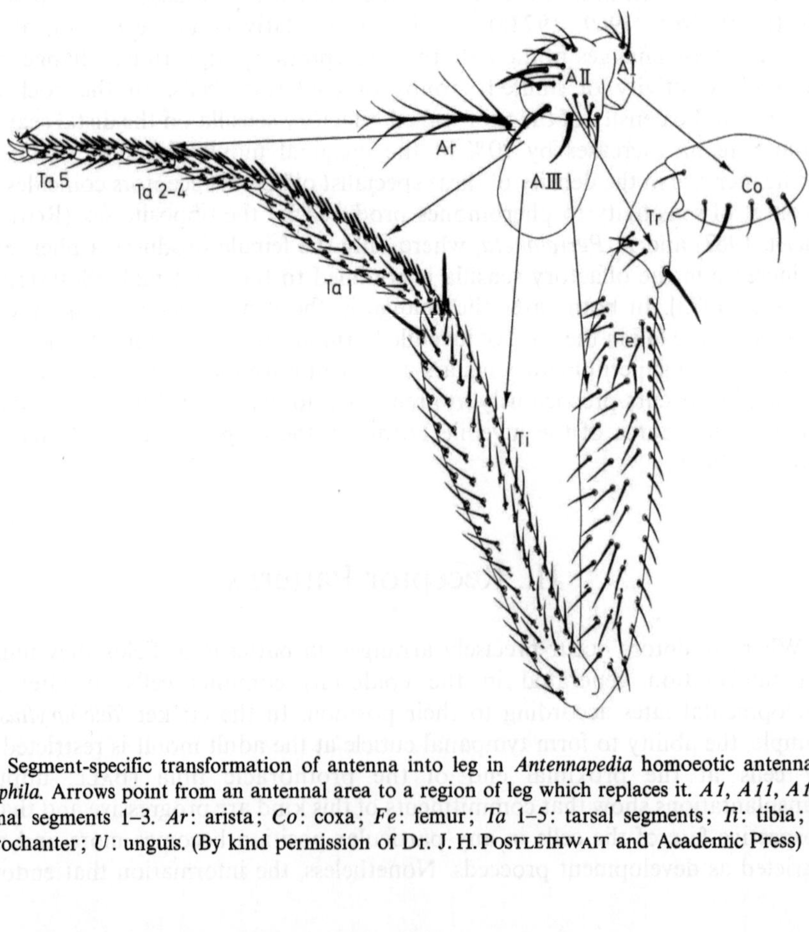

Fig. 1. Segment-specific transformation of antenna into leg in *Antennapedia* homoeotic antenna of *Drosophila*. Arrows point from an antennal area to a region of leg which replaces it. *A1*, *A11*, *A111*: antennal segments 1–3. *Ar*: arista; *Co*: coxa; *Fe*: femur; *Ta* 1–5: tarsal segments; *Ti*: tibia; *Tr*: trochanter; *U*: unguis. (By kind permission of Dr. J. H. POSTLETHWAIT and Academic Press)

cells express patterns autonomously in fields of mesothoracic and metathoracic leg cells (TOKUNAGA and STERN, 1965). The same is true for mosaics of aristapedia (leg converted to arista ROBERTS, 1964), in mosaics of wings and halteres (LEWIS, 1963) and in leg antenna mosaics; GLOOR and KOBEL, 1966; POSTLETHWAIT and SCHNEIDERMAN, 1971). The common significant part of all these observations is that cells have access to the same positional cues as to their proximodistal and mediolateral location no matter what appendage they happen to be in. This information not only assigns them to the appropriate segment (Fig. 1) but locates identifiable patterns of receptors such as campaniform and trichoid sensilla in their proper position despite the fact that the surrounding cells are of another kind of appendage (POSTLETHWAIT and SCHNEIDERMAN, 1971).

In one axis of the epidermis the origin of these cues is in a process that corresponds with a gradient of some substance between the anterior and posterior margins of each segment (LAWRENCE et al., 1972). Experimental interference with the segmental fields shows that cells that make polarized structures such as sensilla refer to the axial gradient during alignment (LAWRENCE, 1966a; LOCKE, 1959; PIEPHO, 1955) and that regionally differentiated cuticles are synthesized in response to the positional information provided by the scalar value of the gradient (MARCUS, 1962; STUMPF, 1968; LAWRENCE, 1973a). Transplants of cells between segments show, like the mosaics of homeotic mutants, that the same cues are repeated in each segment and that cells respond to these cues according to their origins. In this context these cues are one coordinate of a system that generates precise patterns of receptors and marks out the limits of fields.

III. Patterns of Receptor Neurons

In general patterns develop in the insect epidermis by the progressive specialisation of each cell's response to unspecific signals such as bristle-forming substances and positional information. The receptor neurons too are a diverse population of cells but the extent and stability of their individuality are not known. Each sensory neuron is assigned by pattern-forming mechanisms and lineage (see below) to a particular modality, and it is an important question whether each at the same time acquires properties that are peculiar to its position. Certainly during normal growth receptors of the same modality form different connections according to their position at the surface (see Section G.V.2). Whether the sensory neurons are subdivided into separate stable classes by position is open to the same simple experimental analysis by transplant and by the creation of mosaics of homoeotic mutants that has been used to test the individuality of the epidermal cells and the generality of the cues to which they respond. A start has been made in this direction with experiments on homoeotically transformed antennae in *Drosophila* (DEAK 1976; STOCKER et al., 1976). In amphibia retinal ganglion cells acquire locus-specific properties according to their position in a field of cells whose axes are defined during embryogenesis (HUNT and JACOBSON, 1972). Exactly when each cell acquires the properties it later manifests in connections with tectal neurons is not known. In insects the boundaries of segmental fields are also established in embryogenesis (LAWRENCE, 1973b). The cells in each field refer to their relative position during a limited part of the cell cycle (CAVENEY, 1973). Presumably if the sensory neurons acquire locus-specific properties in this field, they do so during the cycle of differentiative divisions by which they are produced from cells of the epidermis.

D. The Development of Receptor Cell Groups

I. Mechanoreceptive, Chemoreceptive Bristles, Innervated Scales and Campaniform Organs

Each sensillum is derived from an enlarged mother cell by a short sequence of oriented mitoses. This is true in general for crustacea (DEBAISIEUX, 1949) and among the insects the pattern of divisions is known for several receptor types in different species (Fig. 2). Uninnervated and innervated organules are produced by similar sequences, with the difference that in the uninnervated groups the prospective sensory cell degenerates (e.g. STOSSBERG, 1938).

In *Limnophilus* (HENKE and RÖNSCH, 1951) and *Galleria* (KRUMINS, 1952) the mother cell of a mechanoreceptive bristle divides perpendicular to the surface of the epidermis. The outer daughter divides again obliquely to produce a bristle-forming cell and a socket-forming cell. In uninnervated bristles the inner daughter of the first division degenerates, but in the sensilla it either differentiates as a sense cell or divides again to produce the several neurons of multiply innervated receptors. RÖNSCH (1954) subsequently modified the sequence in *Limnophilus* to include a preliminary oblique division producing the mother cell and an accessory cell, and suggested that campaniform organs were produced in the same way. Possibly an unanalyzed sequence of vertical and oblique divisions produces scales, bristles, and campaniform organs in *Ephestia* to the same pattern (CLEVER, 1958) but the sequence of divisions is not identical in other species. The suggestion that the neurons of multiply innervated receptors are derived from further divisions of a sensory mother cell has been confirmed in *Calliphora* (PETERS, 1965, and *Oncopeltus* (LAWRENCE, 1966b), but the exact lineage of the neurons in these more complex structures remains in doubt. In *Calliphora* the marginal chemosensory bristles of the adult labellum consist of seven cells—two bristle-forming cells and five sensory neurons, of which one is mechanoreceptive and the rest chemoreceptive (SHIRAISHI and TANABE, 1974). The enlarged mother cell from which all seven are derived divides parallel to the surface, to produce two adjacent daughter cells, one of which divides again obliquely to produce bristle and socket cells, the other being the mother cell of the sensory neurons. It is likely that the lineage of the mechanoreceptive and chemoreceptive cells is separate from an early stage. A similar division sequence has been described for the uninnervated hairs, bristles, and chemoreceptors in *Oncopeltus* (LAWRENCE, 1966b) with, as in

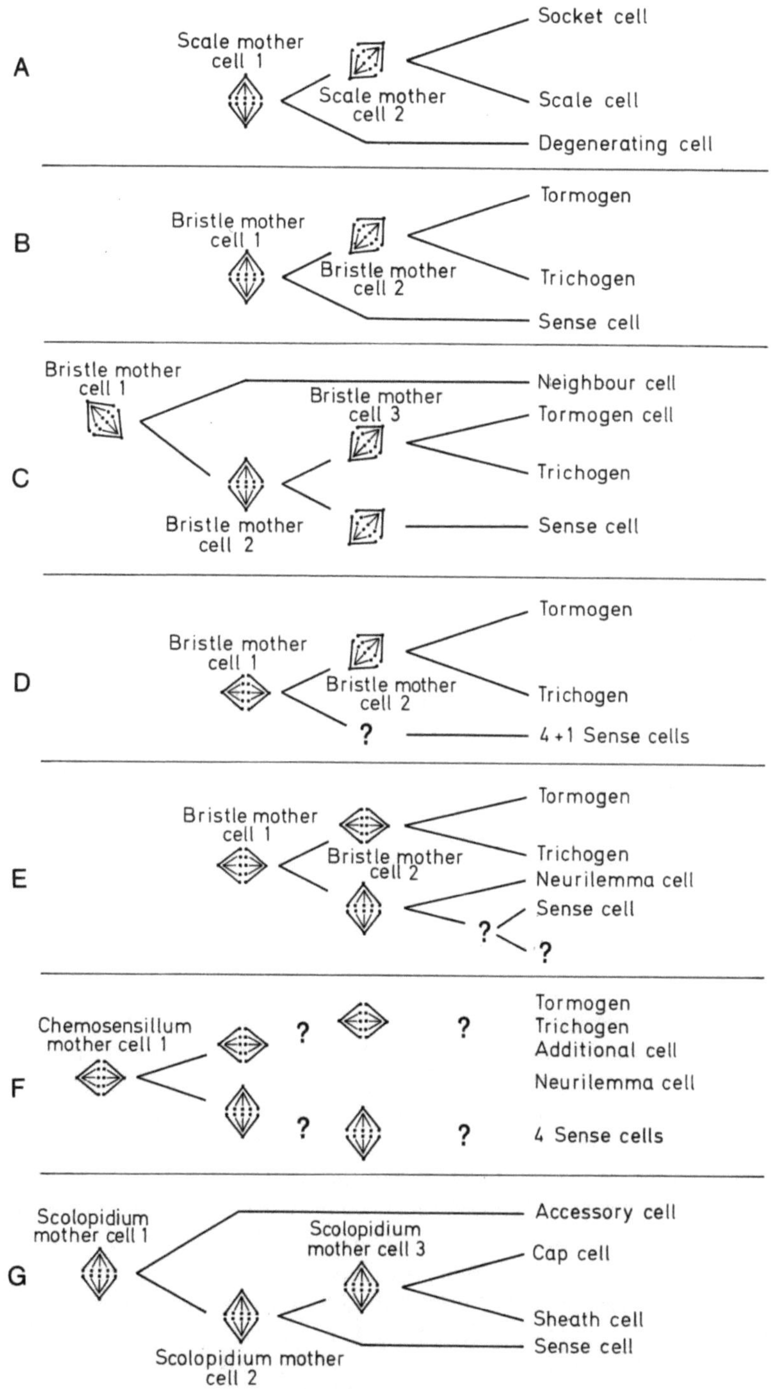

Fig. 2. Division sequences that generate organules and sensilla in insects. A: Uninnervated scales, *Ephestia* (STOSSBERG, 1938); B: Mechanoreceptive bristles *Galleria* (KRUMINS, 1952); C: Bristles, scales, campaniform organs, *Limnophilus* (RÖNSCH, 1954); D: Chemosensilla, *Calliphora* (PETERS, 1965); E: Mechanoreceptive bristles, *Oncopeltus* (LAWRENCE, 1966b); F: Chemosensilla, *Oncopeltus* (LAWRENCE, 1966b); G: Chordotonal organs, *Carausius* (JÄGERS-RÖHR, 1968)

Calliphora, the distinction that the first division of the sequence is parallel to the surface of the epidermis. Taken together, these observations show that although many of the sensilla and organules in the epidermis are generated by a similar pattern of cell divisions, involving an early separation of receptor and sense cells, there is no necessary sequence to which all conform.

The fate of the cells within the group is not entirely determined by their lineage. Position within the group may also be important, so that the orientation of successive mitoses is a necessary prelude to an appropriate arrangement and an aberrant receptor may be formed if the normal pattern is disturbed (LEES and WADDINGTON, 1942). Many authors report a typical orientation of the mitoses with respect to the body axes, so the mother cell has an inherent polarity before division begins and the alignment of the group improves during subsequent development (LAWRENCE, 1966a). This polarity is significant for example in determining the curvature of bristles and the eccentric positioning of the sense-cell dendrite, which gives mechanoreceptors directional sensitivity.

II. Chordotonal Organs

Nothing is known of the development of the various mechanoreceptive strands in arthropods except for the chordotonal organs, which function as proprioceptors in the body wall and at joints and as vibration receptors, often in association with a specialized cuticular *tympanum*. Each chordotonal organ is a compound structure made up of separate *scolopidia* consisting of six cells radiating from an attachment point. The anlagen of the chordotonal organs appear in embryogenesis and in developing imaginal discs (KIM, 1961; VAN RUITEN and SPREIJ, 1974). Earlier descriptions outline the proliferation of the cell masses that form future chordotonal organs (SCHÖN, 1911) and the cell movements that produce regular arrays of oriented *scolopidia* such as Johnston's organ (CHILD, 1894). The linear sequence of cells in each *scolopidium* may be derived from a common mother cell by an orderly sequence of oriented mitoses (Figs. 2) (JÄGERS-RÖHR, 1968; LUKOSCHUS, 1962), although simple observation of division pattern is not proof of a clonal origin for the units of compound sense organs. In *Carausius* (JÄGERS ROHR, 1968) the subgenual organ first appears in the 35-day embryo as a small invagination of the upper surface of the tibia. The cells in this early anlage divide at right angles to the surface, generating two rows of daughter cells, which are pushed into the interior of the developing leg by the encroachment of neighbouring epidermal cells that divide parallel to the surface. The outer daughter remains in contact with the epidermis, where it forms an attachment cell. The inner row of daughter cells divides again at right angles to the axis of the row, producing inner sensory cells and a medial row of cells which will divide again. The anlage now consists of three rows of cells, of which the middle row divides to provide an inner sheath cell and an outer scolopale cell. No further divisions occur during embryogenesis, though it is not clear whether additional units are added to the sense organ during postembryonic life as in *Teleogryllus* (BALL and YOUNG, 1974). In both the subgenual organ and the chordotonal organ the orientation of the anlage depends on the divisions of surrounding epidermal cells, which push the developing *scolopidia* into the interior of the developing leg and in the case of the femoral chordotonal organ construct a chitinous tendon to which the scolopidia become attached at their distal end. In *Melanoplus* the femoral chordotonal organs are constructed in a similar way (SLIFER, 1935). During subsequent development the chordotonal organs elongate passively as their distal (chitinous tendon) and proximal (body wall) attachments become further separated by the growth of the leg. In *Melanoplus* and other Orthoptera (GRABER, 1876; SCHWABE, 1906) the sensilla of the tympanal organ are also produced during embryogenesis and may be complete at hatching. Whether or not the organ is actually fully differentiated can only be determined by a count of the *scolopodia*. In *Teleogryllus* the sensilla of the tympanal organ accumulate progressively from the third instar onwards from an undifferentiated mass of cells present in the first larval instar (YOUNG and BALL, 1974a; BALL and YOUNG, 1974). The same is true of the equivalent tracheal organ of the mesothoracic leg (YOUNG and BALL, 1974b) and the subgenual organs, which are incompletely differentiated in the first instar and continue to develop new scolopidia until the fourth larval instar.

III. Compound Eyes

The compound eye develops from a field of cells in the epithelium of the head, which is crossed by successive waves of cellular proliferation, assembly, and differentiation (Fig. 3); these produce the twenty or so cells of each ommatidial cluster (reviewed by BODENSTEIN, 1953; MEINERTZHAGEN, 1973). The timing of this sequence varies in different groups. In crustacea (MOROFF, 1912; PEABODY, 1939; PARKER, 1891) and many hemimetabolous insects the process is partially completed in embryogenesis, and the larva is equipped with a functional compound eye from which an adult eye develops by the enlargement of existing facets and the incorporation of additional ommatidia at the margins during postembryonic life. In *Schistocerca* for example, about 1200 new ommatidia are added at the advancing margin of the eye during each five day intermoult period (SHELTON, 1976). In some hemimetabola such as dragonflies (LEW, 1933; LERUM, 1968; MOUZE, 1972b), the development of the adult eye involves a reorganization of the larval visual system with the production of specifically adult ommatidia from restricted regions of the eye field. In holometabolous insects the adult eye develops from a specialized imaginal disc, which differentiates gradually during larval life (Diptera, Hymenoptera, some Coleoptera) or suddenly with the onset of metamorphosis (Coleoptera, Lepidoptera). The discs too have various origins: in the higher Diptera they are formed from sac-like protrusions of the embryonic pharyngeal epithelium, which include the prospective antennal field, while in Coleoptera and Lepidoptera they develop as thickenings of the head epithelium, which lie in front of the larval visual organs. Despite the different origins of the retinal field and the various times at which it differentiates in the life cycle, the formation of ommatidia from the cells of the field is remarkably constant in different groups.

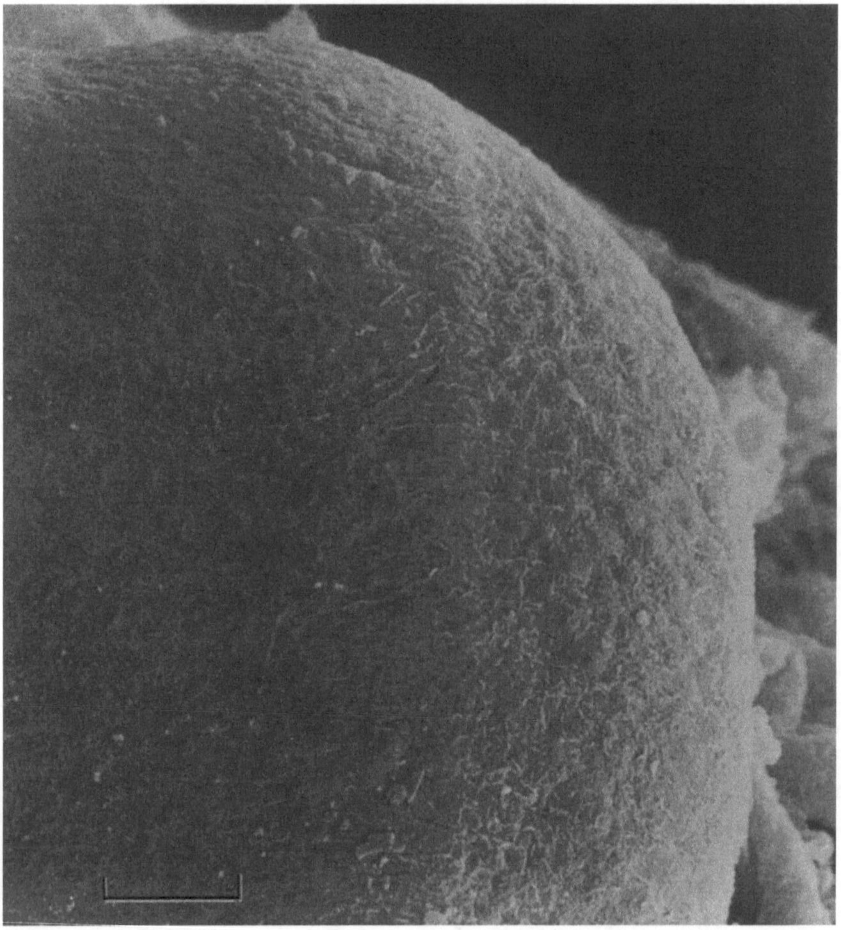

Fig. 3. Differentiation in embryonic eye field of *Locusta*. Wave of proliferation and assembly crosses field from posterior (right) to anterior (left). Proto-ommatidial clusters form as a front, which advances across the field. (Scanning electron micrograph of dorsal part of right eye at 144 hrs of embryonic life at 30° C). Scale: 30 μm

1. Proliferation in the Eye Field

During development the eye field is crossed in a postero-anterior direction by a wave of proliferation, consisting of two bands of mitoses, separated by a strip of non-dividing cells (figure 4) (*Ephestia*, EGELHAAF et al., 1975; *Drosophila*, CAMPOS ORTEGA and HOFBAUER, 1977; READY et al., 1976; embryonic *Locusta*, BATE, unpublished). Proliferation is initiated at the posterior margin of the field and generates cells which cluster together forming protoommatidia. The presence of a differentiation centre at the posterior margin was inferred from ablation experiments (SEIDEL, 1935; WOLSKY, 1956) and its properties were convincingly demonstrated in the mosquito *Aedes* (WHITE, 1961, 1963). In *Aedes* the eye

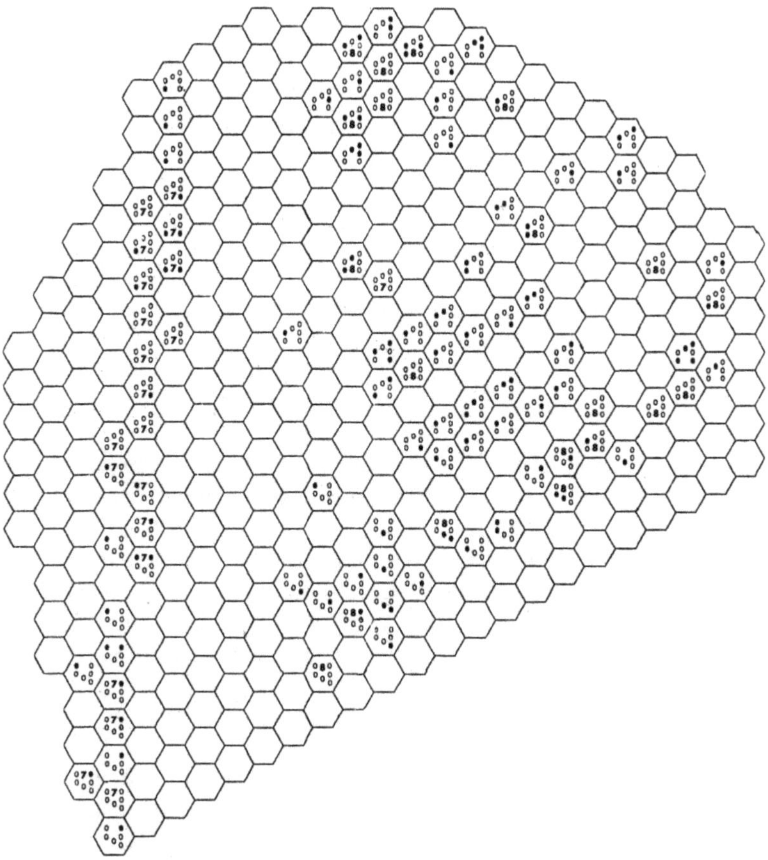

Figure 4. Distribution of labelled retinula cells in mature *Drosophila* eye reconstructed after injection of 3H thymidine in third larval instar. Two vertical waves of mitosis cross the eye field: label is incorporated in a broad anterior (right) zone and in a second, sharply defined posterior band. Serial injections show anterior movement of both waves at one ommatidial row/hr. Labelled retinula cells of all classes R1–8 are present in anterior zone; only equatorial group R1, R6, R7, are labelled in posterior band. Labelled R1–6: black, according to position; R7, 8, numbered. By kind permission of Drs. J. A. CAMPOS ORTEGA and A. HOFBAUER

develops externally from a field of epidermal cells anterior to the larval ocelli A wave of mitosis crosses the field during larval life, producing a thickened optic placode from the single layer of epidermal cells that it encounters. In summary, White's experiments show: a) by ablation that proliferation among the cells of the eye field depends on a property of adjacent placode cells, b) by rotating portions of the field, that the pattern of proliferation is dictated by the wave and is not inherent in the field, c) by transplantation, that the proliferation response is restricted to competent cells and is blocked by cells that lie outside the prospective field. Subsequent experiments on the eye field of *Galleria* (WACHMANN, 1965) have shown that the regenerative stimulus to proliferate is rather nonspecific. If the prospective eye field is completely removed during one of the last three larval instars, no imaginal eye is formed at pupation. However, if fragments of optic anlage are implanted into the wound tissue before pupation, a normal eye develops, indicating that an eye field has been reorganized in the head epidermis and will differentiate provided it receives an appropriate stimulus. A specific signal from the optic anlage is unnecessary, however—grafts of optic or wing anlage from different genera or orders, or fragments of the imaginal retina are competent (even after treatment with acetone) to induce differentiation of the eye. Presumably prospective eye tissue would stimulate differentiation in a field of wing cells. The signal is nonspecific, but the response of the cells is highly individual and reflects their organization as an eye field.

2. The Eye Field

WHITE's experiments show that the cells of the head epidermis in *Aedes* are organised into a prospective eye field whose limits are defined (by the competence of cells to respond to an inductive stimulus) by the fourth larval instar. Cells that normally lie outside this field cannot contribute to the developing eye when grafted into it. Earlier PFLUGFELDER (1936/37) had shown in *Pentatoma* that only cells of the prospective eye field were competent to form a new eye when the developing eye was removed. These results suggest that the form of the adult eye is fixed by a boundary between competent and noncompetent cells, but in other insects the evidence of the range of cells that can contribute to the eye is contradictory. In *Periplaneta* SHELTON et al. (1977) were unable to repeat experiments (HYDE, 1972) in which fragments of prothoracic epidermis were apparently incorporated into the eye and in *Oncopeltus* such grafts never contributed to the eye and frequently rounded up and were rejected, apparently manifesting sorting out properties specific to their normal compartment (GREEN and LAWRENCE, 1975). These results were confirmed by MOUZE (1975) in the dragonfly *Aesehna.* A series of grafts showed that only cells immediately adjacent to the normal eye field can contribute to it. Other cells, as in *Aedes* block the normal wave of proliferation. In a second series of experiments cells of the adult head epidermis, which manifestly had not contributed to the eye during development, were recruited to form ommatidia when grafted into the advancing margin of the eye of third-instar *Oncopeltus* larvae (GREEN and LAWRENCE, 1975). That cells outside the eye field can contribute to it is therefore not in question, but the evidence suggests that competence to form eye is normally restricted to a narrow band of cells adjacent to the advancing margin of the retina.

Wachmann's experiments with *Galleria* show that the eye field has considerable homeostatic properties and will manifest its reorganization from wound tissues at an appropriate position in the head, until removal of the eye anlage is delayed until the stage normally preceding proliferation and differentiation. In crickets too the information necessary to reorganize the eye field from wound tissue is present at least until the third larval instar (HELLER and EDWARDS, 1968). In *Drosophila* fragments of the imaginal eye disc show regulative properties until the middle of the last larval instar (VOGT, 1946) although the disc itself is divided into alternative fields during embryogenesis; the boundaries of these fields reflect the stable commitment of the cells within them to form eye, ocelli, or head cuticle. An analysis of the fractional contribution of mutant ommatidia to the adult eye after X-ray-induced crossing over in larvae heterozygous for defective pigments shows that the cells of the *Drosophila* eye are derived from a field consisting of about 20 cells at the end of the first larval instar (BECKER, 1957, 1966). As in other organs (BRYANT and SCHNEIDERMAN, 1969), a regular pattern of oriented mitoses is important in determining the form of the adult eye which develops from this field. The distribution of the larger clones produced by early irradiation shows that until the end of the first larval instar mitoses are oriented predominantly dorsoventrally with respect to the adult eye. At about the beginning of the second instar the preferred orientation of mitoses swings through 90° and each of the 20 cells gives rise to a strip of about 40 ommatidia radiating out from the posterior margin of the eye. Proliferation becomes restricted in the posterior part of the field but continues anteriorly until the end of the last larval instar. Continuing proliferation depends on the timing of metamorphosis—if it is delayed or brought forward then the number of ommatidia is correspondingly increased or reduced in the adult eye (BODENSTEIN, 1953; CAMPOS ORTEGA and GATEFF, 1976). In normal flies the number of facets varies and this variability is greatest at the anterior (the youngest) end of the eye (READY et al., 1976). In Lepidoptera, on the other hand, where cell division in the eye begins at pupation, precocious metamorphosis causes a smaller adult eye, with the normal number of ommatidia, which are reduced in size (BODENSTEIN, 1953; KOPEC, 1913).

3. Ommatidium Formation

In both insects and crustaceans (BODENSTEIN, 1953; DEBAISIEUX, 1944) proliferation in the eye field produces clusters of cells, each of which is the precursor of a differentiated ommatidium. In *Drosophila* the first band of mitosis to cross the eye field (figure 4) is followed by an advancing morphogenetic furrow (also in *Ephestia* EGELHAAF et al., 1975; *Phaenicia* MELAMED and TRUJILLO CENÓZ, 1975; embryonic *Locusta* BATE unpublished), a region where no division occurs and cells cluster to form protoommatidial groups (figure 5). The cells produced by the second mitotic wave are then added to these clusters, so completing the formation of the immature ommatidial units (READY et al., 1976). How is the fate of each newly recruited cell determined? The choice between alternative destinies might depend on the position which each cell comes to occupy within a cluster. The alternative view, that the fate of the cells is determined by their lineage, is implicit in much of the earlier literature. VIALLANES (1885) describes the derivation of the ommatidial cells from a common "cellule optogène". BERNARD (1937) interpreted sections through the proliferating eye region of *Formicina* as indicating a sequence of divisions in which the progeny of a single mother cell is divided into two groups, a distal one giving rise to the cells of the dioptric apparatus and a more proximal one producing the pigment and retinula cells. This unifying idea is attractive in theory as a simple lineage related to the division sequences that produce other epidermal receptors (HENKE, 1953; WEBER, 1966). More recently

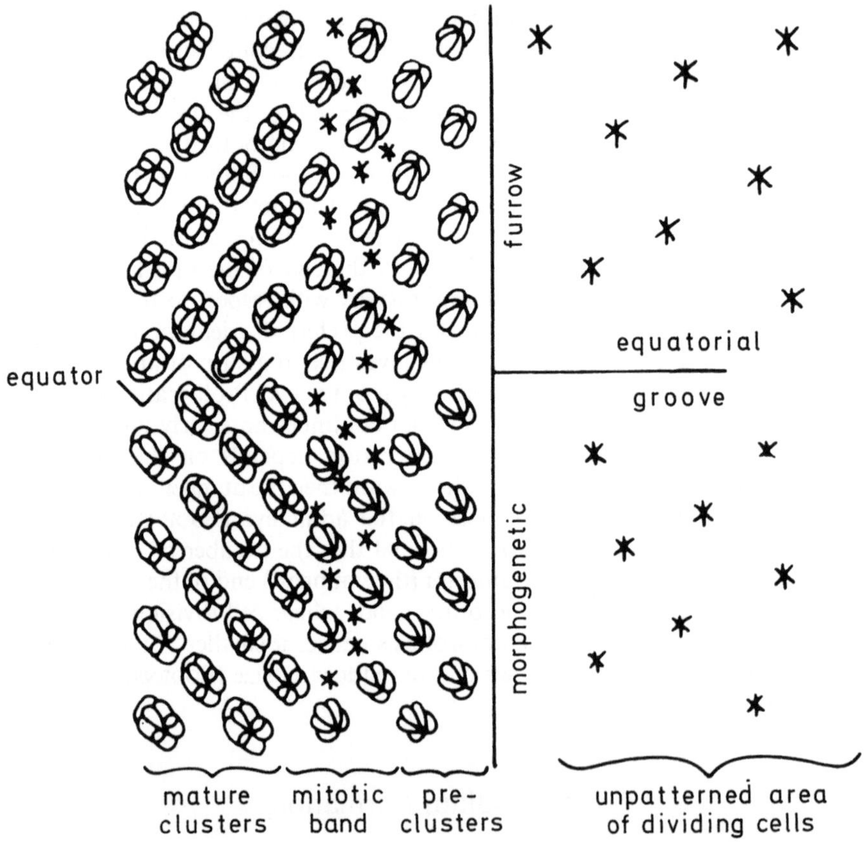

Figure 5. Diagram illustrating sequence of proliferation and assembly in eye imaginal disc of *Drosophila*. A morphogenetic furrow crosses disc from left (posterior) to right (anterior) preceded by equatorial groove and a first broad band of mitoses (asterisks). Protoommatidial clusters accumulate behind advancing furrow and a second narrower band of mitoses crosses the field. As disc develops, area of mature clusters expands anteriorly at the expense of the unpatterned remainder. By kind permission of Drs. D. F. READY, T. E. HANSON, Professor S. BENZER and Academic Press

clonal origin of ommatidial components has been tested experimentally by incorporating genetically marked cells (with a mutant pigment genotype) into the developing eye. In both the open rhabdom eyes of *Drosophila* (HANSON et al., 1972; HOFBAUER and CAMPOS ORTEGA, 1976; READY et al., 1976) and *Oncopeltus* (SHELTON and LAWRENCE, 1974), and the fused rhabdom eye of *Periplaneta* (SHELTON et al., 1977) the cells of single ommatidia at the borders of these marked patches are found to have different origins. The mixed ommatidia are confined to the borders of the mosaic patches, and this, together with the coordinate mixing of pigment and retinula cells, suggests that extensive cell migration is not an important part of the assembly process. Adjacent cells cooperate to form an ommatidium, but not necessarily with their immediate relatives. These experiments also reveal that the horizontal equator (see page 35) in the eye of *Drosophila* does not represent a boundary between two clones.

While these results show that the members of one ommatidial cluster are not necessarily the descendants of a single mother cell, they do not rule out the possibility that regular patterns of mitosis within the eye field relate cells of a particular class (for example pigment cells) by lineage. For this there is now some evidence. In *Drosophila* (CAMPOS ORTEGA and HOFBAUER, 1977) an analysis of mosaics induced in larval eye discs indicates that the lineage of pigment and receptor cells is separate from early in the last larval instar. The lineage of the receptor cells is further subdivided. Each precursor autonomously produces a quota of prospective retinula cells, probably eight (ignoring the deeper of the two central retinula cells, R 8, which was not included in the analysis). The genealogy of this group of cells, judged from an analysis of small marked clones induced late in the last larval instar, indicates that the destiny of retinula cells R 1–R 6 is interchangeable and depends on the position they attain within an ommatidial cluster, whereas the destiny of the second of the two central cells, R 7, is fixed. Each precursor therefore produces a variable number of cells R 1–R 6 and a single R 7, and the progeny of the same or different precursors accumulate in an orderly sequence to form an ommatidial cluster. In this view then, although the ommatidial unit is not a clone, the fate of its component cells is dictated wholly (R 7) or partly (R 1–R 6) by their lineage.

An outline scheme of the process of assembly (figure 5) is suggested by READY et al. (1976). Autoradiographic analyses of the two mitotic steps in the proliferative wave in *Drosophila* (CAMPOS ORTEGA and HOFBAUER, 1977; READY et al., 1976) show that retinula cells 2, 3, 4, 5, and 8 are produced exclusively in the first phase of cell division, and suggest that R 1, 6, and 7 are produced only in the second. Groups of cells including future R 2, R 3, R 4, R 5, and R 8 form 'pre-clusters' behind the first mitotic wave, and in the second, intervening cells between the clusters divide to contribute retinula cells 1, 6, 7, and other non receptor cells to complete the immature ommatidial unit.

The process of assembly however is not simply one in which proliferated cells are added to an advancing template. The orientation of the ommatidia depends on the polarity of the eye field *and* the polarity of the tissue from which cells are recruited (LAWRENCE and SHELTON, 1975).

IV. Simple Eyes

Simple eyes develop from localized groups of enlarged epidermal cells or from an invagination of the epidermis (principle eyes of spiders). The eyes develop at various times in the life cycle but for insects the sequence of differentiation is similar in larval stemmata (holometabola) and adult ocelli (hemimetabola and holometabola). The anlagen of the adult ocelli first appear in the embryo (Orthoptera: MALZACHER, 1968; MOBBS, 1976; LINK, 1909; Ephemeroptera: SEILER, 1905; *Rhodnius*: LASCHAT, 1944) or during larval life (Neuroptera, Lepidoptera: LINK, 1908; Hymenoptera: CAESAR, 1913; CORNELI, 1924; REDIKORZEW, 1900) and differentiate gradually to completion at the adult moult. Earlier descriptions of the origin of these eyes by epidermal invagination (VON REIZENSTEIN, 1905) are incorrect. Different authors confirm that the anlagen consist of small groups of flask-shaped cells, which are apparently derived by simple enlargement from the cells of the epidermis. Each group of cells becomes divided into distal and proximal layers, the outer cells forming the cornea, the inner cells the retina and the pigment. In *Ephestia* the anlage of the six larval stemmata appear in the early embryo (BUSSELMANN, 1935), each group of cells differentiating into three distal and four proximal retinulae. The cells of the dioptric apparatus appear later, migrating inwards from the surrounding epidermis.

The principle eyes (Hauptaugen) of spiders are unusual in that they develop from an invagination of the embryonic prosoma, which folds over to give an anlage consisting of three cell layers (reviewed by HOMANN, 1971). The outer layer forms the vitreous body, the median layer the retinal and pigment cells, and the inner layer a sheathing basal membrane. The more lateral accessory eyes (Nebenaugen), which have an underlying reflective layer of tapetal cells, develop from epidermal thickenings in which cells accumulate by mitosis. The differentiated accessory eye is a bilaterally symmetrical structure and this consistent polarity is expressed during development by oriented cell migrations within the anlage. The prospective retinal cells sink inwards and retain a medial position within the developing cell mass. Cells which migrate to the peripheral pole of the eye develop into the vitreous body, while other cells move towards the inner pole and sink in to produce the pigment cells and the tapetum.

E. The Development of Association Centres

I. Proliferation in the Central Nervous System

In hemimetabolous insects the composition of the central nervous system is largely fixed during embryogenesis (for reviews see EDWARDS, 1969; PIPA, 1973). The full complement of adult neurons (GYMER and EDWARDS, 1967; SBRENNA, 1971) is generated by a sequence of divisions in ganglionic sets of neuroblasts, which degenerate before emergence from the egg or shortly thereafter (BATE, 1976; PANOV, 1966). During postembryonic life ganglion cells are added only to the brain, where neuroblasts persist. With some exceptions (dragonfly optic lobe, for example; LERUM, 1968) the basic structure of the association centres is unchanged during larval life, so that the adult lobes, like the receptor layers that supply them, are the sum of their larval increments. In holometabolous insects, on the other hand, larval and adult association centres are separate (PANOV, 1960, 1961; NORDLANDER and EDWARDS, 1969b, 1970; for reviews see EDWARDS, 1969; MEINERTZHAGEN, 1973) and the brain and thoracic ganglia are partially reorganized during postembryonic life as a prelude to adult emergence. The sequence and extent of the changes depends on the complexity of the larval sense organs and the timing of adult receptor differentiation. Insects with well-developed larval olfactory and visual systems have complex pre-adult olfactory and optic lobes and the adult centres develop alongside, either continuously during postembryonic life or abruptly with the onset of metamorphosis.

In embryogenesis and postembryonic life the anlagen of the association centres are derived from groups of neuroblasts, which divide equally (producing more neuroblasts) or unequally (producing clusters of preganglion cells) (PANOV, 1957, 1960, 1961). The distribution and oriented divisions of these parent cells determine the characteristic disposition of the developing cell cortices and neuropiles. The pattern of growth is known best in the three visual neuropiles of the insect optic lobe (lamina, medulla, lobula), which are derived from the two outer lobes of the embryonic protocerebrum (HEYMONS, 1895; VIALLANES, 1891; WHEELER, 1893). In crustacea the outer visual neuropiles are produced from a common neurogene layer, which distributes cells to the retina, lamina and external medulla (ELOFFSON and DAHL, 1970). The sequence of asymmetric neuroblast divisions typically generates a chain of preganglion cells with a spectrum of ages, the oldest cells of the anlage being the furthest from the parent neuroblasts (NORDLANDER and EDWARDS, 1969b). Differentiation occurs in order according to age and in many cases to the arrival of centripetal axons from outer layers (MEINERTZHAGEN, 1973). Patterns of growth in which neuropile layers are generated more or less in synchrony, a horizontal sequence of differentiation in each producing a wave of vertical axons growing between them, are common.

The detailed ancestry of the central neurons is unknown, so that it is impossible to say what aspects of the fate of any cell are determined by its lineage, its birthday, its position or the innervation it receives, although the coordinate growth of adjacent neuropile and receptor layers suggests that centripetal fibres regulate the extent if not the type of differentiation.

II. Morphogenetic Dependence of Peripheral and Central Nervous Systems

In general, arthropod receptors develop independently of the central nervous system and the coordinate differentiation of receptor layers and their target association centres is ensured by a regulatory effect of incoming sensory axons on the differentiation of central ganglion cells (compare HUGHES and CARR, this volume).

KOPEC's early experiments on *Lymantria* (1922) showed that the eye differentiates autonomously when the optic lobe is destroyed or the eye anlage is transplanted to an abdominal segment. With few exceptions experiments of this kind have repeatedly shown that many kinds of receptors continue to differentiate normally when they are disconnected from the central nervous system (reviews: BODENSTEIN, 1953; EDWARDS, 1969; MEINERTZHAGEN, 1973). In crustacea the capacity to regenerate an eye depends on the level to which the underlying optic lobe is damaged (STEELE, 1907), presumably indicating the extent to which the common proliferation zone for the retina and outer visual neuropiles (ELOFFSON and DAHL, 1970) is removed by the operation. In *Carausius* PFLUGFELDER (1947) found an early embryonic dependence of the developing eye on contact with the optic lobe, but development subsequently becomes autonomous. Many contradictory results with transplanted eyes, or eyes severed from the brain, are probably caused be degeneration of the damaged retinula cells (DRESCHER, 1960; WOLSKY, 1938; WOLSKY and WOLSKY, 1971, 1973). Since adjacent undamaged retinula axons develop normally although their axons never reach the brain (MOUZE, 1972a, 1974) this retrograde degeneration cannot be taken as good evidence of a centrifugal morphogenetic influence from the optic lobe. SCHOELLER's suggestion (1964) that the differentiation of antenna in *Calliphora* depends on centripetal innervation is doubtful as the central origin of the fibres in the partially differentiated antenna was not demonstrated. In the more rigorously investigated *Manduca* antennal sensilla develop in the complete absence of the brain (SANES et al., 1976).

Naturally occurring variations in receptor number are usually correlated with a reduction of the appropriate association centre (HOLMGREN, 1909; DE LATTIN, 1939). Mutants of *Drosophila* with reduced or absent eyes, such as eyeless or bar, also show reduced optic lobes (KRAFKA, 1924; JOHANSSEN, 1924; EL SHATOURY, 1956). POWER (1943) quantified this relationship by means of genetically controlled reductions in facet number. His conclusion that hypoplasia in the outer visual neuropiles is directly related to the facet number has since been confirmed by HINKE (1961). Because the effects of relative eyelessness are exclusively ipsilateral and directly related to the failure in centripetal innervation, they suggest that the development of the visual neuropiles depends on an interaction between ganglion cells and incoming axons from more peripheral layers.

The evidence for a peripheral control of ganglion cell *number* is scanty although some experiments indicate that where a developing receptor layer is removed the number of ganglion cells is reduced (CHIARODO, 1963), and degenerating ganglion cells are found in the normal development of the optic lobes of *Danaus* (NORDLANDER and EDWARDS, 1968) and *Schistocerca* (SHELTON, 1976). These cells occur in layers of ganglion cells that have already begun to put out axons, perhaps indicating that an overproduction of cells occurs and only those cells that make appropriate contact survive, but more precise observations are needed to exclude the possibility that the degenerating cells are committed to die by a normal sequence of differentiative divisions (see SILVER, this volume).

Control of ganglion cell *differentiation* by incoming sensory axons is indicated by the observation that normal neuropile structure fails to develop when the receptor layer is removed. The effect of removing sensory fibres from differentiated neuropile is more equivocal. CAMPOS ORTEGA, and STRAUSFELD (1972) have clearly shown transsynaptic degenerative changes in *Musca* following selective ablation of retinula axons. Earlier work in general shows that the differentiated system is stable in the face of removal of receptors, particularly if the possibility of simultaneous damage to second-order cells is avoided (PFLUGFELDER, 1936/37). Similarly TITSCHAK (1928) and PANOV (1961) found that the antennal lobe was unchanged after removal of antennae from *Cimex* and *Tenebrio*, apart from a reduction in volume associated with the loss of the sensory fibres. If an antenna is removed early in development, however, the characteristic structure of the neuropile fails to develop (PANOV, 1961).

III. Axon Initiation

In the receptor layers the timing of axon outgrowth appears to depend on the developmental age of the differentiating sense cell and not on some extrinsic cue. Axon outgrowth is the first sign of differentiation among groups of cells forming receptors of all kinds. Retinula axons grow out in a sequence which reflects the passage of a wave of differentiation across the developing eye field (IMBERSKI, 1967) and in the epidermis various different stages of axon initiation, outgrowth and bundle formation can be identified in the same preparation. In the embryonic central nervous system too, axons appear in a sequence which depends on the order in which cells are budded off from the parent neuroblasts (BATE, unpubl.). In the optic lobes, on the other hand, there is evidence that axon outgrowth from second-order cells depends on the arrival of the centripetal innervation from the receptor layer.

In *Drosophila* retinula cell and lamina cell fields are matched without a phase of adjustment during which excess cells are eliminated (HANSON, 1972), and in *Calliphora* and *Lucilia* the arrival of retinula axons appears to precede the outgrowth of axons from the lamina (MEINERTZHAGEN, 1973). Anatomical evidence for an inductive relationship between axons growing in from the retina and immature lamina neurons is found in the development of the cyclopic eye in *Daphnia* (LOPRESTI et al., 1973, 1974). In this case, the leading axon from an ommatidial bundle of eight reaches the five cells of its lamina cartridge in a sequence which depends on their staggered positions in the lamina anlage. Each cartridge neuron puts out an axon in turn, following the sequence in which they are touched by the first ommatidial axon. As each cartridge neuron is touched it temporarily embraces the leading axon and in this fold a transient gap junction is formed between the two cells. The wrapping reaction and the formation of the junction occur only between the immature lamina neurons and the lead axon, not with any of its seven followers, which suggests that they are part of a normal differentiative sequence in which the leading ommatidial axon and the immature neurons of the lamina cartridge have a special relation. Specifically LOPRESTI et al. (1974) propose that the lead axon triggers the further differentiation of the lamina neurons, and that the gap junction is the anatomical basis of information transfer between the cells. Such a sequence is of course open to alternative interpretations in which the regulative factors are the age or position of the developing cells, but experimental evidence in the dragonfly *Aesehna* (MOUZE, 1974) shows that undifferentiated lamina neurons accumulate in the lamina proliferation zone of the larval eye when the normal increment of newly differentiated retinula cells is withheld. Although in some insects differentiation of the lamina follows the arrival of axons of retinula cells or is synchronized with it, this is not always the case. In *Culex* (PFLUGFELDER, 1937) and *Danaus* (NORDLANDER and EDWARDS, 1969b) lamina differentiation precedes the ingrowth of retinal axons, ruling out an inductive relation between the two.

F. The Peripheral Nervous System

The axons of the sense cells that differentiate in the epidermis reach the central ganglia by growing along extended peripheral nerve trunks. These trunks are like a wiring harness in that we have to decide how far the accurate connection of the wires in the harness depends on a colour code and how far it depends on the terminals reappearing at the other end in an order and sequence that reflects the way they went in. One important generalization to emerge is that the peripheral nerve trunks develop sequentially from pathways laid down in early embryogenesis. Long-distance exploratory growth is not part of normal development.

I. Growth of Peripheral Nerves

The initial stages in the growth of sensory axons were described by WIGGLESWORTH (1953), who used silver-stained preparations to study the process in the larval epidermis of *Rhodnius*. Each newly differentiated sense cell puts out a proximal process, which grows between the epidermis and the basement membrane and joins with other axons to form a compound bundle that leaves the basement membrane as a peripheral nerve trunk. The early stages of growth are exploratory. Regenerating axons sometimes fail to find a bundle leaving the basement membrane, and these lost axons form a ramifying network between the basement membrane and the epidermis, occasionally growing to form a continuous ring (Fig. 6). Similar endless bundles of growing axons have been described in wounded wing buds of *Galleria* (CLEVER, 1960) and in the regenerating retina of the locust *Chortoicetes* (BATE and KIEN, in MEINERTZHAGEN, 1973), and suggest that it is the bundling reaction (normally leading the axon to a pathway that delivers it to the central nervous system) that is important during the initial phase of growth and not immediate cues directing the processes to their central destination. RÖNSCH's observations on *Limnophilus* (1954) led him to a similar conclusion. He found that the processes of adjacent receptors often grew out in opposite directions and that the first phase of growth might be directed away from the central nervous system. Golgi preparations of the visual system in *Pieris* show that lamina ganglion cells occasionally put out processes that begin to grow in the opposite direction to the one they finally follow (SANCHEZ, 1919a).

That the growing tips of the axons are sensitive to certain cues is not in doubt, and these cues may be important in determining the arrangement of axons within the bundle. CLEVER (1959) found that the axons of sensilla in the wing buds of *Galleria* grew down only those lacunae that subsequently form the veins of the wing, and suggested that signals directed the axons selectively to the prospective vein. In *Limnophilus* the axons of neighbouring campaniform sensilla always associate together before joining the axons of adjacent bristles to form a compound bundle (Fig. 6) even though each campaniform sense cell might be offered a shorter route to the bundle *via* the axon of a bristle (RÖNSCH, 1954). This

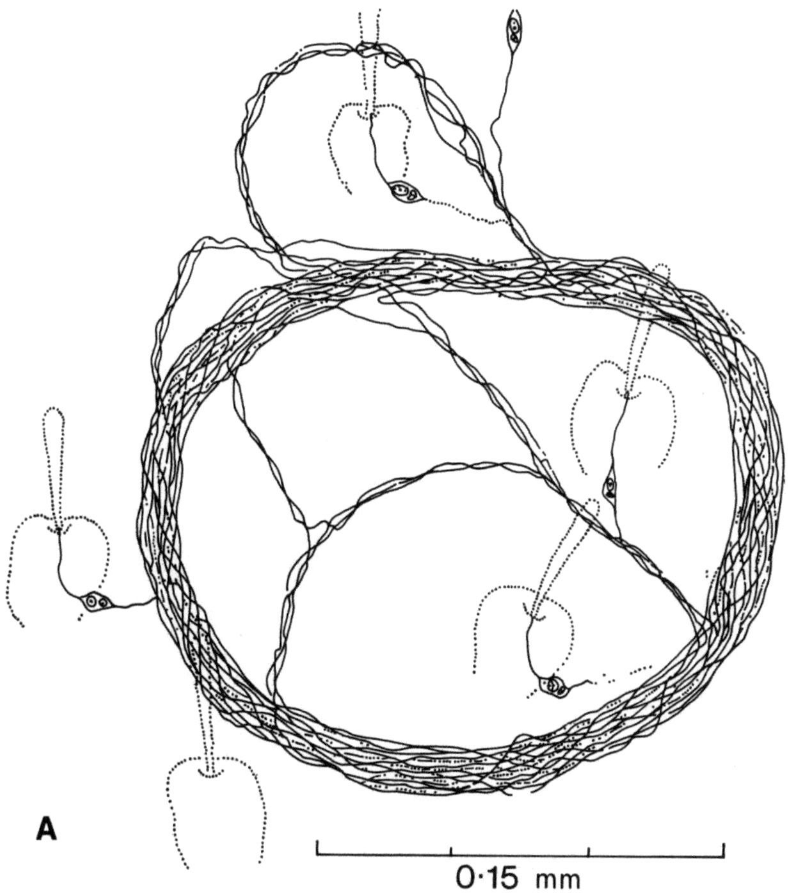

Fig. 6. Patterns of growth in insect sensory neurons. A. Regenerating axons in larval *Rhodnius* associate to form a continuous circular nerve between the epidermis and the basement membrane; B. Selective association between axons in *Limnophilus*: axons of campaniform organs (CO) grow together rather than joining with those of adjacent bristle neurons (B). Scale: 0.1mm; C. Nonselective association between axons from different segments in *Oncopeltus*. Axons of sensilla in an intersegmental graft (shaded) combine with adjacent axons in host segment to form compound bundles. A. By kind permission of Sir VINCENT WIGGLESWORTH, F.R.S.; B. After RÖNSCH (1954); C. By kind permission of Dr. P. A. LAWRENCE

apparent fastidiousness may simply depend on the staggered development of bristles and campaniform organs (campaniform organs slightly later than bristles), or more significantly on a refinement of the mechanism allowing the growing processes of peripheral neurons to recognize those of other nerve cells. It certainly deserves to be looked at again.

The axons are also sensitive to segment boundaries. Maps of peripheral nerves in *Rhyacophila* and *Galleria* (HASSENFUSS, 1973a and b) show that the segment boundaries are a limit to the exploring processes of the differentiating sense cells. Earlier maps show the same restriction on the deployment of sensory axons

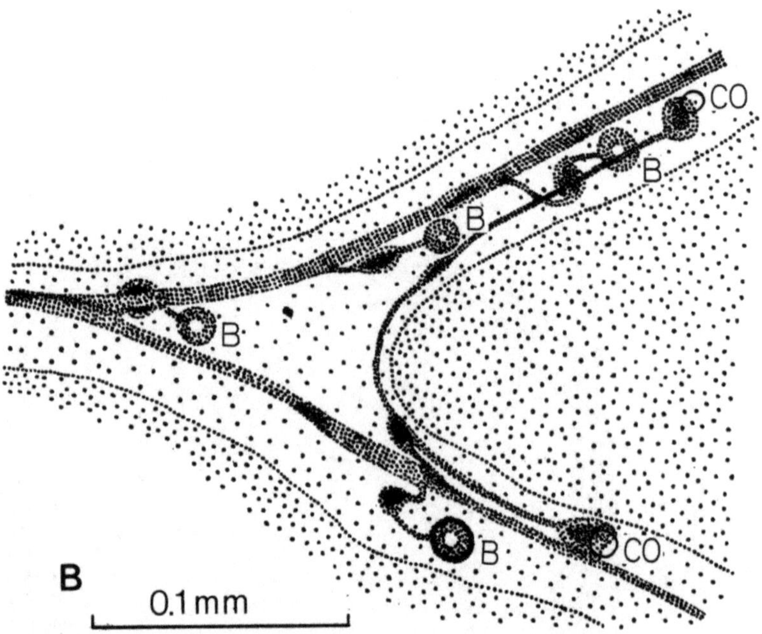
CO
B
B
B
B
CO
B
0.1 mm

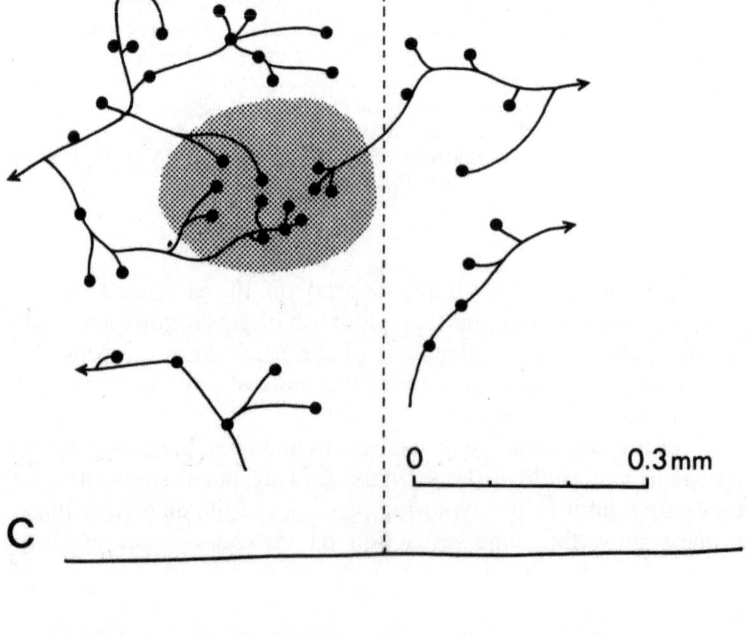
0
0.3 mm
C

(HILTON, 1924; JÖSTING, 1942). The distal processes of multipolar peripheral neurons (type-2 neurons) do cross the segment borders, so the barrier is not simply a mechanical one. Nor is the boundary signalled by a difference between the sense cells of adjacent segments (LAWRENCE, 1975). The axons of sensilla transplanted between segments (Fig. 6) are incorporated into the bundles of the segment to which they are transferred and there is no interruption to growth at the margin of host and graft tissues. The midline is not respected and axons occasionally cross from one side to the other. Like the restricted pathways described by CLEVER in *Galleria* wing buds, these observations suggest that the axons are sensitive to the surface over which they are growing and respond to localized cues, which might, for example, be changes in adhesiveness (LETOURNEAU, 1975) or oriented structures in the base of the epidermal cells.

The extent to which the bundling of the axons is an orderly process according to position or class deserves to be systematically studied. In the visual pathways retinotopic order is precisely conserved and is probably necessary for accurate connection in the underlying neuropiles (see Section G. I.). In less obviously ordered systems it is not clear that it is, but it is certainly too early to reject the possibility. Local groupings may be conserved within the bundle even where the surface arrangement of receptors is not highly regular (EDWARDS and PALKA, 1974; HUGHES and WIERSMA, 1960; LETOURNEAU, 1976b). It is important to determine whether this somatotopic order depends on an active assessment of the origins of neighbouring fibres or more simply on the orderly accumulation of axons into the bundle during normal growth.

II. The Embryonic Origin of Peripheral Nerves

Many authors make brief, unsupported references to the peripheral nervous system in general descriptive embryologies, which will not be reviewed here. A crucial question to be settled is whether the pathways that provide all subsequent axons with a route into the central nervous system are first established by axons growing out from the central ganglia or by an ingrowth of pioneer sensory axons that contribute to the developing neuropile.

Early observations are often coloured by the assumption that the sensory neurons lie in the central ganglia and that the bundles, when they appear, are necessarily growing outwards (VIALLANES, 1891; WHEELER, 1893; BADEN, 1936). In *Locusta* nerves are certainly growing outwards from the ganglia into the appendages early in embryonic life, 12 hours after the beginning of axon initiation within the developing ganglia (BATE, unpubl.). Nevertheless these efferent axons are not responsible for forming the first peripheral pathway for the ingrowing sensory nerves. In the embryonic antenna the first axons to appear are growing inwards (BATE, 1976b), from two pairs of cells located at the tip (Fig. 7). These cells are present from an early stage of antennal development and from about 105 hrs onwards at 30° C they put out axons proximally, so that two pairs of nerves run down either side of the antenna fairly directly towards its base (Fig. 8).

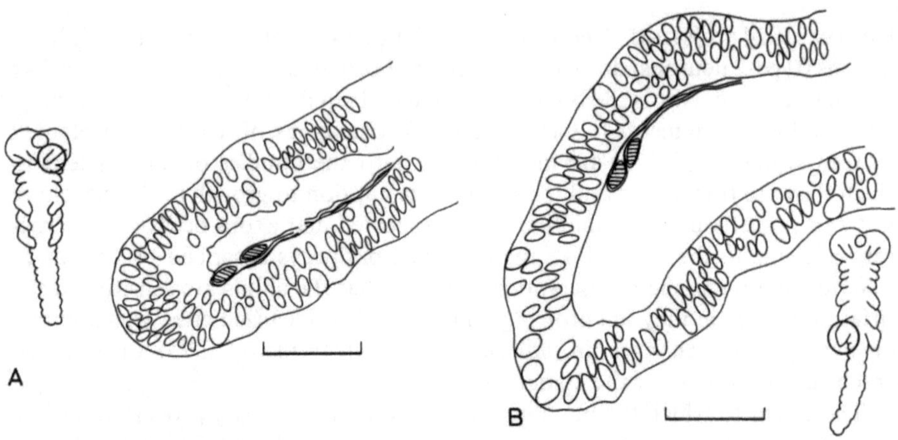

Fig. 7. A. One of two pairs of antennal neurons in *Locusta* at 100 hrs of embryonic life. This pair lies on the central surface, second pair on opposite dorsal surface of antennal lumen; B. Limb bud neurons from the metathorax of *Locusta* at 95 hrs of embryonic life. Scales (both): 40 μm

At first the axons are unsheathed but they become invested by the processes of neurilemma cells that appear at intervals along the wall of the developing antenna. The axons of later developing sensilla accumulate about the pioneering fibres, so forming the twin bundles of the larval antennal nerve. In the embryonic limb buds similar axon pairs grow inwards from peripheral neurons which develop at an early stage. Each limb bud contains a single pair of cells on the anterior margin of its distal tip (Fig. 7). The cells are spindle-shaped and from 90 hrs onwards each member of the pair puts out a process towards the central nervous system. The inward growth of these axons coincides with the appearance of other growing axons in the embryonic ganglia of the thorax, but the sensory fibres reach the base of the limb buds independently before entering the developing neuropile. Fifteen hours after axon initiation the expanded growth cones (Figure 9) of the pioneer afferent neurons enter the ganglia of the thorax from the limb buds (BATE, 1976b). As in the antennae, distal neuron pairs initiate the development of a pathway in the limbs, which subsequently differentiating axons will follow in growth to the centre, and presumably other sensory pathways are established in a similar way. Tracing the central destination of the pioneering axons and their interaction with developing second-order cells must be an important goal in understanding how the foundations of the later sensory projections are laid.

III. Sensory Innervation of Imaginal Discs

An obvious reconstruction of the sensory system occurs at metamorphosis in holometabolous insects with the assembly of adult structures partially or completely from imaginal discs. Nonetheless, existing pathways continue to provide the inward route followed by the new wave of differentiating axons.

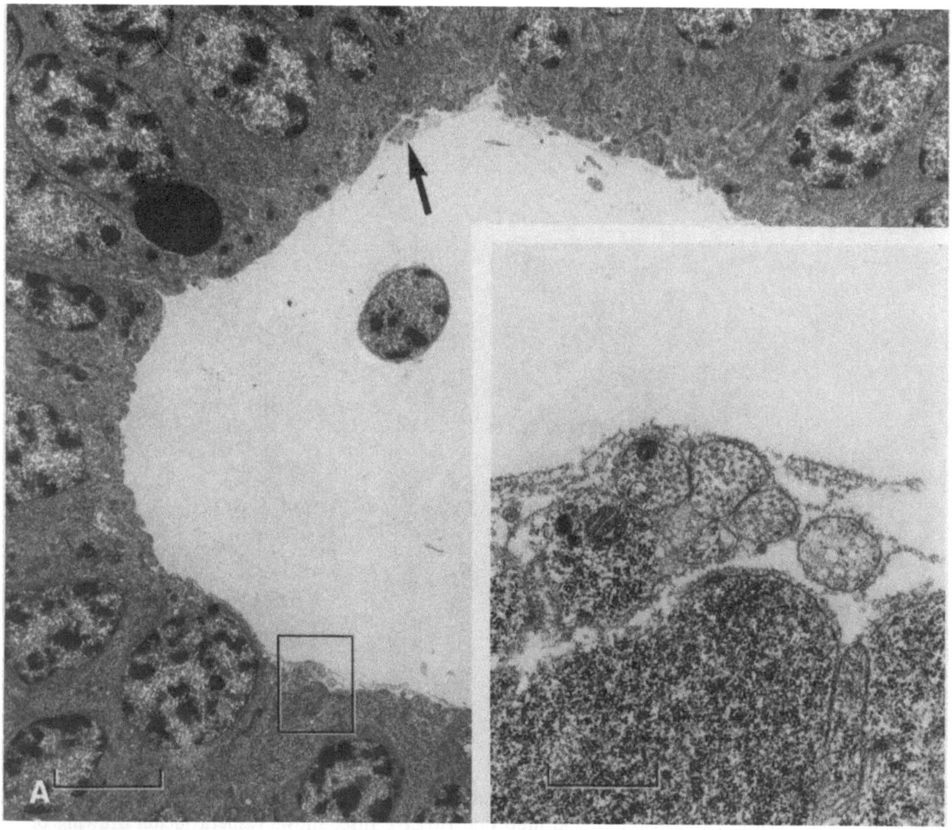

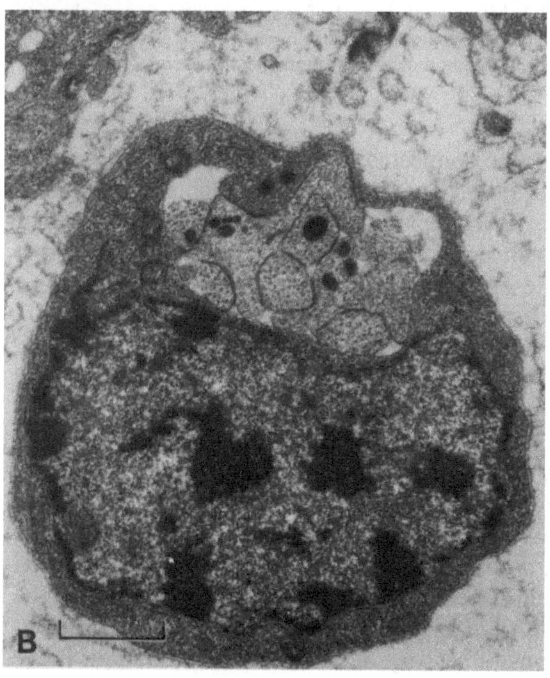

Fig. 8. A. Two sets of ingrowing sensory axons (arrowed and inset) in a cross-section through the lumen of an embryonic antenna of *Locusta* at 115 hrs of embryonic life. Scales: 10 µm and 0.8 µm (inset); B. A bundle of ingrowing axons enclosed by the processes of a glial cell in the antennal lumen at 144 hrs. Scale: 1 µm

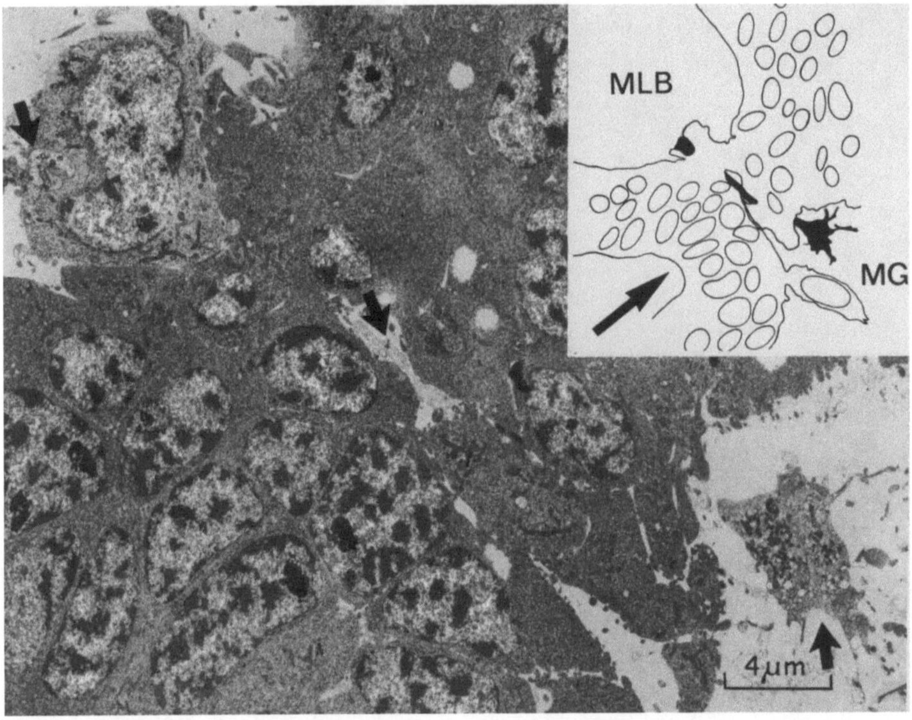

Figure 9. Expanded growth cone of pioneer peripheral axons entering metathoracic ganglion (MG) of embryonic *Locusta* from metathoracic limb bud (MLB). Inset shows camera lucida drawing of base of limb bud (arrow points anteriorly). Micrograph is part of same area resectioned, showing neurilemma cell, (arrowed top left) and growth cone (arrowed bottom right). The growth cone subsequently insinuates between adjacent ganglion cells and enters metathoracic neuropile.

Many imaginal discs are known to be connected with the larval nervous system before metamorphosis (e.g. WAHL, 1901). Contrary to earlier reports (AUERBACH, 1936; LUDWIG, 1949; SCHOELLER, 1964) it is now clear that the imaginal discs of Diptera are in general associated with larval nerves and this connection is probably established in embryogenesis. The eye disc and the anterior two leg discs are innervated in the embryo or at least by the first larval instar (LEWERENZ, 1961; SPREIJ, 1970). The antennal disc (I. I. DEAK personal communication), the wing disc (REINHARDT et al., 1977) and the metathoracic leg disc (NÜBLER JUNG, 1972) are all linked with the larval nervous system before the onset of metamorphosis, as are the ventral set of abdominal histoblasts (M. J. PEARSON personal communication). In *Tenebrio* the early developing adult leg contains bipolar peripheral neurons and nerve bundles (SOROKINA AGAFONOWA, 1924). In the lepidopteran *Calandra* (MURRAY and TIEGS, 1935) the adult limb buds are innervated in the larval stages. In *Pieris*, however, a connection with the developing chordotonal organ in the labial palp may be established by the outgrowth of glial cells from the larval olfactory nerve (KIM, 1961). All descriptions of the developing disc innervation agree that the ingrowing axons of sensory neurons accumulate about nerve bundles that predate the onset of adult differentiation, and if the association between discs and nerves is experimentally disturbed, the outgrowing sensory axons follow the nearest larval nerve into the centre, so that imaginal structures become connected with inappropriate parts of the nervous system (GHYSEN and DEAK in preparation).

In the lepidopteran *Manduca* the adult innervation of the antenna is anticipated by a population of pupal neurons, which probably develops in the antennal disc during the preceding larval instars (SANES and HILDEBRAND, 1975). The guiding function of these neurons appears to be similar to that of the neurons of the embryonic antenna in *Locusta.* At pupation, before the onset of adult development, the prospective antenna already contains a pair of nerves, each consisting of about 160 axons surrounded by a sheath of glial cells. The axons are derived from a population of cells that are periodically distributed in groups of from two to six, concentrated in the distal end of the antenna. During adult development, which is initiated about 25 to 60 hours after pupation, the adult sensory neurons that differentiate from the antennal epidermis contribute their axons to the pre-existing bundles, which gradually enlarge to form the two halves of the antennal nerve (SANES and HILDEBRAND, 1976). A similar population of cells may exist in the pupal antenna of Saturniids (PRÜFFER, 1929). The origin of the pioneering pathway, which connects the antenna with the deutocerebrum and presumably delivers newly differentiating adult axons to the developing adult olfactory lobe, is probably in the innervation of the antennal disc by a branch of the larval antennal nerve (DETHIER, 1941). Stained neurons can be observed in larval antennal discs (SANES and HILDEBRAND, 1975) and the developing pupal innervation is unaffected by the removal of the larval antennae, suggesting that the larval and pupal neurons have a separate origin.

The guiding role of persistent larval neurons is clearly demonstrated in the development of the adult visual system of holometabolous insects. Where the eye develops externally during larval life, the bundle of retinula axons from the larval eye provides a pathway between the surface and the optic lobe, down which the accumulating axons of the differentiating adult eye grow in sequence during postembryonic life (*Culex*, PFLUGFELDER, 1937; *Gyrinus*, BOTT, 1928). In Lepidoptera (MURRAY and TIEGS, 1935; NORDLANDER and EDWARDS, 1969b; UMBACH, 1934) and some Coleoptera (KIRCHHOFFER, 1910) the axons of the adult eye succeed the migrating pigment of the larval stemmata, which moves inwards to the optic lobes at the end of larval life. The stemmatal bundle loses its investing sheath at pupation and splits in two orthogonal planes to provide a radiating framework of fibres between the cells of the differentiating eye field and the optic lobe (NORDLANDER and EDWARDS, 1969a and b).

In *Ephestia* the development of the adult innervation can be followed from the early outgrowth of the stemmatal nerve to the inward growth of imaginal axons along the fragments of this bundle. The anlagen of the stemmata develop early in embryonic life (see above) as groups of flask-shaped cells in the lateral epidermis of the head (BUSSELMANN, 1935), which is still in contact with the embryonic protocerebrum at this stage. Each cell in the differentiating stemma puts out a process that insinuates between the ganglion cells of the adjacent brain lobe and enters the developing neuropile (Fig. 10). In this way the foundation of the

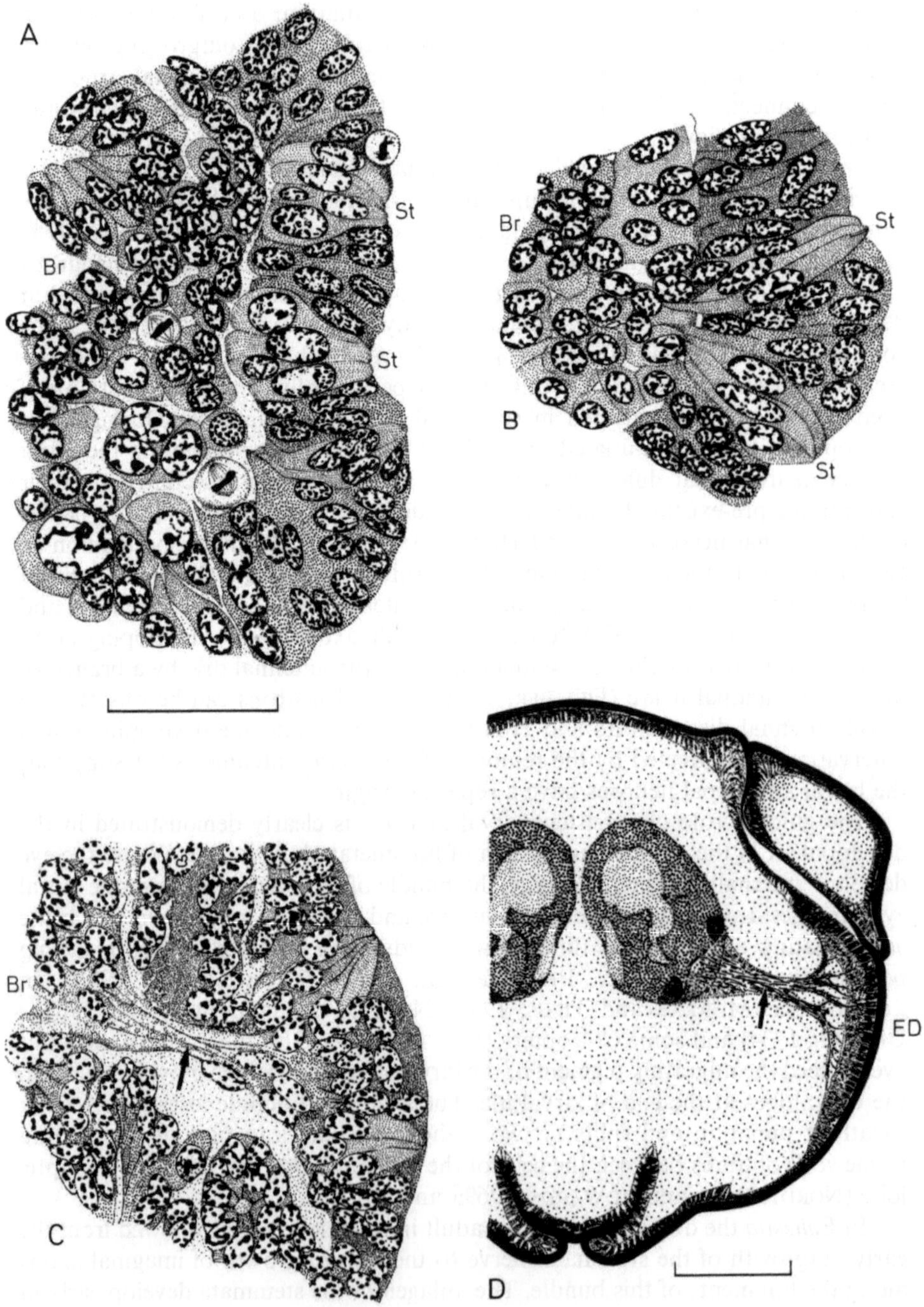

Fig. 10. A, B, C, successive stages in early development of stemmatal nerve (arrowed) in embryo of *Ephestia* (see text). A: 120 hrs, B: 124 hrs, C: 150 hrs. Scale: 20 µm; D. Degenerating stemmatal bundle (arrowed) provides a route to optic lobe for adult axons growing out from differentiating eye disc at pupation. Scale: 100 µm. *St*: Stemma, *Br*: Brain anlage, *ED*: Eye Disc. (A, B, C, after Busselmann, 1935. D after Umbach, 1934)

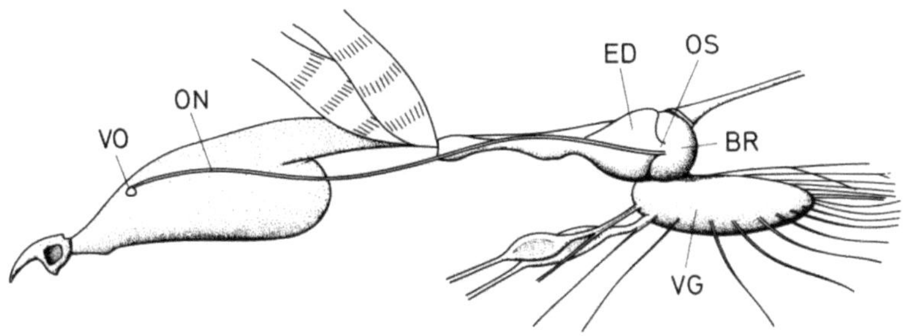

Fig. 11. Larval visual pathway in *Calliphora*. *VO*: larval visual organ. *ON*: Ocellar nerve, *ED*: Eye Disc, *OS*: Optic stalk, *Br*: Brain, *VG*: ventral ganglion. (After ROBERTS, 1971)

stemmatal pathway is established while the differentiating sense cells are still in direct contact with the brain. During the next 24 hrs *after* the initial contact has been made, the developing brain withdraws from the epidermis and the stemmatal nerve grows out from the anlage, sheathed in glial cells that are apparently derived from the differentiating stemmata. The subsequent path of the stemmatal nerve is determined by the increasing separation of the larval eyes from the posteriorly migrating brain and the development of muscles that force the stemmatal bundle through a right-angle turn before it enters the larval optic lobe. As in other Lepidoptera, the axons of immature photoreceptors from the developing imaginal eye travel along the radiating fragments of this bundle to the optic lobe after the inward migration of the stemmatal pigment in the prepupa (UMBACH, 1934).

The importance of these observations is twofold. First they show that the accurate connection of a relatively long peripheral nerve does not depend on exploratory growth by developing axons but on the early differentiation of the bundle when the sense organs and the brain are still in direct contact. The route the nerve subsequently takes is simply a consequence of passive displacement and elongation. Secondly, BUSSELMANN's work shows that careful observation is essential if the earliest stages of sensory innervation are to be observed. The absence of the obvious nerve bundles, which many authors describe as the first sign of innervation, is not good evidence that the differentiating sense cells have not yet produced axons.

In higher Diptera the eye disc is also linked with the developing optic lobe by a nerve which supplies the larval organs of vision. The larval ocelli lie at the anterior tip of the animal and the nerve passes backwards to the optic stalk, which it enters at the caudal margin of the eye disc (Fig. 11). The ocellar nerve is present from the first larval instar (MELAMED and TRUJILLO CENÓZ, 1975; WADDINGTON and PERRY, 1960) and was described by BOLWIG in *Musca* (1946) and in *Calliphora* by ROBERTS (1971) and SPREIJ and DE PRIESTER (1972). Like the ocellar nerve in *Culex* (PFLUGFELDER, 1937) it runs to the larval visual neuropile passing through the developing anlage of the adult visual system (MEINERTZHAGEN, 1973). The composition of the larval nerve, which probably degenerates at pupation (TRUJILLO CENÓZ and MELAMED, 1973), varies in different insects, forming the core of the bundle surrounded by glial cells and a thick outer sheath (MELAMED and TRUJILLO CENÓZ, 1975). The number of axons remains constant through larval life

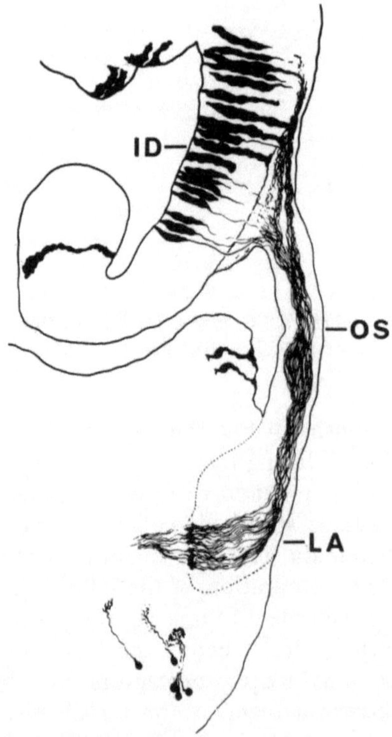

Fig. 12. Golgi preparation of developing eye, optic stalk and lamina in *Phaenicia* 12 hrs after pupariation. *ID:* Eye imaginal disc, *OS*: optic stalk, *LA*: lamina anlage. (By kind permission of Dr. O. TRUJILLO CENÓZ and Academic Press)

until the middle of the third instar, when the differentiation of the eye disc begins and the axons of immature photoreceptors begin to grow out from the imaginal disc, down the optic stalk and into the developing optic lobe (Fig. 12). The axons of the adult cells grow medially to the larval bundle, which they soon outnumber and displace without mixing, so that in *Phaenicia* at the end of the larval period there are two populations of nerves in the optic stalk, a persistent bundle of about 40 larval axons lateral to a growing bundle of about 5000 adult axons (TRUJILLO CENÓZ and MELAMED, 1973), most of which are separated into bundles of eight by intervening glial cells.

IV. Delivery of Axons to the Optic Lobe and the Formation of Chiasmata

Because of the regular nature of the receptor layer and the precise orientation of the underlying neuropiles, the visual system illustrates particularly clearly the combined effect of a pioneering bundle and a temporal sequence of differentiation in the transfer of somatotopic order to the central nervous system.

Because the larval visual pathway passes through the developing adult optic anlage, imaginal axons growing along it are delivered to the proliferating lamina and medulla cortices. Chiasmata develop between the neuropile layers as a consequence of the oriented proliferation of the optic anlage (PFLUGFELDER, 1937; MEINERTZHAGEN, 1973, 1974). In *Culex* (Fig. 13) the imaginal ganglia are derived from separate sets of neuroblasts which bud off cells laterally (lamina), medially

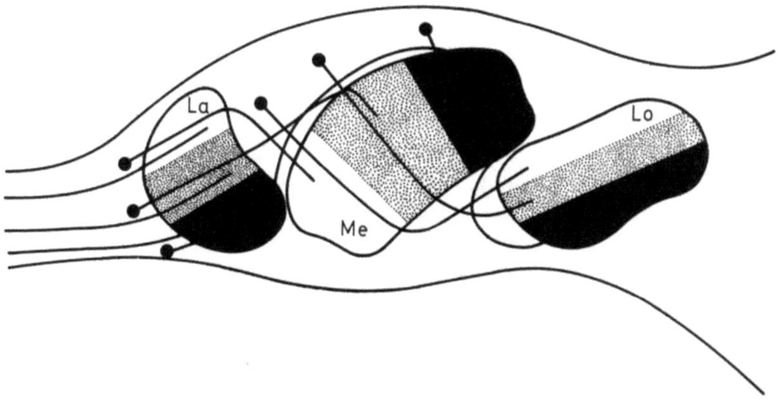

Fig. 13. Diagrammatic representation of proliferation in the optic lobe of *Culex*. Axons grow centripetally in sequence to youngest region of each anlage (unshaded), while older regions (shaded) are progressively displaced by newly proliferated ganglion cells. *Lo*: lobula, *Me*: Medulla, *La*: lamina. (After PFLUGFELDER, 1937)

(medulla) and medio-posteriorly (lobula). The oldest cells in each are posterior (lamina) medial (medulla) medioposterior (lobula). The pathways between the cortices develop necessarily as the axons are delivered in sequence to the newest parts of the anlage which are subsequently displaced by the differentiating daughters of further neuroblast divisions. MEINERTZHAGEN (1973) describes the oriented proliferation of the lamina, medulla and lobula cortices in *Calliphora*, and estimates the timetable to which photoreceptor axons from the eye disc conform in growing to the lamina (assuming a growth rate of 25 μm/hr). The completion of growth in 50 proto-ommatidial rows takes three days in *Calliphora*, so that new rows of axon bundles would be added to the lamina at a rate of one every 90 minutes. Allowing for the greater path length of axons from the dorsal and ventral margins of the eye compared with those from the equator (150 μm in *Calliphora*), there is a time lag of six hrs between the arrival of axons from the margins and the centre of a vertical row (100 bundles in *Calliphora*), so that the axon bundles should arrive in a vertical sequence of one every six minutes. This is a useful exercise, because it indicates that at any one time only a few bundles are forming connections in the lamina, at widely separate points on any dorsoventral row of ommatidia.

G. The Formation of Nerve Connections

Because the arthropod nervous system is divided into two halves, the one continually contributing axons to the other during an extended larval life, it invites an experimental analysis of the formation of connections between the two. This is all the more so because the patterns of receptors are predictable and they terminate on identifiable central neurons, so that the same group of cells can be used repeatedly in experimental tests. Unfortunately the possibilities of this approach are so far strictly limited by our inability to describe in detail even the simplest pattern of connections outside the visual system. It may be that the elusive pattern of connections is not quite what we expect it to be. The arthropod central nervous system consists of a relatively small number of highly stereotyped cells and it is generally assumed that the receptors too make very specific connections with these cells. As it happens some evidence suggests that the central arrangement of receptor endings is less predictable (KENNEDY, 1971, 1974). We may also be misled by the electrophysiological techniques that are used to analyze the input connections of central cells. The discovery that unit A is always connected with receptor B may conceal considerable variation in the anatomy of this stereotyped connection and the growth process that constructs it.

In the visual system, on the other hand, a repetitive pattern of connections between the receptors and the outer visual neuropiles can be precisely defined. As a result it becomes possible to observe a unit process of connection and to describe the assembly of the system at a cellular level (MEINERTZHAGEN, 1975).

I. Connectivity in the Visual System of Flies

Because of its subdivision into sets of parallel visual and integrative units the visual system is wired up by a highly repetitive pattern of connections, which conveys retinotopic order in turn to the lamina and medulla neuropiles. In particular, in those insects with neural superposition eyes (KIRSCHFELD, 1967), visual information is faithfully transferred by a regular interweaving pattern of axons, which connects retinula cells with coincident optical axes with a common unit of the underlying lamina neuropile (BRAITENBERG, 1967; TRUJILLO CENOZ and MELAMED, 1966). The perfection of this connectivity pattern and its precisely inverted projection to the medulla neuropile is a striking demonstration of the reliability of the morphogenetic process that produces it.

Each ommatidium of the fly eye contains eight retinula cells in a precise asymmetrical arrangement (Fig. 14). Six of these cells have short axons that project to the lamina, and surround a central pair of cells whose axons run down the equatorial side of the underlying lamina cartridge to the medulla. Each of the six marginal retinula cells looks at a different point in the visual field, and the axons of retinula cells with coincident optical axes in different ommatidia are combined in a common lamina cartridge. Six axons diverge from each ommatidium and are

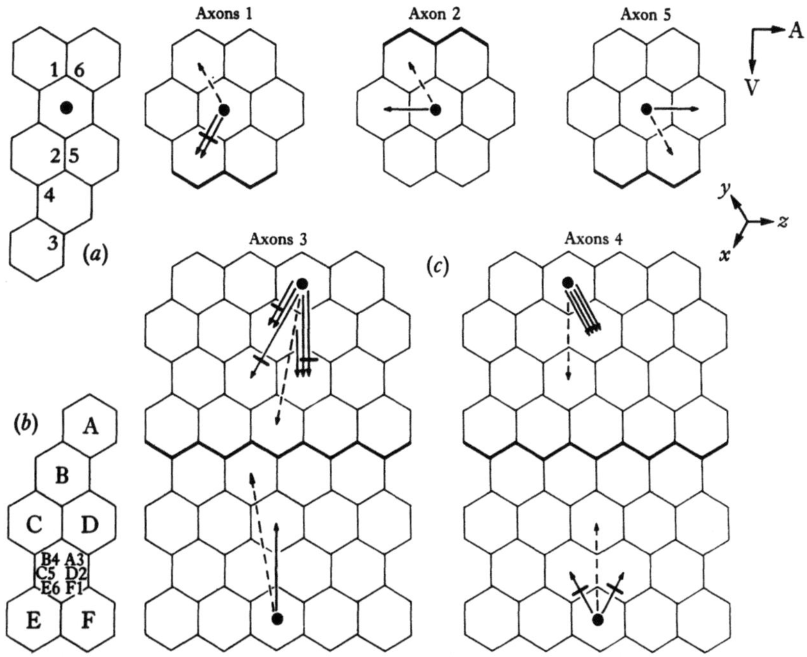

Fig. 14. Organization of first visual projection in adult fly. a) Pattern of lamina cartridges to which divergent axons of a single ommatidial bundle (solid circle) spread in dorsal half of eye. Axons 2–6 grows towards the equator, axons 1 and 6 away from it; b) Convergent pattern of retinula axons on a single lamina cartridge in dorsal half of lamina. Retinula axons from ommatidia *A–F* terminate on cartridge in a precise rotational sequence, 1–6. The sequence is anticlockwise (clockwise for ventral cartridges); c) Summary diagram of projections of 17 axons, which terminated on incorrect cartridges in a region of right eye with a dislocation of retinal equator. Separate patterns are shown for axons 1, 2, 3, 4 and 5. Each axon's origin is shown as a solid circle, solid arrows show actual paths of axons to their cartridge. Broken arrows show predicted path for each class. Heavy horizontal lines represent equator (dorsal above). Arrows crossed by short heavy lines indicate path of an axon that crosses the equator to reach its cartridge. Axes x, y, z, defined by the orientation of the retinal facet rows. *A*: anterior, *V*: ventral. (By kind permission of Dr. I. A. MEINERTZHAGEN)

distributed to a trapezoid array of underlying cartridges, while each lamina cartridge collects axons from six different ommatidia in a trapezoid arrangement in the retina. The interweaving pattern of divergent and convergent axons is repeated throughout the eye and the retinula axons on each lamina cartridge contribute their axons in an exact rotational sequence. The precision of the connection process can be assessed by assigning to each ommatidium an available field of 31 cartridges defined by the length of the longest retinula axon it produces (BRAITENBERG, 1970). Since each cartridge offers six alternative terminals, the number of possible connections is 186, of which a precise arrangement of six is chosen. Errors in the differentiated system are particularly interesting because they are characteristic of the growth processes which produce the normal pattern.

In the normal eye the pattern of asymmetric ommatidia is divided into dorsal and ventral mirror-image halves by a horizontal equator (DIETRICH, 1909). Because of the unequal divergence of axons towards (4) and away from (2) the equator in any ommatidium, retinula cell axons are in excess in the equatorial rows and depleted at the margins of the eye. On optical grounds (the axes of retinula cells in the equatorial ommatidia) lamina cartridges with eight and seven retinula cells are predicted in the rows next to the equator (KIRSCHFELD, 1967) and this prediction is borne out by the anatomy (HORRIDGE and MEINERTZHAGEN, 1970). The regular pattern of convergence of lamina cartridges is faithfully reproduced without error, even though in the equatorial rows four of the axons contributed to each cartridge have grown from ommatidia in the opposite half of the eye, that is to say, through a medium that is the mirror image of their own. Departures from the normal projection pattern do occur, however, when the field of ommatidia is disrupted in some way. HANSON et al. (1972) report mutants in *Drosophila* where the pattern of retinula axons is almost random. At dislocations in the equator (where the normal pattern of dorsal and ventral ommatidia is disturbed at the line of symmetry) in the eye of *Calliphora*, MEINERTZHAGEN (1972) described 17 wrong connections from a traced field of 144 ommatidia (500 axons). This is the sort of manageable figure that can usefully be catalogued so that the common features of the mistakes can be identified, together with the type of error to which the system is most prone.

Firstly all such wrong projections suggest (although they do not prove) that the precise specification of the retina lamina projection does not depend on a chemical labelling of receptor and second-order cells in a way that permits only certain connections. Apart from the extravagance of such a system, it is worth noting that the retinula axons that connect with a single cartridge do not appear to share any common property apart from their connection with that cartridge. The common property is shared by equivalent retinula cells in different ommatidia that send axons to different but equivalent points in the lamina. Secondly, mistaken connections never involve an axon in growth beyond its normal destination. All wrong connections are with cartridges nearer the parent ommatidium than the correct one and are commonest in those axons (3 and 4) that normally have to travel farthest. In addition to these incorrect trajectories, errors in the predicted rotational sequence also occur, at dislocations, occasionally at the equator, and rarely in the rest of the field (HORRIDGE and MEINERTZHAGEN, 1970). The commonest error is a transposition of axons 3 and 4 on any cartridge, these being the axons which in growth to the cartridge have the smallest angular separation. Despite these errors, axons consistently terminate on the side of the cartridge nearest to their parent ommatidium. On opposite sides of the equator the rotational sequence is reversed (BRAITENBERG, 1967) and axons which cross the equator conform to the sequence of their own side (HORRIDGE and MEINERTZHAGEN, 1970), inverting the normal arrangement and in so doing conserving the rule that axons terminate on the side nearest to their parent ommatidium.

Taken together, these characteristics argue against the following systems of generating the normal pattern of connection: axon growth by continuous guidance (signposts inverted at the equator); coordinate labelling of receptors and cartridge neurons (wrong connections); cooperative growth of axons to the same cartridge (no chains of errors); labelling of terminal positions on cartridges (wrong terminals); reference to nearest neighbour on the cartridge (transpositions). The system is most prone to errors when the normal orientation of the ommatidia or cartridges is disturbed or their relative orientation is inverted, and most mistakes are made by the axons that have to grow furthest and between those whose angular separation in growth is least. These characteristics suggest that the general destination of retinula axons is determined only by the spatial order in which they are delivered to the lamina and that a property, such as an orientation, common to equivalent retinula cells in different ommatidia consigns each to a particular cartridge, where it terminates at the first point it reaches.

II. Development of the Retina Lamina Projection in Flies

The arrival of axon terminals at the differentiating lamina has been described in *Drosophila* (HANSON et al., 1972), *Calliphora* (MEINERTZHAGEN, 1973) and *Phaenicia* (TRUJILLO CENOZ and MELAMED, 1973), and the events are similar in all three insects. As we have seen, retinotopic order is conveyed to the lamina by a postero-anterior sequence of axon accumulation in the bundle, which passes down the optic stalk to the developing lamina. At this stage, the lamina consists of a plexus of ommatidial growth cones interspersed by the cell bodies of the immature lamina monopolar neurons. Newly arriving axons grow over the surface of the axons that precede them until they reach the first unoccupied cell group. They grow centrally down this group on the side furthest from the equator, spiralling through 180° to reverse their initial orientation exactly and reach the opposite side of the lamina cell group (HANSON et al., 1972). On arrival at the lamina plexus the growth cones of the retinula axons swell, possibly because of the impedance offered by the epithelial cells, but probably because of a general signal, since the long visual fibres put out a transitory growth cone in the lamina as well. The expanded growth cones put out numerous straight filopodia, which embark on a complex and apparently random pattern of reciprocal invagination with the growth cones of neighbouring cells. Initially the eight axons of any one ommatidium are associated with a single group of monopolar cells, but progressive restriction of the filopodial activity is associated with the oriented growth of each short retinula axon terminal to another cartridge (Fig. 15), so that an orderly transfer of growth cones between neighbouring bundles occurs, to produce a trapezoid distribution of the immature terminals. This crucial phase of growth, in which the final pattern of connections emerges, occurs either by a progressive attrition of redundant connections, possibly as a result of territorial interactions between neighbouring growth cones (MEINERTZHAGEN, 1973), or by the oriented outgrowth of a process from each growth cone (TRUJILLO CENÓZ and MELAMED, 1973). This important difference between the descriptions of growth cone transfer should be resolved, since the two indicate fundamentally different methods of specific connection—either by the gradual elimination of redundant contacts or by the directed outgrowth of a single cell process.

So far these observations bring us no nearer to knowing how the retinula cells select their destinations than the process of deduction from naturally occurring errors. They do provide a firm cellular foundation for the idea that the common

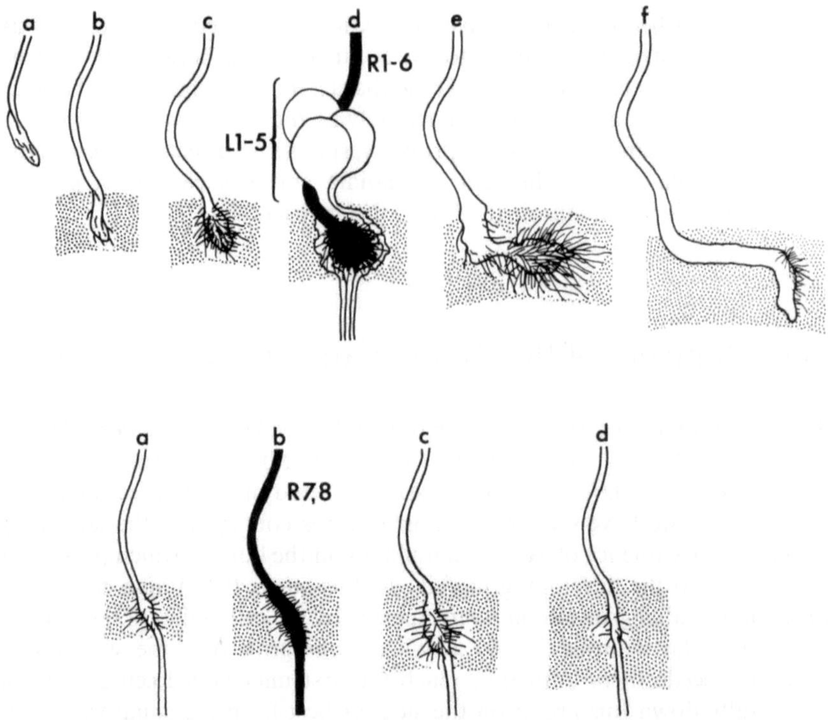

Fig. 15. Summary diagram of morphologies of expanded retinula growth cones in the developing lamina plexus of fly. Each diagram represents idealized appearance of elements R1–6 (top series) and R7, 8 (bottom series), either as seen in Golgi impregnations or reconstructed from serial electron micrographs. Derived from the data of SANCHEZ (1919a and b), MEINERTZHAGEN (1973) and TRUJILLO CENOZ and MELAMED (1973). In both series *a* is the most immature form, *b*, *c*, etc., representing progressively more mature forms. Forms *a–d* for short retinula axons R1–6 and *a–d* for long visual fibres R7, 8 are essentially indistinguishable in species of Lepidoptera (*Pieris*: SANCHEZ, 1919a and b) and Diptera (MEINERTZHAGEN, 1973; TRUJILLO CENOZ and MELAMED, 1973) in which the adult retinula elements have quite different morphologies (MEINERTZHAGEN, 1976). In particular, the retinula divergence pattern of R1–6 arises by the interpolation of an additional lateral growth phase to the growth cone shown in *e* and *f* and illustrated by TRUJILLO CENOZ and MELAMED (1973). Expanded growth cones appear first as swellings of the slender growth cone (top, *a*), (observed in insecs solely in *Pieris*, SANCHEZ, 1919a and b), which enlarge and put out filopods (top *b c*). Filopod activity is restricted to a narrow zone of presumptive lamina neuropile (stippled area in each figure), which also contains expanded growth cones of second-order monopolar neurons L1–5, three of which are illustrated (top, *d*). Figure by kind permission of Dr. I. A. MEINERTZHAGEN

property shared by retinula cells at equivalent positions in different ommatidia is an orientation of axon growth, derived from an interaction with neighbouring cells. In ontogeny and presumably in evolution, the development of the open rhabdom eye passes through a fused rhabdom phase in which adjacent cells are joined by junctions. The formation of the neural superposition pattern coincides with the loss of these junctions and the departure of component axons to alternative destinations which appear to depend on their position in the ommatidial bundle.

III. Connectivity Outside the Visual System

Before dealing with the formation of sensory connections outside the optic ganglia, it is essential to appreciate the difference between this system and equivalent vertebrate systems, such as the amphibian retinotectal projection from which much of the vocabulary and methodology of developmental neurobiology is derived (HUNT and JACOBSON, 1974). Arrays of neurons that terminate in register relaying topographic order from one neuropile to the next are not known outside the visual system. For the most part, arthropod ganglion cells are monopolar units whose cell bodies are arranged as the peripheral rind to a globular neuropile. Each cell contributes an axon to the neuropile, where it forms a branched dendritic tree and here synapses are formed between the first- and second-order neurons. If there is a regular projection to particular cells according to the modality or position of the receptor neurons, no spatial order is imposed on the pattern of connections by the somata of the ganglion cells, which are devoid of synapses. At the same time the number of ganglion cells is relatively small and it is unusual to find interneurons that are devoted to the input of a single receptor, so that a major part of the development of the characteristic pattern of connections involves an accumulation of convergent sensory endings on complex dendritic trees.

The pattern of connections is unknown anatomically. Most of the interneurons act as filters rather than amplifiers, so that consistently heterogeneous patterns of connections are rarely revealed by overt behavioural discriminations between receptors (compare, for example, reflex tests for the origin of skin sensory nerves in the amphibia MINER, 1956). The input connections of different interneurons can be discovered by recording from the axons of single cells between the segmental neuropiles, and more recently by recording from the axons and cell bodies of the larger cells in the ganglia. This *distribution* of receptor terminals to different interneurons is the present limit to our knowledge of the precision with which connections are formed outside the visual system, although electrophysiological data from neuropile recordings and histological data on the shape and extent of the arborizations formed by interneurons and receptor cells give some indication of the *arrangement* of connections and how the pattern develops.

IV. The Distribution of Connections to Second-Order Cells

Patterns of connectivity outside the visual system are best known in the crayfish. WIERSMA and co-workers have analyzed the input connections of units in the ventral cord and circumoesophageal connectives and produced a catalogue of more than 100 highly individual interneurons (WIERSMA, 1958; WIERSMA and HUGHES, 1961; WIERSMA and BUSH, 1963). Each unit shows a specific pattern of connection with receptors according to modality and peripheral fields. The overall pattern is of second-order units arranged in parallel with overlapping receptive fields. Many of the interneurons have multisegmental inputs, i.e., they receive in many ganglia precisely the equivalent input, according to modality and position, which they receive in one. Each cell appears, to be highly selective in its connections with different receptor neurons, which are classed according to type and surface position. The multisegmental pattern of input rules out the possibility that each neuron is specified by the input it receives, and since the same interneurons recur reliably in cross-sections of the ventral cord the specificity of the second-order cells appears to be part of a characteristic pattern of differentiation to which each is committed by its position in the developing nervous system. Interneurons in the sixth abdominal ganglion of the crayfish which receive mechanoreceptive input from hairs on the telson present us with the best analysed material on the type of connections formed between mechanoreceptors and interneurons and to some extent of the way in which these connections develop (KENNEDY, 1971, 1974). The receptors and central cells are connected according to two criteria: position at the surface and directional sensitivity. The first criterion is rather sloppy: the receptive fields of the interneurons are probabilistic (CALABRESE, 1976a; WIESE et al., 1976), with a peak density of connection in one region and the likelihood that any one receptor will be connected declining in the surround. Such a connection system could be generated by 'anatomical labelling', that is the delivery of axons in somatotopic order to the appropriately distributed dendritic trees of the second order cells. Since the field of a single interneuron may include afferents entering by separate roots, the ordering of the afferent fibres must be continuous and compatible from root to root, and this has now been demonstrated (LETOURNEAU, 1976b). The second criterion is precise: many of the telson receptors are dually innervated by afferent neurons of opposite directional sensitivity (WIESE, 1976) and interneurons connect selectively with afferents of one or the other polarity only. There is no anatomical segregation of afferents according to sign, so this subclassification of the connection pattern depends on some other, very specific kind of labelling. In the next two sections we consider further evidence for labelling of this kind.

V. Modality and Position-Specific Connection

1. Modality

Connection according to modality is part of a stable commitment to a particular physiological phenotype, which sensory neurons continue to manifest during regeneration and after transplantation. It is important to be clear about what such a transplant sets out to do. In developmental biology a transplant is often a test of a cell's state of determination, i.e., how far it is committed to a particular developmental fate and how far this programme can be reset by the new position in which it finds itself. In neurobiology, on the other hand, a transplant conventionally tests a neuron's differentiated state. The transplanted neuron reveals the extent to which its normal pattern of comnection depends on some form of specification or on an orderly process of assembly.

The anal cercus of the cricket *Acheta* has been extensively used by EDWARDS and co-workers in such experimental tests of neuronal regeneration (review: EDWARDS and PALKA, 1973, 1976). When the cerci are removed the sensilla, which reappear in sequence in the regenerating blastema, reconnect with the giant fibres, which ascend the ventral cord from the last abdominal ganglion. When the cerci are removed from the abdomen and transplanted to the thorax instead of a thoracic leg the regenerating axons of the cercal sensilla grow inwards to the thoracic ganglia, where again they selectively connect with the giant fibres with which they normally synapse in more posterior parts of the system (EDWARDS and SAHOTA, 1967). The property of specific connection is stable, so that if cerci are removed and prevented from regrowing during postembryonic life normal connections are still reformed when a cercus is allowed to regenerate (EDWARDS and PALKA, 1971). In a similar way, reflex tests of the connections formed by heteromorph antennules in place of an eye in lobsters and crayfish show that growing fibres from the abnormal appendages find their way to appropriate antennular second-order cells in many cases (LISSMAN and WOLSKY, 1933; MAYNARD and COHEN, 1965; SANDEMAN and LUFF, 1974). Apparently major interneuron and receptor classes are coordinately labelled in a way that enables connections to be reformed even when the normal process of development is disturbed.It is important to realize the limits of this experimental approach, however. As in the visual system, errors that occur when the system is disturbed reveal rules that govern the process of connection to a level of specification limited by the available techniques for visualizing the anatomy of sensory and second-order cells. They cannot reveal the mechanism of connection itself.

2. Position

Where sensory neurons of the same modality make consistently different connections according to their position (e.g. CAMHI, 1969; BATE, 1973) they indicate that the location of each neuron is coded and that the normal process of differentiation translates this information into a specific pattern of central connections. As we have seen (p. 40), relative position within a receptor group can be reproduced centrally by nearest neighbour assessments between ingrowing sensory axons. There is as yet no experimental evidence to show that patterns of connection are accurately reconstructed during regeneration in a way that would indicate that the terminals of sensory neurons are labelled centrally according to their position at the surface.

Nonetheless, distinctive connections made by similar receptors at different positions in the same segment have been provisionally interpreted as evidence of locus-specific properties which neurons in the epidemis acquire in response to their level in a segmental gradient (BATE, 1973; BATE and LAWRENCE, 1973). In a recent paper, PALKA and SCHUBIGER (1975) describe the results of operations on the cerci of crickets, which they find consistent with this theory, but their experiments are not an adequate trial. The filiform hairs on the cercus are selectively connected with the giant interneurons according to the orientation of the socket and the consequent plane of their vibration (EDWARDS and PALKA, 1974). Hairs that vibrate transversely to the axis of the cercus are concentrated in its dorsal and ventral sectors, hairs that vibrate longitudinally are located laterally and medially. Only the transverse hairs excite the two largest interneurons in the cord, and interestingly this pattern of selective connection is reproduced by the regenerating axons of the two types of hairs, even when the cerci are rotated or exchanged left and right. After exchange the cerci are necessarily misaligned in one axis and the authors suggest that the ability of the hairs to reconnect faithfully according to their position in the cercus and not according to their new position with respect to the body axis may depend on the presence of cercal gradients to which the sensory neurons refer when making their central connections. While this may be true, the experiment provides no new evidence in favour of the theory. Clearly the transplant that is required is of the first type discussed above (Section G.V.1), i.e., a sheet of epidermis that contains prospective sensilla, since the question to be answered is whether neurons are committed to different development fates by positional cues. The stability of this commitment is another question that can be answered by transplants of differentiated neurons. In these experiments the cercus itself is unchanged and the authors do not state whether new hairs differentiate postoperatively. These hairs, and specifically the hairs that develop at the margin of the host and graft tissues, where intermediate positional values are generated, provide a crucial test of what consigns the terminals of the hairs to different connections within the central nervous system. For the moment, then, it is unproven that positional information generated in the epidermis commits neurons of the same modality to different connections according to their surface location.

VI. The Arrangement of Sensory and Second-Order Terminals

Our knowledge of the arrangement of sensory endings on the dendritic trees of central neurons depends largely on microelectrode recordings from neuropiles and the use of intracellular dyes. Both methods are a fallible guide to the exact pattern of synaptic connections.

Electrophysiological data indicate that the geometry of dendritic trees is relatively stereotyped (KENNEDY, 1966) and although the anatomical evidence reveals some variation, in general it supports the conclusion that the shape of second-order cells and of the central arborizations of receptor cells is consistent. A normal achievement of the morphogenetic process is a characteristically branched neuron, which at a superficial level is recognizably different from other neurons. Because of the reliability with which many identified neurons function, the dendritic tree may possess homeostatic properties that compensate for small variations in shape, or the effects of these small variations are at the noise level for present levels of electrophysiological analysis. The consistency of the receptor and second-order cells argues for an orderly arrangement of sensory endings on the dendritic branches of the interneurons, and in the large scale this is true. Giant interneurons of crickets show ipsilateral and contralateral branches within one ganglion, which are normally exclusively supplied by the terminals of receptors on the respective side (MURPHEY et al., 1975), and in crayfish the dendritic tree of a single interneuron can be subdivided into regions with different receptive fields and separate sites of spike initiation within a single ganglion (CALABRESE and KENNEDY, 1974). How is the development of the sensory projection controlled? There is no evidence that parts of the dendritic tree are labelled and when interneurons are deprived of their normal ipsilateral innervation contralateral fibres may cross and innervate them (PALKA and EDWARDS, 1974; MURPHEY et al., 1975). This finding and the observation that in normal growth in the crayfish tactile efferent fibres never cross the midline (CALABRESE, 1976b) suggest that the ganglion itself is in some way divided into sectors which growing sensory neurons respect unless the system is disturbed. It would still be possible to argue that on a smaller scale some sort of locus specificity is written into the dendritic tree and ensures the orderly arrangement of the sensory terminals in any region. The only evidence on this point is provided by an electrophysiological analysis of the projection of sensory hairs on the telson of the crayfish to interneurons which normally respond to water-borne vibrations (KENNEDY, 1971, 1974). In this case it is likely (page 40) that the order in the sensory projection depends on a sorting out amongst the ingrowing afferent fibres and not on local specificities in the interneuron.

VII. Growth of Dendritic Trees

In many cases (page 18) the number of interneurons is fixed at an early stage of development (perhaps, for the primary cells, by an interaction with early differentiating sensory axons). From then on the continual addition of new receptor elements to a fixed number of central cells demands that the existing dendritic trees expand to accommodate the enlarging input. In the differentiated system each cell has a characteristic pattern of first-order branches, which becomes more variable in the finer ramifications of the tree. For any developmental stage the normal arborization can be defined as an area and density of branching (Fig. 16), a generalization that applies to motor neurons (BURROWS, 1973a and b), interneurons and sensory neurons (ALTMAN and TYRER, 1977a, b, O'SHEA et al., 1974; BURROWS, 1975; MACAGNO et al., 1973; SANDEMAN and OKAJIMA, 1973). An important question is whether the variation within the domain of a single neuron is genetically determined or a consequence of the growth process that produces it, and in this respect the most useful data include a comparison with equivalent cells on oppocite sides of the nervous system. The neurons are genetically identical and differences between them reveal variation in the expression of the phenotype. Comparisons of this kind show that there is no greater variability between cells from different animals than there is between equivalent cells from the same animal (BURROWS, 1975). The unpredictability of the dendritic trees is therefore a consequence of the normal growth process that produces them.

Already it is clear that the pattern of growth is reciprocal and that both sensory neurons (MOUZE, 1974; TRUJILLO CENOZ and MELAMED, 1975) and interneurons (MURPHEY et al., 1975) fail to develop normally when they are deprived of their target cells. The dendrites of giant interneurons in *Acheta* are stunted by chronic removal of their normal cercal input throughout postembryonic life. Unlike the cercal nerves the giant fibres have contralateral as well as ipsilateral projections, and it is particularly interesting that when the

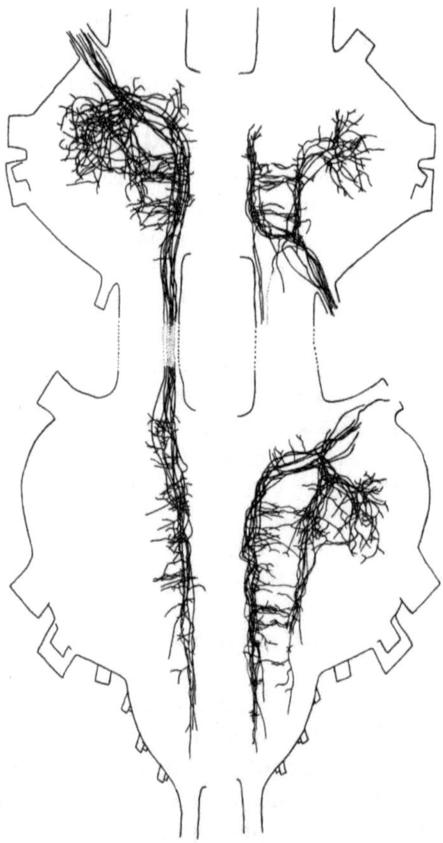

Fig. 16. Variation in branch pattern of central projection of wing-hinge stretch receptors, as revealed by cobalt impregnation in meso- and metathoracic ganglia of *Locusta*. Left: forewing-stretch receptors; Right: hindwing-stretch receptor. In each case 7–10 preparations from different animals are superimposed and normalized. Whole mounts in dorsal view. By kind permission of Drs. J. S. ALTMAN and N. M. TYRER

cercus is removed from one side only, the interneuronal branches on that side are affected, but the dendrites of the same cells are normal on the opposite side. Although these results indicate that the ingrowing axons of the sensory neurons have a local trophic interaction with the dendrites of their target interneurons, they do not show a specific relation. Possibly the normal growth of the dendritic tree depends on a general increase in the volume of the neuropile, which occurs during postembryonic life but is drastically curtailed when a cercus is removed. A general structural change occurs in interneurons that are deprived of sensory input rather than sensory connections (BENTLEY, 1975) but there is no evidence (with the important qualification that we know nothing about synaptogenesis in arthropod nervous systems) that as in the vertebrates regularly repeated patterns of input have a specific effect on the developing nervous systems.

An important point in these observations is that the fundamental branch pattern of the interneurons is unaffected by experimental interference during postembryonic life. Presumably the major branches are established during embryogenesis, but whether this involves an interaction with pioneering sensory fibres is a significant question still to be answered. The consistency of the basic pattern may indicate that the early phases of growth conform to a genetic programme that distributes axon branches to major neuropile areas within which the later sensory projections are formed. Clearly answers to fundamental questions in the development of sensory connections must be sought at this early, relatively uninvestigated stage, when differentiating sensory axons establish the framework of later innervation by selecting between alternative immature interneurons. The commitment to a particular cell is stable and recreated in regeneration, possibly by the recapitulation of a process of exploratory growth and connection according to some identifying tag. The sensory axons that differentiate later arrive at their general destination by following preformed pathways, and insofar as they form a regular projection, this is not because of a regional specification of the target neuron but because the ingrowing neurons can read their position relative to their immediate neighbours. It is probably significant that the cells of the epidermis from which the sensory neurons are derived are known to have properties that enable them to sort out in this way (GARCIA BELLIDO, 1966).

References

Alexandrowicz, J.S., Whitear, M.: Receptor elements in the coxal region of Decapoda Crustacea. J. mar. Biol. Ass. U.K. **36**, 603—628 (1957).

Altman, J.S., Tyrer, N.M.: The Locust Wing Stretch Receptors I Primary sensory neurones with enormous central arborizations. J. Comp. Neur. **172**, 409—430 (1977a)

Altman, J.S., Tyrer, N.M.: The Locust Wing Stretch Receptors II Variation, alternative pathways, and 'mistakes' in the central arborizations. J. Comp. Neur. **172**, 431—440 (1977b)

Auerbach, C.: The development of the legs, wings and halteres in wild type and some mutant strains of *Drosophila melanogaster*. Trans. roy. Soc. Edin. **58**, 787—815 (1936).

Baden, V.: Embryology of the nervous system in the grasshopper *Melanoplus differentialis*. J. Morph. **60**, 159—190 (1936).

Ball, E., Young, D.: Structure and development of the auditory system in the prothoracic leg of the cricket *Teleogryllus commodus* (Walker). II. Postembryonic development. Z. Zellforsch. **147**, 313—324 (1974).

Bate, C.M.: The mechanism of the pupal gin trap. I: Segmental gradients and the connections of the triggering sensilla. J. exp. Biol. **59**, 95—107 (1973).

Bate, C.M.: Embryogenesis of an insect nervous system. 1) A map of the thoracic and abdominal neuroblasts in *Locusta migratoria*. J. Embryol. exp. Morph. **35**, 107—123 (1976a)

Bate, C.M.: Pioneer Neurones in an insect embryo. Nature **260**, 54—56 (1976b)

Bate, C.M., Lawrence, P.A.: Gradients and the developing nervous system. In: Developmental Neurobiology of Arthropods (D. Young, Ed.). London: Cambridge University Press 1973.

Becker, H.J.: Über Röntgenmosaikflecken und Defektmutationen am Auge von *Drosophila* und die Entwicklungsphysiologie des Auges. Z. Vererbung **88**, 333—373 (1957).

Becker, H.J.: Genetic and variegation mosaics in the eye of *Drosophila*. Curr. Top. dev. Biol. **1**, 155—171 (1966).

Bentley, D.: Single Gene Cricket Mutations: Effects on Behavior, Sensilla, Sensory Neurons and Identified Interneurons. Science **187**, 760—764 (1975).

Berlese, A.: Gli Insetti. Milano (1909).

Bernard, F.: Recherches sur la morphogenèse des yeux composés d'arthropodes. Développement, croissance, réduction. Bull. Biol. **66**, 111—148 (1937).

Bodenstein, D.: Postembryonic development. In: Insect Physiology (K. D. Roeder, Ed.). New York: Wiley 1953, pp. 822—865.

Bolwig, N.: Senses and sense organs of the anterior end of the house fly larvae. Vidensk. Medd. dansk. naturh. Forenh Kbh. **109**, 81—217 (1946).

Bott, H.R.: Beiträge zur Kenntnis von *Gyrinus natator substriatus*. I: Lebensweise und Entwicklung. II: Der Sehapparat. Z. Morph. Ökol. Tiere **10**, 207—306 (1928).

Braitenberg, V.: Patterns of projection in the visual system of the fly I. Retina-lamina projections. Exp. Brain Res. **3**, 271—298 (1967).

Braitenberg, V.: Ordnung und Orientierung der Elemente im Sehsystem der Fliege. Kybernetik **7**, 235—242 (1970).
Bryant, P. J.: Cell lineage relationships in the imaginal wing disc of *Drosophila melanogaster*. Dev. Biol. **22**, 389—411 (1970).
Bryant, P. J., Schneiderman, H. A.: Cell lineage, growth and determination in the imaginal leg discs of *Drosophila melanogaster*. Dev. Biol. **20**, 263—290 (1969).
Burrows, M.: Physiological and morphological properties of the metathoracic common inhibitory neuron of the locust. J. comp. Physiol. **82**, 59—78 (1973a).
Burrows, M.: The morphology of an elevator and a depressor motoneuron of the hindwing of a locust. J. comp. Physiol. **83**, 165—178 (1973b).
Burrows, M.: Monosynaptic connexions between wing stretch receptors and flight motoneurons of the locust. J. exp. Biol. **62**, 189—219 (1975).
Busselmann, A.: Bau und Entwicklung der Raupenocellen der Mehlmotte *Ephestia kühniella* Zeller. Z. Morph. Ökol. Tiere **29**, 218—228 (1935).
Caesar, C. J.: Die Stirnaugen der Ameisen. Zool. Jb. (Anat.) **35**, 161—242 (1913).
Calabrese, R. L.: Crayfish Mechanoreceptive Interneurons: I The nature of ipsilateral excitatory inputs. J. comp. Physiol. **105**, 83—102 (1976a).
Calabrese, R. L.: Crayfish Mechanoreceptive Interneurons: II Bilateral interactions and inhibition. J. comp. Physiol. **105**, 103—114 (1976b).
Calabrese, R. L., Kennedy, D.: Multiple sites of spike initiation in a single dendritic system. Brain Res. **82**, 316—321 (1974).
Campos Ortega, J. A., Gateff, E. A.: The development of ommatidial patterning in metamorphosed eye imaginal disc implants of *Drosophila melanogaster*. Wilhelm Roux's Archiv **179**, 373—392 (1976)
Campos Ortega, J. A., Hofbauer, A.: Cell Clones and Pattern Formation: On the Lineage of Photoreceptor Cells in the Compound Eye of *Drosophila*. Wilhelm Roux's Archiv **181**, 227—245 (1977).
Campos Ortega, J. A., Strausfeld, N. J.: The columnar organisation of the second synaptic region of the visual system of *Musca domestica* L. Z. Zellforsch. **124**, 561—585 (1972).
Camhi, J.: Locust wind receptors. II: Interneurones in the cervical connective. J. exp. Biol. **50**, 349—362 (1969).
Caveney, S.: Stability of polarity in the epidermis of a beetle *Tenebrio molitor* L. Dev. Biol. **30**, 321—335 (1973).
Chiarodo, A. J.: The effects of mesothoracic leg disc extirpation on the postembryonic development of the nervous system of the blowfly *Sarcophaga bullata*. J. exp. Zool. **153**, 263—277 (1963).
Child, C. M.: Ein bisher wenig beachtetes antennales Sinnesorgan der Insekten mit besonderer Berücksichtigung der Culiciden und Chironomiden. Z. wiss. Zool. **58**, 475—528 (1894).
Clever, U.: Untersuchungen zur Zelldifferenzierung und Musterbildung der Sinnesorgane und des Nervensystems im Wachsmottenflügel. Z. Morph. Ökol. Tiere **47**, 201—248 (1958).
Clever, U.: Über experimentelle Modifikationen des Geäders und die Beziehung zwischen den Versorgungssystemen im Schmetterlingsflügel an *Galleria mellonella* L. Wilhelm Roux' Arch. Entw. Mech. Org. **151**, 242—279 (1959).
Clever, U.: Der Einfluß der Sinneszellen auf die Borstenentwicklung bei *Galleria mellonella* L. Wilhelm Roux' Arch. Entw. Mech. Org. **152**, 137—159 (1960).
Corneli, W.: Von dem Aufbau des Sehorgans der Blattwespenlarven und der Entwicklung des Netzauges. Zool. Jb. (Anat.) **46**, 573—608 (1924).
Counce, S. J., Waddington, C. H. (Eds.): Developmental Systems: Insects, Vols. I and II. London-New York: Academic 1973.
Deak, I. I.: Demonstration of sensory neurones in the ectopic cuticle of spineless-aristapedia, a homoeotic mutant of *Drosophila*. Nature **260**, 252—254 (1976)
Debaisieux, P.: Les yeux de crustacés: Structure, développement, réactions à éclairement. Cellule **50**, 9—122 (1944).
Debaisieux, P.: Les poils sensoriels d'arthropodes et histologie nerveuse. I: *Praunus flexuosus* Müll. et *Crangon crangon* L. Cellule **52**, 311—360 (1949).
de Lattin, G.: Untersuchungen an Isopodenaugen. Zool. Jb. (Anat.) **65**, 417—468 (1939).
Dethier, V. G.: The antennae of lepidopterous larvae. Bull. Mus. comp. Zool. Harvard **87**, 455—507 (1941).
Dietrich, W.: Die Facettenaugen der Dipteren. Z. wiss. Zool. **92**, 465—539 (1909).

Drescher, W.: Regenerationsversuche am Gehirn von *Periplaneta americana* unter Berücksichtigung von Verhaltensänderung und Neurosekretion. Z. Morph. Ökol. Tiere **48**, 576—649 (1960).
Edwards, J. S.: Postembryonic development and regeneration of the insect nervous system. Adv. Insect Physiol. **6**, 97—137 (1969).
Edwards, J. S., Palka, J.: Neural regeneration: delayed formation of central contacts by insect sensory cells. Science **172**, 591—594 (1971).
Edwards, J. S., Palka, J.: Neural specificity as a game of cricket: some rules for sensory regeneration in *Acheta domesticus*. In: Developmental Neurobiology of Arthropods (D. Young, Ed.). London: Cambridge University Press 1973.
Edwards, J. S., Palka, J.: The cerci and abdominal giant fibres of the house cricket *Acheta donesticus*. I: Anatomy and physiology of normal adults. Proc. roy. Soc. B. **185**, 83—103 (1974).
Edwards, J. S., Palka, J.: Neural Generation and Regeneration in Insects. In: 'Simpler Networks and Behavior', John C. Fentress Ed., Sinauer Associates. Massachusetts (1976).
Edwards, J. S., Sahota, T. S.: Regeneration of a sensory system: the formation of central connections by normal and transplanted cerci of the house cricket *Acheta domesticus*. J. exp. Zool. **166**, 387—396 (1967).
Egelhaaf, A., Berndt, P., Kuthe, H.-W.: Mitosenverteilung und ^{3}H-Thymidin-Einbau in der proliferierenden Augenanlage von *Ephestia kühniella* Zeller. Wilhelm Roux' Archiv Entw. Biol. **178**, 185—202 (1975).
Elofsson, R., Dahl, E.: The optic neuropiles and chiasmata of Crustacea. Z. Zellforsch. **107**, 343—360 (1970).
El Shatoury, H. H.: Differentiation and metamorphosis of the imaginal optic glomeruli in *Drosophila*. J. Embryol. exp. Morph. **4**, 240—247 (1956).
Fielden, A., Hughes, G. M.: Unit activity in the abdominal nerve cord of a dragonfly nymph. J. exp. Biol. **39**, 31—44 (1962).
Franzl, W.: Die Cytogenese der bipolaren Sinneszellen bei der Larve von *Dytiscus* sp. Z. Zellforsch. **31**, 54—59 (1941).
Garcia Bellido, A.: Pattern reconstruction by dissociated imaginal disc cells of *Drosophila melanogaster*. Dev. Biol. **14**, 278—306 (1966).
Gloor, H., Kobel, H.: *Antennapedia* (*SS^{Anq}*) eine homoeotische Mutante bei *Drosophila hydei* Sturtevant. Rev. suisse Zool. **73**, 229—252 (1966).
Graber, V.: Die tympanalen Sinnesapparate der Orthopteren. Denkschr. Akad. Wien. Math.-Naturwiss. **36**, 1—140 (1876).
Green, S. M., Lawrence, P. A.: Recruitment of epidermal cells by the developing eye of *Oncopeltus* (Hemiptera). Wilhelm Roux' Arch. Entw. Biol. **177**, 61—65 (1975).
Gymer, A., Edwards, J. S.: The development of the insect nervous system 1. An analysis of postembryonic growth in the terminal ganglion of *Acheta domesticus*. J. Morphol. **123**, 191—197 (1967).
Haffer, O.: Bau und Funktion der Sternwarzen von *Saturnia pyri* Schiff. und die Haarentwicklung der Saturnid Raupen. Arch. Naturg. **87**, 110—122 (1921).
Hanson, T. E.: Neurogenesis in the eye and optic tracts of *Drosophila*. Ann. Rep. Div. Biol. Caltech., 1972, p. 41.
Hanson, T. E., Jiang, Y.-H., Lee, J. Y.: Growth cone dynamics in the lamina of *Drosophila*. Ann. Rep. Div. Biol. Caltech., 1972, pp. 41—42.
Hanson, T. E., Ready, D. F., Benzer, S.: Use of mosaics in the analysis of pattern formation in the retina of *Drosophila*. Ann. Rep. Div. Biol. Caltech., 1972, p. 40.
Hassenfuss, I.: Vergleichend morphologische Untersuchung der sensorischen Innervierung der Rumpfwand der Larven von *Rhyacophila nubila* Zett. (Trichoptera) und *Galleria mellonella* L. (Lepidiptera). Zool. Jb. (Anat.) **90**, 1—54; 175—253 (1973a).
Hassenfuss, I.: Über die Beziehung zwischen sensorischer Innervierung und primären Segmentgrenzen bei Arthropoden. Verh. dtsch. zool. Ges. **66**, 71—75 (1973b).
Heller, R., Edwards, J. S.: Regeneration of the compound eye in *Acheta domesticus*. Amer. Zool. **8**, 786 (1968).
Henke, K.: Über Zelldifferenzierung im Integument der Insekten und ihre Bedingungen. J. Embryol. exp. Morph. **1**, 217—226 (1953).
Henke, K., Rönsch, G.: Über Bildungsgleichheiten in der Entwicklung epidermaler Organe und die Entstehung des Nervensystems im Flügel der Insekten. Naturwissenschaften **14**, 335—336 (1951).

Heymons, R.: Die Embryonalentwicklung von Dermapteren und Orthopteren unter besonderer Berücksichtigung der Keimblätterbildung. Jena: Fischer 1895.
Hilton, W. A.: The body sense hairs of lepidopterons larvae. Amer. Nat. **36**, 561—578 (1902).
Hilton, W. A.: Afferent and efferent pathways in an abdominal segment of an insect. J. comp. Neurol. **36**, 299—308 (1924).
Hinke, W.: Das relative postembryonale Wachstum der Hirnteile von *Culex pipiens*, *Drosophila melanogaster* und *Drosophila*-Mutanten. Z. Morph. Ökol. Tiere **50**, 81—118 (1961).
Hofbauer, A., Campos Ortega, J. A.: Cell clones and pattern formation: Genetic eye mosaics in *Drosophila melanogaster*. Wilhelm Roux's Archiv **179**, 275—289 (1976).
Holmgren, N.: Termitenstudien. K. svenska vetensk. Akad. Handl. **44**, 1—215 (1909).
Homann, H.: Die Augen der Araneae. Anatomie, Ontogenie und Bedeutung für die Systematik (Chelicerata, Arachnida). Z. Morph. Ökol. Tiere **69**, 201—272 (1971).
Horridge, G. A.: In: Structure and Function in the Nervous Systems of Invertebrates. San Francisco and London: Freeman 1965.
Horridge, G. A., Meinertzhagen, I. A.: The accuracyof the patterns of connexions of the first and second order neurons of the visual system of *Calliphora*. Proc. roy. Soc. B. **175**, 69—82 (1970).
Hughes, G. M., Wiersma, C. A. G.: Neuronal pathways and synaptic connexions in the abdominal cord of the crayfish. J. exp. Biol. **37**, 291—307 (1960).
Hunt, R. K., Jacobson, M.: Development and stability of positional information in *Xenopus* retinal ganglion cells. Proc. nat. Acad. Sci. (Wash.) **69**, 780—783 (1972).
Hunt, R. K., Jacobson, M.: Neuronal specificity revisited. Curr. Top. dev. Biol. **9**, 203—259 (1974).
Hyde, C. A. T.: Regeneration, postembryonic induction and cellular interaction in the eye of *Periplaneta americana*. J. Embryol. exp. Morph. **27**, 367—379 (1972).
Imberski. R. B.: The effect of 5-fluorouracil on the development of the adult eye in *Ephestia kühniella*. J. exp. Zool. **166**, 151—162 (1967).
Jägers-Röhr, E.: Untersuchungen zur Morphologie und Entwicklung der Scolopidialorgane bei der Stabheuschrecke *Carausius morosus* Br. Biol. Zbl. **87**, 393—409 (1968).
Jösting, E. A.: Die Innervierung des Skelettmuskelsystems des Mehlwurms (*Tenebrio molitor* L. Larve). Zool. Jb. (Anat.) **67**, 381—460 (1942).
Johannsen, O. A.: Eye structure in normal and eye mutant *Drosophila*. J. Morph. **39**, 337—349 (1924).
Kennedy, D.: The comparative physiology of invertebrate central neurones. Adv. comp. Physiol. Biochem. **2**, 117—184 (1966).
Kennedy, D.: Crayfish interneurons. The Physiologist **14**, 5—30 (1971).
Kennedy, D.: Connections among neurons of different types in crustacean nervous systems. In: The Neurosciences Third Study Program (F. O. Schmitt and F. G. Worden, Eds.). Cambridge Mass.: MIT Press 1974.
Kennedy, D., Evoy, W. H., Fields, H. L.: The unit basis of some crustacean reflexes. Symp. Soc. exp. Biol. **20**, 75—109 (1966).
Kim, C. W.: Development of the chordotonal organ, olfactory organ and their nerves in the labial palp of *Pieris rapae*. Bull. Dep. Biol. Korea Univ. **3**, 1—8 (1961).
Kirchhoffer, O.: Die Entwicklung des Komplexauges nebst Ganglion opticum von *Dermestes vulpinus* Arch. Naturg. **76**, 1—27 (1910)
Kirschfeld, K.: Die Projektion der optischen Umwelt auf das Raster der Rhabdomere im Komplexauge von *Musca*. Exp. Brain Res. **3**, 248—270 (1967).
Kopec, S.: Untersuchungen über die Regeneration von Larvalorganen und Imaginalscheiben bei Schmetterlingen. Wilhelm Roux' Arch. Entw. Mech. Org. **37**, 440—472 (1913).
Kopec, S.: Mutual relationship in the development of the brain and eyes of Lepidoptera. J. exp. Zool. **36**, 459—468 (1922).
Krafka, J.: Development of the compound eye of *Drosophila melanogaster* and its bar eyed mutant. Biol. Bull. **47**, 143—146 (1924).
Krumins, R.: Die Borstenentwicklung bei der Wachsmotte *Galleria mellonella* L. Biol. Zbl. **71**, 183—210 (1957).
Laschat, F.: Die Embryonale und postembryonale Entwicklung der Netzaugen und Ocellen von *Rhodnius prolixus*. Z. Morph. Ökol. Tiere **40**, 314—347 (1944).
Lawrence, P. A.: Gradients in the insect segment: the orientation of hairs in the milkweed bug, *Oncopeltus fasciatus*. J. exp. Biol. **44**, 607—620 (1966a).

Lawrence,P. A.: Development and determination of hairs and bristles in the milkweed bug, *Oncopeltus fasciatus* (Lygaeidae, Hemiptera). J. Cell. Sci. **1**, 475—498 (1966b).

Lawrence,P. A.: Cellular differentiation and pattern formation during metamorphosis of the milkweed bug *Oncopeltus*. Dev. Biol. **19**, 12—40 (1969).

Lawrence,P. A.: Polarity and patterns in the postembryonic development of insects. Adv. Insect Physiol. **7**, 197—260 (1970).

Lawrence,P. A.: The development of spatial patterns in the integument of insects. In: Developmental Systems: Insects, Vol. 2 (S. J. Counce and C. H. Waddington, Eds.). London-New York: Academic 1973a.

Lawrence,P. A.: A clonal analysis of segment development in *Oncopeltus* Hemiptera. J. Embryol. exp. Morph. **30**, 681—699 (1973b).

Lawrence,P. A.: The structure and properties of a compartment border: the intersegmental boundary in *Oncopeltus*. In: Cell Patterning. Ciba Symposium **29**, 3—23 (1975).

Lawrence, P. A., Crick, F. H. C., Munro, M.: A gradient of positional information in an insect *Rhodnius*. J. Cell. Sci. **11**, 815—854 (1972)

Lawrence,P. A., Shelton,P. M. J.: The determination of polarity in the developing insect retina. J. Embryol. exp. Morph. **33**, 471—486 (1975).

Lees, A. D., Waddington,C. H.: The development of the bristles in normal and some mutant types of *Drosophila melanogaster*. Proc. roy. Soc. B. **131**, 87—110 (1942).

Lerum,J. E.: The postembryonic development of the compound eye and optic ganglia in dragonflies. Proc. Iowa Acad. Sci. **75**, 416—432 (1968).

Letourneau, J. G.: Addition of sensory structures and associated neurons to the crayfish telson during development. J. comp. Physiol. **110**, 13—23 (1976a).

Letourneau, J. G.: Somatotopic organization of afferent axons in peripheral nerves. J. comp. Physiol. **110**, 25—32 (1976b).

Letourneau,P. C.: Cell-to-substratum adhesion and guidance of axonal elongation. Dev. Biol. **44**, 92—101 (1975).

Lew,G. T.-W.: Head characters of the Odonata. Ent. Am. **14**, 41—97 (1975).

Lewerenz,G.: Untersuchungen über Wachstum und Structur der im Thoraxbereich liegenden Imaginalscheiben von *Calliphora erythrocephala*. Dtsch. Ent. Z. **8**, 222—249 (1961).

Lewis,E.: Genes and developmental pathways. Amer. Zool. **3**, 33—56 (1963).

Link,E.: Über die Stirnaugen einiger Lepidopteren und Neuropteren Zool. Anz. **33**, 445—450 (1908).

Link,E.: Über die Stirnaugen der hemimetabolen Insekten. Zool. Jb. (Anat.) **27**, 281—376 (1909).

Lissmann,H. W., Wolsky,A.: Funktion der an Stelle eines Auges regenerierten Antennule bei *Potamobius leptodactylus* Eschh. Z. vergl. Physiol. **19**, 555—573 (1933).

Locke,M.: The cuticular pattern in an insect *Rhodnius prolixus* Stal. J. exp. Biol. **36**, 459—477 (1959).

Lopresti,V. A., Macagno,E. R., Levinthal,C.: Structure and development of neuronal connections in isogenic organisms: Cellular interactions in the development of the optic lamina of *Daphnia*. Proc. nat. Acad Sci. (Wash.) **70**, 433—437 (1973).

Lopresti,V., Macagno,E. R., Levinthal,C.: Structure and development of neuronal connections in isogenic organisms: Transient gap junctions between growing optic axons and lamina neuroblasts. Proc. nat. Acad. Sci. (Wash.) **71**, 1098—1102 (1974).

Ludwig,C. E.: Embryology and morphology of the larval head of *Calliphora erythrocephala* (Meigen). Microent. **14**, 73—111 (1949).

Lukoschus,F.: Uber Bau und Entwicklung des Chordotonalorgans am Tibia-Tarsus Gelenk der Honigbiene. Z. Bienenforsch. **6**, 48—52 (1962).

Macagno,E. R., Lopresti,V., Levinthal,C.: Structure and development of neuronal connections in isogenic organisms: variations and similarities in the optic system of *Daphnia magna*. Proc. nat. Acad. Sci. (Wash.) **70**, 57—61 (1973).

Malzacher,P.: Embryogenese des Gehirns von *Carausius* und *Periplaneta*. Z. Morph. Ökol. Tiere **62**, 103—161 (1968).

Marcus,W.: Untersuchungen über die Polarität der Rumpfhaut der Schmetterlinge. Wilhelm Roux' Arch. Entw. Mech. Org. **154**, 56—102 (1962).

Maynard,D. M., Cohen,M. J.: The function of a hetermorph antennule in a spiny lobster *Panulirus argus*. J. exp. Biol. **43**, 55—78 (1965).

Meinertzhagen, I. A.: Erroneous projection of retinula axons beneath a dislocation in the retinal equator of *Calliphora*. Brain Res. **41**, 39—49 (1972).
Meinertzhagen, I. A.: Development of the compound eye and optic lobe of insects. In: Developmental Neurobiology of Arthropods (D. Young, Ed.). London: Cambridge University Press 1973.
Meinertzhagen, I. A.: The development of neuronal connection patterns in the visual systems of insects. In: Cell Patterning. Ciba Symposium **29**, 265—288 (1974).
Melamed, J., Trujillo Cenoz, O.: The fine structure of the eye imaginal discs in muscoid flies. J. ultrastruct. Res. **51**, 79—93 (1975).
Miner, N.: Integumental specification of sensory fibres in the development of cutaneous local sign. J. comp. Neurol. **105**, 161—170 (1956).
Mobbs, P. G.: Development of the locust ocellus. Nature **264**, 269—271 (1976).
Moroff, T.: Entwicklung und phylogenetische Bedeutung des Medianauges bei Crustaceen. Zool. Anz. **40**, 11—25 (1912).
Mouze, M.: Etude expérimentale des facteurs morphogenetiques et hormonaux réglant la croissance oculaire des Aesehnidae (Odonates, Anisopteres). Odonatol. **1**, 221—232 (1972a).
Mouze, M.: Croissance et métamorphose de l'appareil visuel des *Aeschnidae* (Odonata). Internat. J. Insect Morphol. Embryol. **1**, 181—200 (1972).
Mouze, M.: Interactions de l'oeil et du lobe optique au cours de la croissance postembryonnaire des Insectes odonates. J. Embryol. exp. Morph. **31**, 377—407 (1974).
Mouze, M.: Croissance et régénération de l'oeil de la larve d'*Aesehna cyanea* Müll. (Odonate, Anisoptère). Wilhelm Roux' Arch. Entw. Mech. Org. **176**, 267—283 (1975).
Murphey, R. K., Mendenhall, B., Palka, J., Edwards, J. S.: Deafferentation slows the growth of specific dendrites of identified giant interneurons. J. comp. Neurol. **159**, 407—418 (1975).
Murray, F. V., Tiegs, O. W.: The metamorphosis of *Calandra oryzae*. Quart. J. micr. Sci. **77**, 405—495 (1935).
Nordlander, R. H., Edwards, J. S.: Morphological cell death in the postembryonic development of the insect optic lobes. Nature **218**, 780—781 (1968).
Nordlander, R. H., Edwards, J. S.: Postembryonic brain development in the Monarch Butterfly *Danaus plexippus plexippus* L. I: Cellular events during brain morphogenesis. Wilhelm Roux' Arch. Entw. Mech. Org. **162**, 197—217 (1969a).
Nordlander, R. H., Edwards, J. S.: Postembryonic brain development in the Monarch Butterfly *Danaus plexippus plexippus* L. II: The optic lobes. Wilhelm Roux' Arch. Entw. Mech. Org. **163**, 197—220 (1969b).
Nordlander, R. H., Edwards, J. S.: Postembryonic brain development in the monarch butterfly, *Danaus plexippus plexippus*. III: Morphogenesis of centers other than the optic lobes. Wilhelm Roux' Arch. Entw. Mech. Org. **164**, 247—260 (1970).
Nübler Jung, K.: Diplomarbeit. Universität Freiburg (1972).
Nüesch, H.: Über den Einfluß der Nerven auf die Muskelentwicklung bei *Telea polyphemus* (Lepid.). Rev. suisse Zool. **59**, 294—301 (1952).
Nüesch, H.: The role of the nervous system in insect morphogenesis and regeneration. Ann. Rev. Entomol. **13**, 27—44 (1968).
O'Shea, M., Rowell, C. H. F., Williams, J. L. D.: The anatomy of a locust visual interneurone—the descending contralateral movement detector. J. exp. Biol. **60**, 1—12 (1974).
Pabst, H., Kennedy, D.: Cutaneous mechanoreceptors influencing motor output in the crayfish abdomen. Z. vergl. Physiol. **57**, 190—208 (1967).
Palka, J., Edwards, J. S.: The cerci and abdominal giant fibres of the house cricket *Acheta domesticus*. II: Regeneration and effects of chronic deprivation. Proc. roy. Soc. B. **185**, 105—121 (1974).
Palka, J., Schubiger, M.: Central connections of receptors on rotated and exchanged cerci of crickets. Proc. nat. Acad. Sci. (Wash.) **72**, 966—969 (1975).
Panov, A. A.: Bau des Insektengehirns während der postembryonalen Entwicklung. Ent. Oboz. **36**, 269—284 (1957).
Panov, A. A.: The structure of the insect brain during successive stages of postembryonic development. III: Optic lobes. Ent. Oboz. **39**, 55—68 (1960).
Panov, A. A.: The structure of the insect brain at successive stages in postembryonic development. IV: The olfactory centre. Ent. Oboz. **40**, 140—145 (1961).
Panov, A. A.: Correlations in the ontogenetic development of the central nervous system in the house cricket *Gryllus domesticus* L. and the house cricket *Gryllotalpa gryllotalpa* L. (Orthoptera, Grylloidea). Ent. Oboz. **45**, 179—185 (1966).

Parker, G. H.: The compound eyes of crustaceans. Bull. Mus. comp. Zool. Harvard **21**, 45—140 (1891).

Peabody, E. B.: Development of the eye of the isopod *Idothea*. J. Morph. **64**, 519—553 (1939).

Peters, W.: Die Sinnesorgane an den Labellen von *Calliphora erythrocephala* Mg. (Diptera). Z. Morph. Ökol. Tiere **55**, 259—320 (1965).

Pflugfelder, O.: Vergleichend anatomische, experimentelle und embryologische Untersuchungen über das Nervensystem und die Sinnesorgane der Rhynchoten. Zoologica **34** (93), 1—102 (1936/7).

Pflugfelder, O.: Die Entwicklung der optischen Ganglien von *Culex pipiens*. Zool. Anz. **117**, 31—36 (1937).

Pflugfelder, O.: Die Entwicklung embryonaler Teile von *Carausius* (*Dixippus*) *morosus* in der Kopfkapsel von Larven und Imagines. Biol. Zbl. **66**, 372—387 (1947).

Piepho, H.: Über die polare Orientierung der Bälge und Schuppen auf dem Schmetterlingsrumpf. Biol. Zbl. **74**, 467—474 (1955).

Pipa, R. L.: Proliferation, movement and regression of neurons during the postembryonic development of insects. Developmental Neurobiology of Arthropods (D. Young, Ed.). London: Cambridge University Press 1973.

Postlethwait, J. H., Schneiderman, H. A.: Pattern formation and determination in the antenna of the homoeotic mutant *Antennapedia* of *Drosophila melanogaster*. Dev. Biol. **25**, 606—640 (1971).

Power, M. E.: The effect of reduction in numbers of ommatidia upon the brain of *Drosophila melanogaster*. J. exp. Zool. **94**, 33—72 (1943).

Prüffer, J.: Untersuchungen über die Innervierung der Fühler bei *Saturnia pyri* L. Zool. Jb. (Anat.) **51**, 1—46 (1929).

Ready, D. F., Hanson, T. E., Benzer, S.: Development of the *Drosophila* retina, a neurocrystalline lattice. Dev. Biol. **53**, 217—240 (1976).

Redikorzew, W.: Untersuchungen über der Bau der Ocellen der Insekten. Z. wiss. Zool. **68**, 581—624 (1900).

Reinhardt, C. A., Hodgkin, N. M., Bryant, P. J.: Wound healing in the imaginal discs of *Drosophila*. I Scanning electron microscopy of normal and healing wing discs. Dev. Biol. **60**, 238—257 (1977).

Rice, M. J.: Insect mechanoreceptor mechanisms. In: Sensory Physiology and Behaviour. New York-London: Plenum 1975.

Roberts, M. J.: The structure of the mouth parts of some calypterate dipteran larvae in relation to their feeding habits. Acta zool. (Stockh.) **52**, 171—188 (1971).

Roberts, P.: Mosaics involving *aristapedia* a homoeotic mutant of *Drosophila melanogaster*. Genetics **49**, 593—598 (1964).

Rönsch, G.: Entwicklungsgeschichtliche Untersuchungen zur Zelldifferenzierung am Flügel der Trichoptere *Limnophilus flavicornis* Fabr. Z. Morph. Ökol. Tiere **43**, 1—62 (1954).

Roth, L. M., Barth, R. H.: The sense organs employed by cockroaches in mating behavior. Behavior **28**, 58—94 (1967).

Sánchez, Y., Sánchez, D.: Sobre el desarrollo de los elementos nerviosos en la retina del *Pieris brassicae* L. Trab. Lab. Invest. Biol. **17**, 1—63 (1919a).

Sánchez, Y., Sánchez, D.: Sobre el desarrollo de los elementos nerviosos en la retina del *Pieris brassicae* L. Trab. Lab. Invest. Biol. **17**, 117—180 (1919b).

Sandeman, D. C., Luff, S. E.: Regeneration of the antennules in the Australian freshwater crayfish *Cherax destructor*. J. Neurobiol. **5**, 475—488 (1974).

Sandeman, D. C., Okajima, A.: Statocyst induced eye movements in the crab *Scylla serrata*. III: The anatomical projections of sensory and motor neurons and the responses of the motor neurons. J. exp. Biol. **59**, 17—38 (1973).

Sanes, J. R., Hildebrand, J. G.: Nerves in the antennae of pupal *Manduca sexta* Johannsen (Lepidoptera: Sphingidae). Wilhelm Roux' Arch. Entw. Mech. Org. **178**, 71—78 (1975).

Sanes, J. R., Hildebrand, J. G.: Origin and morphogenesis of sensory neurons in an insect antenna. Dev. Biol. **51**, 300—319 (1976).

Sanes, J. R., Hildebrand, J. G., Prescott, D. J.: Differentiation of insect sensory neurons in the absence of their normal synaptic targets. Dev. Biol. **52**, 121—127.

Sbrenna, G.: Postembryonic growth of the ventral nerve cord in *Schistocerca gregaria* Forsk. (Orthoptera: Acrididae). Boll. Zool. **38**, 49—74 (1971).

Schafer, R.: Postembryonic development in the antenna of the cockroach *Leucophaea maderae*: Growth, regeneration and the development of the adult pattern of sense organs. J. exp. Zool. **183**, 353—364 (1973).

Schafer, R., Sanchez, T. V.: Antennal sensory system of the cockroach *Periplaneta americana*: postembryonic development and morphology of the sense organs. J. comp. Neurol. **149**, 335—354 (1973).
Schafer, R., Sanchez, T. V.: Juvenile hormone inhibits the differentiation of olfactory sense organs during the postembryonic development of cockroaches. J. Insect Physiol. **20**, 965—974 (1974).
Schoeller, J.: Recherches descriptives et expérimentales sur la cephalogenèse de *Calliphora erythrocephala* (Meigen) au cours des développements embryonnaire et postembryonnaire. Arch. Zool. exp. Gén. **103**, 1—216 (1964).
Schön, A.: Bau und Entwicklung des tibialen Chordotonalorgans bei der Honigbiene und bei Ameisen. Zool. Jb. (Anat.) **31**, 439—472 (1912).
Schwabe, J.: Beiträge zur Morphologie und Histologie der tympanalen Sinnesapparate der Orthopteren. Zoologica **50**, 1—154 (1906).
Seidel, F.: Der Anlagenplan in Libellenei. Wilhelm Roux' Arch. Entw. Mech. Org. **132**, 671—751 (1935).
Seiler, W.: Beiträge zur Kenntnis der Ocellen der Ephemeriden. Zool. Jb. (Anat.) **22**, 1—40 (1905).
Shelton, P. M. J.: The development of the compound eye. In 'Insect Development', (P. A. Lawrence, Ed.), Symp. Roy. Ent. Soc. **8**, 152—169 (1976).
Shelton, P. M. J., Anderson, H. J., Eley, S.: Cell lineage and cell determination in the developing compound eye of the cockroach *Periplaneta americana*. J. Embryol. exp. Morphol. **39**, 235—252 (1977).
Shelton, P. M. J., Lawrence, P. A.: Structure and development of ommatidia in *Oncopeltus fasciatus*. J. Embryol. exp. Morph. **32**, 337—353 (1974).
Shiraishi, A., Tanabe, Y.: The proboscis extension response and tarsal labellar chemosensory hairs in the blowfly. J. comp. Physiol. **92**, 161—179 (1974).
Slifer, E. H.: Morphology and development of the femoral chordotonal organs of *Melanoplus differentialis* (Orthoptera, Acrididae). J. Morph. **58**, 615—637 (1935).
Snodgrass, R. E.: The morphology of insect sense organs and the sensory nervous system. Smithson Misc. Coll. **77** (8), 1—80 (1926).
Sorokina Agafanowa, M.: Das Verhalten des peripheren Nervensystems der Insekten in der Metamorphose. Z. Anat. Entw. Gesch. **74**, 318—337 (1924).
Spreij, T. E.: Morphological and histochemical changes during the development of some of the imaginal discs of *Calliphora erythrocephala*. Neth. J. Zool. **20**, 253—275 (1970).
Spreij, T. E., de Priester, W.: Myofilaments in the lemnoblast cells of the larval optic nerve in *Calliphora erythrocephala*. Neth. J. Zool. **21**, 221—264 (1972).
Steele, M. I.: Regeneration in compound eyes of crustacea. J. exp. Zool. **5**, 163—243 (1907).
Stern, C.: Two or three bristles. Amer. Sci. **42**, 213—247 (1954).
Stern, C.: Genetic Mosaics and Other Essays. Cambridge Mass.: Harvard University Press (1968).
Stocker, R. F., Edwards, J. S., Palka, J., Schubiger, G.: Projection of sensory neurons from a homeotic mutant appendage *Antennapedia* in *Drosophila melanogaster*. Dev. Biol. **52**, 210—220 (1976).
Stossberg, M.: Die Zellvorgänge bei der Entwicklung der Flügelschuppen von *Ephestia kühniella* Z. Z. Morph. Ökol. Tiere **34**, 173—206 (1938).
Stumpf, H.: Further studies on gradient dependent diversification in the pupal cuticle of *Galleria mellonella*. J. exp. Biol. **49**, 49—60 (1968).
Titschack, E.: Der Fühlernerv der Bettwanze, *Cimes lectularius* L. und sein zentrales Endgebiet. Zool. Jb. (allg. Zool.) **45**, 437—462 (1928).
Tokunaga, C., Stern, C.: The developmental autonomy of extra sex combs in *Drosophila melanogaster*. Dev. Biol. **11**, 50—81 (1965).
Trujillo Cenoz, O., Melamed, J.: Compound Eye of Dipterans: anatomical basis for integration – an electron microscope study. J. Ultr. Res. **16**, 395—398 (1966).
Trujillo Cenoz, O., Melamed, J.: The development of the retina-lamina complex in muscoid flies. J. ultrastruct. Res. **42**, 554—581 (1973).
Trujillo Cenoz, O., Melamed, J.: Transplanted eyes of Dipterans. Some aspects of their structural organisation. Naturwissenschaften **62**, 42 (1975).
Umbach, W.: Entwicklung und Bau des Komplexauges der Mehlmotte *Ephestia kühniella* Zeller nebst einigen Bemerkungen über die Entstehung der optischen Ganglien. Z. Morph. Ökol. Tiere **28**, 561—594 (1934).
Van Ruiten, Spreij, T. E.: The ultrastructure of the developing leg disk of *Calliphora erythrocephala*. Z. Zellforsch. **147**, 373—450 (1974).

Viallanes, H.: Le ganglion optique de quelques larves de Diptères *(Musca, Eristalis, Stratiomys)*. Ann. Sci. nat. (Zool.) (6) **19**, 2—34 (1885).
Viallanes, H.: Sur quelques points de l'histoire du développement embryonnaire de la mante religieuse (*Mantis religiosa*). Ann. Sci. nat. (Zool.) (7) **11**, 283—328 (1891).
Vogel, R.: Über die Innervierung der Schmetterlingsflügel und über den Bau und die Verbreitung der Sinnesorgane auf denselben. Z. wiss. Zool. **98**, 68—134 (1911).
Vogel, R.: Zur Kenntnis des feineren Baues der Geruchsorgane der Wespen und Bienen. Z. wiss. Zool. **120**, 281—324 (1923).
Vogt, M.: Zur labilen Determination der Imaginalscheiben von *Drosophila*. I: Verhalten verschiedenartiger Imaginalanlagen bei operativer Defektsetzung. Biol. Zbl. **65**, 223—238 (1946).
Vom Rath, O.: Über die Hautsinnesorgane der Insekten. Z. wiss. Zool. **46**, 413—454 (1888).
Von Reizenstein, W.: Untersuchungen über die Entwicklung der Stirnaugen von *Periplaneta orientalis* und *Cloeon*. Zool. Jb. (Anat.) **21**, 161—180 (1905).
Wachmann, E.: Untersuchungen zur Entwicklungsphysiologie des Komplexauges der Wachsmotte *Galleria mellonella* L. Wilhelm Roux' Arch. Entw. Mech. Org. **156**, 145—183 (1965).
Waddington, C. H., Perry, M. M.: The ultrastructure of the developing eye of *Drosophila*. Proc. roy. Soc. B **153**, 155—178 (1960).
Wahl, B.: Über die Entwicklung der hypodermalen Imaginalscheiben im Thorax und Abdomen der Larve von *Eristalis* Latr. Z. Wiss. Zool. **70**, 171—191 (1901).
Weber, H.: Grundriss der Insektenkunde. Jena: Fischer 1966.
Wheeler, W. M.: A contribution to insect embryology. J. Morph. **8**, 1—160 (1893).
White, R. H.: Analysis of the development of the compound eye in the mosquito *Aedes aegypti*. J. exp. Zool. **148**, 223—240 (1961).
White, R. H.: Evidence for the existence of a differentiation center in the developing eye of the mosquito. J. exp. Zool. **152**, 139—148 (1963).
Wiersma, C. A. G.: On the functional connections of single units in the central nervous system of the crayfish *Procambarus clarkii* Girard. J. comp. Neurol. **110**, 421—471 (1958).
Wiersma, C. A. G., Bush, B. M. H.: Functional neuronal connections between the thoracic and abdominal cords of the crayfish, *Procambarus clarkii* Girard. J. comp. Neurol. **121**, 270—235 (1963).
Wiersma, C. A. G., Hughes, G. M.: On the functional anatomy of neuronal units in the abdominal cord of the crayfish *Procambarus clarkii* Girard. J. comp. Neuron. **116**, 209—228 (1961).
Wiese, K.: Mechanoreceptors for near-field water displacements in crayfish. J. Neurophysiol. **39**, 816—831 (1976).
Wiese, K., Calabrese, R. L., Kennedy, D.: Integration of directional mechanosensory input by crayfish interneurons. J. Neurophysiol. **39**, 834—843 (1976).
Wigglesworth, V. B.: Local and general factors in the development of "pattern" in *Rhodnius prolixus* (Hemiptera). J. exp. Biol. **17**, 180—200 (1940).
Wigglesworth, V. B.: The origin of sensory neurones in an insect *Rhodnius prolixus* (Hemiptera). Quart. J. micr. Sci. **94**, 93—112 (1953).
Wigglesworth, V. B.: The Principles of Insect Physiology, 7th ed. London: Chapman and Hall 1972.
Wolpert, L.: Positional information and the spatial pattern of cellular differentiation. J. theoret. Biol. **25**, 1—47 (1969).
Wolpert, L.: Positional information and pattern formation. Curr. Top. dev. Biol. **6**, 183—224 (1971).
Wolsky, A.: Experimentelle Untersuchungen über die Differenzierung der zusammengesetzten Augen des Seidenraupenspinners (*Bombyx mori* L.). Wilhelm Roux's Archiv **138**, 335—344 (19
Wolsky, A.: The analysis of eye development in insects. Trans. N. Y. Acad. Sci. Ser. II **18**, 592—596 (1956).
Wolsky, A., Wolsky, M. de I.: Phase specific and regional differences in the development of the complex eye of the mulberry silkworm (*Bombyx mori* L.) after unilateral removal of the optic lobe of the brain in early pupal stages. Am. Zool. **11**, 679 (1971).
Young, D. (Ed.): Developmental Neurobiology of Arthropods. London: Cambridge University Press 1973.
Young, D., Ball, E.: Structure and development of the auditory system in the prothoracic leg of the cricket *Teleogryllus commodus* (Walker). I: Adult structure. Z. Zellforsch. **147**, 293—312 (1974a).
Young, D., Ball, E.: Structure and development of the tracheal organ in the mesothoracic leg of the cricket *Teleogryllus commodus* (Walker). Z. Zellforsch. **147**, 325—334 (1974b).
Zawarzin, A.: Histologische Studien über Insekten. II: Das sensible Nervensystem der *Aeschna* Larven. Z. wiss. Zool. **100**, 245—286 (1912).

Chapter 2

Continuous Nerve Cell Renewal in the Olfactory System

P. P. C. GRAZIADEI and G. A. MONTI GRAZIADEI, Tallahassee, Florida (USA)

With 23 Figures

Contents

A. Introduction

The olfactory system has been long reported as one of the most phylogenetically primitive (ARIENS KAPPERS et al., 1967), because of the peripheral location of its neurons and the relatively conspicuous development in lower vertebrates.

Recent observations indicate that the olfactory receptor neurons of mammals are continuously renewed. The basal cells of the olfactory neuroepithelium support the neurogenetic process throughout adult life and sustain the replacement of the neurons when extensive destruction of mature elements is experimentally induced. In this latter case a tumultuous neurogenetic process leads to a morphological and functional recovery of the system (GRAZIADEI, 1973a,b; GRAZIADEI and DEHAN, 1973; OLEY et al., 1975).

The persistence of neurogenesis in adult animals challenges our belief that neurons are not replaceable in sexually mature vertebrates (BIZZOZERO, 1894; LEBLOND and WALKER, 1956; MESSIER and LEBLOND, 1960; LEBLOND, 1964). References to regenerative phenomena in the nervous system of adult mammals have so far been limited to the regeneration of axons (CAJAL, 1928; GUTH, 1974; GUTH and CLEMENTE, 1975). In the case of the olfactory primary neuron, by constrast, the neuron as a whole is replaced and the regeneration encompasses both the axon and the perikaryon itself.

This chapter gives a survey of the embryological development, structural organization and regenerative capacity of the peripheral portion of the olfactory system. Emphasis will be placed on observations that illustrate the potential of the olfactory cells for turnover and for regeneration after injury.

B. Embryology

Contrasting views were proposed by early investigators on the origin and development of the olfactory neurons. MARSHALL (1878, 1879), BALFOUR (1877), KOELLIKER (1860, 1882, 1883) and BEARD (1885) maintained that the olfactory neurons originated from the walls of the cerebral hemispheres. CHIARUGI (1891) was of the opinion that both the olfactory placode and the encephalic vesicles might be involved in the formation of the olfactory cells. HIS (1889), CAJAL (1890), and DISSE (1901, 1902), in contrast, reported that the primary olfactory neurons differentiated exclusively from the olfactory placodes. These are visualized as symmetrical thickenings of the ectoderm on the ventrolateral aspect of the head. Further observations confirmed the development of the olfactory receptors from the placodes (HINSBERG, 1901), and also the detailed events of their differentiation (MARCHETTI, 1918; SIGGIA, 1937; MANFREDONIA, 1938). It is currently agreed that the neurons differentiate in the olfactory placodes, send their axons to contact the part of the telencephalon that will differentiate into the olfactory bulb, and establish synaptic contacts with the mitral cells at the level of the glomeruli.

Few authors have described the process of neurons differentiation in the olfactory placode. In human embryos the placode is recognizable at the 4.5-mm stage, but olfactory neurons are not (YOKOH, 1968, 1975). They were observed at $4\frac{1}{2}$ weeks by HIS (1889) and at 9.5 mm by PEARSON (1941a and b). In accordance with these findings, MASY (1955) described axons growing outside the placode at the 9-mm stage. VAN CAMPENHAUT (1956) noticed the first appearance of the placode in pig at 3–4 mm and the growth of axons outside the epithelium at the 7-mm stage. Neurons and supporting cells were previously recognized in the same animal when 6.5 mm in length (BEDFORD, 1904). Complete morphological differentiation of the olfactory neurons and the connection of their axons with the forebrain was observed by VAN CAMPENHOUT at the 6.5 mm stage (1956). This author believes that the Schwann cells of the olfactory nerve in swine differentiate from the olfactory placode at the 8-mm stage. In chicken the Schwann cells develop and migrate from the placode at 16 hs and the nerve reaches the telencephalon at 78 hrs (VAN CAMPENHOUT, 1956). In rabbit, rat and guinea-pig FILOGAMO and ROBECCHI (1969), using the Koelle method, identified the olfactory neuroblasts at 14 days of gestation. In chick embryos bipolar cells can be identified at the ultrastructural level after 3 days' incubation (ROBECCHI, 1970; GRAZIADEI, unpubl.). Our observations in mice (GRAZIADEI, unpubl.) are consistent with those of CUSCHIERI and BANNISTER (1975a and b), which show differentiated axons growing from the placode at 10 days of gestation.

The nervus terminalis and the vomeronasal nerve also originate from the olfactory placode. HIS (1889) demonstrated that the neuroblasts from which the nervus terminalis originate migrate from the placode to form a true ganglion closely associated with the olfactory nerve. The origin of the nervus terminalis neurons from the olfactory placode was also described in many vertebrates by DISSE (1897) and BEDFORD (1904). These ganglion cells supply the fibers of the *nervus terminalis*, which was described in adult animals by FRITSCH (1878) and then by a series of other authors (PINKUS, 1894, 1895; LOCY, 1905; BROOKOVER, 1908, 1914; DOELLKEN, 1909; HERRICK, 1909, 1910; MCKIBBEN, 1911, 1914; MCCOTTER, 1912, 1919; JOHNSTON, 1913; HUBER and GUILD, 1913; HOLMGREN, 1918/20; LARSELL, 1918, 1918/19, 1950; BIANCHI, 1931a and b; PEARSON, 1941a and b; GRUNEBERG, 1973). The vomeronasal organ was first described by JACOBSON (1811). Its embryological development from an invagination of the olfactory placode was established by a number of authors (SEYDEL, 1895; PETER, 1906; ZUCKERKANDL, 1910; ADDISON and RADEMAKER, 1927; HUMPHREY, 1940; PEARSON, 1941a and b; PARSONS, 1959; CUSCHIERI and BANNISTER, 1975).

C. Experimental Embryology

The peculiar embryological origin of the olfactory neurons from the olfactory placode, which is located outside the neural groove, aroused the interest of experimental embryologists, particularly in the classic era of neurology. Studies were concentrated on the effects of the removal or transplantation of the region where the primitive placode differentiates. Many authors agreed that the removal of the placode arrests the differentiation of the olfactory bulb and even of some telencephalic structures (BURR, 1916; CLAIRAMBAULT, 1968, 1971; CHANCOINE and CLAIRAMBAULT, 1975), and that the olfactory placode develops independently of the forebrain (BELL, 1906, 1907; EKMAN, 1924; MAY, 1927). BURR (1916, 1924, 1930) observed the rather interesting phenomenon that the transplanted olfactory placode can send its outgrowing fibers to the *pars dorsalis thalami* or, through the ophthalmic branch of the V nerve, to the ophthalmic ganglion. In both cases, the incoming fibers produce an hypertrophy of the receiving centers.

Connections of the olfactory placode with brain structures other than the olfactory bulb are not uncommon and were also reported by BELL (1906, 1907), MAY (1927), SPIRITO (1930), PIATT (1948), KOSCIUSZKO (1958), HARRISON (1958). The observations of LUNA (1915) are of particular interest. This author observed that the telencephalon is not essential for the differentiation of the olfactory organ; its absence, however, may result in partial atrophy of the organ. He noticed also that the transposed olfactory placode could send its fibers into the mesencephalon, where formations like glomeruli are established by the ramifications of the incoming olfactory fibers.

WADDINGTON and COHEN (1936) suggested that the forebrain might induce the formation of a nasal placode from nonpresumptive nasal epidermis in chick embryos, and ZWILLING (1934) claimed that the transplant of the brain in the ectoderm of the flank could induce an olfactory placode. In a later paper ZWILLING (1940) stated that the olfactory organ was induced by the roof of the archentron. Most of these experiments have been carried out in amphibia. In adult urodeles WEISSFEILER (1924, 1925) has shown that removal of the olfactory bulb and part of the olfactory nerve is followed by regeneration of the olfactory bulb and re-establishment of the connections between periphery and centers. The regenerative phenomenon, according to WEISSFEILER, is supported by the regrowth of the olfactory axons, whose peripheral cell body has not been damaged by the surgery. The bulb itself regenerates from young nerve cells, which migrate centripetally from the olfactory neuroepithelium and differentiate into neurons typical of the olfactory bulb. That the regeneration of the forebrain could be experimentally induced in adult amphibia was also observed by DANIELEWSKI (1890), VINNIKOV (1956), VINNIKOV and TITOTA (1957), and KIRSCHE and KIRSCHE (1963, 1964).

D. Degeneration-Regeneration Experiments

ECKER (1855) was the first to describe the olfactory receptors and to mention incidentally that they could have been replaced by cells lying at the base of the epithelium. When other authors proved the neuronal nature of these receptors (BABUCHIN, 1871; EHRLICH, 1806; ARNSTEIN, 1887; GRASSI and CASTRONUOVO, 1889; VAN GEHUCHTEN, 1890; VON BRUNN, 1892; RETZIUS, 1892, 1894; CAJAL, 1890), the problem of degeneration and regeneration of the olfactory neuroepithelium attracted considerable interest. A complete survey of the literature on the subject until 1970 has been provided by TAKAGI (1971; Vol. IV, Part 1, this series).

Three major experimental approaches to the study of the degeneration of the olfactory neurons will be considered here as more closely related to our present interest: 1) section of the olfactory axons; 2) removal of the olfactory bulb; 3) removal of the peripheral receptors by surgical or chemical means. In the hands of different investigators these experimental procedures yielded inconsistent results.

Section of the olfactory nerves was attempted first by SCHIFF (1859) and COLASANTI (1875), in dog and frog respectively. These authors found that the section did not affect the morphology of the olfactory mucosa. Failure to show degeneration may have been due to the rather unsophisticated histological techniques of the time. Degeneration of the receptors, not followed by regeneration, was obtained in frogs by HOFFMANN (1867), NAKAMURA (1916), TAKAGI and YAGJIMA (1964, 1965), and in mammals by HOFFMANN (1867), BAGINSKY (1894) and TAKATA (1929). In contrast to these results, NAGAHARA (1940), in a well documented work, proved that the section of the *fila olfactoria* in mice induces degeneration of the neurons followed by regeneration of new neural elements from "resting cells" lying at the base of the epithelium.

Removal of the olfactory bulb was reported to produce terminal degeneration of the neurons (EXNER, 1877; LUSTIG, 1924; SEN GUPTA, 1967). Only LEGROS CLARK (1951, 1957) observed a 50% survival of the receptors in similar experiments.

Removal of the olfactory neurons by chemical means was attempted by SMITH (1938), who found that zinc sulfate permanently destroyed the olfactory receptors in rats. He later reported that recovery of the receptors can occur in frogs (1951). JACKSON and LEE (1965), using colchicine, obtained degeneration without recovery of the receptor neuroepithelium. The regenerative capacity of the olfactory mucosa after zinc sulfate treatment was reported by SCHULTZ (1941, 1960), MULVANEY and HEIST (1971), and MATULONIS (1974, 1975). Zinc sulfate has been recently been commonly used for the deafferentation of the olfactory bulbs (for review see ALBERTS, 1974). The difficulty of exact grading of the depth of damage in the neuroepithelium following treatment, however, makes this technique difficult to standardize when regeneration of neurons has to be estimated in detail.

A further approach to the study of olfactory neurons degeneration and regeneration was attempted by FRUHWALD (1935), who closed the nasal cavity in young rabbits. The author found a decreased number of receptors after 12 days and claims that olfactory sensory cells never proliferate or regenerate after birth.

E. Morphology

It is generally agreed that the olfactory mucosa of vertebrates contains three basic cell types: 1) olfactory neurons; 2) basal cells; and 3) supporting cells (Fig. 1).

The olfactory neurons are small, pear-shaped cells, each of which has a perikaryon 5 μm to 7 μm in diameter and a slender, rod-like, unbranched dendrite, provided at the distal end with cilia. The axon is 0.2 μm in diameter, unmyelinated; it reaches the olfactory bulb where, at the level of the glomeruli, it establishes synaptic contacts with the apical dendrite of the mitral cells (Figs. 2, 3, 4). The basal cells are irregular elements situated at the base of the epithelium. Some of them are small, triangular cells, filled with fibrillar material, others are rather globose or pear-shaped. The supporting cells are typical epithelial secretory cells, provided with microvilli at their free surface. The main portion of the cytoplasm, containing the nucleus, is located in the upper third of the epithelium

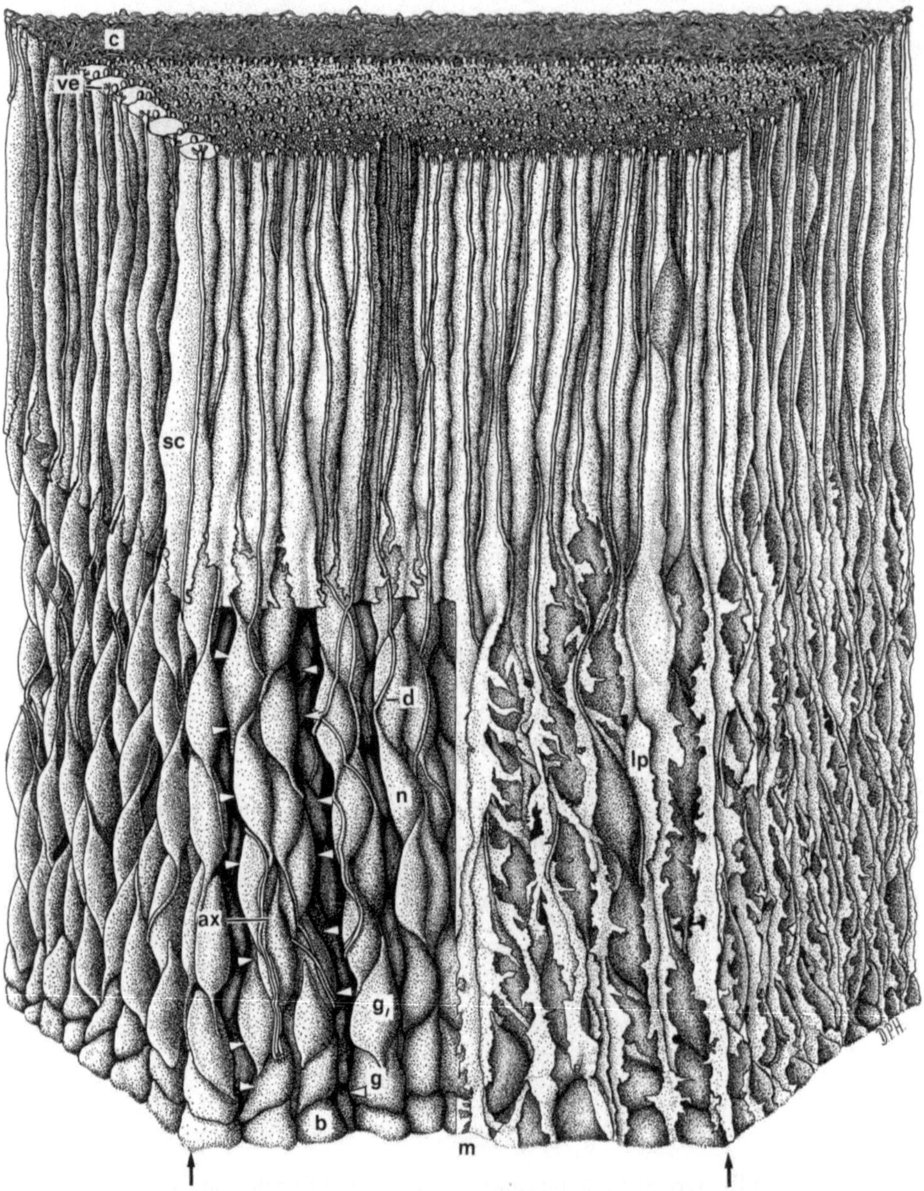

Fig. 1. Three-dimensional diagram of portion of olfactory neuroepithelium as derived from authors' observations in rat. Small basal cells (*b*) are arranged at base of epithelium close to basal lamina (not shown). Globose basal cells (*g*) lie close above basal elements and (g_1) have a clear bipolar shape. Mature neurons (*n*) have long unbranched dendrite (*d*) reaching free surface, and axon running centripetally in *fila olfactoria* to reach olfactory glomeruli (see Fig. 2). Supporting cells (*sc*) have main cell body located in distal third of epithelium. They reach basal lamina with laminar, irregular processes (*lp*) that insinuate between neurons and basal cells. Columnar arrangement of the neurons, as seen in light-microscope pictures of neuroepithelium, preserved in diagram and is indicated at small arrows. Surface of epithelium represented without cilia in anterior portion of block, to show olfactory vesicles of receptors (*ve*). Cilia (*c*) preserved in posterior portion. Two large arrows at base of block indicate its anterior corners. At (*m*), artificial line divides right-hand portion, showing laminar processes (*lp*) of supporting cells, from left-hand portion which has been artifically represented without the supporting cells laminar processes

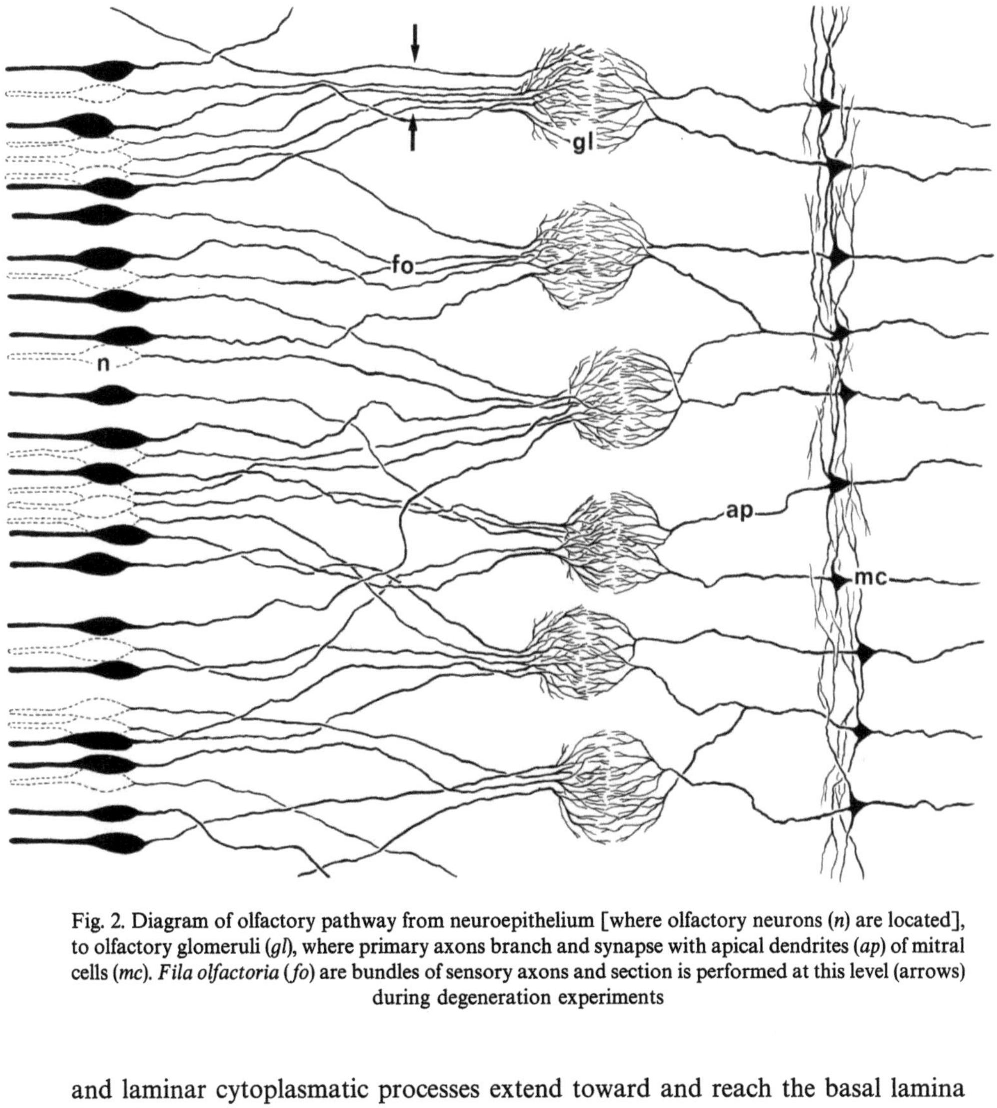

Fig. 2. Diagram of olfactory pathway from neuroepithelium [where olfactory neurons (*n*) are located], to olfactory glomeruli (*gl*), where primary axons branch and synapse with apical dendrites (*ap*) of mitral cells (*mc*). *Fila olfactoria* (*fo*) are bundles of sensory axons and section is performed at this level (arrows) during degeneration experiments

and laminar cytoplasmatic processes extend toward and reach the basal lamina (Fig. 1).

The basal cells can be subdivided into two types, with different morphological and topographical characteristics. Close to the basal lamina of the epithelium and directly adjacent to it there are small irregularly shaped cells with a very dense nucleus. Above these, other cells have a large, rather globose, clear nucleus and more abundant cytoplasm. Transitional morphological forms exist between the two. We consider all these cells as basal, because neither has the characteristics of mature neurons (Figs. 5, 6, 7). Furthermore, our autoradiographic studies have shown that these basal cells are the stem cells and the direct precursors of the neurons, not of the supporting cells. The basal cells are arranged in discrete groups at the base of the columns of neurons (Figs. 7, 8) that make up most of the epithelial layer (GRAZIADEI, 1975a and b).

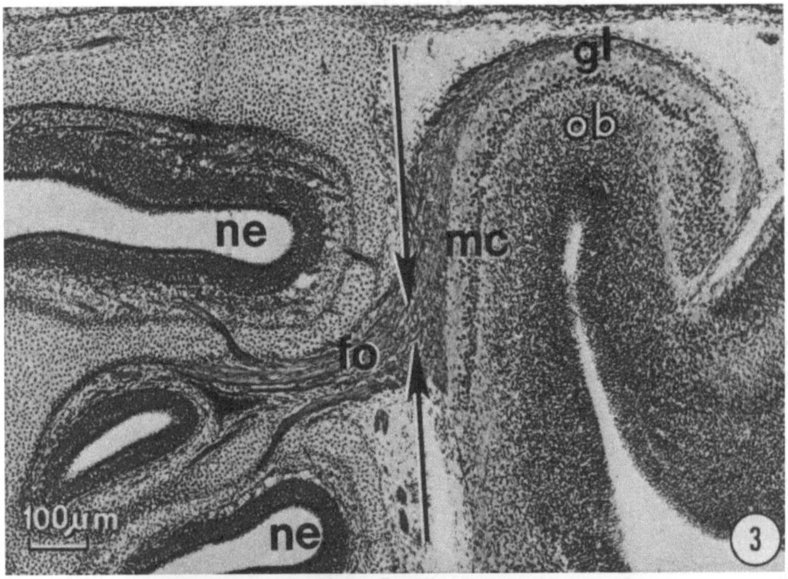

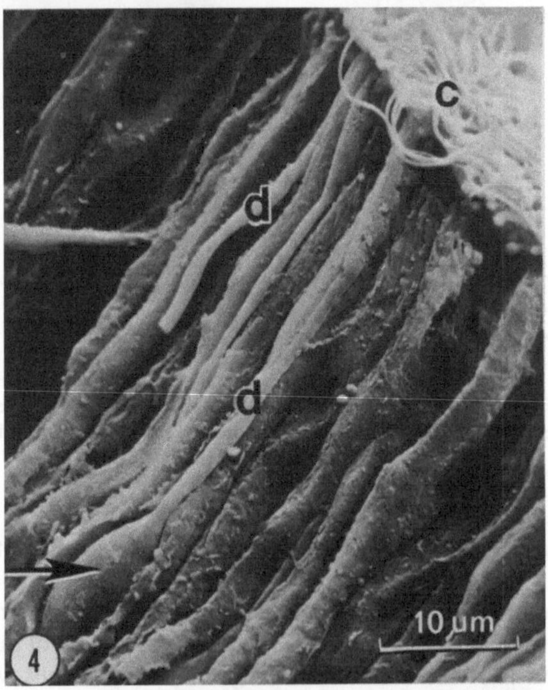

Fig. 3. Parasagittal section of portion of head of mouse embryo at 15 days' gestation. Olfactory bulb (*ob*) is shown with regular arrangement of mitral cells (*mc*), glomeruli (*gl*) and olfactory axons reaching bulb through *fila olfactoria* (*fo*). Axons originate from primary sensory neurons located in nasal neuroepithelium (*ne*), which covers nasal turbinates. Two converging arrows indicate the point of section of olfactory axons. Bodian stain, 10 µm histological section

Fig. 4. Scanning electron micrograph of fractured olfactory mucosa, to show cell surfaces. Free epithelial surface visible in upper right-hand corner, covered with cilia (*c*). Receptor cell body indicated at arrow, its dendrite can be followed until close to surface. Dendrites indicated at (*d*). Other profiles of fractured surface belong to supporting cells

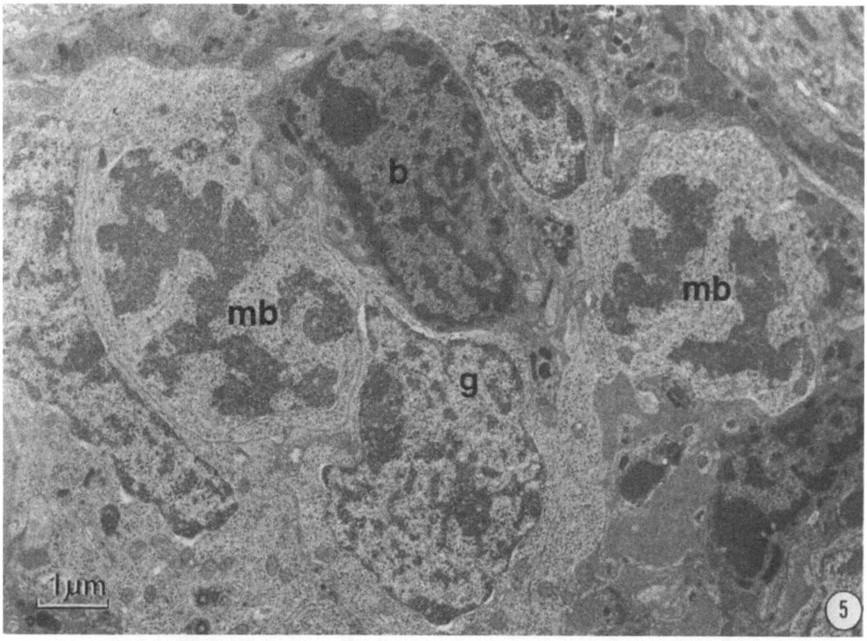

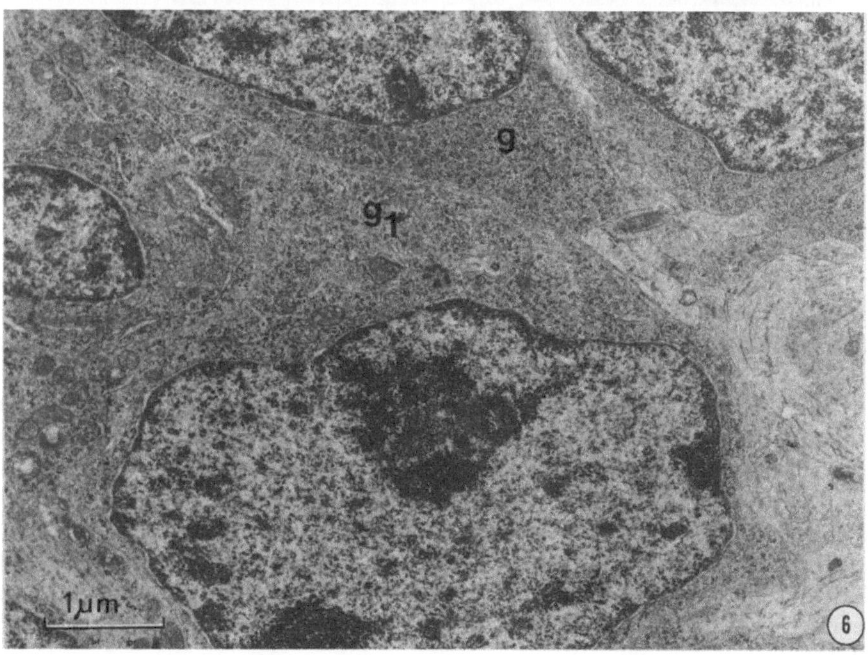

Fig. 5. Basal layer of normal olfactory mucosa of adult rat. Two mitotic figures belonging to basal cell populations (*mb*) close to basal lamina of epithelium. Typical basal cell (*b*) and globose basal cell (*g*) recognized by the dark (*b*) and clearer (*g*) nucleus. In globose cell cytoplasm is filled with free ribosomes

Fig. 6. Two globose basal cells (*g*) show cytoplasm filled with free ribosomes. In (g_1), profiles of *rER* start to differentiate

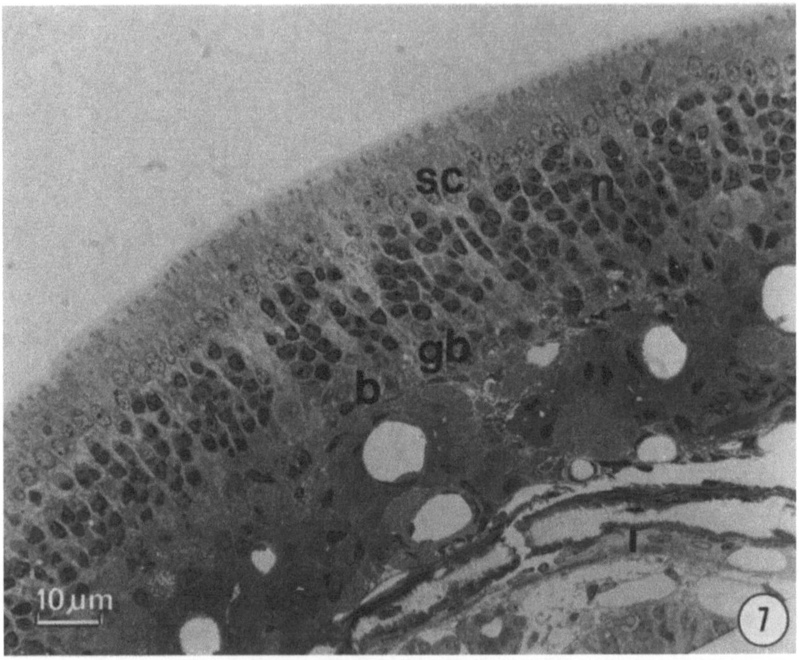

Fig. 7. Histological section of normal olfactory mucosa of rat, showing regular arrangement of supporting cell nuclei (*sc*) and regular columnar arrangement of mature neurons (*n*). Basal (*b*) and globose (*g*) cells indicated at base of columns

The neurons are recognized by their nuclei, which have an irregular outline and a dense chromatin pattern (Figs. 8, 9). The neuron layer varies in thickness in different zones of the olfactory epithelium. We have confirmed these observations in mammals and in lower vertebrates (GRAZIADEI, 1971a and b; GRAZIADEI and GRAZIADEI, 1976). The distribution of neurons appears not to be random when sections are cut perpendicular to the surface of the epithelium. The neurons are arranged in columns (Fig. 7); these are perpendicular to the epithelial surface and can vary in length (2–7 neurons) and in thickness (2–4 neurons). Each columnar unit varies in the number of its single components, and in different areas of the neuroepithelium. Some have a larger number of immature elements, occupying the lower portion of the epithelial layer, while others have a large number of mature elements and the columns are high and made up of a relatively homogeneous population of already differentiated neurons. These observations seem to indicate that in some regions there is an ongoing differentiation and maturation of neurons, while in others an equilibrium of mature elements has been reached (Figs. 10, 11).

The variable morphology of the neural portion of the neuroepithelium contrasts with the organization and cytological characteristics of the supporting cells. These have their nuclei regularly distributed in a single row in the upper third of the epithelium and their morphology is homogeneous and easily distinguished from that of the neurons.

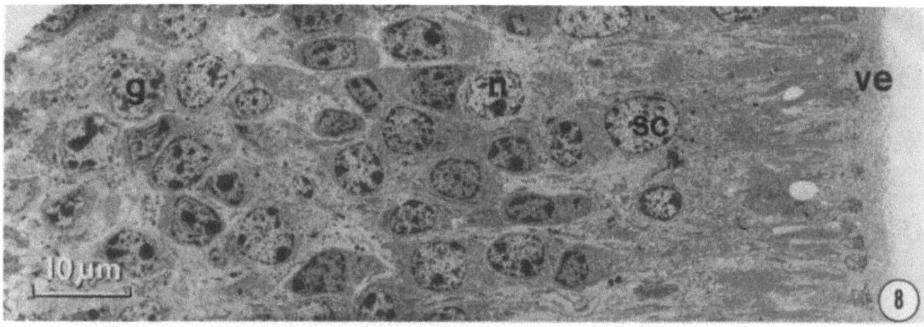

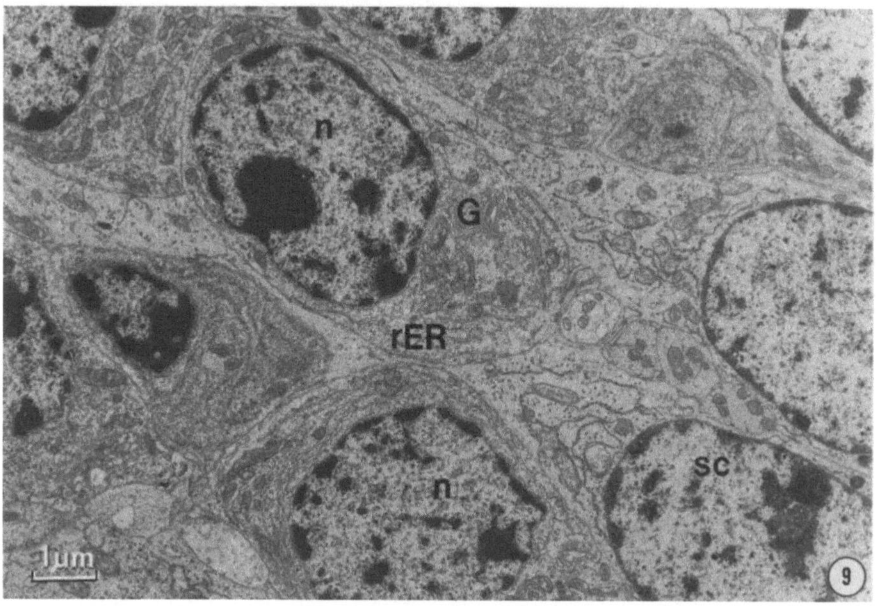

Fig. 8. Low-power electron micrograph showing the entire depth of the olfactory epithelium. *g*, globose basal cells; *n*, mature olfactory neurons; *sc*, supporting cells; *ve*, olfactory vesicle

Fig. 9. Mature neuron perikarya (*n*) show normal component of organelles. *G*, Golgi bodies, *rER*, rough endoplasmic reticulum. Supporting cell nuclei (*sc*) easily differentiated from neuron nuclei

Ultrastructural observations of the olfactory mucosa confirm that the cellular components identified as basal cells and receptors form a unique lineage of transitional stages. In our description we have purposely schematized the most common stages of the neural elements of the olfactory mucosa; these stages, however, represent arbitrary steps in a population of cells undergoing continuous renewal. Five "prominent" morphological types can be recognized, and they have been confirmed in terms of sequential development by autoradiographic studies. These types are described in order of their arrangement from the base of the epihelium upwards.

Close to the basal lamina there are the small basal cells, mentioned above, which correspond to the classic category of basal cells. These elements, in which

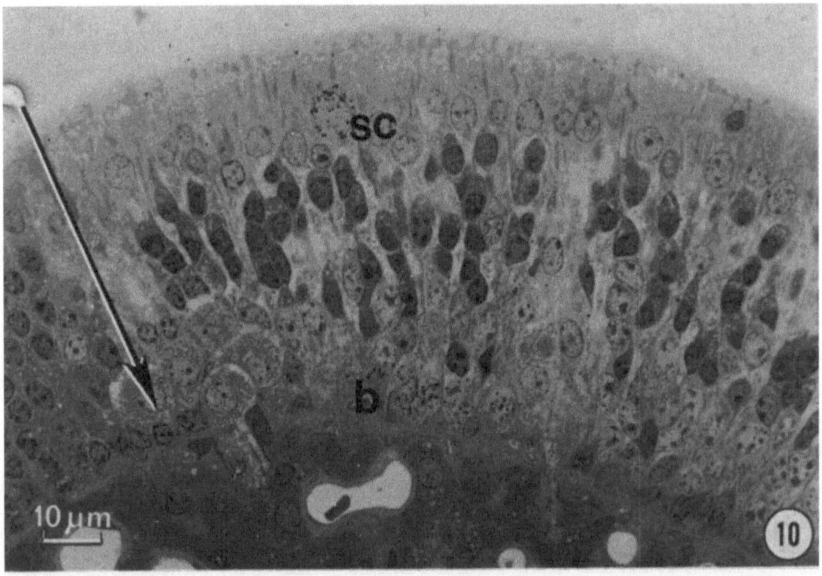

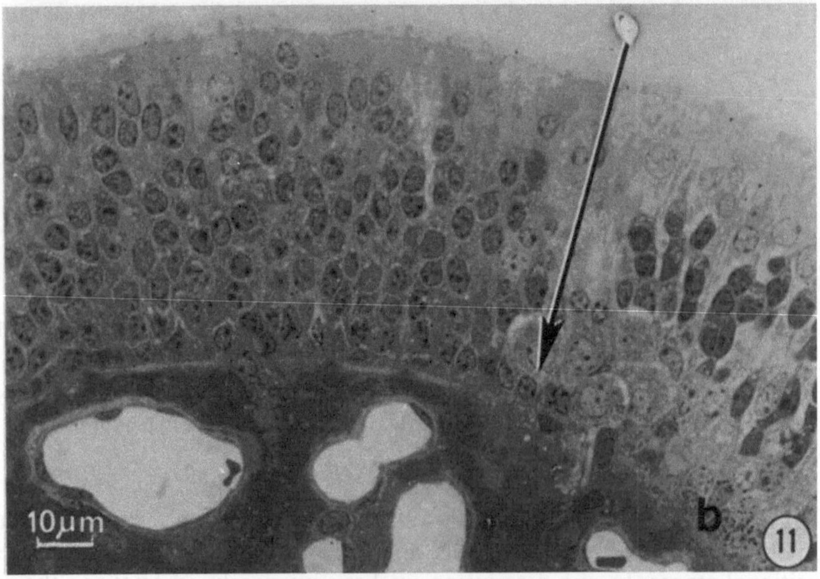

Figs. 10–11. Autoradiographs from normal olfactory mucosa of mouse. Two figures are of two adjoining regions of epithelium, and arrow shows overlapping line common to both photographs. The figures show structural difference of two adjoining regions, with presence in Fig. 10 of nest of labeled cells, while no activity is seen along the base of all remaining portion of epithelium. The 24 hr survival of animal shows also that only basal cells (*b*) and one supporting cell (*sc*) are labeled (see text)

the nuclei are small and irregular, have cytoplasm that adapts to the interstices between the supporting cells' laminar processes and the overlying globose cells. The dense ground cytoplasm with randomly arranged bundles of filaments is characteristic of these small cells. The common organelles are not conspicuous and their quantity and deployment do not characterize the cell. Morphologically they are not different from the basal cells of a common covering epithelium (Figs. 5, 6).

Above and between the small basal cells there are globose elements with a round nucleus recognizable by the homogeneous, sparse pattern of chromatin and obvious nucleoli. The cytoplasm is abundant, characterized by a paucity of organelles, with the exception of numerous free ribosomes which are evenly distributed and give the cell an overall homogeneous appearence (Figs. 5, 6).

In the basal layer of the epithelium among both basal and globose cells it is common to find mitotic figures (Fig. 5). Mitosis has been observed in sexually mature mammals as well as in adult lower vertebrates (GRAZIADEI and METCALF, 1970, 1971, 1972; GRAZIADEI, 1971a and b, 1973; GRAZIADEI and GRAZIADEI, 1976). We consider the basal globose cells homologous to the "resting cells" of NAGAHARA (1940).

Among and slightly above the globose basal cells there are others that can be classified as neurons on the basis of their shape and organelle content. They are more ovoid, often with an apical process recognizable as a developing dendrite and a centripetal small axon. The cell is clearly bipolar. The nucleus is also ovoid, with a denser chromatin pattern than the globose cells. In the cytoplasm are abundant free ribosomes, rough endoplasmic reticulum (rER), Golgi profiles and several mitochondria. In some of these elements the rER is arranged in parallel cisternae. Basal bodies tend to be concentrated at the distal pole of the cell. We consider these elements are the young developing neurons.

The mature neuron is characterized by the complete extension of the dendrite to the free surface, where the bulbous end is provided with cilia, and by a fine axon running outside the epithelium. The cell body is located in the midthird of the epithelium. The perikaryon contains a prominent display of rER and Golgi, most evident in the supranuclear position. There are a few dense bodies of different electron density and irregular pattern, resembling lysosomes (Fig. 9).

The lysosomes become the most conspicuous organelles in the neurons closer to the surface. These cells have an unusually dense ground cytoplasm with swollen profiles of the Golgi and ER. It is not unusual to find some of these elements in process of degeneration, or fragments of them, still recognizable, in the cytoplasm of macrophages that have penetrated into the epithelium.

The use of ^{3}H-thymidine provides definite evidence that in adult animals a continuous process of proliferation and maturation of neurons occurs in the olfactory neuroepithelium and that there is turnover of these neurons.

While our autoradiographic experiments have been conducted in several vertebrates (frogs, turtles, pigeons, mice and rats), the data reported here are primarily derived from observations in mice and frogs. ^{3}H-thymidine has been administered intraperitoneally in single or repeated pulses at doses of 5 μCi/gbw.

When animals are sacrificed 12 and 24 hrs after the first pulse of ^{3}H-thymidine, only basal cells proper, supporting cells, and globose basal cells nuclei are labeled. The percentage of nuclei labeled is proportional to the number of

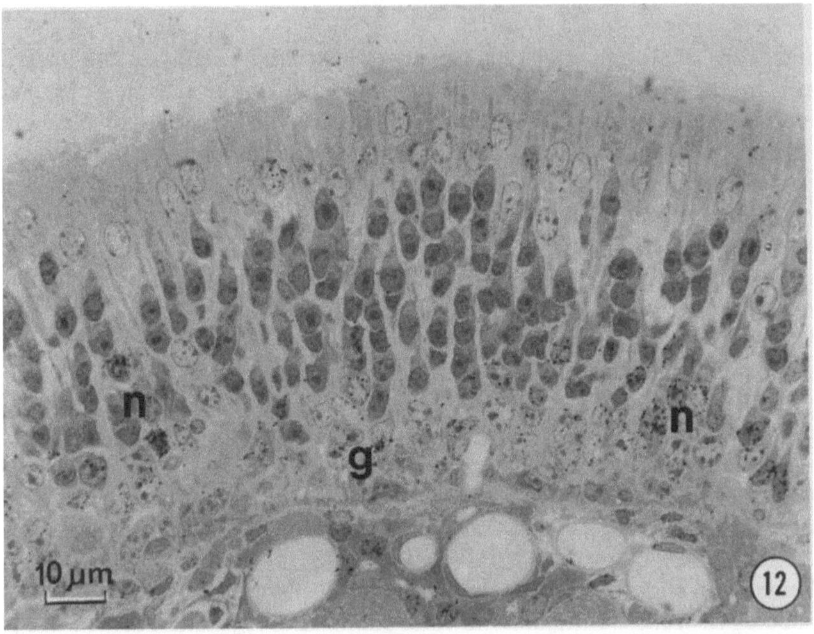

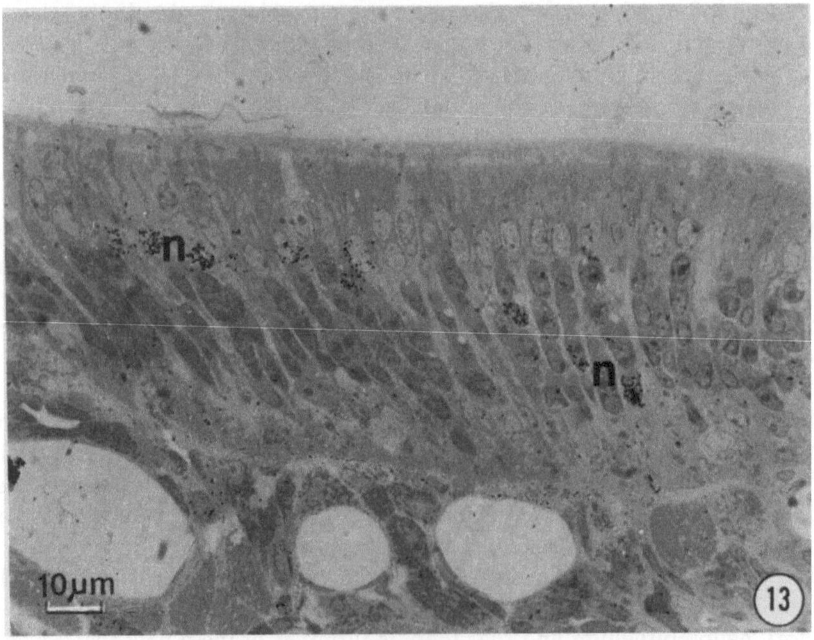

Fig. 12. Autoradiograph of normal olfactory mucosa of mouse, sacrificed 12 days from first of 3 days of labeling with ^{3}H-thymidine in 3 doses daily. Some mature neurons already labeled (*n*). Other labeled cells globose (*g*) and basal cells

Fig. 13. Autoradiograph of normal olfactory mucosa of mouse sacrificed 23 days from first of 5 days of labeling with ^{3}H-thymidine in 3 doses daily. All labeled elements are mature neurons (*n*) and most are in upper portion of the neuron layer, indicating an advanced age of the element. At 23 days' survival no more basal cells are labeled, indicating either that all have differentiated into neurons or that their label has been diluted to a noise level pulses given in these time intervals. When three pulses are administered in 24 hrs of survival time from the first dose of tritiated thymidine, an estimated 20% of basal cells are labeled, as against less than 1% of the supporting cell nuclei. In the brief survival time of 12 and 24 hrs, labeled nuclei belong to basal and supporting cells only, and not to mature neurons. The shape, size, and chromatin density of the labeled nuclei in the upper third of the epithelium make them easily identifiable as belonging to the supporting cells. The cells of the Bowman's glands, distinguishable from the other epithelial elements by their typical secretory characteristics, are seldom labeled.

When animals treated with the above mentioned three pulses of isotope are sacrificed after 7 days, the label is concentrated in the globose basal cells and in some of the differentiated neurons lying above the basal layer. After 15 to 20 days the label is recognizable in the elements described as mature neurons (Figs. 10–13). The labeled neurons disappear from the olfactory mucosa over 30 days.

This sequence of observations indicates that in 8 days the basal cells can differentiate into young mature neurons, and in 30 approximately days these neurons have been eliminated and replaced by new elements. The number of the labeled supporting cells and the number of granules in their nuclei does not change during this period.

In comprehensive histological autoradiographs we have observed that the label is not uniformly distributed, but there are zones of high activity adjoining others where there are very few labeled elements. Where the label is concentrated, the globose cells are numerous, in contrast to the areas where the label is minimal and the columns of mature neurons reach close to the basal lamina (Figs. 10, 11). In the zones with higher activity we have observed that the labeled cells seem to be arranged in groups or nests. It cannot be proved with our present techniques that the columnar units observed morphologically represent functional entities.

F. Experimental Studies

The anatomical organization of the peripheral segment of the olfactory system provides a simple model of neuronal organization well suited to the study of degenerative and regenerative phenomena. The primary olfactory neurons send an unbranched, unmyelinated bundle of axons to directly contact, at the level of the glomeruli, the apical dendrites of the mitral cells. Section of the primary axons allows the study of the degeneration of the glomerular terminals and of the retrograde reaction of the peripheral perikaryon (Figs. 2, 3). We will describe here the reaction of the peripheral cell bodies to the section of their centripetal axons, in frog, rat, and mouse.

The severance of the olfactory nerves is easily performed in frogs because these animals have a long, discrete, and surgically accessible olfactory nerve. In mammals, where the short *fila olfactoria* make section of the olfactory axon more

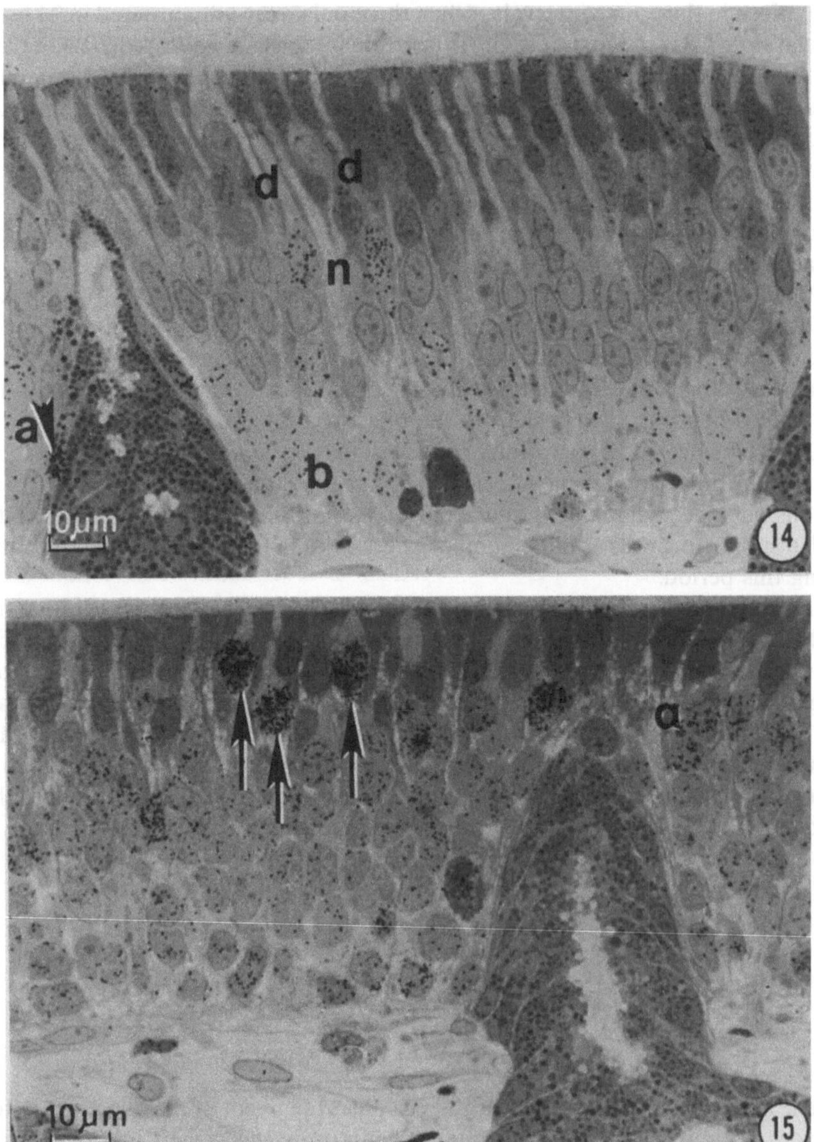

Figs. 14 and 15. Autoradiographs of olfactory mucosa of frog taken from the unoperated (Fig. 14) and operated (Fig. 15) side. Animal sacrificed 40 days after olfactory nerve section and injected with ^{3}H-thymidine from 5th to 15th postoperative day. On unoperated side (Fig. 14) many basal cells (*b*) and a few mature neurons (*n*), recognizable by their clear dendrite (*d*), are labeled. On operated side (Fig. 15) all elements have a labeled nucleus, but number of granules varies. Heavily labeled elements close to surface (arrows) are those that differentiated after first mitotic division, to replace the neurons degenerated after axotomy (see text). Note that none of supporting cell nuclei in both Figs. 14 and 15 are labeled, nor any of the Bowman's glands secretory cells, except for that indicated in (*a*) by arrow difficult, we have succeeded in cutting an estimated 80—90% of the fibers without appreciable damage to the blood supply of either the forebrain or the sensory neuroepithelium. Section of the olfactory axons could have resulted in: a) regeneration of the axons; b) permanent retrograde degeneration of the neurons; c) retrograde degeneration of the neurons followed by regeneration of new ones and reestablishment of their connections with the olfactory bulb glomeruli. The last hypothesis implies that the basal cells assume the role of nerve cell precursors even when total destruction of the existing mature neurons is experimentally induced. Our results provide evidence of a complete regeneration of neurons following the total degeneration of the axotomized elements.

The time sequence of neuronal degeneration and regeneration varies in different animals, being slower in amphibia than in mammals, but the modalities are similar. In rat signs of degeneration are seen with light microscopy three days after section of the *fila olfactoria*. The layers of neurons in the epithelium are unchanged but some of the nuclei of the mature elements show signs of pyknosis. The number of macrophages in the epithelium increase as they begin to participate in the removal of the degenerating neurons. The supporting cells are unaffected by the neuronal degeneration and the basal cells are normal in number and morphology. The degenerative process involving the neurons extends to their axons in the *fila olfactoria* (Figs. 16, 17) both proximally and distally to the point of section and it affects the terminals in the bulb glomeruli where the degeneration of the synaptic endings is easily noticeable at the third day (ESTABLE-PUIG et al., 1969; BERGER, 1971). The process of degeneration is maximal in the olfactory mucosa at 8 days postoperative, when there are no more mature neurons in the epithelium (Figs. 18, 19). The basal cells show considerable mitotic activity and many are of the globose type. These elements alone fill the epithelial layer below the supporting cells.

Throughout the degeneration period some immature neurons, with the characteristics of bipolar cells, persist in the olfactory mucosa. These elements are the ones that were in a developmental stage at the moment of the axotomy and therefore did not have the axon severed, consequently their process of differentiation into neurons was unaffected by the surgery. This circumstance explains why, at any time after surgery, a small percent of elements may be seen in relatively advanced stages of maturation in the olfactory mucosa. At 15 days survival the mitotic activity of the basal cells has decreased to near-normal values and the more numerous elements are globose cells and differentiating young neurons (Figs. 20, 21). Some of the neurons send the new dendrites to reach the epithelial surface. Maturation of these elements proceeds very fast and the olfactory mucosa regains its normal appearence in a period of 25 to 30 days.

These degenerative—regenerative phenomena observed by means of light microscopy are confirmed at the ultrastructural level, where the cytological characteristics of degenerating and regenerating neurons can be seen in finer detail. At the ultrastructural level we observe that the process of degeneration induced by axotomy is similar, although faster, to the one observed in the aging process of the normal neurons (Figs. 22, 23). Many lysosome-like bodies fill the cytoplasm of the receptors; these then disintegrate and are rapidly engulfed by macrophages. The supporting cells do not play any major role in the disposal of the neurons and their morphology is unchanged during the entire period of the

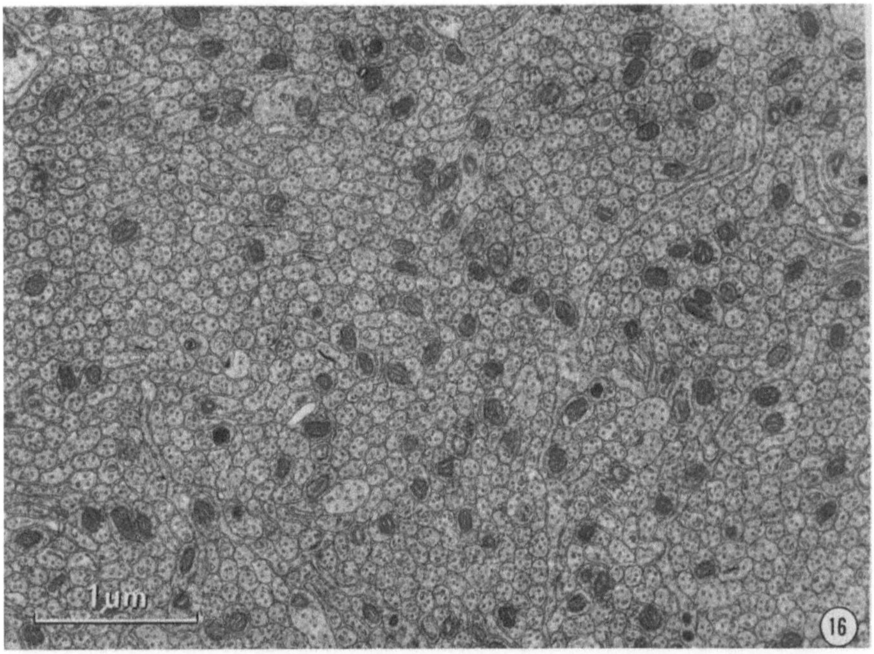

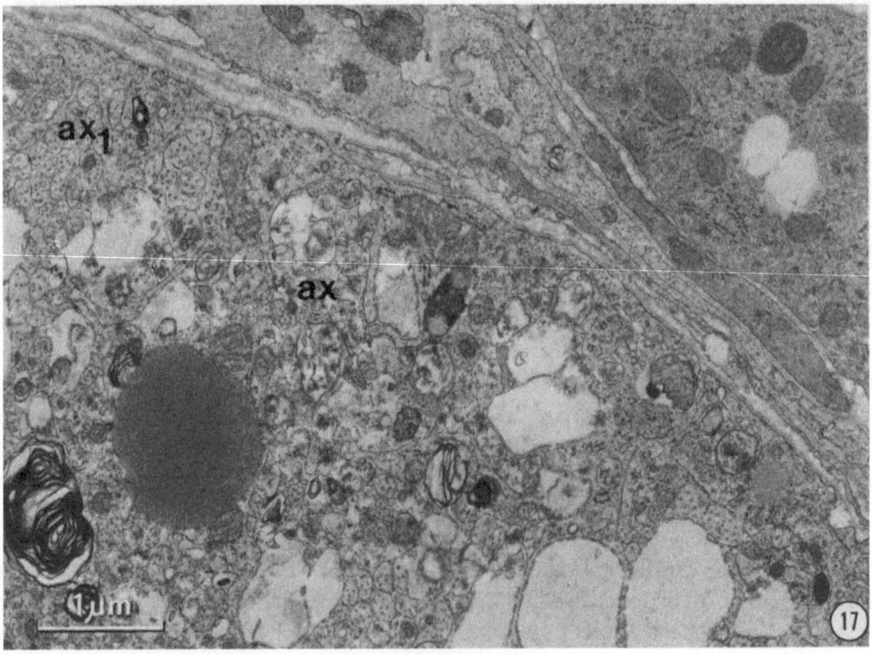

Figs. 16 and 17. Portion of normal olfactory nerve from rat (Fig. 16), showing intact axons to compare with degenerating axons of rat operated upon 3 days before sacrifice (Fig. 17). While some axons remain intact (ax_1) at 3rd postoperative day, most are degenerated (*ax*)

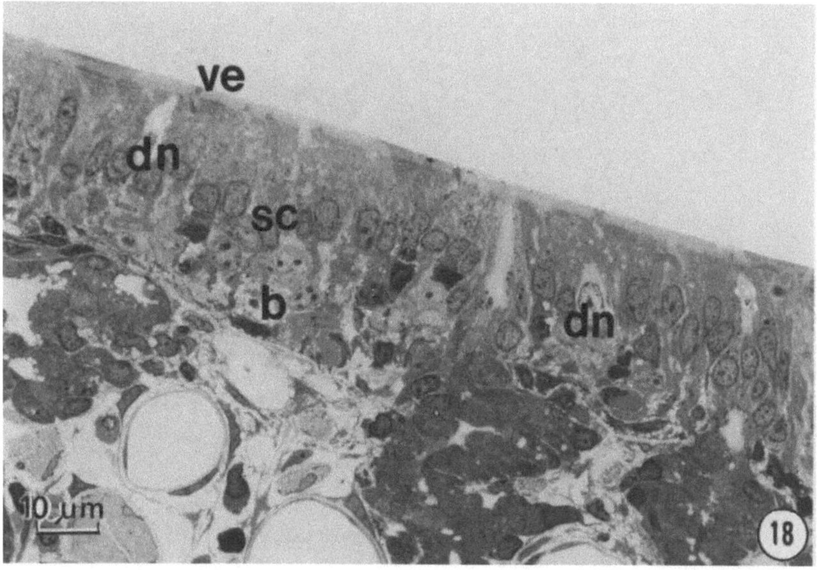

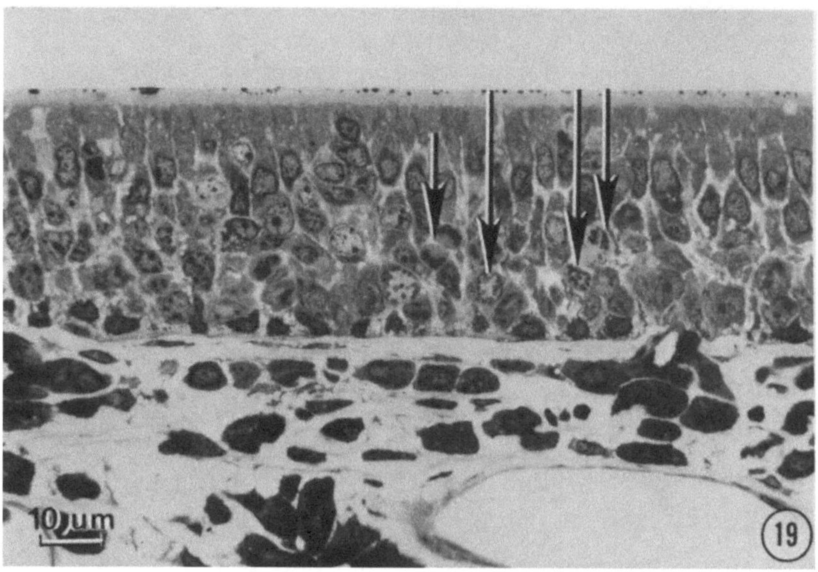

Fig. 18. Histological section of olfactory mucosa of rat 6 days after section of *fila olfactoria*. Supporting cell nuclei (*sc*) numerically and morphologically unchanged. Neuron layer has disappeared and two clear, degenerating receptors are indicated (*dn*). Below supporting cells, layer of irregular elements of the basal type (*b*). Surface of epithelium shows few remaining olfactory vesicles (*ve*); however their number is minimal compared to both normal (Fig. 7) and regenerated (Fig. 21) olfactory mucosa

Fig. 19. Histological section of olfactory mucosa of rat 8 days after section of *fila olfactoria*. No olfactory vesicles at free surface of epithelium. Neurons not yet differentiated, but in layer of epithelium occupied by basal and globose cells, a large number of mitotic figures is typical. Four cells in mitosis indicated by arrows

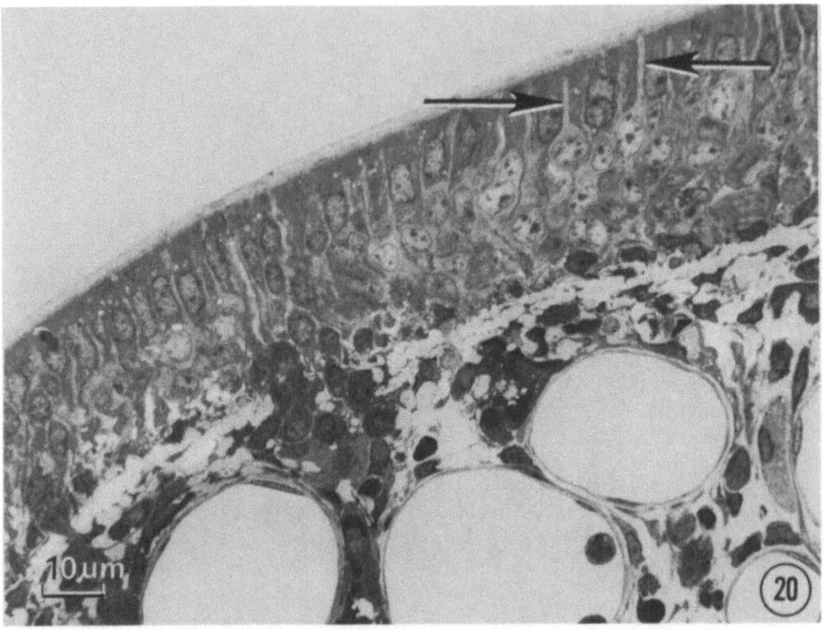

Fig. 20. Histological section of olfactory mucosa of rat 20 days after section of *fila olfactoria*. Basal layer higher than in previous Figures (18, 19), and many globose cells and young neurons can be recognized. Some mature neurons already extending dendrites to reach surface (arrows)

Fig. 21. Histological section of olfactory mucosa of rat 33 days after section of *fila olfactoria*. While vesicles not so numerous as in normal olfactory mucosa (*ve*) (see Fig. 7), neuron layer (*n*) is present and rather discretely arranged from supporting cells (*sc*) and basal cells (*b*). Olfactory mucosa by this time has reaquired near-normal appearance. However, the columnar arrangement of neurons is not as obvious as normal (Fig. 7) experiment. At the ultrastructural level the basal cells show no signs of damage. They react to the severance of the *fila olfactoria* with an increased mitotic activity and the process of differentiation follows the same modalities of maturation observed in the normal olfactory mucosa. Autoradiographic controls were established in the animals submitted to surgery, to prove that all the mature neurons degenerated after severance of the nerves and all the new neurons were the product of the mitotic activity of the basal cells. The experiments were conducted in frogs, where it is possible to cut 100% of the olfactory axons, and in mice. All the neural elements of the epithelium are labeled (Figs. 14, 15) when ^{3}H-thymidine is administered postoperatively during the period corresponding to maximal mitotic activity, which spans days 5 to 15 and 5 to 10 after surgery in frogs and mice respectively. Examination of autoradiographs reveals some heavily labeled mature neurons and others with a proportionally reduced number of granules toward the base of the epithelium. The latter derive from the basal cells that have undergone more than one division after treatment with ^{3}H-thymidine has been discontinued. The label in the supporting cells is minimal and reaches the usual percentage; the cells of the Bowman's glands are not significantly labeled and they are not a source of neurons even in the experimental conditions we have studied.

G. Discussion

Recent experimental evidence indicates that, when proper procedure is followed, the degeneration of neurons is followed by total regeneration in the olfactory epithelium. The controversial results obtained by many research workers, who used zinc sulfate to establish regeneration in the olfactory epithelium, may be attributed to the uneven action of the chemical on the tissue, due to variability of the concentration, the flow rate, the time of exposure to the solution, and to the extreme compartmentalization of the nose of the mammals commonly used. Consequently, the damage to the epithelium may be restricted to the superficial epithelial layer or may reach and affect its basal neurogenetic portion. Routine examination at light microscopy level does not allow adequate evaluation of the cytological changes in the neuroepithelium and of the effectiveness of the procedure for the specific study of neuronal regeneration. Section of the *fila olfactoria* and removal of the olfactory bulb in mammals also deserves some criticism. Legros Clark and Warwick (1946) and Legros Clark (1951, 1957) reported that in the rabbit 50% of the olfactory receptors are unaffected by the bulb removal. From the illustrations provided we inferred that the authors did not document the early total degeneration of the mature receptors and only observed the partial regeneration of these elements that can occur, according to our observations, even when the bulb is totally removed. We should also consider the possibility that some of the receptors could have survived owing to the incomplete bulb removal. The use of thick paraffin sections could have partially impaired the exact evaluation of the cytological data, and we know that

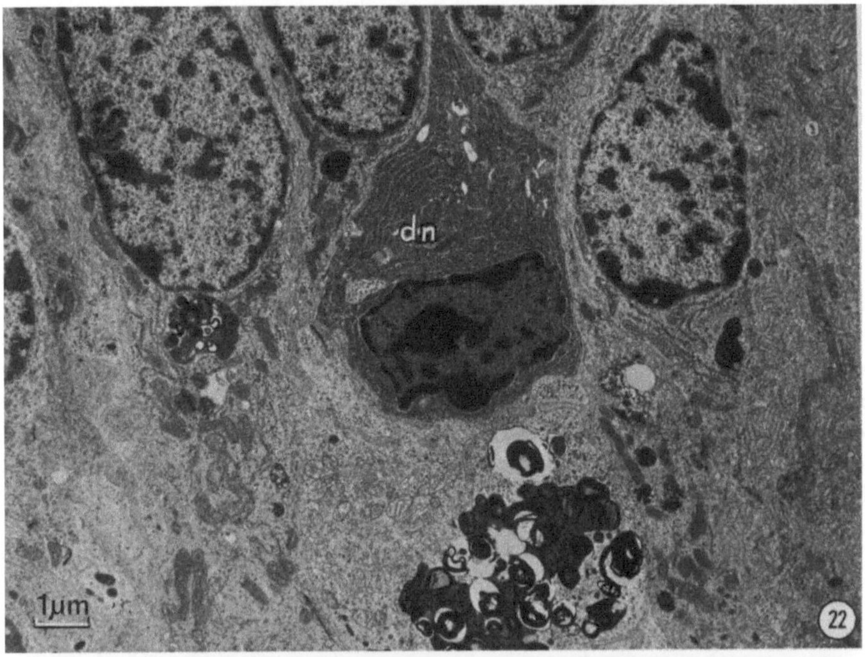

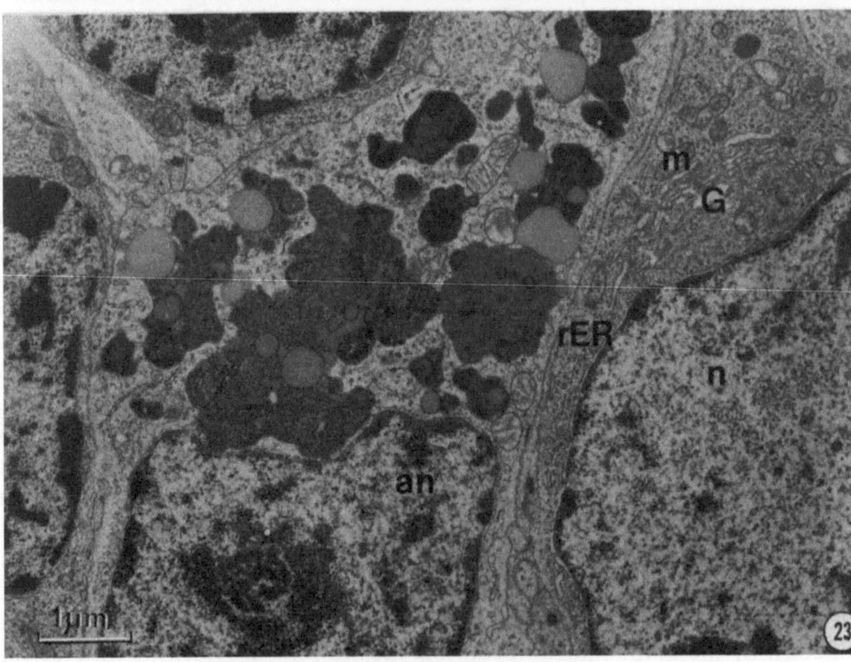

Fig. 22. Olfactory mucosa from rat sacrificed 3 days after section of *fila olfactoria*. Degenerating neurons (*dn*) show several electron-dense bodies in cytoplasm matrix

Fig. 23. Normal olfactory mucosa of rat showing an aging neuron (*an*) with cytoplasm filled with vacuoles and polymorphous electron-dense bodies. Normal olfactory neuron (*n*) shows usual organelles; *G*, Golgi; *m*, mitochondria; *rEF*, rough endoplasmic reticulum the olfactory epithelium varies in thickness from region to region, making comparisons on the basis of height alone an unreliable criterion. NAGHARA (1940), who was the first and only worker to perform successful surgery on the olfactory axons in mice, described the regeneration of new neurons, which we have recently confirmed.

Our observations have shown that the basal cells and the neurons of the olfactory neuroepithelium vary widely in number and arrangement. On extensive sampling of the organ, some zones appear to be *quiescent*. They are characterized by highly regular columnar arrangements of mature neurons and by the presence of few basal and immature cells. Other areas, conversely, display many globose basal cells and young differentiating neurons, with a relatively reduced number of mature elements. These zones, which could be termed *active*, seem to be randomly distributed among the quiescent ones. They do not extend to large areas of the nose but are in the range of some 100 µm in diameter. These observations are confirmed by autoradiographic results, which show that areas of intense labeling (active zones) are mixed with others where there are few labeled elements (quiescent zones). We do not yet have data indicating that these zones of different neurogenetic activity are functionally diverse, but the variability of the neurogenetic process in discrete areas of the olfactory neuroepithelium poses the problem of the existence of a regulatory mechanism controlling the continuous renewal of the neurons in the sensory area as a whole.

Pulse labeling of the epithelium shows not only the zonal distribution of the labeled elements but also the time course of their lifespan. Our data indicate that the immature cells differentiate into bipolar, typical neurons in a period of 7 to 8 days, and that the labeled elements degenerate and disappear from the epithelium in no longer than 30 to 35 days. During this short time each element undergoes dramatic morphological and functional changes, establishes synaptic relationship with the olfactory bulb, degenerates, and is eventually removed by a macrophage. The rapid sequence of recognizable developmental stages of these neural elements provides very useful material for the study of the chemical makeup of this peculiar neuron, which progresses from birth to death in the short period of few weeks.

The different levels of neurogenetic activity in the normal olfactory epithelium must be considered, and have a direct bearing on the interpretation of our results of the experimental degeneration-regeneration of the olfactory neurons. We have observed that a few days after axotomy all mature receptors degenerate. Some developing neurons, however, are still present. They represent the product of the differentiation of those immature elements that have not been affected by the surgical procedure.

The use of ^{3}H-thymidine has proved that the basal cells are the stem cell elements of the new neurons and that these elements differentiate locally, giving origin only to neurons. Their upward migration into the neuron layer can be followed morphologically and autoradiographically as a wave of labeled elements into the epithelium.

Morphological and autoradiographic evidence does not support the hypothesis that basal cells are common progenitors for both neurons and supporting cells. The possibility that the neurogenetic process may follow the modality observed by FUJITA (1962, 1963, 1964, 1966) in the developing central nervous system does not seem to apply to the olfactory neuroepithelium of adult animals.

The regenerative capacity of the olfactory neurons suggests that their continuous replacement may be an adaptation of the system to injury from the environment during the normal life of the animal. However, we tend to believe that the short lifespan of these neurons is not the result of chemical and physical damage from the environment but is a predetermined, and possibly acquired, genetic characteristic. We support our view with the observation that these neurons do not show signs of injury and are not eliminated at an early stage of their development, but undergo a process of morphological differentiation, maturation, and senescence that is typical of, and similar to, that of the cells of our central nervous system. If these neurons are programmed to live such a short time, with their capacity to be replaced, they are indeed a unique and exceptional population in our nervous system.

H. List of Abbreviations

an	ageing neuron
ap	apical process
ax	axon
b	basal cell
bl	blood vessel
bla	basal lamina
c	cilia
d	dendrite
dn	degenerating neuron
fo	fila olfactoria
G	Golgi
g	globose basal cell
gl	olfactory glomerulus
lp	laminar process
m	mitochondria
mb	mitosis of basal cell
mc	mitral cell
n	mature olfactory neuron
ne	olfactory neuroepithelium
ob	olfactory bulb
rER	rough endoplasmic reticulum
sc	supporting cell
ve	olfactory vesicle

I. Acknowledgements

The authors wish to thank H. M. Sloan for electron microscopic and photographic preparation; D. Harlos for the illustrations; C. F. Lewis for manuscript preparation.

This research was supported in part by USPHS NS08943 and USPHS NS11668.

References

Addison, W. H. F., Rademaker, L. A.: The postnatal growth of the vomeronasal organ of the Albino Rat (*Mus norvegicus albinus*). J. comp. Neurol. **44**, 69—86 (1927).

Alberts, J. R.: Producing and interpreting experimental olfactory deficits. (Theoretical review). Physiol. and Behavior **12**, 657—670 (1974).

Allison, A. C.: The morphology of the olfactory system in vertebrates. Biol. Rev. **28**, 195—244 (1953).

Ariens Kappers, C. U., Huber, G. C., Crosby, E. C.: The comparative anatomy of the nervous system of vertebrates, including man. New York: Hafner 1967.

Arnstein, C.: Die Methylenblaufärbung als histologische Methode. Anat. Anz. **2**, 125—135 (1887).

Babuchin, A.: Das Geruchsorgan. In: Stricker S. Handbuch der Lehre von den Geweben des Menschen und der Tiere, Vol. 2 (S. Stricker, Ed.). Leipzig: Engelman 1871, pp. 964—976.

Baginsky, B.: Über das Verhalten von Nervenendorganen nach Durchschneidung der zugehörigen Nerven. Virchows Arch. path. Anat. **137**, 389—404 (1894).

Balfour, F. M.: The development of elasmobranch fishes. J. Anat. Physiol. **11**, 406—490 (1877).

Beard, J.: The system of branchial sense organs and their associated ganglia in Icthyopsida. Quart. J. micr. Sci. **26**, 95—156 (1885).

Bedford, E. A.: The early history of the olfactory nerve in swine. J. comp. Neurol. **14**, 390—410 (1904).

Bell, E. T.: Experimental studies on the development of the eye and the nasal cavities in Frog Embryos. Anat. Anz. **29**, 185—194 (1906).

Bell, E. T.: Some experiments on the development and regeneration of the eye and nasal organ in frog embryos. Arch. Entw. Mech. d. Org. **23**, 645—678 (1907).

Berger, B.: Formes diverses de dégénerescence des boutons synaptiques dans le glomerule olfactif de lapins après lesion du nerf olfactif. Brain Res. **33**, 218—222 (1971).

Bianchi, L.: Il cosidetto ganglio olfattivo e i suoi rapporti d'origine coi nervi olfattivo e terminale. Arch. It. Anat. Embriol. **29**, 187—209 (1931a).

Bianchi, L.: Sui caratteri e sulla evoluzione del cosidetto ganglio olfattivo. Boll. Soc. ital. Biol. sper. **6**, 153—155 (1931b).

Bizzozero, G.: Accrescimento e rigenerazione nell'organismo. Arch. Sci. Med. **18**, 245—287 (1894).

Brookover, C.: Pinkus nerve in Amia and Lepidosteus. Science **27**, 913—914 (1908).

Brookover, C.: The development of the olfactory nerve and its associated ganglion in Lepidosteus. J. comp. Neurol. **24**, 113—130 (1914).

Burr, H. S.: The effects of the removal of the nasal pits in amblystoma embryos. J. exp. Zool. **20**, 27—57 (1916).

Burr, H. S.: Some experiments on the transplantation of the olfactory placode in amblystoma. I. An experimentally produced aberrant cranial nerve. J. comp. Neurol. **37**, 455—479 (1924).

Burr, H. S.: Hyperplasia in the brain of Amblystoma. J. exp. Zool. **55**, 171—191 (1930).

Burr, H. S.: An electro-dynamic theory of development suggested by studies of proliferation rates in the brain of Amblystoma. J. comp. Neurol. **56**, 347—371 (1932).

Cajal, S. R.: Origen y terminación de las fibras nerviosas olfactorias. Gac. médica Catalana **12**, 6 (1890).

Cajal, S. R.: In: Degeneration and Regeneration of the Nervous System (R. M. May, Ed.). London: Hafner 1928.

Cameron,J., Milligan,W.: The development of the olfactory nerves in vertebrates. Trans. internat. Congr. Med. London, Section 15: Rhinology and Laryngology, Part 2, pp. 451—465 (1914).
Chancoine,M., Clairambault,M.P.: L'influence morphogénétique de la placode nasale sur le développement du télencéphale de la Truite (Salmo irideus Gib.) C. R. Acad. Sci. **280**, pp. 475—778 (1975).
Chiarugi,G.: Observations sur les premieres phases de développement des nerves encéphaliques chez les Mammifères, et, en particulier, sur la formation du nerf olfactif. Monit. zool. It. **11**, 418—425 (1891).
Clairambault,P.: Morphogénèse du télencéphale: Morphogénèse expérimentale du télencéphale des Ranides. Arch. Biol. (Liège) **79**, 537—578 (1968).
Clairambault,P.: Les effects de l'ablation bilaterale de la placode nasale sur la morphogénèse du télencéphale des anoures. Acta Embryol. exp. 61—92 (1971).
Colasanti,G.: Untersuchungen über die Durchschneidung des Nervus Olfactorius bei Fröschen. Arch. physiol. u. wiss. Med. **23**, 469—476 (1875).
Cuschieri,A., Bannister,L.H.: The development of the olfactory mucosa in the mouse: light microscopy. J. Anat. **119**, 277—286 (1975a).
Cuschieri,A., Bannister,L.H.: The development of the olfactory mucosa in the mouse: electron microscopy. J. Anat. **119**, 471—498 (1975b).
Danielewsky,B.: Über die Regeneration der Großhirn-Hemisphären bei Fröschen. Verh. 10. Internat. med. Kongr. **2**, 2 (1890).
Disse,J.: Die erste Entwicklung der Riechnerven. Anat. Hefte **9**, 255—300 (1897).
Disse,J.: Riechschleim und Riechnerv bei den Wirbeltieren. Ergebn. anat. Entwick. **10**, 487—523 (1901).
Disse,J.: Riechschleimhaut und Riechnerv bei den Wirbeltieren. Ergeb. anat. Entwick. **11**, 407—436 (1902).
Doellken,A.: Ursprung und Zentren des Nervus terminalis. Mschr. Psychiat. Neurol. **26**, 10—33 (1909).
Ecker,A.: Über das Epithelium der Riechschleimhaut und die wahrscheinliche Endigung der Geruchsnerven. Z. wiss. Zool. **8**, 303—306 (1855).
Ehrlich,P.: Über die Methylenblaureaktion der lebenden Nervensubstanz. Dtsch. med. Wschr. **12**, 49—52 (1896).
Ekman,G.: Über die Regeneration der Nasenanlage bei Rana fusca. Soc. Sci. fennica. Comment. Biol. **6**, 1—23 (1924).
Estable-Puig,J.F., DeEstable,R.F.: Acute ultrastructural changes in the rat olfactory glomeruli after peripheral deafferentation. Exp. Neurol. **24**, 592—602 (1969).
Exner,S.: Fortgesetzte Studien über die Endigungsweise des Geruchnerven. Will. Wien Akad. **76**, 171—220 (1877).
Filogamo,G., Robecchi,M.G.: Neuroblasts in the olfactory pits in mammals. Acta Anat. **73** (Suppl.) 56, 182—187 (1969).
Fritsch,G.: Untersuchungen über den feineren Bau des Fischgehirns, Berlin 1878; cited by W.A.Locy, 1905.
Fruhwald,V.: Die Folgen des einseitigen Nasenverschlusses auf die Riechschleimhaut und auf den Bulbus und Tractus olfactorius. Arch. Ohr-, Nas.- u. Kehlk.-Heilk. **139**, 153—173 (1935).
Fujita,S.: Kinetics of cellular proliferation. Exp. Cell Res. **28**, 52—60 (1962).
Fujita,S.: The matrix cell and cytogenesis in the developing central nervous system. J. comp. Neurol. **120**, 37—42 (1963).
Fujita,S.: Analysis of neuron differentiation in the central nervous system by tritiated thymidine autoradiography. J. comp. Neurol. **122**, 311—328 (1964).
Fujita,S.: In: Evolution of the Forebrain (R.Hassler and H.Stephan, Eds.). New York: Plenum 1966.
Grassi,V., Castronuovo,A.: Beitrag zur Kenntnis des Geruchsorgans des Hundes. Arch. mikr. Anat. **34**, 385—390 (1889).
Graziadei,P.P.C.: The olfactory mucosa of vertebrates. In: Handbook of Sensory Physiology, Vol. IV (L. M. Beidler, Ed.). Berlin-Heidelberg-New York: Springer 1971a, p. 27—58.
Graziadei,P.P.C.: Topological relations between olfactory neurons. Z. Zellforsch. **118**, 449—466 (1971b).
Graziadei,P.P.C.: Cell dynamics in the olfactory mucosa. Tissue and Cell **5**, 113—131 (1973).
Graziadei,P.P.C.: The olfactory organ of vertebrates: A survey. In: Essays on the Nervous System (R. Bellairs and E. G. Gray, Eds.). Oxford: Clarendon 1974, pp. 191—222.

Graziadei,P.P.C.: Neuronal plasticity in the vertebrate olfactory receptor organ. 10th Int. Cong. Anat. Tokyo, pp. 45 (1975a).

Graziadei,P.P.C.: Structural evidence of functional units in the olfactory mucosa of mice. Soc. for Neuroscience. 5th Annual meeting, p. 1143 (1975b).

Graziadei,P.P.C., DeHan,R.S.: Neuronal regeneration in frog olfactory system. J. Cell Biol. **59**, 525—530 (1973).

Graziadei,P.P.C., Graziadei,G.A.: Olfactory epithelium of Necturus maculosus and Ambystoma tigrinum. J. Neurocytol. **5**, 11—32 (1976).

Graziadei,P.P.C., Metcalf,J.F.: Autoradiographic study of frog's olfactory mucosa. Amer. Zool. **10**, No. 4, 716 (1970).

Graziadei,P.P.C., Metcalf,J.F.: Autoradiographic and ultrastructural observations on the frog's olfactory mucosa. Z. Zellforsch. **116**, 305—318 (1971).

Graziadei,P.P.C., Metcalf,J.F., DeHan,R.S.: Cell dynamics in the olfactory mucosa of vertebrates. Soc. for Neuroscience. 2nd Annual Meeting (1972).

Gruneberg,H.: A ganglion probably belonging to the N. terminalis system in the nasal mucosa of the mouse. Z. Anat. Entwickl.-Gesch. **140**, 39—52 (1973).

Guth,L.: Axonal regeneration and functional plasticity in the central nervous system. Exp. Neurol. **45**, 606—654 (1974).

Guth,L., Clemente,W.F.: Growth and regeneration in the central nervous system. Exp. Neurol. **48**, 3: p. 2, 1—251 (1975).

Harrison,J.L.: Some hypoplastic modifications on the telencephalon following unilateral excision of the nasal placode in Rana pipiens embryos. Dissert. Abstr. **19**, 604 (1958).

Herrick,C.J.: The nervus terminalis (Nerve of Pinkus) in the frog. J. comp. Neurol. **19**, 175—190 (1909).

Herrick,C.J.: The morphology of the forebrain in Amphibia and Reptilia. J. comp. Neurol. **20**, 413—547 (1910).

Hinsberg,V.: Die Entwicklung der Nasenhole bei Amphibien. Arch. mikr. Anat. **58**, 411—482 (1901).

His,W.: Über die Entwicklung des Riechlappens und des Riechganglions und über diejenige des verlängerten Markes. Anat. Anz. **4**, 63—66 (1889).

Hoffmann,C.K.: Über die Membrana olfactoria. Inaugural dissertation, Amsterdam, 1867.

Holmgren,N.: Zur Kenntnis des Nervus terminalis. Fol. neurobiol. **11**, 16—36 (1918/20).

Huber,G.C., Guild,S.R.: Observations on the peripheral distribution of the nervus terminalis in mammalia. Anat. Rec. **7**, 253—272 (1913).

Humphrey,T.: The development of the olfactory and the accessory olfactory formations in human embryos and fetuses. J. comp. Neurol. **73**, 431—468 (1940).

Jackson,R.T., Lee,C.C.: Degeneration of olfactory receptors induced by colchicine. Exp. Neurol. **11**, 483—492 (1965).

Jacobson,L.: Description anatomique d'un organe observé dans les mammifères. Ann. Mus. Hist. nat. Paris **18**, 412—431 (1811).

Johnston,J.B.: Nervus terminalis in Reptiles and Mammals. J. comp. Neurol. **23**, 97—120 (1913).

Kirsche,K., Kirsche,W.: Regenerative Vorgänge im Telencephalon, von Regeneration des Nervus olfactorius auf die Vorderhirngeneration von Ambistoma mexicanus. J. Hirnforsch. **7**, 315—333 (1963).

Kirsche,K., Kirsche,W.: Experimentelle Untersuchung über den Einfluß der Regeneration des Nervus olfactorius auf die Vorderhirngeneration von Ambistoma mexicanus. J. Hirnforsch. **7**, 315—333 (1964).

Koelliker,A. von: Über die Entwicklung des Geruchsorganes beim Menschen und beim Hühnchen Würzburger. med. Zschr. **1**, 425—435 (1860).

Koelliker,A. von: Der Lobus olfactorius und die Nervi olfactorii bei jungen menschlichen Embryonen. S.-B. phys.-med. Ges. Würzb. **36**, 68—72 (1882).

Koelliker,A. von: Zur Entwicklung des Auges und Geruchsorganes menschlicher Embryonen. Verh. phys.-med. Ges. Würzb. **17**, 229—257 (1883).

Kosciuszko,H.: The influence of the olfactory nerve on the regeneration of the endbrain in Xenopus tadpoles. Fol. Biol. **6**, 117—130 (1958).

Larsell,O.: Studies on the nervus terminalis: mammals. J. comp. Neurol. **30**, 3—68 (1918).

Larsell,O.: Studies on the nervus terminalis: turtle. J. comp. Neurol. **30**, 423—433 (1918/19).

Larsell,O.: The nervus terminalis. Ann. Oto-laryng. (Paris) **59**, 414—438 (1950).

LeBlond, C.P.: Classification of cell populations on the basis of their proliferative behavior. Nat. Cancer Monogr. **14**, 119 (1964).
LeBlond, C.P., Walker, B.E.: Renewal of cell populations. Physiol. Rev. **36**, 255—275 (1956).
Le Gros Clark, W.E.: The projection of the olfactory epithelium on the olfactory bulb in the rabbit. J. Neurol. Neurosurg. Psychiat. **14**, 1—10 (1951).
Le Gros Clark, W.E.: Observations on the structure and organization of olfactory receptors in the rabbit. Yale J. biol. Med. **29**, 83—95 (1956).
Le Gros Clark, W.E.: Inquiries into the anatomical basis of olfactory discrimination. Proc. roy. Soc. B. **146**, 299—319 (1957).
Le Gros Clark, W.E., Warwick, R.T.T.: The pattern of olfactory innervation. J. Neurol. Neurosurg. Psychiat. **9**, 101—111 (1946).
Locy, W.A.: On a newly recognized nerve connected with the forebrain of Selachians. Anat. Anz. **26**, 33—63 (1905).
Luna, E.: Ricerche sperimentali sullà morfologia dell'organo dell'olfatto negli amfibi., Arch. ital. Anat. Embriol. **14**, 609—628 (1915).
Lustig, A.: Die Degeneration des Epithels der Riechschleimhaut des Kaninchens nach Zerstörung der Riechlappen desselben. S.-B. Wien. Akad. **89**, (Abst.). (1924).
McCotter, R.E.: The connection of the vomeronasal nerves with the accessory olfactory bulb in the opossum and other mammals. Anat. Rec. **6**, 299—318 (1912).
McCotter, R.E.: The vomero-nasal apparatus in Christemys punctata and Rana catesbiana. Anat. Rec. **13**, 51—67 (1919).
McKibben, P.S.: The nervus terminalis in urodele amphibia. J. comp. Neurol. **21**, 261—309 (1911).
McKibben, P.S.: Ganglion cells of the nervus terminalis in the dogfish (Mustelus canis). J. comp. Neurol. **24**, 431—433 (1914).
Manfredonia, M.: Localizzazione della zona presuntiva del placode' olfattivo in Discoglossus pict. Arch. It. Anat. Embriol. **40**, 356—368 (1938).
Marchetti, L.: Sui primi momenti dello sviluppo di alcuni organi primitivi nel germe di Bufo vulgaris. Arch. It. Anat. Embriol. **16**, 175—245 (1918).
Marshall, M.: The development of the cranial nerves in the Chick. Quart. J. micr. Sc. **18**, 10—40 (1878).
Marshall, M.: Morphology of the vertebrate olfactory organ. Quart. J. micr. Sc. **19** (1979).
Masy, A.: (1955); quoted by E. Van Campenhout (1956).
Matulonis, D.H.: Ultrastructural morphology of the normal nasal respiratory epithelium of the mouse. Anat. Rec. **176**, 65—84 (1973).
Matulonis, D.H.: Ultrastructure of olfactory epithelia in mice after smoke exposure. Ann. Otol. Rhinol. Laryngol. **83**, 192—202 (1974).
Matulonis, D.H.: Ultrastructural study of mouse olfactory epithelium following destruction by $ZnSO_4$ and its subsequent regeneration. Amer. J. Anat. **142**, 67—90 (1975).
May, R.M.: Modifications des centres nerveux dues à la transplantation de l'oeuil et de l'organe olfactif chez les embryones d'Anoures. Arch. Biol. (Liège) **37**, 336—396 (1927).
Messier, B., LeBlond, C.P.: Cell proliferation and migration as revealed by radioautography after injection of thymidine H3 in male rats and mice. Amer. J. Anat. **106**, 247—285 (1960).
Mulvaney, B.D., Heist, H.E.: Regeneration of rabbit olfactory epithelium. Amer. J. Anat. **131**, 241—252 (1971).
Nagahara, Y.: Experimentelle Studien über die histologischen Veränderungen des Geruchsorgans nach der Olfactoriusdurchschneidung. Beiträge zur Kenntnis des feineren Baus des Geruchsorgans. Jap. J. Med. Sci., V: Pathol. **5**, 165—199 (1940).
Nakamura, Y.: Über die Veränderung des Geruchsorgans nach der Olfactorius-Durchschneidung bei Fröschen. Dainihon Jiji Ih. **22** (1916); cited by Nagahara, Y. (1940).
Oley, N., DeHan, R.S., Tucker, D., Smith, J.C., Graziadei, P.P.C.: Recovery of structure and function following transection of the primary olfactory nerves in pigeons. **88**, 477—495 (1975).
Parsons, T.S.: Studies on the comparative embryology of the reptilian nose. Bull. Mus. comparative Zool. **120**, 101—277 (1959).
Pearson, A.A.: The development of the nervus terminalis in man. J. comp. Neurol. **75**, 39—66 (1941a).
Pearson, A.A.: The development of the olfactory nerve in man. J. comp. Neurol. **75**, 199—217 (1941b).
Peter, K.: Die Entwicklung des Geruchsorgans und Jacobsonschen Organs in der Reihe der Wirbeltiere. Bildung der äußeren Nase und des Gaumens. In: Handbuch der Entwicklung der Wirbeltiere. Jena: Fischer 1906, pp. 1—82.

Piatt, J.: Form and causality in neurogenesis. Biol. Rev. **23**, 1—45 (1948).
Pinkus, F.: Über einen noch nicht beschriebenen Hirnnerven des Protopterus annectens. Anat. Anz. **9**, 562—566 (1894).
Pinkus, F.: Die Hirnnerven des Protopterus annectens. Morph. Arb. **4**, 275—346 (1895).
Read, E. A.: A contribution to the knowledge of the olfactory apparatus in dog, cat, and man. Amer. J. Anat. **8**, 17—47 (1908).
Retzius, G.: Zur Kenntnis der Nervendigungen in der Riechschleimhaut. Biol. Unters. N. F. **4**, 62—64 (1892).
Retzius, G.: Die Riechzellen der Ophidier. Biol. Unters., N. F. **6**, 48—51 (1894).
Robecchi, M. G.: Données ultrastructurelles sur la différentiation des cellules neurosensorielles dans le placode olfactive de poulet. C. R. Ass. Anat. **55**, 515—519 (1970).
Schiff, M.: Der erste Hirnnerv ist der Geruchsnerv. Moleschotts Untersuchungen **6**, 254—424 (1859).
Schultz, E. W.: Regeneration of olfactory cells. Proc. Soc. exp. Biol. **46**, 41—43 (1941).
Schultz, E. W.: Repair of the olfactory mucosa. Amer. J. Pathol. **37**, 1—19 (1960).
Sen Gupta, P.: Olfactory receptor reaction to the lesion of the olfactory bulb. In: Olfaction and Taste, Vol. 2 (T. Hayashi, Ed.). New York-London: Pergamon 1967, pp. 193—201.
Seydel, O.: Über die Nasenhöhle und das Jacobson'sche Organ der Amphibien. Eine vergleichend-anatomische Untersuchung. Morph. Jb. **23**, 453—543 (1895).
Siggia, S.: Contributi allo studio della determinazione del placode olfattivo in Discoglussus Pictus. Monit. zool. It. **47**, 116—117 (1937).
Smith, G. C.: Changes in the olfactory mucosa and the olfactory nerves following intranasal treatment with one per cent zinc sulfate. Canad. med. Ass. J. **39**, 138—140 (1938).
Smith, G. C.: Regeneration of sensory olfactory epithelium and nerves in adult frogs. Anat. Rec. **109**, 661—671 (1951).
Spirito, A.: Rigenerazione e regolazione nell'encefalo degli Anfibi. Roux Arch. Entw. Org. **122**, 152—178 (1930).
Takagi, S. F.: Degeneration and regeneration of the olfactory epithelium. In: Handbook of Sensory Physiology, Vol. IV (L. M. Beidler, Ed.). Berlin-Heidelberg-New York: Springer 1971, pp. 76—94.
Takagi, S. F., Yajima, T.: Electrical responses to odours of degenerating olfactory epithelium. Nature (Lond.) **202**, 1220 (1964).
Takagi, S. F., Yajima, T.: Electrical activity and histological change in the degenerating olfactory epithelium. J. gen. Physiol. **48**, 559—569 (1965).
Takata, N.: Riechnerv und Geruchsorgan. Arch. Ohren-, Nasen- u. Kehlkopf-Heilk. **121**, 43—78 (1929).
Van Campenhout, E.: Le développement embryonnaire comparé des nerfs olfactif et auditif. Acta otorhino-laryngologica belgica **11**, 279—287 (1956).
Van Gehuchten, A.: Contributions a l'étude de la muqueuse olfactive chez les mammifères. Cellule **6**, 393—407 (1890).
Vinnikov, A.: Processus de dégénérescence et de régeneration de l'organe olfactif chez les poissons et les amphibiens (in Russian). Bjull. Eksp. Biol. i. Med. **11**, 875—877 (1956).
Vinnikov, J. A., Titova, L. K.: Morphology of olfactory organs. Moscow: Medgiz 1957.
von Brunn, A.: Beiträge zur mikroskopischen Anatomie der menschlischen Nasenhöhle. Arch. mikr. Anat. **39**, 632—651 (1892).
Waddington, C. H., Cohen, A.: Experiments on the development of the head of the chick embryo. J. exp. Biol. **13**, 219—236 (1936).
Weissfeiler, J.: Régénération des lobes olfactifs et des hémisphères cérébraux. C. R. Soc. Biol. **91**, 543—544 (1924).
Weissfeiler, J.: Régénération du cerveau et du nerf olfactif chez les Batraciens urodèles. Rev. suisse Zool. **32**, 1—43 (1925).
Yokoh, Y.: The early development of the nervous system in man. Acta anat. **71**, 492—518 (1968).
Yokoh, Y.: Early development of the cerebral vesicles in man. Acta anat. **91**, 455—461 (1975).
Zuckerkandl, E.: Über die Wechselbeziehung in der Ausbildung des Jacobsonschen Organs und des Riechlappens. Anat. Hefte **41**, 3—73 (1910).
Zwilling, E.: Induction of the olfactory placode by the forebrain in Rana pipiens. Proc. Soc. exp. Biol. **31**, 933—935 (1934).
Zwilling, E.: An experimental analysis of the development of the nuran olfactory organ. J. exp. Zool. **84**, 291—323 (1940).

Chapter 3

The Interaction of Periphery and Center in the Development of Dorsal Root Ganglia

ARTHUR F. HUGHES, Cleveland, Ohio (USA) and VIRGINIA MCMILLAN CARR, Evanston, Illinois (USA)

With 4 Figures and 1 Table

Contents

A. Introduction

In this chapter we shall be concerned with the interactions that occur between the dorsal root ganglia (DRGs) and their peripheral fields of innervation in the course of vertebrate development. The term "center", as used here, refers to the neuronal cell bodies comprising the dorsal root ganglia. "Periphery" refers to the organ or field of innervation from which sensory nerve impulses originate.

The dorsal root ganglia and their peripheral fields lie some distance apart, and the link between them can be readily severed. Consequently, much of the work to be described has been concerned with the effects of experimental manipulations of the periphery, such as the removal or addition of limb rudiments, upon the developing ganglia. There is a corresponding and more extensive body of literature analyzing the effects of such procedures on the motor systems. In addition, there is an analogous branch of sensory research concerned with the effects of the removal of an eye on the development of the midbrain roof. These various systems have all been reviewed from time to time, notable by DETWILER (1926a, 1933, 1936), PIATT (1948), HUGHES (1968), JACOBSON (1969, 1970), and COWAN (1971).

The dorsal root ganglia and their periphery may be causally linked in the opposite direction also. It has been found, for example, that in the later stages of development a sensory nerve supply is required for normal development of specialized sensory organs such as encapsulated nerve endings and muscle spindles (JACOBSON, 1971; SAXOD, this volume). Such central-peripheral relationships may be akin to other instances where neurons exert trophic effects, as with regeneration of the amphibian limb (SINGER, 1965).

B. The Early Development of the Dorsal Root Ganglia

The dorsal root ganglia are derived from the neural crest, which is located at the margins of the neural plate of the early embryo (HIS, 1868; HÖRSTADIUS, 1950). The first experimental study on the neural crest was by HARRISON (1904) and involved extirpation of neural crest material from frog embryos. HARRISON demonstrated that the crest gave rise by migration and dispersal of its component cells to both the dorsal root ganglia and the Schwann cells of the peripheral nervous system. Since those early experiments, numerous other neural crest derivatives, including the autonomic ganglia, various types of pigment cells, glial elements, and some skeletal and connective tissue elements, have been identified (WESTON, 1970).

Migration of the neural crest cells was followed by WESTON (1963) in early chick embryos whose own crest he had exchanged for that from donor embryos previously labeled with tritiated thymidine. WESTON's study demonstrated that neural crest cells migrate with no apparent impediment through the closely packed tissues of the ectoderm and somitic mesenchyme. In further experiments (WESTON and BUTLER, 1966) involving grafting of labeled crest material between embryos of different ages, it was shown that the direction of the neural crest migration is controlled by the surrounding tissues and that a developmentally "changing embryonic environment affects the extent of migration".

The dorsal root ganglia are no exception to this observation. Those neural crest cells that give rise to the dorsal root ganglia lodge in the spaces between the somites. Thus, the segmentation of the DRGs is secondarily derived. This was first demonstrated by DETWILER in 1934 in experiments involving removal or addition of somitic units to the axes of *Amblystoma* embryos. DETWILER showed that disturbances of the serial arrangement of the axis resulted in corresponding alterations in the number and arrangement of the dorsal root ganglia.

A reciprocal relationship between somites and ganglia may also exist, for it may be that the presence of neural crest derivatives is required for the maintenance of segmentation of the somitic mesenchyme. This possibility is indicated by the Splotch mutant in mice. In Splotch embryos the neural crest is deficient (AUERBACH, 1954), and the boundaries between the somites can be seen to break down prematurely.

Once the individual dorsal root ganglia are recognizable, they grow rapidly. Ganglionic growth is perhaps most rapid in the opossum embryo (HUGHES, 1973). In this animal, as soon as the cervical and anterior thoracic ganglia are recognizable, approximately two days before birth, each ganglion gains some 200 cells per day. The mitotic rates in these ganglia are correspondingly high, approximately five percent, which is sufficient to account for the increased cell numbers without additional neural crest cell migration into the ganglia.

At some stage in this developmental process the dorsal root ganglia come under the influence of their peripheral fields and begin to make adjustments to the requirements of the periphery. The adjustments are by way of alterations in cell sizes and numbers within the ganglia. In some species the peripheral control over cell number may include effects on proliferation and differentiation of neuronal precursors. In all species examined it most certainly involves maintenance of developing neurons.

It is generally assumed that the control the periphery exerts on differentiating ganglia is chemical in nature and is transmitted to the cell bodies by retrograde axonal flow in peripheral nerve fibers. This notion is supported by the observation of retrograde axonal flow in explanted ganglia (HUGHES, 1953) and in isolated DRG neurons (NAKAI, 1956). The chemical identity of the controlling factor(s), however, is totally unknown.

C. Peripheral Effects on the Size, Cell Number, and Differentiation of Dorsal Root Ganglia

The influence of the periphery on the development of the dorsal root ganglia was first examined by SHOREY (1909) in a study of the effect of limb amputations on the size of the ganglia in embryos of chicks and in embryos and larvae of three amphibian species, *Amblystoma tigrinum*, *Rana pipiens*, and *Bufo americanus*. SHOREY found that the ganglia that had innervated the amputated limbs became reduced in size following the operations in all four species. She considered these reductions to be due to a "failure of certain neuroblasts to differentiate normally" and discussed the implications of these findings in the light of the concepts of self-differentiation and dependent differentiation that had been enunciated by ROUX (1894). Thus, one possible explanation for her results was that there were two classes of neuroblasts or young neurones, "one of which is self-differentiating, and the other not". There was no evidence for this hypothesis. The only alternative explanation for the results was that in development all differentiating neurons are dependent upon some sort of stimulation from the end organs, which had been removed in the course of the amputations.

Subsequent workers in this field have chosen to confine their studies to a single genus or species, and further discussion of these studies will be according to the animal class examined.

Before progressing further, however, it first ought to be noted that the analysis of cellular events within dorsal root ganglia and of the mechanisms controlling these events is hindered by a lack of information regarding the degree of fiber outgrowth toward the periphery at any given time during development. Caution is especially required in dealing with the earlier stages. Until a few years ago the absence of silver-stainable nerve fibers from early limb buds examined at the light microscope level was held to indicate that neurons had not yet grown into the limb buds or made contact with the periphery. However, the application of electron microscopy to the examination of the developing nervous system at once made it clear that most embryonic neurites are not sufficiently myelinated to take up silver stains and that, even if they were, they are below the limits of resolution of the light microscope (PRESTIGE and WILSON, 1972). Furthermore, even with electron microscopy it is extremely difficult to recognize single fine neurites among a mass of mesenchymal cells that have their own fine interlocking processes.

More recently, use of the marker protein horseradish peroxidase (HRP) has provided another tool for the examination of the degree of fiber outgrowth from ganglionic neurons. Use of HRP overcomes the difficulties associated with electron microscopy of recognizing single fine neurites and also permits localization within the central structures of neurons innervating any particular peripheral region. For example, examination of the uptake and retrograde transport of HRP from the limb buds into early ventral horn cells of *Xenopus* revealed that these motor cells had already made connection with the periphery at the earliest stage of limb development (LAMB, 1974). Prior to this study the source of fibers observed in the limb buds was unknown.

This note of caution having been sounded, the discussion can proceed.

I. Urodela

In 1920 SAMUEL DETWILER began publishing a series of neuroembryological studies on *Amblystoma*, which continued to appear for over thirty years. These investigations proceeded alongside many on the physiology of vision. Of Detwiler's many papers on neuroembryology, only a few concerned sensory ganglia. The rest were devoted to experimental studies on the spinal cord, the function of grafted limbs, and, towards the end of the series, research on the neural crest.

His first neuroembryological paper (1920) concerned the effects on dorsal root ganglia of the heterotopic grafting of a forelimb bud several segments caudal to its normal position. The ganglia thus deprived of their normal periphery showed cellular hypoplasia and reduction in the overall ganglionic size, while ganglia innervating the limb bud in its new position showed cellular hyperplasia and ganglionic hypertrophy.

SCHWIND (1931) studied the relationship between the size of the forelimb and the cell number in brachial ganglia by means of reciprocal grafts between different sized species of *Amblystoma*. He found that grafting limbs of the smaller *A. punctatum* in place of those of the larger *A. tigrinum* resulted in hypoplasia of the host *tigrinum* ganglia that innervated the *punctatum* limbs. The reciprocal arrangement, the larger *tigrinum* limbs replacing the smaller *punctatum* limbs, resulted in hyperplasia of some of the *punctatum* brachial ganglia. DETWILER (1930) had also noted such effects in the course of studies on the function and innervation of heteroplastic grafts.

II. Anura

SHOREY's experimental studies on anura have been continued by several workers on a variety of species. DÜRKEN (1911) found that extirpation of the hindlimb bud of *Rana fusca* and *Bufo vulgaris* resulted in decreases of over 50% in the diameters of the lumbar ganglia. MAY (1933) reported a hyperplasia of neurons of the dorsal root ganglia in *Discoglossus pictus* following grafting of a supernumerary limb. Reduced ganglionic size was observed in *D. pictus* by MAY (1930) following a different sort of reduction in the periphery. MAY grafted spinal cord segments to the region between normal cord and a hindlimb bud of the host animal. In those animals whose limb innervation came to be largely from the grafted cord, the host dorsal root ganglia and that half of the host spinal cord on the side of the graft were deprived of their normal fields of innervation. Subsequently, these ipsilateral ganglia and halves of the host spinal cords were reduced in size.

Neuroembryological studies on *Xenopus laevis* were first carried out by HUGHES and TSCHUMI (1958). These workers examined development of the lumbar dorsal root ganglia and ventral horn in *Xenopus* under normal conditions and following limb bud ablation or isolation at larval stages. In the normal situation they found that ganglia at limb and nonlimb levels differed from each other in two respects. First, limb ganglia were found to contain a greater number of cells than were found in nonlimb ganglia. Second, it was demonstrated that while limb ganglia are comprised of a uniform population of small cells, nonlimb ganglia contain a few cells that are much larger than the rest. Because these differences exist even before the hindlimb buds make their appearance, HUGHES and TSCHUMI concluded that control of the *initial* production of cells and the size of neurons at the various levels must be intrinsic to the ganglia themselves and free from any influence of the peripheral field of innervation.

Following hindlimb bud ablation HUGHES and TSCHUMI observed a hypoplasia in the lumbar ganglia but no change in the sizes of the constituent neurons. They concluded that the periphery must function in some way to control cell numbers at the center.

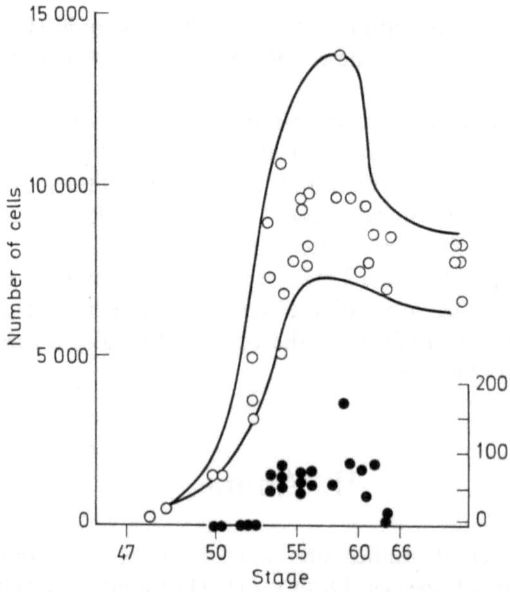

Fig. 1. Counts of living and degenerating cells in lumbar ganglia of *Xenopus laevis* tadpoles during development. Living cells, open circles (left-hand scale). Degenerating cells, closed circles (right-hand scale). Adapted from HUGHES (1967a), based on PRESTIGE (1967a)

HUGHES and TSCHUMI's study was continued and greatly expanded by those of PRESTIGE (1965, 1967a). PRESTIGE made separate counts of the numbers of viable and degenerating cells in normal animals and in experimental animals following hindlimb ablation at various developmental stages. He found that in the normal animal at stage 50 (NIEUWKOOP and FABER, 1956) the three lumbar ganglia are made up of a total of about 1500 cells (Fig. 1). This total increases at an exponential rate until stage 53, when massive cell degeneration begins within the ganglia. As a result of this degenerative activity, the rate of increase of cell numbers is reduced. The total number of cells reaches a peak of about 10000 at stage 59 and then falls to a somewhat smaller number where it levels off. During this period of cell turnover, the amount of cell degeneration is so large that many more cells degenerate than are eventually maintained. Using HUGHES's estimate (1961) of approximately three hours for the duration of the degeneration of a cell in the ventral horn, PRESTIGE calculated that the number of ganglionic cells produced during larval life is approximately two or three times the number actually found at metamorphosis.

PRESTIGE then analyzed the effects imposed on the lumbar ganglia by ablation of the hindlimb at various stages (1967a). His results are summarized in Table 1. He observed three successive phases in the reaction of ganglionic cells to the operation. HUGHES and TSCHUMI (1958) had shown that when amputation is performed before cell turnover begins at stage 53, the effect of the operation is to slow the rate of increase of cell numbers in the ganglia by reducing the rate of cell production. No change in the rate of degeneration occurs. This response cor-

Table 1. The effect of hindlimb amputation on cells of the ipsilateral lumbar dorsal root ganglia of *Xenopus laevis*. The rates of cell production referred to were calculated from the rate of change in cell number of the ipsilateral ganglia relative to the contralateral. See text for details. From PRESTIGE (1967a)

	Phase I	Phase II		Phase III
Amputation at stage	50 from (HUGHES and TSCHUMI, 1958)	53–55 (57) 61		Juvenile
Days	0	20 30 40		80
Rate of production	Slowed 6 days→	Depressed 1 day →	Depressed 1–2 days Elevated 5–25 days	Elevated 11–25 days
Rate of degeneration	Unaffected	Depressed 1 day→ with 2 intense excess periods in 1st and 3rd weeks, superimposed	Depressed 1–2 days Elevated 5–62 days	Elevated 11 days→ (at least 4 months)
Total cell numbers	Rate of increase slowed	50% lost in 2 periods in 1st and 3rd weeks	50% lost between 25 and 62 days	Decline 25 days→ (still in progress at 4 months)
Dorsal root ganglia in control animals		Turnover		
	Cell production			
		Cell degeneration		
	Cell numbers rising-falling-unchanging			

responds to Phase I of the amputation reactions of ventral horn cells (PRESTIGE, 1967b). PRESTIGE (1967a) found that once cell turnover begins, amputations have a striking effect on cell degeneration. Amputations performed between stage 53 and the onset of metamorphosis result in a two-stage response. Following an initial, brief period of reduced cell degeneration, a much longer period of markedly increased degeneration occurs. This response corresponds to the Phase II reaction to amputation. The rate of production of cells also appears to be affected in the Phase II response, for counts of total cell numbers reveal no changes in cell numbers for extended periods even though the rate of degeneration is known to be elevated.

The third phase of amputation reaction of ganglionic cells occurs in juvenile toads after cell turnover has ceased at metamorphosis. Amputations performed after this phase is reached cause no degeneration for a relatively long period following the operation. However, once cell degeneration does get under way, it continues for at least several months. As in Phase II, cell production in Phase III appears to be affected by the operation, for once again total cell numbers remain constant during certain periods despite an increase in the rate of degeneration.

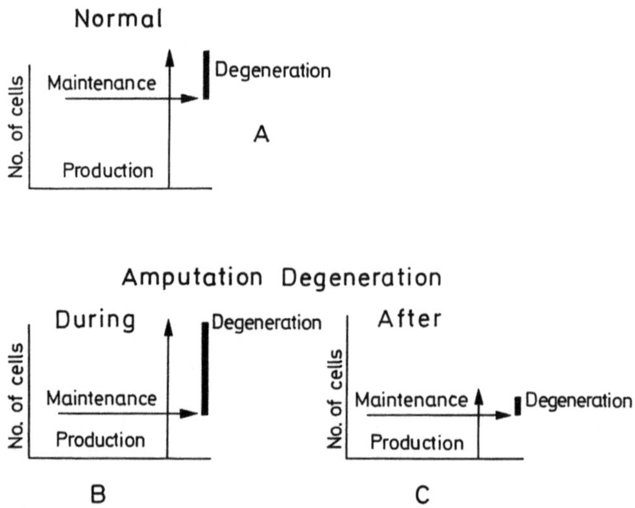

Fig. 2. Diagram of processes controlling number of cells in lumbar dorsal root ganglia. A. Normally, overproduction of cells causes neurons in excess of maintenance capacity of periphery to degenerate (histiogenetic degeneration of GLÜCKSMANN, 1951). B. Amputation reduces the maintenance capacity so additional cells degenerate during the period of cell turnover. C. Amputation also reduces cell production so amount of normal histiogenetic degeneration is reduced. Adapted from PRESTIGE (1967a)

PRESTIGE concluded that these events are governed by an overall control of total ganglionic cell number by the periphery. Axotomy and loss of the peripheral field *per se* do not cause a cell to degenerate, for death may be postponed until long after the initial injury. Rather, the cell body is assumed to store a "maintenance factor" derived from the periphery and transported to the cell body by retrograde axonal flow. Axotomy and removal of the periphery cuts off the supply of this substance. When the supply of maintenance factor stored in the cell body becomes exhausted, the cell dies. Larger, more mature cells are presumed to store a greater amount of maintenance factor than smaller, less mature cells. Thus, degeneration in Phase III is considerably postponed relative to Phase II. On the other hand, PRESTIGE suggests, Phase I cells do not show a degenerative response because they have not yet been exposed to large amounts of maintenance factor. Maintenance factor is presumed to have an addictive quality such that only after neurons have been exposed to it do they require its continued presence to insure their survival. As they mature, young neurons are exposed to increasing amounts of maintenance factor. By the time they have matured to Phase II cells, they are dependent and will degenerate in the absence of sufficient amounts of this factor.

In addition to a maintenance factor, PRESTIGE's observations also require the activity of a "production factor", also produced by periphery. Such a production factor would act to cause an increase in the number of neurons in the ganglia and would thus work in concert with maintenance factor to match the number of lumbar ganglionic neurons to the demands of the hind leg. The relationship between production, maintenance, and degeneration is represented schematically in Figure 2.

A series of neuroembryological studies on species of the anuran genus *Eleutherodactylus* was begun by HUGHES and his colleagues in 1959. Species of *Eleutherodactylus* are among the smallest living tetrapods, and thus offer advantages in studies involving any cell counting. The number of cells in lumbar dorsal root ganglia in *Eleutherodactylus* is about one-sixth that in *Xenopus*. Furthermore, there is no larval stage following hatching in *Eleutherodactylus*; development is wholly embryonic. There is also much less thyroid control over developmental processes than is the case in anurans undergoing a larval stage in the course of their development (HUGHES, 1974).

In *Eleutherodactylus*, as in *Xenopus*, a greater number of cells are seen in the dorsal root ganglia at limb than at nonlimb levels even before the limb buds have formed. In *Eleutherodactylus*, however, there are no early intrinsic differences in the sizes of neurons of ganglia at the various levels (HUGHES, 1959). Studies on *E. martinicensis* indicate that the total number of ganglionic cells normally increases during almost the entire course of development (HUGHES and EGAR, 1972). In the lumbar ganglia this total cell number increases to a peak on day 13, but then falls sharply during the 14th day. The fall in cell number is accompanied by increased amounts of cellular degeneration during the last 12 hours prior to hatching on day 14.

Following limb amputation in *Eleutherodactylus*, extensive degeneration occurs promptly in the innervating ganglia, regardless of the age of the embryo at the time of the operation. Hindlimb amputation on day 5, for example, results in large amounts of degeneration in the lumbar ganglia within a few hours of the operation (HUGHES, 1964), i.e., several days before normally occurring degeneration. This is in striking contrast to the situation in *Xenopus*, where degeneration in response to early amputation does not occur until the time of normally occurring cell death, at stage 53. HUGHES further noticed in *Eleutherodactylus* that even if the amputated limb bud is replaced immediately after its excision, early cell degeneration occurs despite the fact that such autografts are usually re-innervated. However, there do seem to be some adjustments following such immediate replacements of amputated limbs, for the final cell numbers in the ganglia involved are higher than those seen following simple ablation.

Unfortunately, relatively few of the experimental studies carried out on the dorsal root ganglia of *Xenopus* have yet been applied to those of *Eleutherodactylus*. Thus, explanations for the events in *Eleutherodactylus* and for the discrepencies between *Eleutherodactylus* and *Xenopus* are not forthcoming.

III. Reptilia

The one investigation concerning central and peripheral relationships in the dorsal root ganglia of reptiles is of a unique kind. It concerns the re-innervation of regenerated tails of lizards following autotomy or amputation of the original appendages. The regenerate does not reproduce all the features of the original tail. Some structures, including the dorsal root ganglia and the gray matter of the spinal cord, are not found in the regenerate. Rather, as FRAISSE illustrated in 1885, the regenerate is re-innervated by downgrowth of nerve fibers from the spinal segments of the stump. The spinal segments themselves become enlarged in response to the increased size of their periphery. HOOKER (1912) described the descent of branches of two pairs of spinal nerves into the new tail. TERNI (1920) found in *Gongylus ocellatus* that there were contributions from three segments. During tail regeneration histological (HUGHES and NEW, 1959) and ultrastructural (PANNESE, 1963) changes indicative of new fiber growth can be seen in the perikarya of the ganglia in these segments. TERNI observed that large neurons of these overloaded dorsal root ganglia had undergone a cellular hypertrophy, enlarging in diameter to three times their normal size. PANNESE found that hypertrophy occurs in small neurons as well. Such effects on ganglionic neurons have very rarely been observed.

IV. Aves: The Chick

The development of the dorsal root ganglia and their interactions with the periphery in the course of development have been studied rather extensively in the chick embryo at various times since the publication of SHOREY's (1909) observations. One of the first studies on normal chick DRG development was that of LEVI-MONTALCINI and LEVI (1943). According to these workers, the first sign of ganglionic neuronal differentiation occurs at about three days of incubation as the migrating neural crest cells cease their migratory activity and begin sending out neurites. By the end of the third or the beginning of the fourth day of incubation, the distal neurites have innervated the wing bud. Innervation of the hindlimb bud occurs slightly later. The proximal neurites enter the spinal cord to form the dorsal roots during the fifth day.

Differentiation at the ultrastructural level has been closely followed by PANNESE (1968a and b, 1969). He has shown that up to the fourth day of incubation the dorsal root ganglia are composed largely of rounded cells containing ovoid or spherical nuclei that are surrounded by a thin rim of cytoplasm containing free ribosomes and a few mitochrondria. These so-called "indifferent" cells mature into what PANNESE calls "primitive neuroblasts", which have the characteristic ultrastructure of young neurons and are quite numerous during the fourth to the sixth days. In the cytoplasm of these primitive neuroblasts, cisternae are found just beneath the cell membrane. At deeper levels there appears a rough endoplasmic reticulum made up of long interconnecting channels, and also a large number of free ribosomes.

The next stage in neuronal differentiation, that of the so-called "intermediate neuroblast", is first seen during the fifth day of incubation. Cells that have matured to this stage have much more abundant cytoplasm and also an eccentrically located nucleus. The cytoplasm close to the inner edge of the nucleus contains mitochondria and Golgi bodies. Deeper in the cytoplasm is a conspicuous area of endoplasmic reticulum, of both the rough and smooth varieties.

At about this same time, the fifth day of incubation, star-shaped satellite cells first come into contact with the young neurons. Prior to this time, the developing neurons have been closely packed together with both *adhesion plaques* and *fasciae occulentes* type junctions between them. As satellite cells begin to move between the developing neurons and encircle individual cells, these junctions disappear. By day 10 each neuron has become completely ensheated by cytoplasmic expansions of the satellite cells surrounding it.

The final stage of differentiation of the young neuron is that of the rounded pseudo-unipolar neuron. By this stage, the proximal and distal axons have been displaced from the immediate vicinity of the cell body to the end of a long cytoplasmic process that has developed between them and the cell body. Hence the description "pseudo-unipolar". Within such cells, rough endoplasmic reticulum has become concentrated into the Nissl bodies described in light microscopy. Many ribosomes elsewhere are aggregated into polysomes containing six to 13 ribosomal units. Cells of this stage do not appear until the second half of incubation.

The apparent rates of neuronal development are not the same throughout the chick dorsal root ganglia. By day 8 there is a clear boundary between the larger, more mature cells in the ventrolateral region and the smaller cells in more dorsal and medial regions. The two regions are distinguishable until the 15th day of incubation (LEVI-MONTALCINI and LEVI, 1943). Such a sharp distinction between different regions is not generally seen in the developing ganglia of other tetrapods, although in opossum DRGs there does seem to be a greater proportion of large cells in the more ventral than in the more dorsal regions (HUGHES, 1973). There is some evidence that this regional segregation may be peripherally controlled, for YATES (1961) observed no segregation in ganglia isolated on the chorioallantoic membrane. However, a similar absence of segregation has not been noted in ganglia *in situ* following removal of their peripheral fields of innervation (LEVI-MONTALCINI and LEVI, 1942; HAMBURGER and LEVI-MONTALCINI, 1949; CARR, 1976).

Cell numbers in the developing chick dorsal root ganglia were counted first by LEVI-MONTALCINI and LEVI (1942, 1943). Their studies mark the earliest attempt to count all the cells within a particular tetrapod ganglion. The formidable task was multiplied several times by the repetition of the counts at several stages, both in the normal situation and following limb bud amputation. In the normal situation they found that in ganglion 25, the main lumbar ganglion, the total number of the cells increases until day 8 of incubation when it levels off at 8000 to 12000 cells.

The first experimental study on chick ganglia after that of SHOREY was by HAMBURGER in 1934. HAMBURGER found that amputation of the right wing buds of 72-hr chicks (i.e., prior to wing bud innervation) resulted in reductions in volumes of the affected brachial ganglia by 37 to 54%. LEVI-MONTALCINI and LEVI's (1942) cell counts indicated that by four days after amputation of the hindlimb bud the numbers of both total cells and differentiated neurons in the ipsilateral 25th ganglion had been reduced considerably. HAMBURGER (1939) also carried out grafting of limb buds early in development. He observed increases of 15 to 200% in the volumes of ganglia innervating supernumerary limb buds.

These studies were greatly expanded by that of HAMBURGER and LEVI-MONTALCINI (1949), who recorded, in addition to the number of cells, the number of mitotic figures and estimates of the abundance of degenerating cells in dorsal root ganglia at various levels in both control and experimental situations. This landmark study was the first in which cellular degeneration, originally noticed among developing neurons by COLLIN in 1906, was recognized as a general factor in ganglionic development. HAMBURGER and LEVI-MONTALCINI not only observed reductions in both the total number of cells and the number of large, differentiated neurons in ganglia affected by wing bud removal; they also observed the converse, namely an increase in both of these populations, following limb bud addition. Such alterations, these authors felt, were mediated through changes in the rates of proliferation, degeneration, and initial differentiation of ganglionic neurons.

HAMBURGER and LEVI-MONTALCINI found that normally there is a peak in the absolute number of mitotic figures in a ganglion, which they termed the mitotic activity, between days 5 and 6 and that such peaks are higher in limb than in nonlimb ganglia. Mitotic figures decrease in number from the 7th day and can no longer be found by the 9th. Following experimental alterations in the size of the peripheral field of innervation, corresponding changes in the mitotic activity of affected ganglia were observed: increases of 3 to 45% were observed in ganglia innervating supernumerary limbs and decreases of 4 to 55% were observed in those whose field of innervation had been removed by amputation. It was also noticed that these responses to alterations in the periphery occurred with the same temporal sequence as was seen in the normal situation. This observation was later corroborated by YATES (1961) in ganglia isolated on the chorioallantoic membrane.

In a recent study on the cellular effects of wing bud ablation and limb bud addition, CARR (1975, 1976) measured cell production in the dorsal root ganglia of chicks at $4^1/_2$ to $9^1/_2$ days of incubation by autoradiography with tritiated thymidine. Cell production was expressed as the labeling index, that is, the percentage of cells counted that were observed to be labeled following a one-hour exposure to tritiated thymidine. All brachial ganglia, 14, 15, and 16, and nonbrachial ganglia 12, 13, and 17 were examined. Separate counts were made for the ventrolateral and mediodorsal cells in each at all times except $4^1/_2$ days.

At $4^1/_2$ days of incubation CARR found labeling indices of approximately 12% in both brachial and nonbrachial ganglia. In the mediodorsal region the labeling indices of all ganglia decrease from 11–13% at $5^1/_2$ days to one percent or less by $7^1/_2$ days, where they remain for the rest of the period examined (Fig. 3). In the ventrolateral region of all ganglia, labeling indices decrease from approximately 8% at $5^1/_2$ days. In the nonbrachial ganglia these labeling indices decrease to 2 to 3% by $7^1/_2$ days and in the brachial ganglia to 4% by as early as $6^1/_2$ days. However, rather than continuing to decline, as observed for the mitotic activity (HAMBURGER and LEVI-MONTALCINI, 1949), the ventrolateral labeling indices increase again slightly and are still elevated at $9^1/_2$ days. Furthermore, given CARR's high labeling indices at $4^1/_2$ days, her data also differs from that of the earlier studies in showing no peak in proliferative activity at days 5 to 6 but, rather, elevated values through $5^1/_2$ days. Explanations for these discrepancies may be

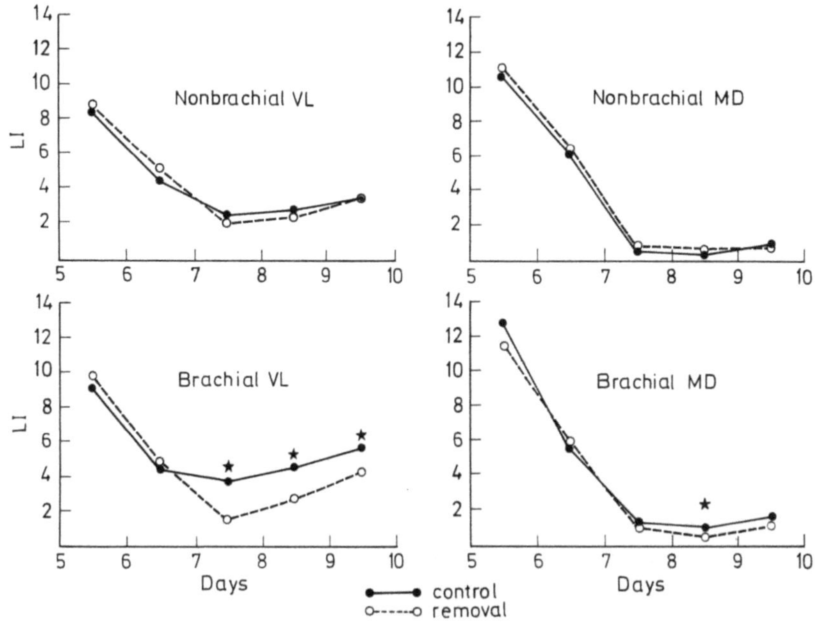

Fig. 3. Labeling indices (*LI*) in ventrolateral (*VL*) and mediodorsal (*MD*) regions of dorsal root ganglia of chick embryos at various times after right wing bud amputation at $2^1/_2$ days of incubation. Star indicates significant response to amputation as analyzed by the paired *t*-test. From CARR (1976)

found in the results of YATES (1961). YATES found that when the proliferative rate was expressed as the absolute number of mitotic figures, as in HAMBURGER and LEVI-MONTALCINI's mitotic activity, a peak was seen at 5 and 6 days of incubation. Proliferative activity expressed in this way varies with the number of cells in different ganglia and at different times. However, when YATES expressed the proliferative rate as a ratio of the mitotic to nonmitotic cells, a parameter not unlike CARR's labeling index and one that does not vary with cell number, it was seen to decrease continuously from days 4 to 6 when it levelled off.

In further autoradiographic studies, CARR has also attempted to analyze the nature of the proliferative population represented by the labeling indices. Her results indicate that neuronal production has virtually ceased by $6^1/_2$ days in the ventrolateral regions and by $7^1/_2$ days in the mediodorsal regions of both brachial and nonbrachial ganglia. Glial proliferation, however, remains elevated. Thus labeling after these times must be primarily glial in nature. Furthermore, given that in the chick DRGs the ventrolateral cells are the larger and the mediodorsal cells the smaller young neurons, these data are in agreement with the observations made in dorsal root ganglia of rats (LAWSON et al., 1974) and elsewhere in the nervous systems of various species (e.g. ANGEVINE, 1970) that precursors of large neurons tend to cease their proliferation before those of smaller neurons.

Following amputations of right wing buds at $2^1/_2$ days of incubation, CARR found no change in the labeling indices in any ganglia on the operated side until after $6^1/_2$ days of incubation (Fig. 3). After this time significantly lower labeling indices were observed, but only in the ventrolateral regions of brachial ganglia. No proliferative responses to amputation were seen in the mediodorsal region. However, as shown by the birthdate studies mentioned above, labeling activity in the ventrolateral regions after $6^1/_2$ days is glial in nature. Therefore, CARR's results indicate that the only labeling response to wing bud removal in the chick dorsal root ganglia must be glial rather than neuronal in nature.

In striking contrast to HAMBURGER and LEVI-MONTALCINI, CARR found no proliferative response to early limb bud addition. Despite the fact that counts were made in only those specimens (34 in all) that showed brachial innervation of both graft and host limbs, as revealed in tracings made of all nerves visible in serial cross sections, CARR found very little evidence of either ganglionic hypertrophy or altered labeling indices at the ages examined. HAMBURGER (pers. comm.) feels the explanation for the differences between the two sets of results lies in the nature and quality of both the grafts and their sensory innervation. He feels that unless the grafts and their innervation, as well as the innervation of the ipsilateral host wing bud, are almost the same as in the normal situation, no responses such as those he observed can be expected. Unfortunately, beyond either histological or behavioral examination, the quality of the sensory innervation was not determined in either set of experiments. Until such time as procedures are devised to define the optimum sensory innervation while leaving the dorsal root ganglia structurally intact for cell counts, the problem will remain unresolved.

As mentioned above, HAMBURGER and LEVI-MONTALCINI also analyzed degenerative activity in dorsal root ganglia. Their study consisted of estimates of the relative amount of degenerative activity within ganglia at various levels in control and experimental animals. Normally, they found cellular degeneration was limited to nonlimb ganglia and, thus, appeared to play a large role in establishing size differences between limb and nonlimb ganglia. Moreover, within the nonlimb ganglia degeneration appeared to be limited almost exclusively to the large cells in the ventrolateral region. Degenerating cells were not seen earlier than days 4 to 5 of development; but, once degeneration began, it rapidly reached a peak in activity at days 5 to 6. After day 7 it was virtually absent. At later times only a slowly progressing atrophy was apparent, primarily in the mediodorsal region. Following wing bud amputation, large numbers of degenerating cells were seen in the affected brachial as well as the nonbrachial ganglia. However, this increased degenerative activity was again limited to the ventrolateral regions, and the temporal sequence of its appearance was the same as in the controls.

CARR carried out a more quantitative analysis of degenerative activity among the dorsal root ganglia. Her data (Fig. 4) are expressed in terms of a degeneration index, that is, the percentage of the cells counted that are observed to be degenerating. In agreement with the earlier studies, CARR found little or no degenerative activity until after $4^1/_2$ days of incubation in either brachial or nonbrachial, control or experimental ganglia following alterations in the size of the peripheral field. However, in no case was degenerative activity absent after day 7. Furthermore, degenerating cells were observed in the mediodorsal regions, and not only of nonbrachial ganglia but of brachial ganglia as well. Mediodorsal degeneration indices rise slowly over the course of the period examined, reaching

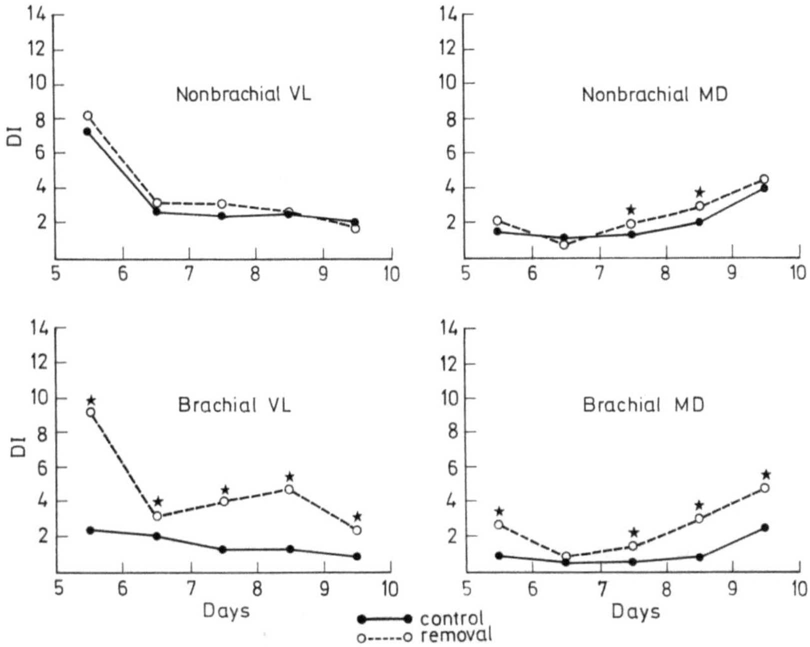

Fig. 4. Degeneration indices (*DI*) in ventrolateral (*VL*) and mediodorsal (*MD*) regions of dorsal root ganglia of chick embryos at various times after right wing bud amputation at $2^1/_2$ days of incubation. Star indicates significant response to amputation as analyzed by the paired *t*-test. From CARR (1976)

values of 2 to 4% by $9^1/_2$ days. In ventrolateral regions, where degenerating cells had been previously described only in nonbrachial ganglia, degeneration was found by CARR to be present in brachial ganglia as well. In both brachial and nonbrachial ganglia, peaks in ventrolateral degeneration indices were observed at $5^1/_2$ days, although the brachial peak was considerably smaller than the nonbrachial.

Following wing bud amputation, CARR, in agreement with the previous investigators, found no degenerative response until the time when normally occurring degeneration begins. In addition, she also found a ventrolateral degenerative response to amputation only in the brachial ganglia. However, this ventrolateral response was observed to consist not of one, but of two peaks of relatively high degenerative activity, one of 9% at $5^1/_2$ days and another of 5% at $8^1/_2$ days. Furthermore, experimental degeneration indices did not return to normal levels but, instead, remained significantly higher than those of the controls throughout the rest of the period examined.

Degenerative responses to wing bud ablation were also observed in the mediodorsal regions, in both brachial and nonbrachial ganglia. In the nonbrachial ganglia the degenerative response is slight and most probably reflects removal of some of the shoulder musculature along with the wing bud. In the brachial ganglia, however, the ipsilateral degeneration indices must reflect the absence of the wing bud. These results are most interesting in view of the fact, mentioned previously, that the small mediodorsal cells appear to be much less mature than the ventrolateral cells at the times covered in the above studies. Indeed, LEVI-MONTALCINI and LEVI (1943) felt that differentiation of the smaller mediodorsal cells did not even begin until day 10 to 12 of incubation. CARR's results suggest, however, that, despite their immature appearance, at least some of these young mediodorsal neurons must be in contact with the periphery. Recent experiments by OPPENHEIM and HEATON (1975), tracing uptake of the marker protein horseradish peroxidase from leg buds to lumbar spinal cords and ganglia lead to the same conclusion, for HRP could be seen in mediodorsal cells as early as day 4 of incubation.

V. Mammalia

TENNYSON (1965) has described the differentiation of neurons of the dorsal root ganglia of the rabbit embryo by light, phase-contrast, and electron microscopy. Her stages and description are not unlike those of PANNESE (1968a, 1968b, 1971) for the chick embryo.

Because of the difficulties of intrauterine experimental procedures, studies on the interaction between nerve centers and their peripheral fields of innervation in mammals have not been as numerous as in other animal classes. Surgical techniques necessary for intrauterine operations on mammalian embryos were described by BORS (1925) and NICHOLAS (1925) and have been applied by workers such as BARRON (1945) and HALL and SCHNEIDERHAN (1945). BARRON amputated limbs of sheep embryos *in utero* and found that by birth there was an 11 to 93% reduction in the volume of the affected dorsal root ganglia.

HALL and SCHNEIDERHAN carried out a more extensive study on rat embryos. They operated at a relatively late stage of fetal life, 15 to 18 days of gestation, which is only seven to four days before birth. By this time limb development is well advanced. The difficulties of the operations are quite apparent from the results: from 50 pregnant rats only 12 living newborns lacking a limb were obtained. Most of these animals were fixed at birth. Cells were counted in every fourth or fifth section through the ganglia, and the volumes of the dorsal root ganglia were estimated by planimetric measurements on sections. Hypoplasia was observed in all ganglia from C_5 to T_1. The most marked hypoplasia was seen in C_8, where the number of cells on the ipsilateral side was less than half that on the control side. In most instances there was close agreement between the number of cells and the volume of a particular ganglion. This suggests that a normally occurring increase in cell volume was unaffected by the operation.

Recent work by LAWSON et al. (1974) indicates that the effects of these operations must have been on postmitotic cells: LAWSON and her co-workers found in labeling studies that neuronal production in rat DRGs has ceased by 13 days of gestation, two days prior to the earliest of HALL and SCHNEIDERHAN's operations. Thus, the effect of these operations must have been to produce the death of postmitotic cells. Since HALL and SCHNEIDERHAN did not report any cell death in animals fixed at birth, such degeneration presumably must have occurred on a massive scale in the few days before birth. However, in the two instances in which a fetus had been examined shortly before birth, HALL and SCHNEIDERHAN did not report seeing any pycnosis in brachial ganglia at that time either. Indeed, in one specimen hypoplasia was not yet apparent. Thus, the exact nature and timing of ganglionic responses to reduced periphery in rats are not at all clear.

MCBRIDE (1974) has examined the ganglionic hypoplasia accompanying thalidomide-induced phocomelia in rabbits. The teratogen was administered to the mothers daily from an early stage of pregnancy until day 29, three days before birth, when each fetus was examined and fixed. The limbs suffered various degrees of reduction, which were correlated with changes in the brachial and lumbar ganglia. These changes included such degenerative features as cytoplasmic vacuolation and paucity of Nissl substance, as well as an "edematous dissociation of the cells, with wide intercellular spaces". In extreme cases few normal cells were observed. MCBRIDE counted all neurons in serial sections of ganglia C_6 and made separate tallies of normal and abnormal cells. A specimen with apparently normal forelimbs showed only a small proportion of abnormal cells and a slight degree of hypoplasia in the C_6 ganglia. However, a much greater degree of hypoplasia was seen in an embryo with almost complete amelia. The hypoplasia was so severe, in fact, that on one side of the embryo C_6 could not be found. It has not yet been determined for certain whether these results are due to the influence of the periphery on the center or to the converse. However, if the results were due to an effect of the center on the periphery, one would have to postulate a special intrinsic sensitivity of the limb ganglia to thalidomide, for degenerative changes and hypoplasia were observed only in brachial and lumbar ganglia.

Marsupial mammals offer particular opportunities to the investigator of the relationships between limb and neural centers. At birth these animals are in quite an immature condition, and the relationships of center to periphery can thus be studied more readily and over a greater course of development than in placental mammals. Much scope exists for inquiry regarding interaction between the limbs and their neural centers in particular, for while limb function and development at birth are quite premature in the forelimbs (LANGWORTHY, 1925), which are used in the climb to the pouch, it is quite delayed in the hindlimbs. Furthermore, early pouch young of the opossum have been found to suffer amputation without prejudice to further development (MCCRADY, 1938). Finally, the cell numbers in the motor centers and sensory ganglia are much lower in the opossum than in placental mammals, making the task of cell counting much less formidable.

HUGHES (1973) has described the cellular events in the normal limb ganglia. His study brings to light the fact that degeneration on a massive scale appears in the main brachial ganglia shortly after birth, leaving large spaces filled with pycnotic debris towards the ventral pole. This degeneration begins only after migration of the fetus to the pouch has been accomplished, and one could theorize that it results from the sensorimotor circuits concerned with the migratory activity having become redundant subsequent to the migration. A second period of degenerative activity can be seen in the brachial DRGs of older pouch young. This later degenerative period probably corresponds to that which results in the decrease in ganglionic cell numbers seen in the anuran tadpole before and immediately after metamorphosis.

D. The Nature of the Degenerative Response

The visible changes which a cell undergoes during the course of degeneration have been studied with the electron microscope by various workers including Bellairs (1961), Goldsmith et al. (1967), Birks and Weldon (1971), Schweichel and Merker (1973), and O'Connor and Wyttenbach (1974). The changes observed, for example the extent to which cytoplasmic lysosomes are involved, vary from cell type to cell type. Schweichel and Merker (1973) report that in rat and mouse fetuses necrosis in dorsal root ganglia is of a type that involves condensation of the nuclear and cytoplasmic material followed by cell fragmentation and lysosomal disintegration of the fragments in neighboring phagocytizing cells. Goldsmith et al. (1967) showed cytoplasmic crystalline bodies, previously described by Bellairs (1961), to be associated with degenerating cells in 6- to 7-day chick embryo DRGs. In cultures of chick dorsal root ganglia, many cells, especially those that had already sent out fibers prior to explantation, undergo degeneration (Birks and Weldon, 1971). In such cells the pycnotic nuclear changes are preceded by a breakdown of endoplasmic reticulum and the formation of the large crystalline bodies in the cytoplasm. Such crystalline arrays are thought to represent ordered arrangements of ribosomes (Crain et al., 1964).

A question of considerable interest and importance is the timing of degeneration and whether cells that degenerate are still undergoing mitosis or have already withdrawn from the mitotic cycle. Carr (1976) made special counts in chick DRGs of degenerating cells that were also labeled. Counts were made at various times after a single dose of tritiated thymidine given at $5^1/_2$ days of incubation. Labeled degenerating cells were found as early as two hours after administration of the isotope, and comprised 15% of the degenerating cells. None had been seen at one hour after a thymidine dose. It has been generally held that neuronal degeneration occurs only after the neuroblast has ceased its proliferative activity. The developing neuron must then send out neurities to the periphery and be "informed" by some sort of chemical "message" from the periphery *via* retrograde axonal flow that inappropriate or insufficient connections have been made [e.g. by transport of insufficient "maintenance factor" (Prestige, 1967a)]. Two hours would seem hardly sufficient for the completion of these processes. Carr's data suggest that either not all of these processes need occur within *each* neuroblast or young neuron prior to degeneration or that they occur at a much faster rate than previously thought.

However, the death of the early labeled degenerating cells nevertheless appears to be under control of the periphery. When the same labeling experiment is performed on embryos that had been subjected to wing bud amputation at $2^1/_2$ days of incubation, no change in the percentage of labeled degenerating cells is observed (Carr and Simpson, in preparation). This is despite the greatly increased overall degenerative activity in the affected brachial ganglia (Hamburger and Levi-Montalcini, 1949; Carr, 1976). Were the death of early labeled degenerating cells independent of peripheral control, one would expect a decline in the percentage of degenerating cells labeled. These findings suggest the possibility of two types of cellular degeneration in the chick DRGs, one under direct peripheral control, and another under more indirect peripheral control. The former would occur in cells that have sent fibers to the periphery, the latter in less mature cells that have not established their own direct contact with the periphery. This observation is especially interesting in light of the report by PILAR and LANDMESSER (1975, 1976) of a difference in the ultrastructural appearance of neurons in normal and peripherally deprived ciliary ganglia prior to the normal period of cell death. PILAR and LANDMESSER suggest that there may be a different cause for normally occurring degeneration than for degeneration occurring as a result of reduction in the periphery.

E. Peripheral Effects and Sensory Modalities

I. Sensory Modalities of Ganglionic Neurons as Revealed by Alterations in the Periphery

In the studies on the effects of limb amputations and additions on dorsal root ganglia development that were discussed in Section C, little attention was paid to whether the neurons affected by the operations belonged to exteroceptive or proprioceptive categories. One of the first experimenters to attempt to analyze this problem was DETWILER. In an ingenious experiment on the effects of peripheral ablation, DETWILER (1926b) made separate estimates of the hypoplasia due to losses in the exteroceptive and proprioceptive fields. He did this by grafting pairs of very young *Amblystoma* embryos in parabiosis and removing a variable amount of somatopleure along with the overlying skin at the time of the union. In some instances the operated area included the forelimb. As a result, the animals were deprived of varying amounts of both skin and muscle at the site of union. The parabiosed pairs were fixed and sectioned 40 to 50 days after the operation and wax plate reconstructions were prepared from the sections extending from the 3rd to the 8th body segments. From these wax reconstructions DETWILER determined the volume of striated muscle and the surface area of skin on the joined and free sides. The numbers of cells in ipsi- and contralateral dorsal root ganglia were also determined. From such data obtained from several parabiosed twins, DETWILER set up pairs of simultaneous equations and attempted to solve them for the proportions of ganglionic cells related to exteroceptive and to proprioceptive functions. He found that at thoracic levels 60% of the ganglionic cells were related to exteroceptive function and 40% to proprioceptive function. These same proportions of 60% and 40% were obtained consistently from all sets of equations when the area examined was confined to thoracic levels. However, when the operated area included the limb field, consistent proportions could not be obtained from the equations. This observation suggested to DETWILER "the possibility of some regulatory adjustment occurring in the limb region in such experiments."

DETWILER's calculated proportions of 60% for exteroceptive and 40% for proprioceptive cells were found to apply to *Amblystoma* ganglia innervating heterotopically grafted forelimbs (CARPENTER and CARPENTER, 1932). Following grafting of the limbs to 21–30 mm larvae, these workers measured the parameters analogous to those measured by DETWILER: increases in skin and muscle and in ganglionic cell numbers. They then proceeded to work DETWILER's equations backwards by substituting in DETWILER's proportions of 60% and 40% for the respective sensory components along with their own measurements of increases in skin and muscle and then solving for the number of cells supplying each of the sensory functions. The CARPENTERS were able to show that the sum of the cell numbers calculated to supply each of the two sensory functions was very close to the actual number of neurons counted in the ganglia on the operated side. Thus, they demonstrated that DETWILER's calculated values for the proportions of proprioceptive and exteroceptive ganglionic neurons do apply to the limb level ganglia in the case of limb bud addition.

In 1927 DETWILER was able to selectively reduce the amount of muscle in *Amblystoma* larvae and study the effects on the number of dorsal root ganglionic cells. He found that following unilateral removal of lateral portions of somites hypoplasia occurred in the ipsilateral DRGs. Furthermore, when simultaneous equations derived in the same manner as described in the 1926 paper are set up for these animals, the same proportions are also obtained: 60% of the DRG cells are devoted to exteroceptive function and 40% to proprioceptive function.

II. Sensory Modalities, Neuronal Sizes, and Peripheral Control

In the fully developed tetrapod nervous system, size differences among the dorsal root ganglionic neurons are correlated with various factors, including the modality of sensation conveyed along the proximal fibers in the spinal cord. Other factors include differences in fiber caliber and in the sites at which the dorsal root fibers terminate and synapse with their respective second-order neurons. For the developmental neurobiologist, profound questions exist as to how features such as these are established in embryonic life, the extent to which control is exercised by central and peripheral factors, and the degree to which these size differences in embryonic ganglia are related to neuronal functional differences in the adult ganglia.

According to MÜLLER's (1833) doctrine of specific nerve energy, the modality of a given sensory neuron ultimately depends upon the connections made by its proximal fibers in the central nervous system. The factors controlling the formation of appropriate central connections are still of an unknown nature. However, whatever these factors may be, it now appears that impulses from specialized peripheral endings are not among them. Some evidence for this comes from observations regarding the time of formation of the sensory tracts in the spinal cord compared with that of the specialized sensory structures at the periphery. Other evidence has been obtained from organ culture studies.

Anatomical and behavioral studies show that cutaneous sensibility and reflexes are established long before specialized cutaneous structures are seen. Much of this work has been concerned with human development. It has been correlated by HOOKER (1952) and HUMPHREY (1964) and reviewed more recently by BRADLEY and MISTRETTA (1975). Humphrey compared the development of cutaneous sensibility with that of sensory structures, as described by a number of earlier authors. In the hand she had found that reflexive bending of the fingers in response to tactile stimulation first occurs at 10.5 weeks (menstrual age). At this stage nerve branches are closely related to the epidermis (CAUNA and MANNAN, 1961). However, definitive Meissner's corpuscles, the fine tactile receptors, do not appear on the palmar surfaces until 24 weeks (SZYMONOWICZ, 1933). Pacinian corpuscles, the deep pressure receptors, are not seen until 12 to 12.5 weeks (CAUNA and MANNAN, 1959). Similarly, primitive nerve endings appear in muscles several weeks prior to the appearance of spindle fibers during the fourth month (See HUMPHREY, 1964, Table XI).

The dorsal columns of the spinal cord, to which the proximal fibers of the dorsal root ganglia contribute, are also present in the fetus before the specialized sensory endings appear. HUMPHREY (1955) directly associated the primitive dorsal funiculus of the embryonic cord with the future ascending sensory columns, the cuneate and gracile columns, of the adult. This primitive dorsal funiculus was first seen by her at 7.5 weeks (menstrual age). HUGHES (1976) has recently identified a different fiber tract appearing slightly later as the future gracile and cuneate columns. This new and denser fiber tract first appears on either side of the midline of the cord at 8 weeks (25–31 mm C. R. length). Because 8 weeks also marks the first appearance of size differences among the neurons of the cervical dorsal root ganglia, HUGHES attaches special significance to this new addition to the white matter as the probable origin of the gracile and cuneate columns. If the new tract does indeed prove to be the origin of the adult dorsal columns, its appearance still antedates those of the specialized sensory endings mentioned above. Thus, it would seem that, whatever the factors are that control the differentiation of sensory modalities, specific impulses from specialized sensory endings cannot be among them.

Recent experimental evidence from organ culture also points to the same conclusion. CRAIN and PETERSON (1975) combined dorsal root ganglia of mice with slices of spinal cord and brainstem at the level of the gracile and cuneate nuclei. Electrophysiological examination showed that outgoing fibers from the ganglia established synaptic connections with "target neurons" and did so with a "remarkable degree of characteristic pharmacologic sensitivity and regional specificity." If these DRG fibers correspond with dorsal funicular elements of the intact fetus, their growth into the appropriate centers of second-order neurons clearly occurred in the absence of impulses from specific sensory terminals.

The modalities of the DRG neurons may be related to early size differences between them. This appears to be the case for chick embryos. According to VISINTINI and LEVI-MONTALCINI (1939), the first exteroceptive reflexes in the chick are seen at day 6 of incubation. These authors reported that this period coincides with the earliest ganglionic fiber outgrowth, which is first seen among the ventrolateral cells. Accordingly, they regarded at least some of these larger ventrolateral cells as belonging to the exteroceptive modality. The first proprioceptive reflexes were not seen until day 11 of incubation, which coincides not only with the beginning of neurofibrillar differentiation in the small mediodorsal cells but also with the beginning of muscle spindle formation in the limb muscles (TELLO, 1922). Thus, some of the small cells were assumed to belong to the proprioceptive modality.

More recent experiments, however, now show that the first peripheral contacts are not made only by cells in the ventrolateral region. As mentioned earlier, OPPENHEIM and HEATON (1975) have shown that by $4^1/_2$ days of incubation horseradish peroxidase injected into leg buds of chick embryos can be transported to the cell bodies of mediodorsal as well as of ventrolateral neurons. CARR's findings (1975, 1976) of degenerative responses to wing bud removal in the mediodorsal cells at $5^1/_2$ days offer additional evidence for the early presence in peripheral tissues of fibers from these cells. Presumably, then, some of the mediodorsal cells, as well as some ventrolateral cells, may serve an exteroceptive function. These results suggest that there may be less correlation between neuronal size and function in chick DRGs than previously believed.

In the rat an observation has been made recently by LAWSON (pers. comm.) that may permit direct evidence to be obtained as to the time of establishment of adult relationships between cell size and sensory modalities. LAWSON has observed that DRG neurons in the newborn rat vary in diameter between 7 and 30 µm. This range is so wide that it ought to be possible to trace the peripheral course of fibers from the larger cells and directly observe the nature of their sensory endings, thereby establishing the modalities of the cells and analyzing the relation between cell size and modality.

F. Trophic Influence of Sensory Fibers on the Differentiation and Maintenance of Specialized Sensory Structures

The differentiation and maintenance of specialized sensory structures belongs to a second class of peripheral-central relationships, that in which the center influences development at the periphery. For example, ZELENÁ has demonstrated the necessity of sensory innervation for the maintenance of the structural integrity of muscle spindles. ZELENÁ (1957) noticed that transection of the sciatic nerve in newborn rats resulted in the disintegration of spindles that had already developed in muscle of the hindlimb. This finding was extended by ZELENÁ and SOUKUP (1974), who showed that removal of motor innervation alone at birth does not prevent or impair further development of the spindles and that, therefore, it is the sensory innervation alone that provides the neural trophic influence required by the muscle-spindle fibers.

A similar dependency on sensory innervation has been demonstrated for other sensory receptor structures. The lateral-line system in amphibian tadpoles requires innervation by the lateral-line nerve, a branch of the vagus nerve, for long term survival (SPEIDEL, 1948; WRIGHT, 1951). JONES and SINGER (1969) have demonstrated that in adult newts, the presence of fibers from the dorsal root ganglia is most probably also involved. The presence of sensory innervation is an absolute necessity for the development of the Herbst and Grandry corpuscles of duck and chick skin (SAXOD, 1972). SAXOD's study further revealed a certain degree of species specificity in this requirement; for while corpuscles in skin from either avian source will differentiate in the presence of avian DRGs in organ cultures, in the presence of mouse DRGs they will not (see SAXOD, Chapter 7, this volume).

G. Specification of Sensory Neurons—the Cutaneous Local Sign

A final topic that ought to be considered, at least briefly, in a discussion of the peripheral-central interactions in dorsal root ganglia is the problem of the specification of the nerve cells that comprise the DRGs, especially the exteroceptive neurons responsible for the cutaneous local sign. This topic has been reviewed extensively, and the reader is referred to the reviews, such as those of GAZE (1970) and JACOBSON (1969, 1970, 1978).

At the moment there are two viewpoints concerning the nature of the control over the specification of sensory nerves. SPERRY (1951) and MINER (1956) have advanced the hypothesis that the central connections made by the proximal fibers of sensory neurons are determined by the topographical location of the cutaneous terminals on the body surface and that specification is mediated through some sort of matching of characteristic cell surface molecules. Alternatively, SZÉKELY (1974) has suggested as an explanation a combination of differences in somatotopic projections to the spinal cord, coupled with specific impulse patterns from receptors in various surface locations.

These hypotheses were put forward to explain the now well-known misdirected responses to cutaneous stimulation of dorsoventrally reversed skin grafts in frogs (MINER, 1956). In these experiments a rectangle of skin had been cut from the flank of *Rana pipiens* tadpoles and regrafted in the same site after back-to-belly rotation. At metamorphosis the regrafted skin self-differentiated so that the metamorphosed animals displayed an area of white belly skin on the back and pigmented back skin on the belly. Behavioral studies carried out subsequent to metamorphosis showed that responses to cutaneous stimulation of the grafted regions were reversed. When the grafted white belly skin on the back was stimulated, wiping movements were directed toward the belly. Conversely, when the darker back skin on the belly was stimulated, the wiping movements were directed toward the back. The responses were directed toward the original sites of the rotated skin instead of toward the surface stimulated.

SPERRY suggested that in the frog the cutaneous local sign of each area of body surface is established through the operation of two gradients, one anterior-posterior, the other dorsal-ventral. These gradients were envisaged in chemical terms. Once established in normal development, the corresponding local specificity is transferred to cutaneous fibers as a result of their contact with the skin. In the reversed grafts, the local sign persists within the skin and is imposed upon the ingrowing, regenerating nerve fibers. Consequently, the nerve terminals become specified according to the original location of the grafted region. Finally, the central connections of the proximal fibers in the spinal cord become altered in some way so that they too correspond to the site of origin of the grafted material. Hence the "misdirected response" to stimulation.

The second hypothesis regarding the nature of the control of specification of sensory nerves has been advanced by SZÉKELY (1974). SZÉKELY notes that specification of local sign through chemical gradients would require an enormous number of specific molecules. He suggests instead that there may be differences between the patterns of impulses generated by receptors in belly skin and those in back skin. He feels that differences in impulse patterns coupled with differences in the somatotopic projections of dorsal and ventral sensory nerves onto the spinal column could account for specification of sensory nerves and also for the misdirected responses following skin rotation.

At the moment both SPERRY'S postulated chemical gradients and SZÉKELY'S postulated differences in impulse patterns await definitive proof.

Somewhat more recent studies in this field have revealed new data, some of which, unfortunately, make interpretation of the misdirected response more difficult. JACOBSON and BAKER (1969) found that the responses of newly metamorphosed frogs bearing large dorsoventrally reversed grafts are at first like those of normal frogs. The misdirected responses develop only after a few days. Furthermore, misdirected responses were shown to develop only in grafts made prior to larval stage XV, indicating the involvement of some sort of temporal factor.

By means of electrophysiological mapping of the receptive fields of the cutaneous nerves, JACOBSON and BAKER were also able to show that grafts are innervated by local nerves and that "each nerve entered the skin within its own receptive field". Thus, misdirected responses cannot arise as a result of disturbance of the terminal patterns of innervation or of the nerves growing for long distances in the skin before making terminal connections. Furthermore, the mapping studies also revealed a difference in the degree of overlap between adjacent receptive fields across the borders of the graft that depends on the orientation plane of rotation of the graft. In normal animals a certain amount of overlap occurs between fields. At the borders of grafts with reversed dorsoventral orientation, however, a sharp boundary occurs between the adjacent receptive fields of the normal and grafted skin. Such a sharp boundary is not seen in grafts reversed in an anteroposterior orientation. In these grafts a normal amount of overlap between adjacent cutaneous fields occurs, even across the borders of the graft. An explanation for these observations is not yet at hand. Interestingly enough, however, anteroposteriorly reversed grafts, which have normal field overlap across their borders, never develop misdirected responses.

BAKER (1970) introduced another level of complexity into the problem of the specification of cutaneous nerves with his report that misdirected responses are not seen in *Rana clamitans* or *R. catesbeiana.* Both of these species have a more extended larval life than does *R. pipiens.* It is not known, however, whether the differences actually reflect a diversity in some feature of larval life or are merely due to species differences.

H. Concluding Remarks

The morphological and cytological influences of the periphery on the center were at one time a dominant theme in neuroembryology. This has now yielded to other studies, leaving many problems unsolved and some hardly attempted. The original question of self-differentiation and dependent differentiation posed by SHOREY in 1909 has been only partly answered. The factors that determine whether a newborn neuron will become exteroceptive or proprioceptive can as yet only be surmised, though it now seems that peripheral influences may be less than might have been suspected at an earlier time. We now see that there are considerable differences in the detailed reactions of ganglia to peripheral influences.

There is little doubt that much further information could be obtained by the use of modern methods such as cell labeling, degenerative techniques of neurohistology, and electrical recording, particularly if these were used in conjunction. It is perhaps characteristic of modern progress in science that much attention has been devoted to the effects of nerve growth factor on nerve cultures, an artificial situation readily accessible to investigation, while very much less attention has been accorded to comparable phenomena within the embryo itself. It is unlikely, however, that this situation will persist indefinitely. The way is now wide open for new discoveries concerning the interactions between the dorsal root ganglia and their peripheral fields of innervation.

I. Summary

1. Developing limbs exert several quantitative effects on their related dorsal root ganglia with respect to cell production, cell differentiation, and cell degeneration.

2. These effects have been demonstrated experimentally in both the hypoplasia caused by limb ablation and the hyperplasia resulting from the grafting of supernumerary limbs.

3. The factors that determine which DRG cells become exteroceptive or proprioceptive neurons are not known, but it seems that peripheral input is not the primary causative factor.

4. Developing sensory neurons exert their own trophic effect on the peripheral sensory receptors.

5. The experimental analysis of the phenomenon of cutaneous local sign in anura is reviewed briefly. It is likely that more than one factor is concerned, most probably in a combination of local influences, either chemically or electrically derived, and somatotopic projection to the spinal cord.

J. Acknowledgements

The authors are deeply grateful to Dr. SARAH LAWSON of Bristol, England, for the privilege of seeing her unpublished material and for permission to quote from her results.

V. McM. C. is indebted to Prof. SIDNEY B. SIMPSON, JR. for reviewing this manuscript and for his helpful suggestions.

References

Angevine, J. B., Jr.: Time of neuron origin in the diencephalon of the mouse. An autoradiographic study. J. comp. Neurol. **139**, 129—188 (1970).

Auerbach, R.: Analysis of the developmental effects of a lethal mutation in the house mouse. J. exp. Zool. **127**, 305—330 (1954).

Baker, R. E.: Behavioral reflexes in *R. catesbeiana* and *R. clamitans* with large skin grafts. J. exp. Zool. **173**, 129—136 (1970).

Barron, D. H.: The role of the sensory fibers in the differentiation of spinal cord in sheep. J. exp. Zool. **100**, 431—443 (1945).

Bellairs, R.: Cell death in chick embryos as studied by electron microscopy. J. Anat. (Lond.) **95**, 54—60 (1961).

Birks, R. I., Weldon, P. R.: Formation of crystalline ribosomal arrays in cultured chick embryo dorsal root ganglia. J. Anat. (Lond.) **109**, 143—156 (1971).

Bors, E.: Die Methodik der intrauterinen Operation im überlebenden Säugetierfoetus. Arch. Entw. Mech. Org. **105**, 655—666 (1925).

Bradley, R. M., Mistretta, C. M.: Fetal sensory receptors. Physiol. Rev. **55**, 352—382 (1975).

Carpenter, R. L., Carpenter, E.: The effect of retarded body growth on the response of spinal ganglia subjected to excessive peripheral overloading. J. exp. Zool. **64**, 187—208 (1932).

Carr, V. McM., Simpson, S. B., Jr.: Peripheral effects on early development of chick spinal ganglia. Neurosci. Abs. **1**, 749 (1975).

Carr, V.McM.: Intrinsic and peripheral factors in the early development of chick dorsal root ganglia. Ph. D. Thesis, Northwestern University, 1976.

Cauna, N., Mannan, G.: Development and postnatal changes of digital Pacinian corpuscles *(corpuscula lamellosa)* in the human hand. J. Anat. (Lond.) **93**, 271—286 (1959).

Cauna, N., Mannan, G.: Organization and development of the preterminal nerve pattern in the palmar digital tissues of man. J. comp. Neurol. **117**, 309—328 (1961).

Collin, R.: Recherches cytologiques sur le développement de la cellule nerveuse. Névraxe **8**, 181—308 (1906).

Cowan, W.M.: Studies on the development of the avian visual system. In: Cellular Aspects of Neural Growth and Differentiation (D. C. Pease, Ed.). Los Angeles: University of California 1971, pp. 177—222.

Crain, S.M., Peterson, E.R.: Development of specific sensory-evoked synaptic networks in fetal mouse cord-brainstem cultures. Science **188**, 275—277 (1975).

Crain, S.M., Benitez, H., Vatter, A.E.: Some cytological effects of salivary nerve-growth factor on tissue cultures of peripheral ganglia. Ann. N. Y. Acad. Sci. **118**, 206—231 (1964).

Detwiler, S.R.: On the hyperplasia of nerve centers resulting from excessive peripheral loading. Proc. nat. Acad. Sci. Wash. **6**, 96—101 (1920).

Detwiler, S.R.: Experimental studies on morphogenesis in the nervous system. Quart. Rev. Biol. **1**, 61—86 (1926a).

Detwiler, S.R.: The effect of reduction of skin and muscle on the development of spinal ganglia. J. exp. Zool. **45**, 399—414 (1926b).

Detwiler, S.R.: The effects of extensive muscle loss upon the development of spinal ganglia in *Amblystoma*. J. exp. Zool. **48**, 1—14 (1927).

Detwiler, S.R.: Some observations upon the growth, innervation, and function of heteroplastic limbs. J. exp. Zool. **57**, 183—203 (1930).

Detwiler, S.R.: Experimental studies upon the development of the amphibian nervous system. Biol. Rev. **8**, 269—310 (1933).

Detwiler, S.R.: An experimental study of spinal nerve segmentation in *Amblystoma* with special reference to the plurisegmental contribution to the brachial plexus. J. exp. Zool. **67**, 395—441 (1934).

Detwiler, S.R.: Neuroembryology. New York: Macmillan 1936.

Dürken, B.: Über frühzeitige Extirpation von Extremitätenanlagen beim Frosch. Z. Wiss. Zool. **99**, 189—355 (1911).

Fraisse, P.: Die Regeneration von Geweben und Organen bei Wirbeltieren, besonders Amphibien and Reptilien. Cassel: Fischer 1885.

Gaze, R.M.: The Formation of Nerve Connections. New York: Academic Press 1970.

Glücksmann, A.: Cell death in normal vertebrate ontogeny. Biol. Rev. **26**, 59—86 (1951).

Goldsmith, M., Weston, J.A., Cowell, L.: Crystalline bodies associated with morphogenetic cell death in embryonic sensory ganglia. Amer. Zool. **7**, 755 (1967).

Hall, E.K., Schneiderhan, M.A.: Spinal ganglion hypoplasia after limb amputation in the fetal rat. J. comp. Neurol. **82**, 19—34 (1945).

Hamburger, V.: The effects of wing bud extirpation on the development of the central nervous system in chick embryos. J. exp. Zool. **68**, 449—494 (1934).

Hamburger, V.: Motor and sensory hyperplasia following limb-bud transplantations in chick embryos. Physiol. Zool. **13**, 268—289 (1939).

Hamburger, V., Levi-Montalcini, R.: Proliferation, differentiation, and degeneration of the spinal ganglia of the chick embryo under normal and experimental conditions. J. exp. Zool. **111**, 457—501 (1949).

Harrison, R.G.: An experimental study of the relation of the nervous system to the developing musculature in the embryo of the frog. Amer. J. Anat. **3**, 197—220 (1904).

His, W.: Untersuchungen über die erste Anlage des Wirbeltierleibes. Die erste Entwicklung des Hühnchens im Ei. Leipzig: F. C. W. Vogel 1868.

Hörstadius, S.: The Neural Crest. London: Oxford University Press 1950.

Hooker, D.: Die Nerven im regenerierten Schwanz der Eidechsen. Arch. mikr. Anat. **80**, 217—222 (1912).

Hooker, D.: The Prenatal Origin of Behavior. Lawrence: University of Kansas Press 1952.

Hughes, A. F.: The growth of embryonic neurites. A study on cultures of chick neural tissues. J. Anat. (Lond.) **87**, 150—162 (1953).
Hughes, A. F.: Studies in embryonic and larval development in amphibia. I. The embryology of *Eleutherodactylus recordii*, with special reference to the spinal cord. J. Embryol. exp. Morph. **7**, 22—38 (1959).
Hughes, A. F.: Cell degeneration in the larval ventral horn of *Xenopus laevis* (Daudin). J. Embryol. exp. Morph. **9**, 269—284 (1961).
Hughes, A. F.: Further experiments on the innervation and function of supernumerary limbs in the embryo of *Eleutherodactylus martinicensis*. J. Embryol. exp. Morph. **12**, 229—245 (1964).
Hughes, A. F.: Aspects of Neural Ontogeny. London: Logos 1968.
Hughes, A. F.: The development of dorsal root ganglia and ventral horns in the opossum. A quantitative study. J. Embryol. exp. Morph. **30**, 359—376 (1973).
Hughes, A. F.: Endocrines, neural development and behavior. In: Studies in the Development of Behavior and the Nervous System, Vol. 2: Aspects of Neurogenesis. G. Gottlieb (Ed.). New York: Academic 1974.
Hughes, A. F.: The development of the dorsal funiculus in the human spinal cord. J. Anat. (Lond.) **122**, 169—175 (1976).
Hughes, A. F., Egar, M.: The innervation of the hind limb of *Eleutherodactylus martinicensis*: further comparison of cell and fiber numbers during development. J. Embryol. exp. Morph. **27**, 389—412 (1972).
Hughes, A. F., New, D.: Tail regeneration in the geckonic lizard, *Sphaerodactylus*. J. Embryol. exp. Morph. **7**, 281—302 (1959).
Hughes, A. F., Tschumi, P. A.: The factors controlling the development of the dorsal root ganglia and ventral horn in *Xenopus laevis* (Daud.). J. Anat. (Lond.) **92**, 498—527 (1958).
Humphrey, T.: Pattern formed at upper cervical spinal cord levels by sensory fibers of spinal and cranial nerves. Arch. Neurol. Psychiatr. **73**, 36—46 (1955).
Humphrey, T.: Some correlations between the appearance of human fetal reflexes and the development of the nervous system. Prog. Brain Res. **4**, 93—135 (1964).
Jacobson, M.: Development of specific neuronal connections. Science **163**, 543—547 (1969).
Jacobson, M.: Developmental Neurobiology. New York: Holt, Rinehart and Winston 1970.
Jacobson, M.: Formation of neural connections in sensory systems. In: Handbook of Sensory Physiology, Vol. 1 (W. R. Loewenstein, Ed.). Berlin-Heidelberg-New York 1971, pp. 166—190.
Jacobson, M.: Developmental Neurobiology, 2nd ed. New York: Plenum 1978.
Jacobson, M., Baker, R. E.: Development of neuronal connections with skin grafts of frogs: behavioral and electrophysiological studies. J. comp. Neurol. **137**, 121—142 (1969).
Jones, D. P., Singer, M.: Neurotrophic dependence of the lateral-line sensory organs of the newt, *Triturus viridescens*. J. exp. Zool. **171**, 433—442 (1969).
Lamb, A. H.: The timing of the earliest motor innervation to the hind limb in the *Xenopus* tadpole. Brain Res. **67**, 527—530 (1974).
Langworthy, O. R.: The development of progression and posture in young opossums. Amer. J. Physiol. **74**, 1—13 (1925).
Lawson, S. H., Caddy, K. W., Biscoe, T. J.: Development of rat dorsal root ganglion neurones. Studies of cell birthdays and changes in mean cell diameter. Cell Tiss. Res. **153**, 399—413 (1974).
Levi-Montalcini, R., Levi, G.: Les conséquences de la destruction d'un territoire d'innervation périphérique sur le développement des centres nerveux correspondants dans l'embryon de poulet. Arch. Biol. **53**, 537—545 (1942).
Levi-Montalcini, R., Levi, G.: Recherches quantitatives sur la marche du processus de différentiation des neurones dans les ganglions spinaux de l'embryon de poulet. Arch. Biol. **54**, 198—206 (1943).
May, R. M.: Répercussions de la greffe de moelle sur le système nerveux chez l'embryon de l'anoure, *Discoglossus pictus* Otth. Bull. Biol. Fr. Belg. **64**, 355—387 (1930).
May, R. M.: Réactions neurogéniques de la moelle à la greffe en surnombre ou à l'ablation d'une ébauche de patte postérieure chez l'embryon de l'anoure *Discoglossus pictus* Otth. Bull. Biol. Fr. Belg. **67**, 327—349 (1933).
McBride, W. G.: Fetal nerve cell degeneration produced by thalidomide in rabbits. Teratology **10**, 283—292 (1974).
McCrady, E.: The Embryology of the Opossum (Amer. anat. Mem. 16). Philadelphia: Wistar Institute 1938.

Miner,N.: Integumental specification of sensory fibers in the development of cutaneous local sign. J. comp. Neurol. **105**, 161—170 (1956).

Müller,J.: Handbuch der Physiologie des Menschen für Vorlesungen, 2 vol. Coblenz: I. Hölscher, 1833—1840. As quoted in: Selected Readings in the History of Physiology. J. F. Fulton and L. G. Wilson (Eds.). Springfield: Charles C. Thomas 1966, pp. 289—291.

Nakai,J.: Dissociated dorsal root ganglia in tissue culture. Amer. J. Anat. **99**, 81—129 (1956).

Nicholas,J.S.: Notes on the application of experimental methods upon mammalian embryos. Anat. Rec. **31**, 385—394 (1925).

Nieuwkoop,P.D., Faber,J.: Normal Table of *Xenopus laevis* (Daudin). Amsterdam: North-Holland 1956.

O'Connor,T.M., Wyttenbach,C.R.: Cell death in the embryonic chick spinal cord. J. Cell Biol. **60**, 448—459 (1974).

Oppenheim,R.W., Heaton,M.B.: The retrograde transport of horseradish peroxidase from the developing limb of the chick embryo. Brain Res. **98**, 291—302 (1975).

Pannese,E.: Investigations on the ultrastructural changes of the spinal ganglion neurons in the course of axon regeneration and cell hypertrophy. II: Changes during cell hypertrophy and comparison between the ultrastructure of nerve cells of the same type under different functional conditions. Z. Zellforsch. **61**, 561—586 (1963).

Pannese,E.: Developmental changes of the endoplasmic reticulum and ribosomes in nerve cells of the spinal ganglia of the domestic fowl. J. comp. Neurol. **132**, 331—364 (1968a).

Pannese,E.: Temporary junctions between neuroblasts in the developing spinal ganglia of the domestic fowl. J. ultrastruct. Res. **21**, 233—250 (1968b).

Pannese,E.: Electron microscopical study on the development of the satellite cell sheath in spinal ganglia. J. comp. Neurol. **135**, 381—422 (1969).

Piatt,J.: Form and causality in neurogenesis. Biol. Rev. **23**, 1—45 (1948).

Pilar,G., Landmesser,L.: Normal and induced neuron death in ganglia during embryogenesis. Neurosci. Abs. **1**, 750 (1975).

Pilar,G., Landmesser,L.: Ultrastructural differences during embryonic cell death in normal and peripherally deprived ciliary ganglia. J. Cell Biol. **68**, 339—356 (1976).

Prestige,M.C.: Cell turnover in the spinal ganglia of *Xenopus laevis* tadpoles. J. Embryol. exp. Morph. **13**, 63—72 (1965).

Prestige,M.C.: The control of cell number in the lumbar spinal ganglia during the development of *Xenopus laevis* tadpoles. J. Embryol. exp. Morph. **17**, 453—471 (1967a).

Prestige,M.C.: The control of cell number in the lumbar ventral horns during the development of *Xenopus laevis* tadpoles. J. Embryol. exp. Morph. **18**, 359—381 (1967b).

Prestige,M.C., Wilson,M.A.: Loss of axons from ventral roots during development. Brain Res. **41**, 467—470 (1972).

Roux,W.: Einleitung. Arch. Entw. Mech. Org. **1**, 1—42 (1894).

Saxod,R.: Rôle du nerf et du territoire cutané dans le développement des corpuscles de Herbst et de Grandry. J. Embryol. exp. Morph. **27**, 277—300 (1972).

Schweichel,J.-U., Merker,H.-J.: The morphology of various types of cell death in prenatal tissues. Teratology **7**, 253—266 (1973).

Schwind,J.L.: Heteroplastic experiments on the limb and shoulder girdle of *Amblystoma*. J. exp. Zool. **59**, 265—295 (1931).

Shorey,M.L.: The effect of the destruction of peripheral areas on the differentiation of the neuroblasts. J. exp. Zool. **7**, 25—63 (1909).

Singer,M.: A theory of the trophic nervous control of amphibian limb regeneration. including a re-evaluation of quantitative nerve requirements. In: Regeneration in Animals and Related Problems (V. Kiortsis and H. A. Trampusch, Eds.). Amsterdam: North-Holland 1965.

Speidel,C.C.: Correlated studies of sense organs and nerves of the lateral-line in living frog tadpoles. II: The trophic influence of specific nerve supply as revealed by prolonged observation of denervated and renervated organs. Amer. J. Anat. **82**, 277—320 (1948).

Sperry,R.W.: Mechanisms of neural maturation. In: Handbook of Experimental Psychology (S. S. Stevens, Ed.). New York: Wiley 1951.

Székely,G.: Problems of neuronal specificity in the development of some behavioral patterns in amphibia. In: Studies in the Development of Behavior and the Nervous System, Vol. 2: Aspects of Neurogenesis (G. Gottlieb, Ed.). New York: Academic 1974.

Szymonowicz, W.: Über die Entwicklung der Nervenendigungen in der Haut des Menschen. Z. Zellforsch. **19**, 356—382 (1933).
Tello, J. F.: Die Entstehung der motorischen und sensiblen Nervenendigungen. I: In dem lokomotorischen System der höheren Wirbeltiere. Z. Anat. Entw. **64**, 348—440 (1922).
Tennyson, V. M.: Electron microscopic study of the developing neuroblast of the dorsal root ganglion of the rabbit embryo. J. comp. Neurol. **124**, 267—318 (1965).
Terni, T.: Sulla correlazioni fra ampiezza del territorio di innervazioni e grandezza della cellule gangliari. II: Ricerche sui gangli spinali che innervano la coda rigenerata, nei Sauri *(Gangylus ocellatus)*. Arch. ital. Anat. Embriol. **17**, 507—543 (1920).
Visintini, F., Levi-Montalcini, R.: Relazione tra differenziazione structurale e funzionale dei centri e delle vie nervose nell'embrione di pollo. Arch. suisse Neur. et Psych. **44**, 119—150 (1939).
Weston, J. A.: A radiographic analysis of the migration and localization of trunk neural crest cells in the chick. Dev. Biol. **6**, 279—310 (1963).
Weston, J. A.: The migration and differentiation of neural crest cells. Adv. Morphogen. **8**, 41—114 (1970).
Weston, J. A., Butler, S. L.: Temporal factors affecting localization of neural crest cells in the chicken embryo. Dev. Biol. **14**, 246—266 (1966).
Wright, M. R.: The lateral line system of sense organs. Quart. Rev. Biol. **26**, 264—280 (1951).
Yates, R. D.: A study of cell division in chick embryonic ganglia. J. exp. Zool. **147**, 167—182 (1961).
Zelená, J.: The morphogenetic influence of innervation on the ontogenetic development of muscle spindles. J. Embryol. exp. Morph. **5**, 283—292 (1957).
Zelená, J., Soukup, T.: The differentiation of intrafusal fibre types in rat spindles after motor denervation. Cell Tiss. Res. **153**, 115—136 (1974).

Chapter 4

Visual Behavior Development in Nonmammalian Vertebrates

DAVID INGLE, Belmont, Massachusetts (USA)

Contents

A. Introduction

Although the study of behavioral development in various vertebrate species has become a popular area for ethologists and for students of animal perception, it is replete with so many inherent methodological problems that few claims stand free of controversy. Within recent years much of the drama has unfolded at the mammalian level, where both behavioral and physiological effects of visual deprivation or of abnormal environments have been reported. The recent emphasis upon mammalian visual studies naturally follows the greatly increased interest in determinants of normal and abnormal development of human infants. However, the tendency to limit such inquiries to mammals has been (perhaps unwisely) influenced by the tacit assumption that developmental plasticity is not found to a comparable extent in other organisms. Students of behavior are still likely to get the impression from textbooks that perception is "innate" for fishes and birds, but is almost entirely "aquired" by kittens and monkeys. However, for two reasons, the extent to which "developmental plasticity" becomes manifest in nonmammals is an important unresolved question. First, if there is a clear hiatus between mammals and their "ancestral" cousins in the extent to which visual behavior depends upon experience, then inter-species comparisons of neurochemical and ultrastructural components of visual systems may provide insight into the mammalian specializations which allow plasticity. If the differences are not so sharp as is commonly believed, then the *simpler* brains of nonmammals (such as frogs) or the experimental *convenience* of certain species (such as chicks) should provide useful model systems for a neurobehavioral approach to plasticity. Although we cannot yet estimate the extent of environmental-dependence for comparable behaviors of fish, amphibians, reptiles and birds, we can point out exemplary experimental approaches which offer promise for future behavioral and physiological studies.

B. Behavioral Taxonomy as a Problem

The classic dichotomy between innate and acquired components of behavior remains a central issue in this review. In approaching this issue, we usually lack a precise definition as to just what a mature organism is able to *do* with visual information, beyond the solution of simple discrimination tests and a few basic visuomotor skills. Since the more complex visual abilities are likely to depend most upon specific visual experiences, we may describe behavior as innate in nonmammals simply because we fail to appreciate their full capacities. As a second limitation of method, we usually fail to distinguish the various *uses* of vision by speaking of a global perceptual ability, such as "color vision". A given species might show innate color preferences for feeding, but show strong effects of experience in using color to identify parents or enemies. The problems raised above point out the need for a taxonomy of behavior, so that distinct classes can be isolated. Success in this will also maximize the physiologists' ability to localize the visual neurons required to mediate a particular type of behavior. Moreover, the aim of comparative studies is blunted if behaviorists cannot identify "homologous" classes of visual behavior shared by mammals and nonmammals. For these reasons, I provide a short introduction to visually guided responses that seem comparable in kind from fish to primate.

I. The Visual Grasp Reflex

All visually developed vertebrates turn head or eyes sharply toward objects for "evaluation" prior to a direct approach. The tendency of eyes or head to "lock on" to a small target is analogous to a grasp with hand or jaws (which may directly follow). As discussed in a recent review (INGLE and SPRAGUE, 1975) this response is associated with the optic tectum at all levels of vertebrate phylogeny. The movements is open-loop for frog or monkey and requires less than half of one second for these species. For other species (e.g. rodents and reptiles) turning can be fractionated into two or more successive movements.

II. Visually Elicited Biting

Animals such as the frog can strike directly at a nearby target with the tongue without prior orientation. The pecking ability of a bird can also be elicited directly by a stimulus within the "strike zone" (although this is much smaller than that of a frog). In reptiles, birds, and mammals the head may be precisely rotated prior to biting or grasping with the mouth—a complexity not seen in the visual-grasp response. It is likely that visual experience would play a larger role in this more subtle "fitting" of mouth to the object configuration than would be required merely to fixate the target.

III. Anticipatory Tracking of Moving Targets

There is little hard evidence that animals other than primates can show anticipatory tracking by head of eye movement. Fishes or frogs can make a series of brief saccades, but each is dependent upon momentary object location. We now have evidence that the gerbil (a rodent) has evolued such predictive tracking (INGLE, 1977). This mechanism is probably mediated by different visual circuitry than the visual grasp reflex and seems more likely to depend upon visual experience.

IV. Orientation Toward Apertures or Around Barriers

Although the motor coordination required to avoid a barrier seems identical to that involved in the visual-grasp response, there is evidence from studies of frogs (INGLE, 1973a) and of tree-shrews (CASAGRANDE and DIAMOND, 1975) that the ability negotiate barriers survives ablation of the optic tectum, but that the visual-grasp response does not. In the frog, the "shape" of an aperture can be distinguished in terms of spontaneous preference tests (INGLE, 1973b).

V. Visual Placing Responses

In reptiles and mammals the ability to "visually place" the forelegs upon a surface during descent may be derived from this primitive mode of guidance. In the kitten, HEIN and DIAMOND (1971) have distinguished triggered-extention from directed paw placement, since the second response is dependent upon active visual experience. The first discrimination is based upon depth estimation, and is probably related to the ability of many newly born animals to select the shallow side of a visual cliff (WALK and GIBSON, 1961). The exact ontogenetic and phylogenetic evolution from simple reactions to visual surfaces to complex locomotor abilities has scarcely been discussed, and remains an important problem for behaviorists.

VI. Phototactic Orientation

Various species show a strong compulsion to move toward (or away from) light sources—a reflex shown in tadpoles, fish hatchlings and newly hatched turtles. Phototaxis in the adult frog can be mediated *via* the ipsilateral retinal projection system, which does not support detection of prey, predators or barriers (unpublished study). A more complex derivation of light-source detection may be the sun-compass orientation shown by fishes, amphibians and birds. In these cases, early experience appears necessary for migratory behavior, as we shall see below.

VII. Avoidance of Visual Threat

The behavior of most vertebrates to rapidly approaching objects of substantial size includes blinking, head aversion, or jumping away. While these startle reflexes may be innate, the animal's selectivity in avoidance depends to some extent upon identification of predators, and upon the choice of an appropriate direction for flight. These behaviors (so easily elicited in some species) have been only occasionally studied, but seem likely to reveal the influence of early experience.

VIII. Visually-Elicited "Communicative" Postures

In the course of courtship or combat, animals often turn such that their own body is aligned parallel to or perpendicular to the antagonist as part of a display (rather than in direct attack). The role of experience in guiding body postures in fish and in reptiles is not known.

IX. Shape Recognition

Beyond the skills involved in maneuvering through the environment, many vertebrates have the ability to recognize a variety of objects and individual conspecifics by visual cues. Elsewhere, the author has argued that shape recognition in fishes and frogs may still be related to configurational discriminations used for the visual-grasp and barrier-detection responses (INGLE, 1978). But abilities to recognize objects despite variation in their size or spatial orientation (as higher mammals, at least, can do) suggest a major dependence upon experience.

C. The Problem of "Innate" Feeding Schemata

It is naturally important for the young of many species to begin adaptive feeding as soon as possible. Although chemosensory cues are important for food identification by many groups, vision is the predominant modality for food-selection or food-begging behaviors of newly hatched chicks. Most studies on feeding behavior of "naive" animals have involved various bird species: chickens, ducklings, gulls, and terns in particular. Analytical studies have presented young birds with artifical models, for example of the parental head and beak, or simple small figures affixed to a vertical wall. The targets vary in color, size, or shape, and they can be moved in various directions. The experimenter examines the number of pecks or bites given to stimuli presented either singly or in a preference test to determine the factors that yield various degrees of salience. In this section we shall review some of the current evidence for both genetic snd experential factors in food-seeking behavior.

I. Pecking Accuracy

Early observers of newly hatched domestic chicks noticed that pecking accuracy during the first day or two of life was relatively low, although persistent, by contrast with pecking accuracy in older birds. A well-controlled study by CRUZE (1935) showed clearly that both experience in pecking and age-related maturation of coordination were factors in normal improvement. The number of misses in pecking at small grains decreases among successive groups of chicks kept in the dark for one to five days before their initial visual tests. However, with one to three days' practice, a better performance level is reached than has occurred spontaneously with dark-reared chicks tested at three to five days. HESS (1956b) also noted that chicks notably decrease their "scatter" in pecking at a seed, even while wearing prisms that prevent their actually hitting the target. HAILMAN (1967) also observed that both maturation and experience were involved in improvement in pecking accuracy of the herring gull chick towards a model of the parents bill (an act that induces regurgitation of food by the parent). HAILMAN added the important observation that the chick's anticipatory head rotation in grasping the parent's bill depended upon appropriate visuomotor experience and did not appear spontaneously.

The implication that experience in hitting the target provides reinforcement for pecking accuracy leads one to ask whether a chick can learn an abnormal direction of pecking if its spatial array is systematically displaced by wearing prisms. There is, of course, a rich literature describing the adaptability of human and monkey subjects in such "visuomotor rearrangement" studies. Although an initial study by HESS (1956b) has produced negative evidence for "adaptation" to displacing prisms by newly hatched chicks, further studies by ROSSI (1968) have shown that negative after-effects (pecking in the opposite direction from prismatic displacement once prisms are removed) do appear after a few days of distorted viewing. In fact chicks that received no practice in pecking at seeds during prism exposure showed the same adaptation (ROSSI, 1971). It would be important to know whether adaptation in the visually naive chick has more lasting after-effects than adaptation that might be obtained in older chicks.

II. Innate Form Preferences

Studies of goldfish and lizards (see INGLE, 1978) reveal spontaneous shape discrimination during feeding tests but there is no evidence that these behaviors are innate. Studies by FANTZ (1957) and by GOODWIN and HESS (1969) have demonstrated that some form preferences appear immediately among naive chicks during unreinforced multiple-choice tests. If stimulus sizes are equated, chicks always prefer small circles to either squares or triangles (as do goldfish, frogs, and lizards). Although it is unlikely that these animals normally find perfectly round objects in this environment, it is possible that a round stimulus best "fits" the typically round receptive fields of retinal ganglion cells that project to the tectum. Unfortunately, no quantitative recording studies have yet been carried out with such shapes in any species. As an alternative explanation, one cannot rule out an attraction toward circular objects as novel, since animals normally see corners and points in the experimental environment and may rapidly habituate to these features. A considerable body of unpublished data from the laboratory of HESS (including an experiment by the reviewer) shows that a few minutes' pre-exposure to circular shapes is sufficient to momentarily reverse the usual circle-over-triangle pecking preference. In fact, FANTZ's own data show that the circle/triangle discrimination is significantly less pronounced after only 30 min of visual experience than later in the test session. Testing of shape preferences in a translucent globe (with collars to occlude the chicks' own pointed toes) might be required to make a definitive test of innate shape preferences. Whatever the basis of the overt preferences, the quoted studies do indicate that chicks have the ability to discriminate certain forms immediately upon emerging into the light.

Using the gull chick's tendency to peck at a model of the parents' bill, HAILMAN (1971) demonstrated a spontaneous tendency to prefer a vertically oriented bar to diagonal or horizontal orientations. Furthermore, a bar descending within the upper field is more potent than a vertical bar ascending from below: this roughly corresponds toward the natural approach toward the parental bill rather than toward their legs. Tests with monocular chicks would be required to rule out the reviewer's suggestion that the preferred upper-field vertical bar is simply the easiest to fixate with binocular visibility. For example, a prey object placed on the frontal midline is prepotent for binocular frogs (INGLE, 1973b) and there is evidence that a bird's beak can occlude part of a frontal stimulus during discriminative pecking (CORBALLIS and BEALE, 1970). However, this type of "visibility" explanation would not explain why QUINE and CULLEN (1964) found a preference for pecking at a horizontal (rather than vertical) beak model by the young Artic tern. This innate preference might reflect a predominance of oriented "edge-detecting" neurons in the retina of CNS of a given species, since a preponderence of horizontal edge detectors in the pigeon's retina was reported by MATURANA and FRANK (1963).

On the whole, HAILMAN's (1967) exemplary study of laughing gull and herring gull chick's perception of their respective parents indicates that the innate schema is rather crude. An isolated bill is as effective as the entire head for naive chicks, and neither type of bird can even discriminate its own species pattern without some days of experience. The earlier claim of TINBERGEN and PERDECK (1950), that the location of a red spot on the bill triggered a "configurational schema" in the chick's brain, seems incorrect: HAILMAN finds that spot-velocity rather than its relative location is the critical factor in its salience for the chick as the beak is moved back and forth. It is worth noting that movement direction is not critically important for the gull chick, as it has been found to be for visual elicitation of feeding in the toad (EWERT, 1968), where motion of a rectangle orthogonal to its long axis sharply inhibits feeding. We should raise the question as to whether the toad, with a more specific innate prey schema, could learn to select such an unnatural "worm".

III. Preference for Solid Objects

The observation of FANTZ (1958), that dark-reared chicks immediately preferred to peck at hemispheres rather than at flat discs, seemed to indicate that detection of "solidarity" was innate for this species. However, studies by HESS (1961) indicated that one important cue to solidarity of a hemisphere in normal light—*i.e.* the shadow cast over the bottom half—is at least partially learned. If chicks were raised with light emanating from *below*, a preference for pecking at photographs of hemispheres with top shadows rather than bottom shadows could be induced.

Further studies by DAWKINS (1968) showed that this developmental problem is still more complex. He confirmed the finding of FANTZ that dark-reared chicks prefer solid over flat objects, and added the information that monocular viewing is sufficient for this discrimination in the chick. However, dark rearing abolishes the preference for solid over flat objects as tested with photographs. This finding might seem to confirm the conclusion of HESS, that solidarity as judged by light-and-shadow orientation is learned. However, an additional test by DAWKINS showed that chicks reared with light from below did *not* prefer the inverted-shadow photographs, but rather the "correct" illustrations. This result contradicts the interpretation of HESS, but not his actual data, since in the former study the grain photos showed conspicuous shadows on the background surface, which provided extra cues. Apparently some experience of movement within a lighted incubator is required for the discrimination to emerge, but the preference among shadow-directions reflects an innate bias. The ability of naive chicks to discriminate solidarity of real objects with no visual experience at all must depend upon other cues, such as motion parallax, to enhance the object's salience.

IV. Color Preferences for Pecking

Various experiments agree that dark-reared chicks or ducklings of various species show spontaneous preferences for pecking at small spots of particular hues placed on an achromatic background (white or grey). HESS (1956) observed that ducklings strongly preferred to peck at green spots, and this conclusion was confirmed by OPPENHEIM (1968), who used careful experimental controls for differential intensity of various colors. Indeed, relative intensity had a negligible effect on the duckling's dominant tendency to peck at green objects. This was an absolute, not simply a relative preference, in that ducklings gave few pecks to an array of colored objects when the green stimuli were removed. Not only was this preference for green spots innate, but it survived the rearing of ducklings in monochromatic environments of other colors. This fact contrasts with data reviewed below based on other kinds of stimulus discrimination behavior. OPPENHEIM further showed that the animals did not simply approach any green surface: colored spots on a green wall had little attraction.

When gull chicks peck at small spots, they show an entirely different behavior according to HAILMAN (1964): they ignore green and yellow in favor of red or blue targets. In a further study of this species, HAILMAN (1966) noted that this bimodal preference curve was exactly inverted when white spots were placed on different colored backgrounds. The green background apparently made the white spot appear reddish, by analogy with human color-contrast effects. This effect may reflect an opponent red-green color coding process within the bird's visual system such as has been described for primates and for fishes. From this perspective, it would be useful to know whether the ducklings tested by OPPENHEIM would have pecked readily at a white spot on a red background. Electrophysiological studies of gulls and ducklings might reveal whether the simplest peripheral coding hypothesis (on-responses from ganglion cells with red and green centers respectively) might hold.

D. Preferences for Large Visual Objects

The tendency of young chicks to approach any large moving objects seems to be the basis of rapid formation of social attachments, which have generally been termed "imprinting". However much this process resembles conventional discrimination learning, there have been many difficulties in demonstrating the exact contributions of innate and acquired preferences and the role of actively following the imprinting object. There have been no definitive studies of shape preferences (if they exist at all) with most students manipulating object size, color and "conspicuousness" as variables. The results of color preference tests make an interesting contrast to conclusions derived from food pecking tests.

I. Color Preferences for Approaching a Model

IMPEKOVEN (1969) is the only worker to examine color preferences in a situation where chicks (gulls) were intentionally biased to seek different types of visual stimulation: *hungry* chicks tend to peck at the bill of a parental model, while *cold* chicks tend to move toward the parental body in order to snuggle under it. Although red and blue models are preferred over others under both motivational states, other rank-order differences are state-dependent. For example, the dominance of black over yellow models for cold chicks reversed for the hungry group. The cold animals also chose black over grey more than did the hungry chicks. The dark body seems to have more salience for the cold chicks, since they all preferred black body plus white head to white body plus black head. However, 80% of choices by the hungry chicks were in the reverse direction. We might explain these data by suggesting that the visual filters for food-begging and crouching-under-body have somewhat different color or luminance optima, and that the relative threshold of the visuomotor connections varies with motivational state. Further evidence that different responses are differently color-biased was provided by HESS (1956), who showed that ducklings also imprint well on red and blue spheres, but not upon green objects.

In general, behaviorists have assumed that naive chicks will acquire a preference for following any distinctive visual object (within a certain size range) which is presented to it. For example, POLT (1969) reports that chicks can be imprinted upon either a blue or a yellow sphere such that subsequent forced-choice tests will reveal a preference for a familiar object. However, KLOPFER and HAILMAN (1964) have shown that this direct imprinting effect is not automatic if the choice test involves stimuli that are quite different in salience. Their experiment showed that although chicks did not follow a plain model less that they followed a "striking" model, they significantly preferred the latter model in a test about one day later, regardless of which model they had initially followed. Since a second control group of the same age did not show such a preference as a function of maturation, the authors concluded that one following experience had primed the following response, but to the more salient model. The possibility that discriminative capacity *had* matured and that only response threshold was affected by the first imprinting experience cannot be excluded. Another study, by GRAY (1961), also shows that general experience plus maturation can produce changes in following preferences among chicks. For example, chicks showed progressively greater preference for a red ball (as against other colors) during five daily test sessions, even though they had equal exposure to all other colors. The physiological basis for this phenomenon is not known, but such shifting biases complicate analysis of the interaction between the innate and acquired preferences of young animals.

Another striking effect of restricted rearing experience has been shown by KOVACH (1961), who measured the preference of chicks for approaching lighted panels of different hues in a Y-maze. As others have surmised from imprinting studies with colored spheres, chicks most readily choose blue and red in paired comparison tests, even if reared in the dark. After rearing in achromatic light such preferences were blunted. On the other hand, exposure to red or blue monocromatic light further enhanced the natural preferences for these colors. KOVACH used the term "canalization" to describe the effect of environment in sharpening natural preferences. It should be noted that effects of pre-exposure to these uniform colors did not "habituate" the chick's interest in such hues, whereas exposure to spots of one color can induce a momentary decline in pecking at spots of that particular color. It would be important, however, to expose chicks to the dominant spot color for much longer periods of time in order to make a test comparable to that of KOVACH.

The effects of similar monochromatic rearing environments on a bird's *conditioned* pecking behavior toward various colored stimuli has also been examined by two investigators. PETERSON (1962) first reared ducklings in either monochromatic light or in white light, and then trained each group to peck at a key illumined with that particular hue. The resulting generalization gradient (proportion of responses made to novel colors as well) was much flatter for the monochromatically reared group. It appeared that an imbalance in chromatic stimulation had somehow affected the immediate discriminability of the new colors. However, an extension of PETERSON's approach to domestic chicks by MALLOTT (1968) failed to obtain overall differences in "sharpness of generalization" when dark-reared, and achromatically and monochromatically reared chicks were trained to peck at a key illumined by a narrow band peaking at 589 μ. The fact that chicks reared in white light showed a better discrimination of 589 μ than of 620 μ is suggestive, but is not conclusive due to the small number of subjects used.

The general hypothesis that restricted stimulation should result in reduced discrimability seems intuitively plausible, but has no compelling support as yet. In fact, a report by BATESON and CHANTREY (1972) suggests "that discriminations between two objects can become more difficult for an animal after it has been exposed to both of them". For example, chicks that were exposed to two different-colored cylinders for one hour each day took longer to learn a food-reinforced discrimination between these two than did control chicks that saw only one of the objects each day. Similar results were obtained with monkeys exposed to letter forms in their cage. These authors suggest by way of interpretation that stimuli can be classified together since they occurred in the same context. The biological importance of such "contextual binding" of stimuli is that a naive animal will associate different views of the same object. A chick might thus lump together different views of its mother as the same real object. However, it would seem more likely that association would occur more readily between like elements that could be perspective transformations of one another, rather than between quite dissimilar objects.

E. Development of Visual Behavior in Amphibians

I. Feeding Behavior of Frogs

The common frog should provide a good model system for developmental correlations of form and function, since recent studies (POMERANZ and CHUNG, 1970; POMERANZ, 1972) report that the axons of various classes of retinal ganglion cells (as first described by LETTVIN et al., 1959) innervate the tectum at different stages of development. First, the edge-detecting class-1 terminals are not to be found at all in the larval stage: the free-moving tadpole may have no need to attend to stationary edges, as does the adult who sits and awaits his prey. Secondly, the so-called "convex-edge detectors" or class-2 fibers are found within a restricted region of tectum, corresponding to the projection of the horizontal meridian. On the other hand, the class-3 (changing contrast detector) and class-4 (dimming detector) units are widely distributed within the tectal neuropil.

Although a definitive study of the onset of visually elicited behaviors in Ranid frogs has not been reported, the present reviewer can add the observation that shadow avoidance, but not visual feeding, is easily seen in species such as *Rana clamitans* and *Rana catesbeiana.* This fact seems to correlate with the maturity of class-3 and class-4 projections in tadpoles, and with the relative incompleteness of the "bug-detector" fibers projection at this stage. When frogs of both species were brought into the laboratory, a rather sudden appearance of visually elicited feeding (toward real or dummy prey objects) was witnessed at the time that froglets emerged from the water. The majority of frogs in a batch would begin to strike at prey on the same day, as if some phase of retinal or central connectivity were suddenly complete. During these first feeding attempts, the froglets strike accurately in all directions, and estimate stimulus distance correctly. Furthermore, when the worm is partly occluded by a grid barrier, the frog is immediately able to sidestep the far edge in order to get an unimpeded view of the moving target. These facts suggest that selectivity and central integration of movements are "hard-wired" in the frog. However, we do not know whether preferences for specific prey objects, detection of particular predators, or ability to negotiate a complex terrain might be altered by selective exposure to atypical early enviroments.

II. Phototaxis in the Frog

Another behavior that appears just at metamorphosis is the frog's tendency to approach blue light in preference to light of other hues. Although young tadpoles do show positive phototaxis, their spectral sensitivity peaks in the green region (MUNTZ, 1963). Only after metamorphosis do they resemble the adult, whose phototactic preference is in the blue range. MUNTZ suggests that development of an inhibitory effect *via* red rods could explain the acquisition of the adult color preference. There is no evidence as to whether or not rearing in the dark or in monochromatic light would modify relative spectral preferences in the standard Y-maze test.

III. Sun-Compass Orientation

For many amphibians (as for fishes and birds) there is strong evidence that celestial cues are sufficient for discrimination of direction during migratory or homing maneuvers. For example, froglets trapped at their shoreline during postmetamorphic dispersal and transferred to a test arena will continue their initial orientation when the sun is in view, but not on a cloudy day (DOLE, 1972). The same frogs will reverse their preferred direction if confined and tested a few days later, as if they were now prepared to head homeward (as normally happens). The direction taken in respect to the sun is determined by the time of day at which the frogs are tested; therefore the frogs do not simply learn to move at one angle to the sun, but rather to "compensate" for its relative position during the day. It is not known whether prior experience with the sun's movement is itself required to make this discrimination. It might be possible for a frog to associate a few distinct sun positions with a few internal states (such as hunger, warmth, sleepiness, etc.), which also vary during the day, and thereafter to seek the appropriate combination of internal and external cues. This is probably not the correct explanation, inasmuch as toads can be trained at a given time to seek food and water by moving within a training arena in respect to an artificial light, and yet show time-compensated direction shift when tested at another time of day (LANDRETH and FERGUSON, 1968).

The problem of sun-compass orientation is complicated by additional evidence that salamanders can orient in reference to sun-position by using either the eyes or an extra-optic receptor, which is probably part of the pineal complex (TAYLOR, 1972). Photosensitive neural elements within the frog's dorsal diencephalon might also mediate this behavior (DODT and JACOBSON, 1963). The extra-optic channel is sensitive to the plane of polarized light, which varies systematically with sun position, so that salamanders trained with an artificial shoreline and polarization cues will later show appropriate directional orientation to the polarization plane in absence of the eyes (TAYLOR and ADLER, 1973). Since both the optic and extra-optic types of orientation are time-compensated, it is possible that both afferent channels converge upon a common thalamic or midbrain visual center. It is not yet known whether an innate mechanism exists for translation between the two sets of coordinates or whether each type of cue must be learned separately. It would be possible to train a salamander in a pond with the pineal surface covered and eyes open, and later to test for sun-compass orientation after optic blinding.

The plane of polarization must be inferred from an illumination intensity gradient or from the ratio between outputs of cells with differently polarized filters. Therefore one should investigate the possibility that optic cues are also encoded *via* intensity ratios, such as those that must exist between the two eyes. The convergence of ipsilateral and contralateral projections within the rostral thalamus of frogs and salamanders could be related to such a "comparator" mechanism. For one thing, the frog's ipsilateral projections do carry information used for localization of light sources (KICLITER, 1973; INGLE, 1976). And secondly, experiments on phototactic orientation of sea-turtle hatchlings (MROSOVSKY and SHETTLEWORTH, 1974) clearly demonstrate that binocular use is required for these animals to move straight towards the brightest region of the horizon. It is perhaps easier to understand how an "optimal ratio" can be systematically shifted during a circadian rhythm than how a two-dimensional map can be rotated.

IV. Substrate Pattern Selection by Tadpoles

In order for amphibians to acquire a "familiar" sun-compass orientation, it is probably necessary for them to use the water-land interface as a reference cue. A recent set of studies by WIENS (1970, 1972) indicates that frog tadpoles learn still more about their surroundings. For example, *Rana aurora* reared in an artificial vertically-striped environment for 14 to 17 days showed a preference for approaching this stimulus at a later age when they would normally seek out substrate near the pond's edge. Controls reared in a featureless environment showed no such preference when first exposed to that pattern. The opposite behavior was shown by *Rana cascadae* tadpoles exposed to the same regime: they became attached to a square pattern if present earlier, but did not become imprinted on a vertical-stripe pattern. WIENS interprets this species-specific capacity for visual imprinting in terms of environmental adaptations: *R. cascadae* lives in a barren pond, which is often splotchy, whereas *R. aurora* breeds in shallow ponds with plants such as weed stems, cattails, floating grass and other linear contours. The results of this study should be replicated, since they are among the first to imply that the brain of certain species may be preadapted to develop attachments to certain stimuli and not others, even though some specific visual experience is required for formation of the attatchments. Further documentation of the degree of specificity in the tadpole's imprinting capacity might encourage the physiologist to investigate the mechanism of substrate-selection at the single-unit level.

F. Stellar Orientation in Birds

In addition to sun-compass cues, some bird species appear to use the pattern of certain stars in the night sky to guide their annual migrations. Recently EMLEN (1975) has summarized his own studies of indigo buntings, which include critical developmental data. In order to investigate the influence of early experience the young birds were raised either in a windowless room or under a planetarium. Those exposed to a natural sky (rotating about the north star) or to a modified pattern (rotating about another star) were able to orient toward the southern sky. Their orientation was based upon the use of the axis of rotation of discernable star patterns, but was not genetically linked to any particular pattern. Once the rotational pattern is learned, however, time-compensation is used to compute the southerly direction. Early exposure to sky rotation is sufficient to impart the knowledge used later when the motivation to migrate arises.

G. Social Behaviors of Fishes

I. Orientation Toward the Parent or Towards a School

Newly hatched fish fry often tend to group together or to swim very near their parents, either for protection or for finding of food. SHAW (1960) has observed the development of schooling in the fry of the silverside. The initial attraction of a fish to another fish is gradual, but not necessarily "reinforced", since isolated fry also approached fish schools with normal intensity. However, these naive fry were at first ineffective in maintaining position within the group: they bumped into other fish and often left the group. Within about four hours their behavior had become indistinguishable from that of socially reared fish. It is not clear whether their errors were due to over-reaction to slight convergences with other fish, confusion resulting from the sudden sensory overload, or deficits in the use of motion and velocity cues to maintain the proper distance. It should be possible to use a mechanical school with preprogrammed movements to study how a naive fish attempts to "track" the movements of its group. It would also be possible to study the relative effects of active or passive movement of the test fish in development of effective schooling.

BAERENDS (1971) has reviewed the variety of species-specific releasers of orientation to parents among various species of the cichlid family. He notes that attraction to red or yellow colors among certain species seems to be innate: isolated fish immediately show a preference for models with red. However, tendencies of species to approach black models or those with short-wavelength colors seem to be learned: isolated young have no such preferences. Some species (*e.g. Hemichromis*) that normally prefer red models will nevertheless become imprinted upon blue models and will later prefer to accompany these, much like chicks or ducklings. The mouthbreeding cichlids (*e.g. Tilapia*) must specifically orient to the parents' mouth (for protection), and these fry show a remarkable tendency to cluster near the *bottom* of any object or cutout figure with high contrast. It would be interesting to determine whether the retina is regionally specialized in such a way that upper-field stimulation is preferred by the fry.

II. Specific Recognition During Sexual and Aggressive Encounters

A large variety of fish will show male-to-male aggressive behavior immediately after prolonged isolation, including mouthbreeders (NEIL, 1964) and platyfishes (SHAW, 1962). A number of other examples are given by BAERENDS (1971), but in nearly all cases the actual basis of the visual discrimination is not well characterized, so that it is not possible to estimate the complexity of the adequate stimulus signifying a "rival". SEITZ (1940) notes that schooling experience is necessary for one species of *Tilapia* to limit its courtship response to conspecific females. Isolated males will court any simple model, including a round disc, but schooling males will only select realistic females that have colors similar to their former juvenile schoolmates. However, even isolated males will show specific recognition of the brightly colored males. The best studies of specific recognition among fish have been carried out by LEONG (1969) and HEILIGENBERG et al. (1972). The first study demonstrated that the presence of a dark eye-bar on *Haplochromis* models enhanced aggression in mature males, while the presence of an orange patch inhibited attack. The second article further showed that the exact angular orientation of the eye-bar was critical to facilitation of aggression. However when a fish model was placed in a head-down posture, the critical feature was the relative angle of the eye-bar to the model's orientation (i.e. parallel to the forehead). The ability of goldfish to detect the "parallelness" of two line segments has been emphasized in a recent review (INGLE, 1978), but the physiological basis of this "gestalt relationship" is not at all obvious. There is a good possibility that this particular feature of the "threatening male" is learned, but no decisive experiments have yet been reported.

H. Innate Depth Discrimination Ability

An important study by WALK and GIBSON (1961) introduced the "visual cliff" as a useful tool for comparative and ontogenetic studies of depth discrimination capacity. In addition to several studies on mammalian species, numerous experiments have used the newly hatched, dark-reared chick, which has the advantage that large numbers of subjects can be conveniently used for important parametric variations in rearing or stimulus conditions. Although the initial observations of WALK and GIBSON left no doubt that some depth discrimination was present in the initial trials of dark-reared chicks, and that this ability did not depend upon the use of binocular vision, there has been uncertainty concerning the exact type of visual information guiding the choice. WALK and GIBSON suggested that the motion parallax produced by differential velocities of the deep vs. shallow surfaces in respect to the edge of the starting runway (between alternative sides) would be a sufficient cue for avoiding the deep side, since the motion-contrast would be greater on that side. This hypothesis was supported by a further study by SOMERVILL (1971), who used a novel three-dimensional striped pattern that provided a combination of velocity contrast (the deep stripes moving in respect to the shallow stripes) and a shallow surface that the chicks could realistically step upon. The presence of the extra deep stripes completely inhibited any approach to the shallow surface, even though controls would choose to approach the shallow stripes alone. In this experiment, the differential motion *within* a given surface area may have appeared to the chick as an "unsteady" surface rather than as a "deep" void. It would be interesting to know whether chicks would step on an oscillating pattern at a shallow depth in preference to a stationary deep surface.

Although the parallax theory of innate depth discrimination seems plausible, we should note that SCHIFFMAN (1968) observed a consistent avoidance of naive chicks to the "optical void" produced by placing the animal on a glass with a deep surface below. In this situation, the chick's own body contours would have to serve for parallax comparison; if these could be reduced in visibility with a yellow background directly below, one might better test the "velocity contrast" hypothesis. Whatever the exact mechanism by which chicks detect deepness, other studies show that they quickly lose their fear of the deep side of a visual cliff when simply left exposed for one day over a deep surface (TALLARICO and FARRELL, 1964). A further study by ZEIER (1970) replicated this habituation effect, and added the interesting observation that chicks pre-exposed to a deep surface monocularly showed no transfer of their learning when later exposed to a visual cliff *via* the second eye. HAILMAN (1968) also suggested that fear of depth could be modified in different ways in cliff-dwelling gull chicks (kittiwakes) from that seen in their ground-dwelling cousins (herring gulls). Although all chicks of both species showed good visual cliff discrimination, the former species learned to stay very still in their nests when disturbed, while herring gull chicks in the cliff-nests of kittiwakes would usually jump out when disturbed. From that behavior alone, one would have guessed that they had not evolved a fear of falling. Such a species difference was reported by YERKES (1904), who noted that water tortoises but not land tortoises would readily step off a high platform. Unpublished studies in our own laboratory indicate that pond frogs (*Rana pipiens* or *Rana climitans*) will readily leap into a void during prey pursuit, while toads (*Bufo*) and tree-frogs (Hyla) will not jump until a surface is present behind the prey object. However, with a modified visual-cliff device the same frogs are about 90% accurate in jumping to a shallow rather than a deep surface, even when the former subtends only about 30° within the lower field. In the latter study, it was surprising to find that frogs strongly avoided jumping to a platform with vertical stripes, when opposed to either horizontal strips or a random-pattern (INGLE, 1978). Apparently these frogs regard vertical contours as barriers. In view of the studies of WIENS (1970, 1972), one should examine the possibility that these orientation-specific preferences are acquired through experience.

I. Concluding Remarks

Our best evidence from the study of newly hatched birds indicates that even fairly complex organisms enter the world with at least a simple set of stimulus preferences for following or pecking behaviors. Certain distinctions such as horizontal-vertical, upper-lower shadows, round-pointed or red-green are already salient to chicks of many species. Unfortunately, comparable data would be harder to obtain on those fish and amphibian species amenable to detailed visual testing, simply because prolonged periods of deprivation would be required before tests were feasible. But the problem is not insuperable, and would be worth the effort because of the convenience of neurophysiological investigation in these species. However, it would be a mistake to rely upon spontaneous preference data alone to characterize the chief dimensions of "discriminability" among objects that may be innate. For example, one cannot conclude that two objects are not easily discriminable, if they are equally salient as potential food even at first look. One promising approach (especially with chicks) would involve selective pre-exposure to a single object (or a restricted class of objects) such that stimulus habituation occurs; then enhanced choice of a novel object in a two-choice test would prove a measure of inherent discriminability.

Although HAILMAN (1967) failed to observe immediate preferences among complex models of gull parents, one cannot conclude that chicks are not endowed with latent abilities to make these discriminations better than classes of "ecologically irrelevant" discriminations. Do birds naturally discriminate better among birds than do other visually developed animals, and are fish specialized for discriminating among other fishes? Although we lack answers to such questions, we have noted above some hints that innate biases may be sharpened *via* experience. In particular we have a demonstration of priming effects of imprinting experience by KLOPFER and HAILMAN (1964), the "canalization" notion applied by KOVACH (1971) to chick color preferences, and the limited imprinting capacities of different tadpole species by WIENS (1972). These suggestive data indicate the importance of studying the specificity of genetic-environmental *interactions*, as well as the immediate behavior of young animals facing the world for the first time.

Although it seems clear that visual experience plays an important and sometime critical role in the shaping of normal species-specific behaviors, we do not usually know whether the unfolding of behavior is simply a manifestation of discrimination *learning* processes as observed in adults or whether stimulus variation directly *induces* more specific stimulus-filtering properties within afferent visual neuronal networks. One method which has been insufficiently exploited for such a distinction involves the training of visual discriminations via an "experienced" eye, and later testing for transfer of training via the second "visually deprived" eye—a method easily carried out with certain fish and bird species. Successful transfer would eliminate the possibility that retina or direct retinofugal target neurons are critically modified by visual experience, since these are entirely unilateral in such species. If fishes or birds should not show direct afferent-induction effects of the environment they become the more valuable for study of "acquired distinctiveness" of visual objects, which must reflect central mechanisms.

References

Baerends, G. P.: The ethological analysis of fish behavior. In: Fish Physiology, Vol. 6. (eds. Hoar, W. S., and Randall, D. J.). New York: Academic Press, 1971.

Bateson, P. P. G., Chantrey, D. F.: Retardation of discrimination learning in monkeys and chicks previously exposed to both stimuli. Nature **237**, 173—174 (1972).

Casagrande, V. A., Diamond, I. T.: Ablation study of the superior colliculus in the tree shrew (*Tupaia glis*). J. comp. Neurol. **156**, 207—238 (1974).

Corballis, M. C., Beale, I. L.: Bilateral symmetry and behaviour. Psychol. Rev. **17**, 451—464 (1970).

Cruze, W. W.: Maturation and learning in chicks. J. comp. Psychol. **19**, 371—409 (1935).

Dawkins, R.: The ontogeny of a pecking preference in domestic chicks. Z. Tierpsychol. **25**, 170—186 (1968).

Dodt, E., Jacobson, M.: Photosensitivity of a localized region of the frog diencephalon. J. Neurophysiol. **26**, 752—759 (1963)

Dole, J. W.: Evidence of celestial orientation in newly-metamorphosed *Rana pipiens*. Herpetol. **28**, 273—276 (1972).

Emlen, S. T.: The stellar-orientation system of a migratory bird. Sci. Amer. **233**, 102—111 (1975).

Ewert, J.-P.: Der Einfluß von Zwischenhirndefekten auf die Visuomotorik im Beute- und Fluchtverhalten der Erdkröte. Z. vergl. Physiol. **61**, 41—70 (1968).

Fantz, R. L.: Form preferences in newly hatched chicks. J. comp. physiol. Psychol. **50**, 422—430 (1957).

Fantz, R. L.: Development discrimination in dark-hatched chicks. Percept. Motor Skills **8**, 47—50 (1958).

Goodwin, E. B., Hess, E. H.: Innate visual form preferences in the pecking behavior of young chicks. Behavior **34**, 223—237 (1966).

Gray, P. H.: The releasers of imprinting: differential reactions to colors as a function of maturation. J. comp. physiol. Psychol. **54**, 597—601 (1961).

Hailman, J. P.: Coding of the colous preference of the gull chick. Nature **204**, 710 (1964).

Hailman, J. P.: Mirror-image color-preferences for background and stimulus object in gull-chicks. Experientia **22**, 257 (1966).

Hailman, J. P.: The ontogeny of an instinct. The pecking response in chicks of the laughing gull (*Larus atricilla* L.) and related species. Behaviour **15**, 159 (1967).

Hailman, J. P.: Visual-cliff responses of newly-hatched chicks of the laughing gull *Larus atricilla*. Ibis **110**, 197—200 (1968).

Hailman, J. P.: The role of stimulus-orientation in eliciting the begging response from newly-hatched chicks of the laughing gull (*Larus atricilla*). Anim. Behav. **19**, 328—335 (1971).

Heiligenberg, W., Kramer, U., Schultz, V.: The angular orientation of the black eye-bar in *Haplochromis burtoni* (Cichlidae, Pisces) and its relevance to aggressivity. Z. vergl. Physiol. **76**, 168—176 (1972).

Hein, A., Diamond, R. M.: Contrasting development of visually triggered and guided movements in kittens with respect to interocular and interlimb equivalence. J. comp. physiol. Psychol. **76**, 219—224 (1971).

Hess, E. H.: Development of the chick's response to light and shade cues of depth. J. comp. physiol. Psychol. **43**, 112—122 (1950).

Hess, E. H.: Natural preferences of chicks and ducklings for objects of different colors. Psychol. Rep. **2**, 477—483 (1956a).

Hess, E. H.: Space perception in the chick. Sci. Amer. **195**, 71—74, 76, 78, 80 (1956b).

Hess, E. H.: Shadows and depth perception. Sci. Amer. **204**, 138—148 (1961).

Impekoven, M.: Motivationally controlled stimulus preferences in chicks of the black-headed gull (*Larus ridibundus* L.). Anim. Behav. **17**, 252—270 (1969).

Ingle, D.: Two visual systems in the frog. Science **181**, 1053—1055 (1973a).

Ingle, D.: Selective choice between double prey objects by frogs. Brain. Behav. Evol. **7**, 127—144 (1973b).

Ingle, D.: Spatial vision in Anurans. In: The Amphibian Visual System (ed. K. Fite). Academic Press, New York (1976)

Ingle, D.: Mechanisms of shape recognition among vertebrates. In: Hdbk. Physiol. voll. VIII, pp. 267—295 (H.-L. Teuber et al., Eds). Berlin: Springer 1978.

Ingle, D.: Role of visual cortex in anticipatory orientation toward moving targets by the gerbil. Soc. Neurosci Abstr. 3: 68 (1977).

Ingle, D., Sprague, J. M.: Sensorimotor functions of the midbrain tectum. Neurosci. Res. Progr. Bull. **13**, 169—288 (1975).

Kicliter, E.: Flux, wavelength and movement discrimination in frogs: forebrain and midbrain contributions. Brain Behav. Evol. **8**, 340—365 (1973).

Klopfer, P. H., Hailman, J. P.: Perceptual preferences and imprinting in chicks. Science **145**, 1333—1334 (1964).

Kovach, J. K.: Interaction of innate and acquired: color preferences and early exposure learning in chicks. J. comp. physiol. Psychol. **75**, 386—398 (1971).

Landreth, H. F., Ferguson, D. E.: The sun compass of Fowler's toad, *Bufo Woodhousei Fowleri*. Behaviour **30**, 27—43 (1968).

Leong, C.-Y.: The quantitative effect of releasers on the attack readiness of the fish *Haplochromis burtoni* (Cichlidae, Pisces). Z. vergl. Physiol. **65**, 29—50 (1969).

Lettvin, J. Y., Maturana, H. R., McCulloch, W. S., Pitts, W. H.: What the frog's eye tells the frog's brain. Proc. IRE, **47**, 1940—1951 (1959)

Mallott, M. K.: Stimulus control in stimulus-deprived chickens. J. comp. physiol. Psychol. **66**, 276—282 (1968).

Maturana, H. R., Frank, S.: Directional movement and horizontal edge detectors in the pigeon retina. Science **142**, 977—979 (1963).
Mrosovsky, N., Shettleworth, S. J.: Further studies of the sea-finding mechanism in green turtle hatchlings. Behaviour **51**, 194—208 (1974).
Muntz, W. R. A.: The development of phototaxis in the frog (*Rana temporaria*). J. exp. Biol. **40**, 371—379 (1963).
Neil, E. H.: An analysis of color changes and social behavior of *Tilapia mossambic*. Univ. Calif. Publ. Zool. **76**, 1—58 (1964).
Oppenheim, R. W.: Color preference in the pecking response of newly hatched chicks (*Anas platyrhynchos*). J. Comp. physiol. Psychol. Monogr. Suppl. part 2, pp. 1—17 (1968)
Peterson, N.: Effects of monochromatic rearing on the control of responding by wave length. Science **136**, 774—775 (1962).
Polt, J. M.: Effect of imprinting experience and discrimination learning in chicks. J. comp. physiol. Psychol. **69**, 514—518 (1969).
Pomeranz, B.: Metamorphosis of frog vision: changes in ganglion cell physiology and anatomy. Exp. Neurol. **34**, 187—199 (1972).
Pomeranz, B., Chung, S. H.: Dendritic tree anatomy codes form vision in tadpole retina. Science **170**, 983—984 (1970).
Quine, D. A., Cullen, J. M.: The pecking response of young Arctic terns (*Sterna macrura*) and the adaptiveness of the releasing mechanism. Ibis **106**, 145—173 (1964).
Rossi, P. J.: Adaptation and negative aftereffect to lateral optical displacement in newly hatched chicks. Science **160**, 430—432 (1968).
Rossi, P. J.: Prism-induced negative after effects without food-reinforced feedback in newly hatched chicks. Psychonom. Sci. **24**, 141—142 (1971)
Schaeffer, T. H., Hess, E. H.: Color preferences in imprinting objects. Z. Tierpsychol. **16**, 161—172 (1961).
Schiffman, H. R.: Physical support with and without optical support: Reaction to apparent depth by chicks and rats. Science **159**, 892—894 (1968).
Seitz, A.: Die Paarbildung bei einigen Cichliden, II. Z. Tierpsychol. **5**, 74—101 (1940).
Shaw, E.: The development of schooling behavior in fishes. Physiol. Zool. **33**, 79—86 (1960).
Shaw, E.: Environmental conditions and the appearance of sexual behavior in the platyfish. In: Roots of Behavior. New York: Harper and Row 1962.
Somervill, J. W.: Motion parallax in the visual cliff situation. Percept. Motor Skills **32**, 43—53 (1971).
Tallarico, R. B., Farrell, W. M.: Studies of visual depth perception: an effect of early experience on chicks on a visual cliff. J. comp. physiol. Psychol. **57**, 94—96 (1964).
Taylor, D. H.: Extra-optic photoreception and compass orientation in larval and adult salamanders (*Ambystoma tigrinum*). Anim. Behav. **20**, 233—236 (1972).
Taylor, D. H., Adler, K.: Spatial orientation by salamanders using plane-polarized light. Science **181**, 285—287 (1973).
Tinbergen, N., Perdeck, A. C.: On the stimulus situation releasing the begging response in the newly hatched Herring Gull Chick (*Larus argentatus* Pont.). Behavior **3**, 1—39 (1950).
Walk, R. D., Gibson, E.: A comparative and analytical study of visual depth perception. Psychol. Monogr. **75**, 519: (whole issue) (1961).
Walk, R. D., Walters, C. P.: Importance of texture-density preferences and motion parallax for visual depth discrimination by rats and chicks. J. comp. physiol. Psychol. **86**, 309—315 (1974).
Wiens, J. A.: Effects of early experience on substrate pattern selection in *Rana aurora* tadpoles. Copeia No. 3, 543—548 (1970).
Wiens, J. A.: Anuran habitat selection: Early experience and substrate selection in *Rana cascadae* tadpoles. Anim. Behav. **20**, 218—220 (1972).
Yerkes, R. M.: The instincts, habits and reactions of the frog. Psych. Rev. Monogr. **4**, 579—638 (1903).
Zeier, H.: Lack of eye-to-eye transfer of an early response modification in birds. Nature **225**, 708—709 (1970).

Chapter 5

Ontogeny of Structure and Function in the Vertebrate Auditory System

EDWIN W RUBEL, Charlottesville, Virginia (USA)

With 15 Figures and 4 Tables

Contents

A. Introduction

Why study the vertebrate auditory system? Throughout the ages, when scholars have attempted to define objectively the qualities that distinguish man from other animals, the power of human communication systems inevitably comes to the fore. Normal development of auditory perception is essential for the establishment of both expressive and receptive aspects of language. Thus, increased understanding, leading to eventual prevention or treatment of the various conditions that cause failures in the normal processing of acoustic information, has important clinical relevance. Furthermore, the auditory system is one of the primary sense modalities, involving both a highly specialized peripheral receptor organ and a complex set of central pathways.

Thus, ontogenetic studies of the vertebrate auditory system are of general importance. For instance, studies of the sensory epithelium and associated peripheral specializations are important for understanding the origins, inductive interactions, and polarization gradients of receptor specializations. Analyses of central nervous system pathways will lead to increasing understanding of the ontogeny of neural coding mechanisms and the role of afferent innervation and/or stimulation in the ontogeny of neural networks. Finally, investigations of an organism's behavior with respect to its acoustic environment will expand our general knowledge of how organisms come to appropriately organize and categorize their sensory environment.

The present review attempts two tasks related to the functions noted above. First, a large portion of the literature on the ontogeny of auditory system structure and function is brought together and summarized in Tables 1–4. The tables, forming the body of this review, are meant to provide access to the various areas of literature and to provide brief summaries of what I believe to be the major findings derived from each group of studies. The text and figures accompanying the tables are intended to point out areas of controversy and agreement, and areas where future research will be most beneficial. It should be pointed out that the tables and text do not exhaustively cover all the literature in this field. Several areas of research have been intentionally omitted or receive only abbreviated treatment because they have been adequately summarized elsewhere, or due to limitations of space and time. Where possible I have attempted to point out these omissions in the text and to cite a few general references in these areas. In addition, in some cases where the same data appear in many sources by the same author(s), only the most inclusive paper has been cited. Finally, there are undoubtedly omissions in areas that are intended to be exhaustive[1].

The second purpose of this review is to point to some examples of research areas where investigations of auditory system development either have had, or can have, relevance for broader problems of behavior or brain development. Thus, the second section gives brief summaries of three research programs in which the vertebrate auditory system is used to approach general problems of neural and perceptual ontogeny. The examples chosen are the research of G. GOTTLIEB, D. KENT MOREST, and the present author.

[1] I would appreciate being informed of any such omissions, as I will attempt to maintain and make generally available a complete list of citations relevant to the area of auditory system development.

B. Ontogeny of the Auditory System

I. Embryology of the Auditory Receptor[2]

1. Early Development of the Inner Ear

Table 1 summarizes the literature on the ontogeny of the vertebrate inner ear. The early literature was reviewed by MINOT (1892), while later reviews concentrated on selected topics. BAST and ANSON (1949) present a detailed account of normal morphogenesis and histogenesis in mammals, especially man. ALTMAN (1950) presents a briefer but excellent account of normal development, and includes much of the comparative and experimental literature. YNTEMA (1955) summarizes the work on amphibians, dealing primarily with inductive interactions and polarization of the ear and surrounding tissue, and VAN DE WATER and RUBEN (1976) present an up-to-date account of the experimental research in this area. Table 1 is restricted to inner ear and ganglion cell development, and does not include the following related topics: induction and differentiation of otic capsule and surrounding tissues (YNTEMA, 1955; VAN DE WATER and RUBEN, 1976); ontogeny of the middle ear (BAST and ANSON, 1949; CANDIOLLO and LEVI, 1969; STEPHENS, 1972); and pathological abnormalities, including those of known genetic origin (BAST and ANSON, 1949; HARDY, 1975; VAN DE WATER and RUBEN, 1976).

The earliest embryological sign of an auditory apparatus is a thickening of the ectoderm on either side of the head at a rostrocaudal level approximately at the middle of the hindbrain. This thickening, first recognized as the anlagen of the inner ear by HUSCHKE (1831), is the auditory placode. It will eventually become the entire otic labyrinth, a continuous series of epithelium-lined tubes housing the auditory and vestibular organs. The placodal region invaginates to form the auditory pit, which then splits off from the overlying ectoderm to form the auditory vesicle or otocyst. These processes and the ensuing evagination of the otocyst into three primary ducts (endolymphatic duct, utriculus, and sacculus) have been described for many species, and timetables are often given (see Table 1, A.). Perhaps the best description is provided by BAST and ANSON (1949). The cochlear duct forms as an outgrowth of the rostromedial portion of the sacculus. Although it has not been firmly established whether cell addition occurs preferentially at the saccular (basal) or apical end during cochlear duct outgrowth, RUBEN (1967), using the ^{3}H-thymidine negative-labeling method in mice, found that cells that had gone through their terminal mitosis earliest (oldest cells) were located at the apical end of the mature cochlea, while younger cells were located at progressively more basal portions. These data led RUBEN to hypothesize that "the growth area might be at the junction of the cochlear duct and the primitive saccule" (p. 39). To the knowledge of the present author, these results have not been replicated in other species, nor has the hypothesis been adequately tested by

[2] Phylogeny of the auditory receptor is not discussed in this chapter. The reader is referred to the excellent discussion by WEVER (1976).

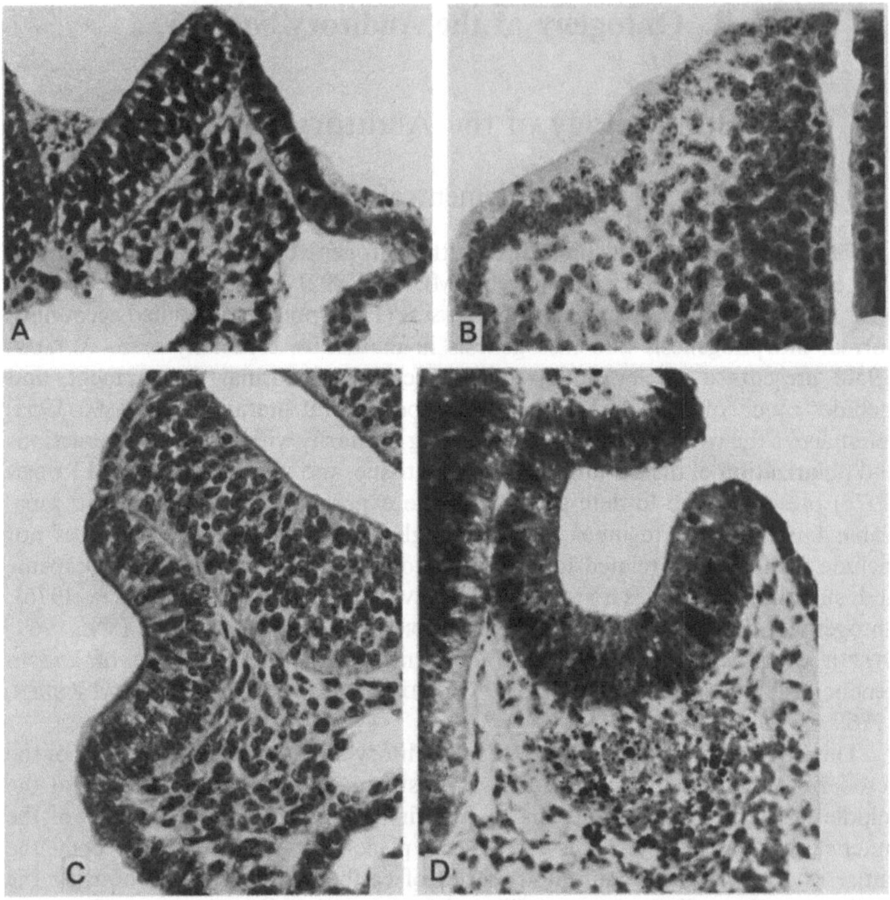

Fig. 1A–D. Taken from O'RAHILLY (1963) this figure shows the early stages of otocyst formation in the human embryo. Numbers at the lower right of each micrograph indicate embryonic stage. Magnification of all photomicrographs is about ×250. (A) Two-somite (about 20th post-ovulatory day) embryo. Section through right rhombencephalon. Otic region seen as lateral (right side) continuation of rhombencephalon. Mesenchymal tissue lies between otic region and neural ectoderm; (B) 10-somite (about 22-day) embryo. Section through left neural tube. Cell bodies of otic plate occupy basal position, no thickening of plate has occurred; (C) 16-somite (about 24-day) embryo. View through left neural tube and left otic region. Extensive cell division has led to thickening and partial invagination of placodal region, forming an otic pit. Note juxtaposition of otic pit and neural tube; (D) 24-somite (about 26-day) embryo, right otocyst and edge of neural tube. Continued cell division and invagination has produced an otic vesicle. Only a narrow pore extends between the cavity and surface of the embryo. Vesicle in close contact with rhombencephalon (left). Note cells below vesicle; these are probably early neuroblasts of the VIIIth nerve ganglia

investigating proliferative zones at various stages of cochlear duct outgrowth. In addition to providing additional information concerning cell addition, such investigations might shed light on the mechanics of spiral formation, e.g. cell proliferation around the duct may occur at unequal rates.

Inductive interactions involved in the early stages of otocyst development have been extensively studied in amphibians, and to a lesser extent in chick and mouse embryos. The first interaction noted in the literature involves the placodal region and the chordamesoderm, which during gastrulation comes to lie under the presumptive auditory field. Transplantation and *in vitro* experiments have firmly established that the underlying mesoderm plays an important role in differentiation of the placodal region and in the early stages of vesiculation.

Following invagination, the vesicle lies in very close apposition to the rapidly developing neural folds. At this stage, the rhombencephalon exerts a powerful influence on further development. In amphibians, the experimentally induced absence of a hindbrain causes the vesicle to remain cystic, and histogenesis is minimal. The inductive role of the rhombencephalon in mammalian otocyst differentiation is more difficult to determine experimentally, but is strongly supported by two lines of research. First, the remarkably close apposition between the developing otocyst and the neural folds is present in man and other mammalian species as well as amphibian and avian embryos (Fig. 1). Second, in several strains of mutant mice, where a primary etiology is thought to be a malformation of the rhombencephalon, the inner ear fails to form normally (VAN DE WATER and RUBEN, 1976). The next phase of investigation in this area will undoubtedly be to determine the biochemical nature of the above tissue interactions. Unfortunately, in the ontogeny of the otic labyrinth, the interactions of biochemical, mechanical, and hydrostatic factors will be unusually difficult to work out. On the other hand, the re-emergence of *in vitro* analyses, especially in conjunction with mutant strains and microsurgical manipulation, holds great promise for this area of research (see VAN DE WATER and RUBEN, 1974).

Investigations of polarization of the ear have been essentially nonexistent in the last 30 years. Earlier studies have shown that rotation or disharmonic transplantation of the placodal ectoderm in amphibians can lead to a variety of results, including dwarfed structures, reversal of normal polarity, or mirror-image reduplication of either the rostral or the caudal half, depending on the age at which the manipulation takes place or the ages of donor and host embryo. In light of the current interest in embryological gradients, it is somewhat surprising that no recent studies have utilized *in vitro* methods in combination with biochemical analyses and/or manipulations in attempts at further study of the nature of the polarization gradients. Furthermore, since the inner ear is composed of several different organs with distinct central connections, investigations of the nature of the central connections, influences on target cells, or behavior effects of manipulating the polarization are in order.

There is also a relative paucity of literature concerning the proliferation of cells in the organ of Corti. An elegant contribution by RUBEN (1967) is the only investigation of proliferation of the various cell types. Two aspects of this study merit additional discussion and future attention. First, it was found that the majority of cell types in the organ of Corti of mice all underwent their terminal mitoses on the 14th day of gestation. These included spiral ganglion cells, inner hair cells, outer hair cells, inner pillar cells, outer pillar cells, Deiter's cells, Hensen's cells, Claudius' cells, inner supporting cells, and external sulcus cells. This remarkable synchrony of structurally and functionally divergent cell types suggests to this author either a common origin or some active process regulating the synchronization of proliferation. Also of considerable interest is the finding (noted above) that the organ of Corti demonstrates an apical-to-basal proliferation gradient; that is, the first cells to go through their terminal mitosis are found near the apex, while later-dividing cells are distributed at progressively basal locations. This finding is somewhat surprising when considered in conjunction with the facts that the spiral ganglion cells show an opposite (basal-to-apical) proliferation gradient, and virtually all other aspects of cochlear and central nervous system differentiation show a basal-to-apical sequence (see below). In later investigations, RUBEN and his colleagues have utilized short-term exposure to ^{3}H-thymidine to investigate cellular kinetics in the inner ear of normal and mutant mice. These investigations led to the demonstration that the pattern of nuclear migration is similar to that found in the neural tube; synthetic activity occurs near the basement membrane and mitosis at or near the lumen. In addition, observations on the Kreisler mouse (Kr/Kr), which is known to suffer aberrant rhombencephalon development, suggest a neural influence on cellular proliferation in the inner ear. These pioneering studies by RUBEN have, therefore, raised many important questions regarding the cellular dynamics responsible for normal and abnormal cochlear development. Additional studies, focusing on the neural regulation of cell proliferation, the origins of the various cochlear cell types, and cellular migration within the cochlear duct will yield valuable information for understanding developmental anomalies of the inner ear.

2. Differentiation[3] of the Organ of Corti

Although adequate means for identifying proliferation times of unidentified cell types became available only recently, examination of tissues during development has been a chief pastime of both embryologists and comparative anatomists for centuries. Thus it is not surprising that within 10 years of the time when Corti provided the first detailed description of the organ that bears his name, KÖLLIKER (1861) published the first description of its development in man. Since that time well over 100 papers have described the differentiation of the organ of Corti in most vertebrates. What may be surprising, in view of the technical advances in fixation, microscopy, and staining over the last 100 years, is that the most detailed descriptions of the final stages of this differentiation are still those in RETZIUS's two-volume opus, *Das Gehörorgan der Wirbelthiere* (1884). In Fig. 2 one of his magnificent plates has been reprinted, showing the basal, middle, and apical turns at three stages of development in the cat. Figure 3 is included for identification of the major cell types.

Inspection of Table 1, Section D, quickly reveals the principal findings on histological differentiation of the organ of Corti. The majority of studies present an ontogenetic timetable of the normal appearance of the organ for the species in question. The fundamental principles shown by these studies are: i) Histological differentiation of all elements at a given cochlear location occurs in synchrony; ii) Differentiation occurs first in the basal turn and proceeds apically; and iii) Differentiation of inner hair cells slightly precedes that of outer hair cells.

[3] "Differentiation" in this chapter refers to an ongoing histological or morphological process, not a particular state or stage of development.

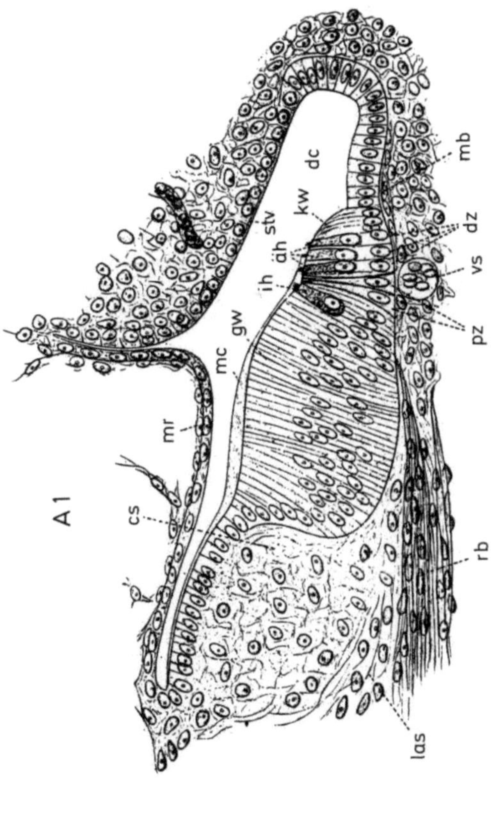

Fig. 2 A1

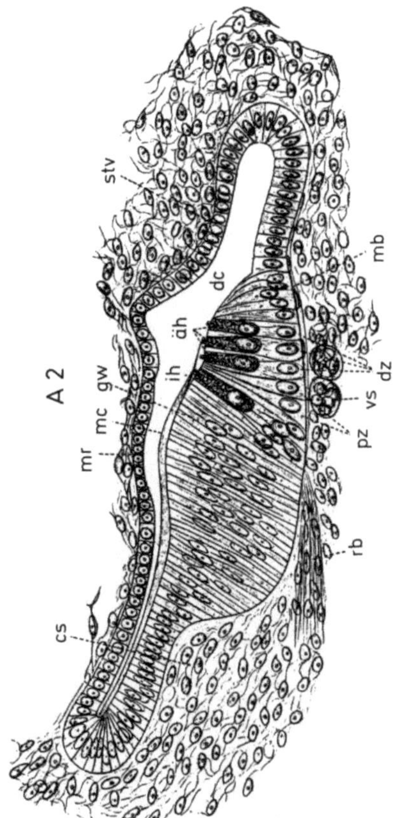

Fig. 2 A2

(Legends see p. 148)

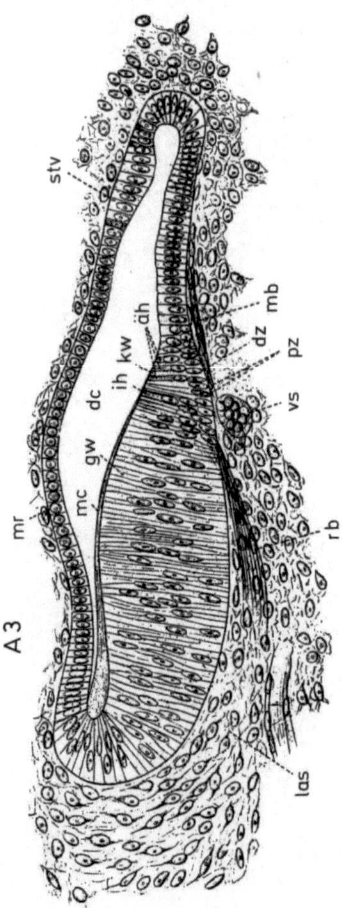

Fig. 2 A3

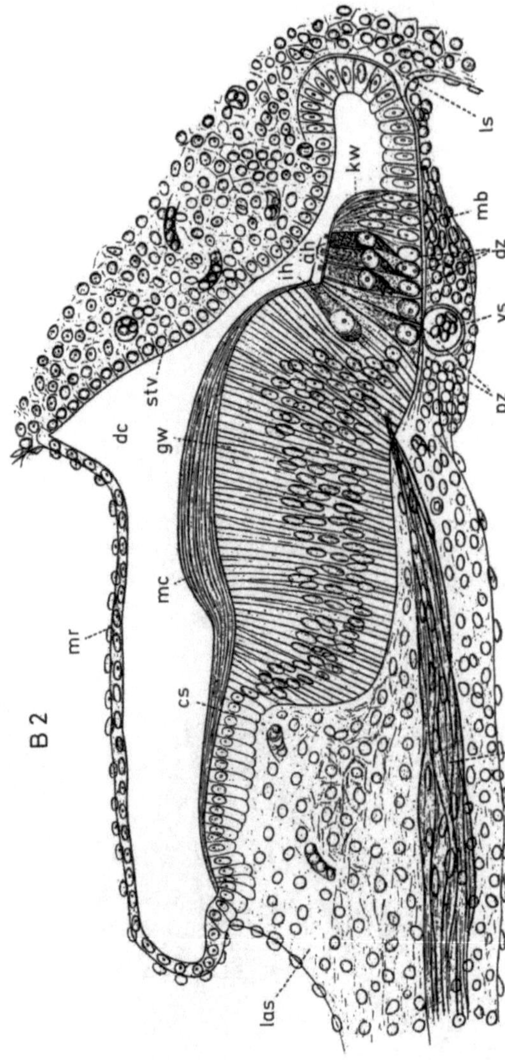

Fig. 2B2

(Legends see p. 148)

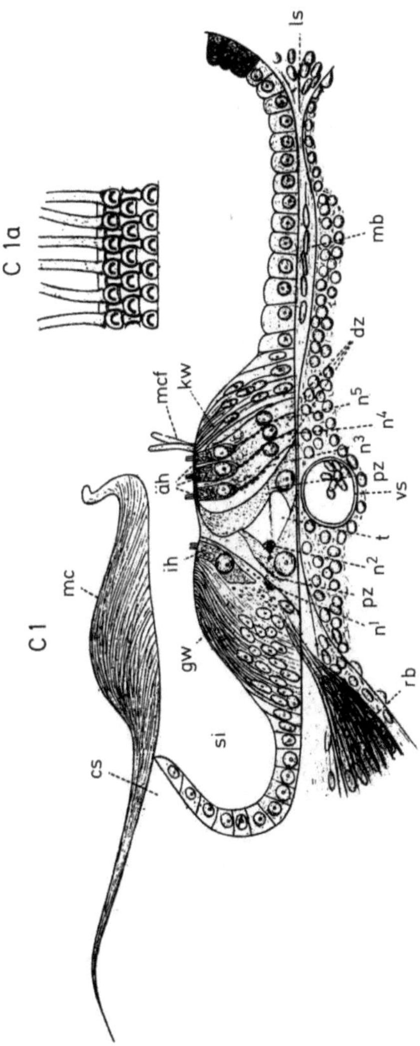

Fig. 2 C1a

Fig. 2 C1

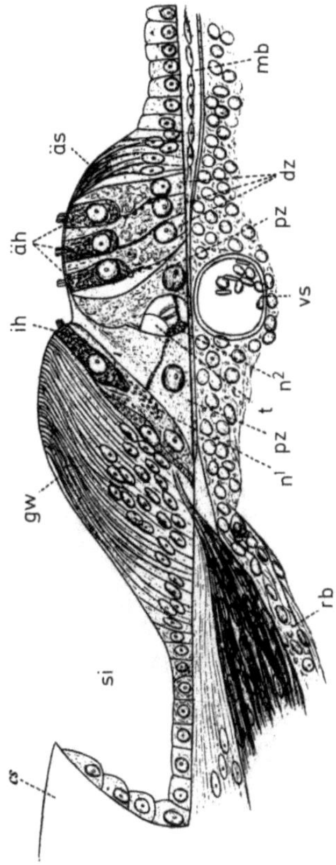

Fig. 2 C2

(Legends see p. 148)

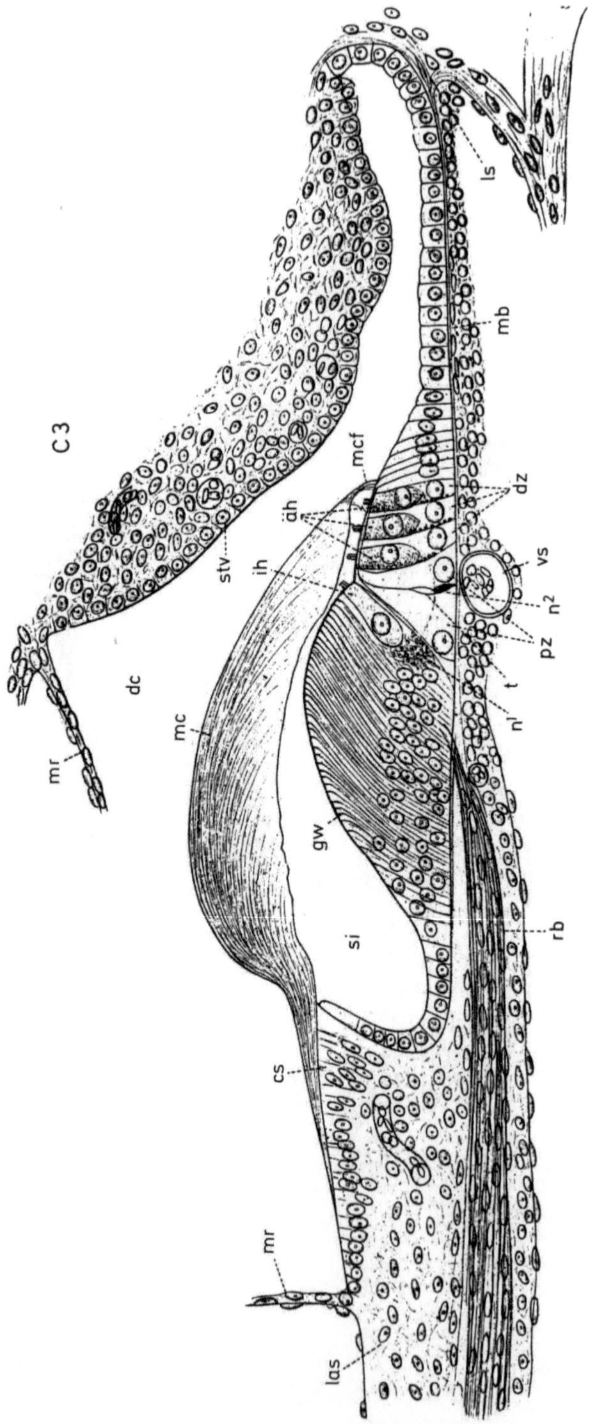

Fig. 2 C3
(Legends see p. 148)

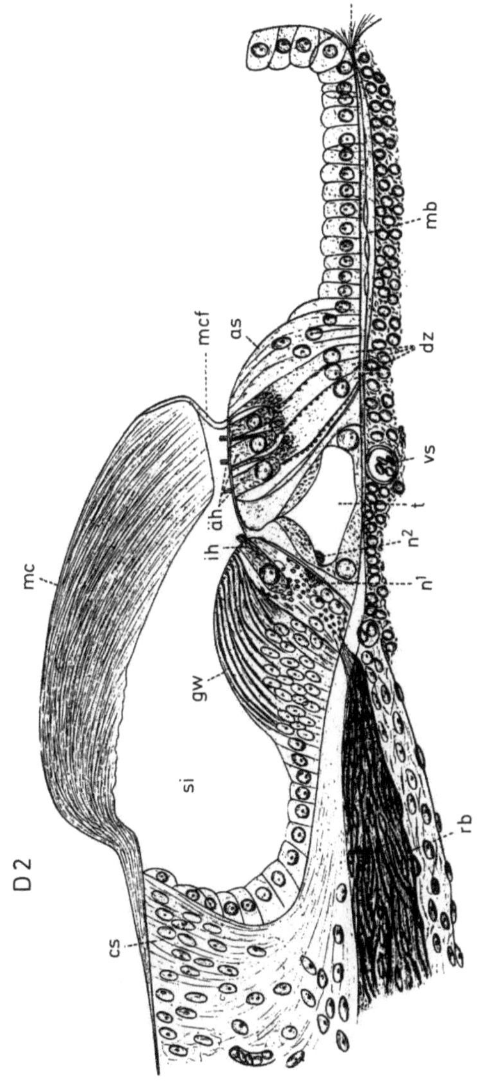

Fig. 2 D2

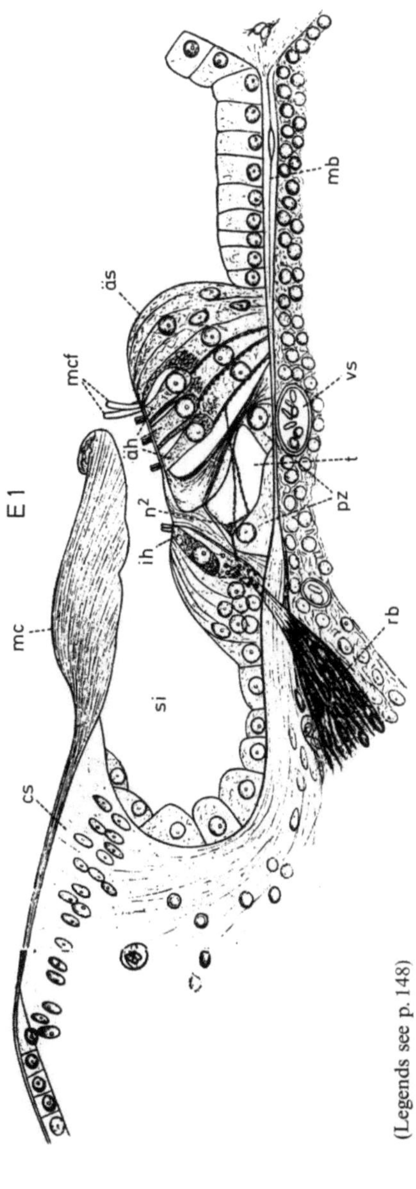

Fig. 2 E1

(Legends see p. 148)

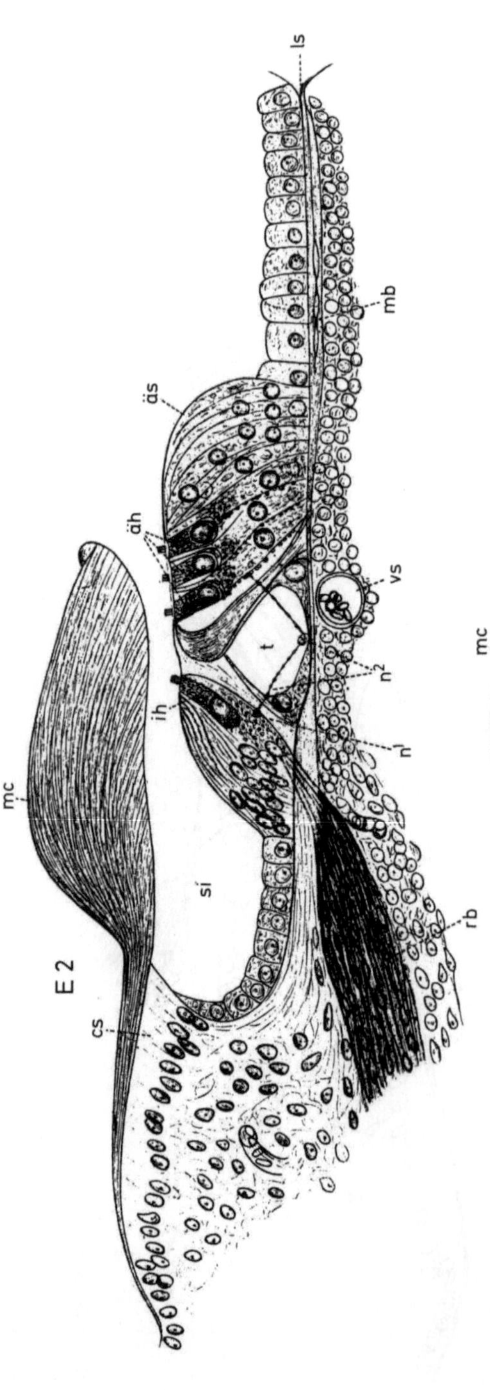

Fig. 2 E2

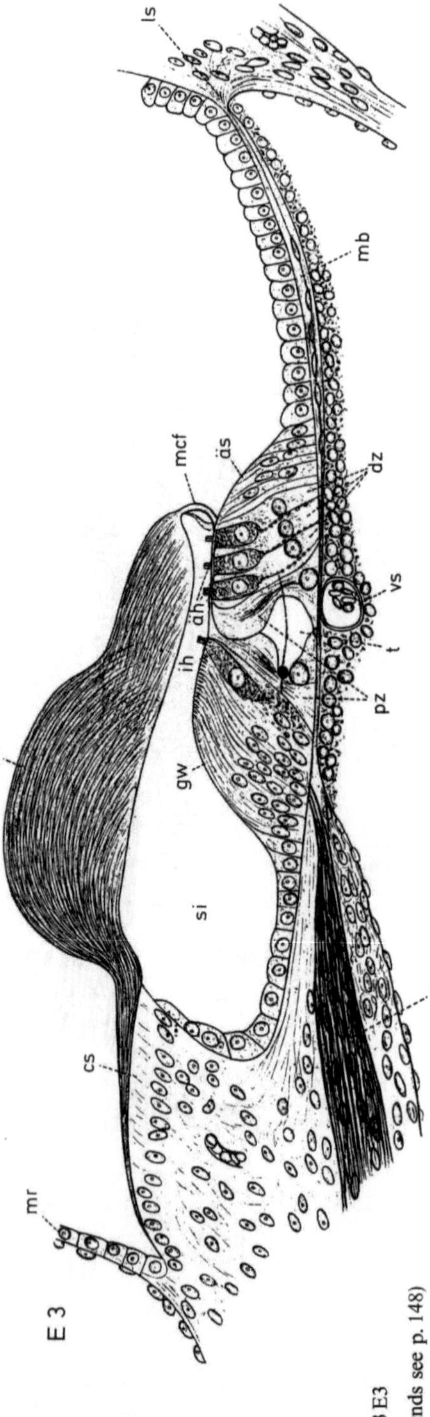

Fig. 3 E3
(Legends see p. 148)

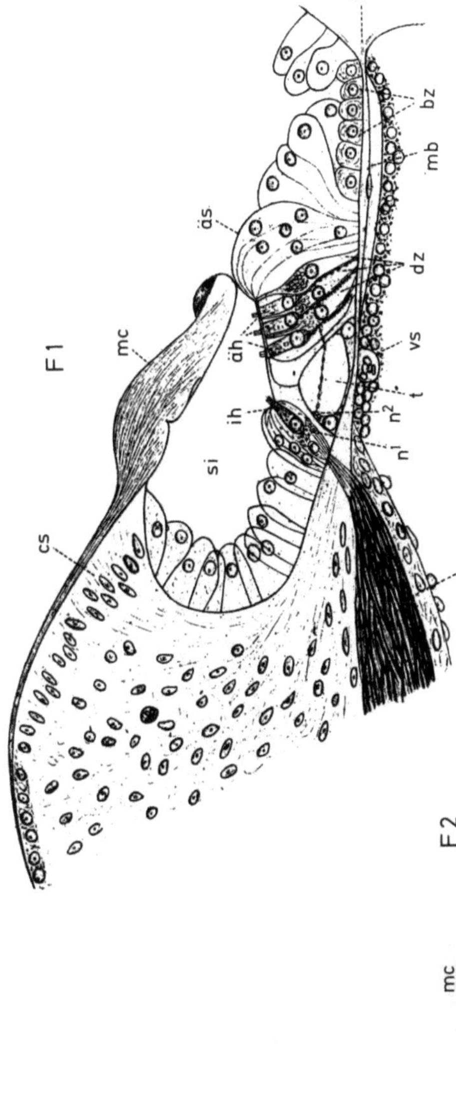

Fig. 2F1

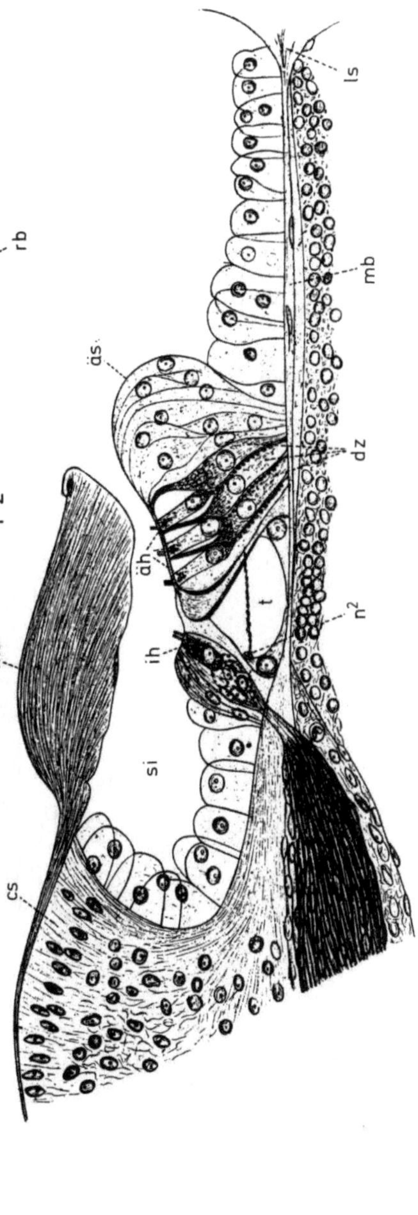

Fig. 2F2

(Legends see p. 148)

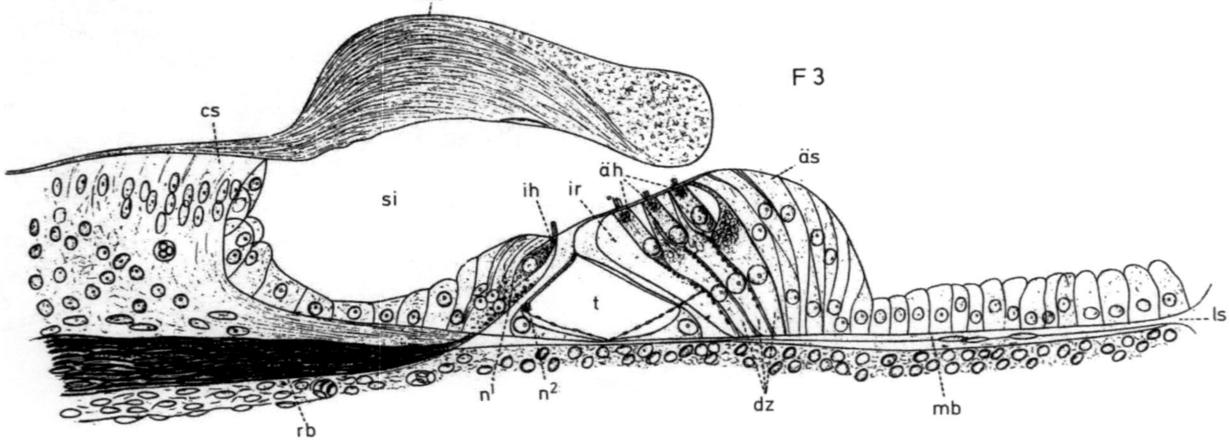

Fig. 2 F3

Fig. 2. Reproduction of lithographs (Nos. 30 and 31) from RETZIUS (1884), showing differentiation of kitten organ of Corti. Figure labeled "A" from 8-cm embryo; "B" from 12-cm embryo; "C" from newborn kitten; "D", "E", and "F" from 7-day-, 11-day-, and 30-day-old kittens, respectively. Numbers (1, 2, or 3) next to each letter (e.g. Fig. A1) refer to basal turn, middle turn, and apical turn respectively. For identification of individual structures see Fig. 3

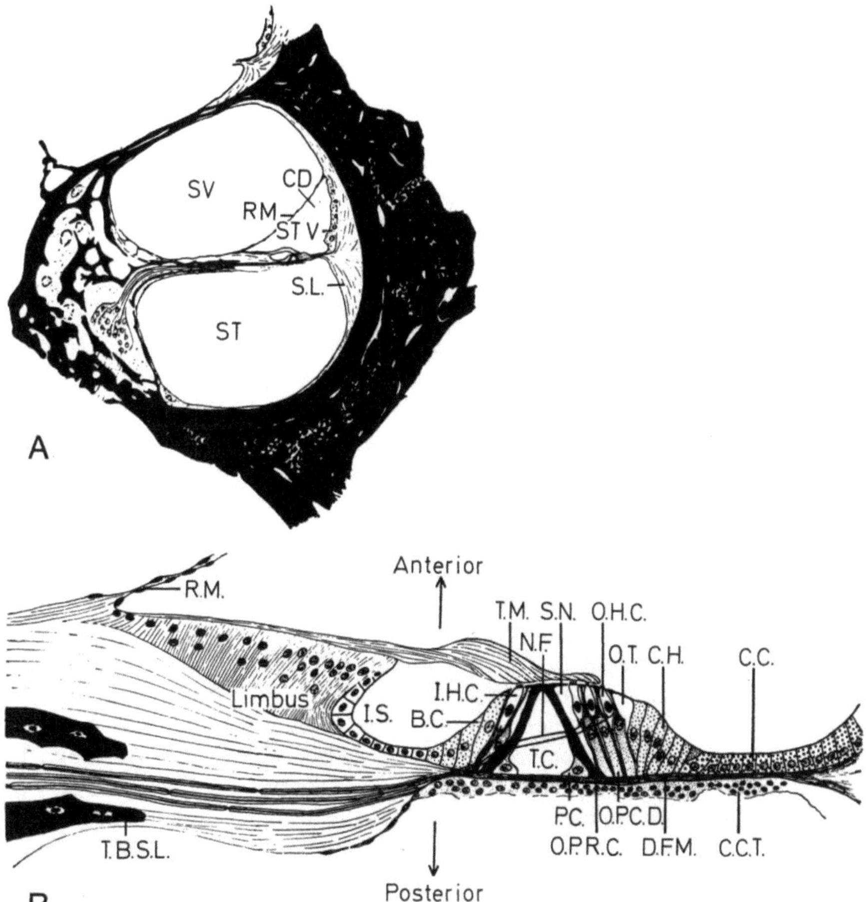

Fig. 3A and B. From BAST and ANSON (1949), Fig. 46. Drawings of horizontal section through middle turn of cochlea from 6-month-old human infant. (A) General structure of cochlea (magnification about ×16). Abbreviations: CD, cochlear duct; RM, Reissner's membrane; SL, spiral ligament; ST, scala tympani; ST V, stria vascularis; SV, scala vestibuli; (B) Detailed structure of organ of Corti, semidiagrammatic (approximately ×220). Abbreviations: B.C., border cell; C.C., cells of Claudius; C.C.T., cellular connective tissue; C.H., cells of Hensen; D.F.M., dense fibrous membrane; I.H.C., inner hair cell; I.S., inner sulcus; N.F., nerve fiber; O.H.C., outer hair cell; O.P.C.D., outer phalangeal cell of Deiters; O.P.R.C., outer pillar rod of Corti; O.T., outer tunnel; P.C., pillar cell; R.M., Reissner's membrane; S.N., space of Nuel; T.B.S.L., tip of bony spiral lamina; T.C., tunnel of Corti; T.M., tectorial membrane

The synchronous development of cellular and morphological aspects of the organ of Corti is evident throughout this literature, but perhaps the best example is to be found in WADA (1923). In this extensive quantitative study on the ontogeny of the organ of Corti in the rat, WADA demonstrated remarkable developmental synchrony of hair cells, supporting cells, and numerous morphogenetic measures. Since WADA's work, histochemical measures on the organ of Corti and *stria vascularis* of the rat have provided additional data supporting this observation.

The finding of synchronous differentiation of functionally related cell types is not new in itself, but does suggest that either intercellular regulation of development or some additional controlling factor may be present. Furthermore, it is interesting to note that the maximal period of histological differentiation also coincides with the onset of auditory function. The possible causal relations between the structural and functional differentiation will be more fully discussed below, but it is important to point out that developmental synchronies of functionally related cell groups are common (cf. BURR, 1932; RUBEL et al., 1976). Often a relationship with functional ontogeny is suggested.

Without doubt, the most frequent conclusion of studies on differentiation of the organ of Corti is that differentiation begins in the basal portion and proceeds apically. This relationship can be clearly seen in Figure 2, and appears true for each aspect of cellular differentiation that has been investigated. Several authors (cf. BREDBERG, 1968) have noted upon close examination that this gradient is not unidirectional; the differentiation begins near the middle of the basal turn rather than at the basal pole. In any case, the unanimous observation of a developmental gradient, in conjunction with the finding that the endocochlear potential is uniform throughout the cochlea at all ages, strongly suggests that neither hormonal changes nor a change in the general ionic environment are regulating the developmental synchrony noted above.

The slight developmental precedence of inner hair cell differentiation over that of outer hair cells has been confirmed by many investigators, and in many mammalian species. This sequence is also true of innervation; inner hair cells receive innervation slightly before outer hair cells (Table 1, E.). Although the significance, if any, of this finding is unclear, several additional differences between these two populations can be related, namely: only about 10% of the nerve fibers innervating the cochlea synapse on outer hair cells, whereas 90–100% innervate the inner hair cells [note that the mammalian cochlea has three or four times as many outer hair cells as inner hair cells (BREDBERG, 1968)]; outer hair cells are more susceptible to chemical or acoustical trauma than inner hair cells; and inner hair cells are sustained longer in cochleopathic mutants and during old age than outer hair cells. It is interesting that the later-developing outer hair cells appear considerably more susceptible to insult and receive a small percentage of the total cochlear innervation. When the functional differences between inner and outer hair cells become firmly established, it may be possible to sort out these relationships.

The application of technical developments in histochemistry and electron microscopy to the problems of cochlear differentiation have primarily served to reaffirm the conclusions stated above. Oxidative metabolism, as expected, follows the sequence, and is temporally coincident with differentiation. Until more is known regarding the synaptic chemistry of the adult cochlea, little further progress regarding its developmental regulation is possible. Electron microscopic investigations, primarily by HILDING (1969) and his colleagues, have yielded information on the time at which efferent and afferent terminals are first observed. In addition, KIKUCHI and HILDING (1965, 1966) provided the important observation that hair cells, supporting cells, and possibly cells of the *stria vascularis* contain a single kinocilium and numerous cilia in the newborn mouse. Around the time of onset of cochlear function, supporting cells lose these structural elements and hair cells lose the kinocilium, but retain the basal body. This observation with regard to hair cells reaffirms their essential structural, and probably functional, similarities to other mechanoreceptors.

The most promising development for advancement in our understanding of the mechanisms involved in cochlear differentiaton is the resurgence of research on cochlear development *in vitro*. Since the classic studies of FELL (1928–29) it has been well known that the otocyst will show relatively complete morphological and histological differentiation *in vitro*. Several laboratories are now utilizing *in vitro* preparations of the chick or mouse cochlea to investigate normal and pathological cochlear development (FRIEDMANN and BIRD, 1961, 1967; ORR, 1968, 1975; VAN DE WATER and RUBEN, 1974; SOBKOWICZ et al., 1975; ROSE et al., 1977). In addition to demonstrating that most of the ontogenetic relationships described above occur in culture, these investigators have begun to unravel the influence of intercellular environmental factors on cochlear differentiation. Most important for the present purpose is the repeated demonstration, in both mutant mouse and normal chick cultures, that the presence of nerve fibers from the ganglion cells is necessary for complete differentiation of the hair cells. In areas where nerve fibers are absent, differentiation of the organ of Corti is arrested at an immature stage. In the future it should be possible to utilize *in vitro* preparations to determine the mechanisms underlying such cellular interactions and temporal gradients in the organ of Corti.

3. Origin of Cochlear Ganglion Cells

The origin of cochlear ganglion cells has been the subject of much debate during the 20th century. From examination of histological sections and organ dissection, most classical anatomists were of the opinion that the acousticovestibular and geniculate ganglia were initially fused, and of neural crest origin. After migration from the neural tube, further proliferation and then eventual subdividing of the separate ganglionic masses were thought to occur (ADELMAN, 1925). STREETER (1906a, 1912) is usually credited with initiating the view that the VIIth and VIIIth nerve ganglion cells are not derived from a common primordium, the geniculate ganglion being derived principally from neural crest and the acousticovestibular ganglion cells arising from the ectoderm of the otic placode. Largely through the influence of ADELMAN (1925), the former view found its way into most textbooks, while during the same period noted embryologists were proving it fallacious in amphibians. Thus a large number of transplantation and extirpation studies (see Table 1, E), culminating in the Nile blue marking experiments of YNTEMA (1937), conclusively showed that in amphibians, the acousticovestibular ganglion cells are derived from the medial wall of the otic vesicle. As these cells split off from the vesicular pouch, they come in close apposition to the crest cells, which will later form the geniculate (VIIth) ganglionic mass; hence the earlier confusion. Therefore the problem appeared resolved, at least for amphibians, and subsequent detailed examination of normal tissue in developing chick (PROCTOR and LAWRENCE, 1959), cat (HALLEY, 1955), sheep (BATTEN, 1958), and human embryos (POLITZER, 1956) all confirm that cochlear and vestibular ganglion cells are of placodal origin. As each of these authors shows an elegant series of micrographs of presumptive ganglion cells budding off from the developing otocyst (Figs. 4 and 5), general resolution of the problem was apparent.

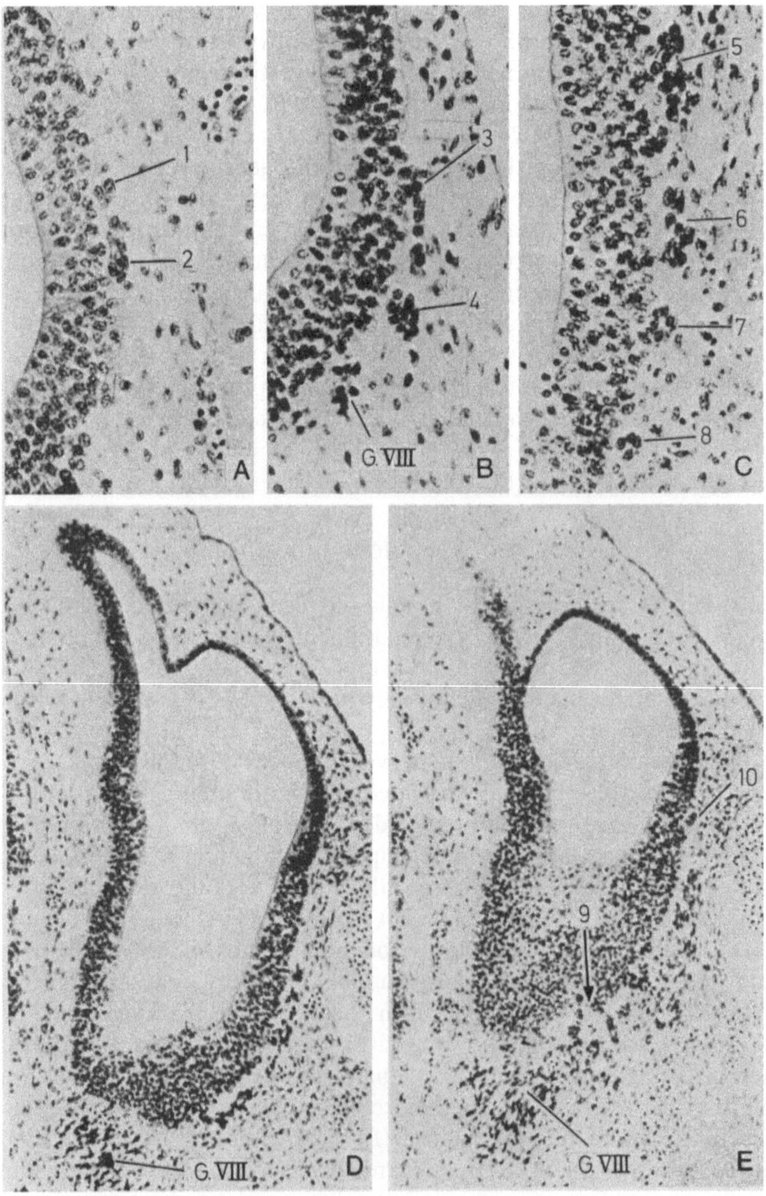

Recently, the accepted view of placodal origin has been challenged on two grounds. On the basis of electron-micrographic evidence that nerve fibers first invade the region of the hair cells at 6—8 days in the chick embryo (*in vivo* and *in vitro*), FRIEDMANN (1969) claims that the ganglion cells are not of placodal origin; otherwise fibers would be apparent at the earlier stage suggested by PROCTOR and his colleagues. In addition to the obvious weakness of this negative argument, FRIEDMANN presents no evidence that his cultures were devoid of ganglion cells, and he clearly states that the afferent fibers are critical for hair cell differentiation. The second, and equally suspect, challenge to the view of placodal origin arises from an extensive series of investigations of cochlear differentiation in mutant mice by DEOL (1967, 1970). Examination of the inner ears of mutants known to suffer from abnormal neural crest development revealed gross malformations of the organ of Corti. From these results, DEOL reasoned that the cochlear ganglion cells were at least partially of neural crest origin, in spite of the fact that the ganglion cell degeneration was less severe and appeared secondary to that of the sensory epithelium. The primary etiology of the mutations is not known, and it is well known that the differentiation of neural crest derivatives is highly dependent on their migratory and postmigratory environment (WESTON, 1971). Thus it is highly possible that the neural crest abnormalities in these animals are secondary in nature. Furthermore, the fact that the ganglion cells themselves were largely spared in many of these animals casts serious doubt on DEOL's conclusion. At present, therefore, the present author feels that there is little or no evidence supporting a neural crest contribution to the neurons of the cochlear ganglion of any vertebrate. However, in view of the evidence suggesting two types of ganglion cells (SPOENDLIN, 1966, 1970, 1974) and the previous suggestion of dual placodal and crest origin of the trigeminal ganglion (HAMBURGER, 1961), this possibility must not be excluded until the critical experiments have been done. Such experiments, which would involve transplantation of otocysts from ^{3}H-thymidine labeled donors to unlabeled host embryos prior to the stage of neural crest migration, are now quite feasibile in avian embryos and would presumably lay this issue to rest.

◁ Fig. 4A—E. VIIIth nerve ganglion cells budding off right otocyst of sheep embryos. From BATTEN (1958). (A) Transverse section of 7.2-mm sheep embryo showing two early placodal buds (1 and 2) on medial wall of otic vesicle. Both buds appear to be sliding ventrally between parent epithelium and basement membrane. × approx. 270; (B) Three placodal spurs associated with rostromedial wall of otic vesicle in 8.0-mm sheep embryo. Spur 3 still enclosed by tubular evagination of basement membrane, but spur 4 completely detached from epithelium. Immediately below lies diffuse tip of mass of detached placodal cells that constitute rudiment of acoustic ganglion (G. VIII). × approx. 270; (C) Transverse section of 8.2-mm sheep embryo showing lateral wall of otic vesicle with four groups of placodal cells apparently migrating towards acoustic ganglion. Upper two spurs (5 and 6) already free of otic epithelium. Of lower spurs, 7 consists of cluster of cells probably about to detach, and 8 is small free spur attached to otic epithelium in previous section. × approx. 270; (D) Coronal section through cranial third of otic vesicle in 8.2-mm sheep embryo. Note groups of small deeply staining placodal cells apparently migrating over lateral face of vesicle towards rudiment of acoustic ganglion (G. VIII). × approx. 120; (E) Coronal section through same 8.2-mm embryo as Fig. 4D but passing through cranial edge of otic vesicle. Lateral groups of placodal cells have almost reached acoustic ganglion and make contact with it in next section. Arrow (9) indicates two spurs of placodal cells presumed to be streaming into ganglion. Small bud of placodal cells probably about to detach from lateral wall of vesicle is visible at 10. × approx. 120

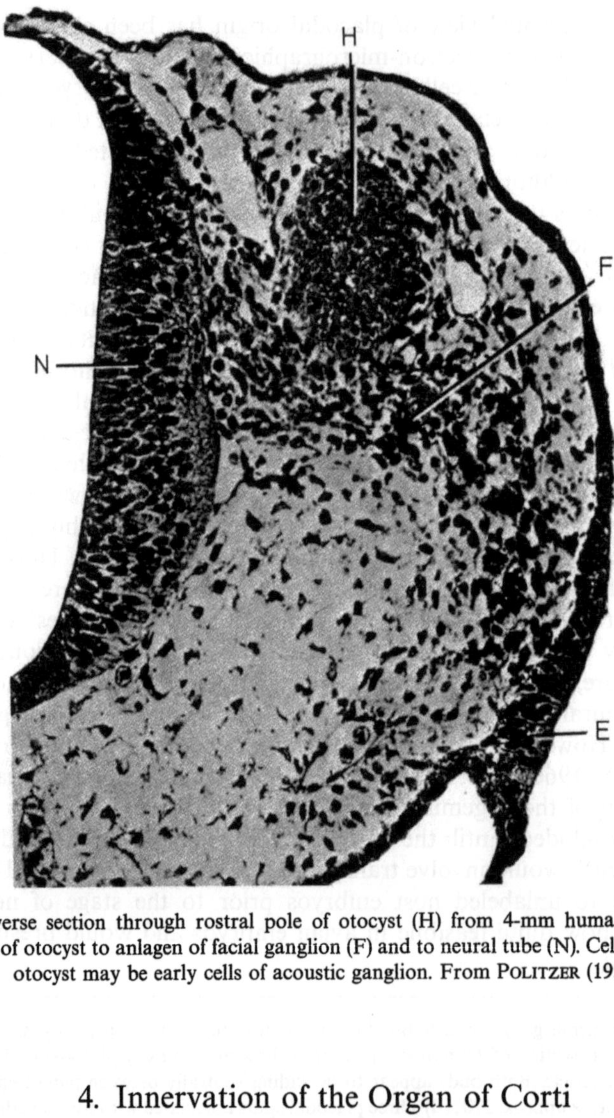

Fig. 5. Transverse section through rostral pole of otocyst (H) from 4-mm human embryo. Shows juxtaposition of otocyst to anlagen of facial ganglion (F) and to neural tube (N). Cells ventromedial to otocyst may be early cells of acoustic ganglion. From POLITZER (1956)

4. Innervation of the Organ of Corti

With osmium-fixed and reduced silver preparations, it has been repeatedly demonstrated that nerve fibers (presumably distal processes of the ganglion cells) are present in the developing cochlea at remarkably early stages. In the human, for example, fibers entering the vesicle have been reported as early as $4^1/_2$ weeks of gestation. In cat, nerve fibers are apparent in the developing organ of Corti at 30 days' gestation (approximately midterm).

In both mice and chickens, where the relevant data are available, the presence of nerve fibers has been reported at times prior to the period of terminal mitosis of ganglion cells. Thus it is entirely plausible that the ganglion cells spin out their peripheral process as they migrate from the lumen of the otic vesicle, such as has been demonstrated for the axons of cerebellar granular cells. In contrast to the

Table 1. Development of the auditory portion of the inner ear

Subject of study	Species (age)	Method(s)	Comments	Relevant references
A. Formation of vesicle and growth of inner ear				
Normal growth	Human (6–16 weeks)[a]	Light microscopy[b]	Cavitation of otic cavity involves chondrolysis by "histocyle" cells	ANDERSON et al. (1969)
	(3–20 weeks)	Light microscopy	Timetable of otocyst formation and duct division and expansion	BOETTCHER (1869), BAGINSKY (1886), STREETER (1906a), WILSON (1914), STREETER (1918), BARTELMEZ (1922), ANSON (1934), BAST and ANSON (1949), COOPER (1948), OMEROD (1960), O'RAHILLY (1963)
	Rabbit (postnatal)	Light microscopy	Canals fully grown at birth	HOYTE (1961)
	Guinea pig (embryogenesis)	Light microscopy and histochemical	Timetable of normal morphological and histochemical development	CHONDYNICKI (1968)
	Sheep and goat (to 50-day embryos)	Light microscopy	Timetable of normal morphogenesis	TITOVA (1970)
	Mice, CBA-J/CBA-J (11-day embryo-adult)	Light microscopy	Cochlear duct first seen on 12th day; $1\frac{3}{4}$ turns seen by 17–18 days	SHER (1971)
	Hamster (embryonic day 1-postnatal day 20)	Light microscopy	Timetable of normal inner and middle inner ear development	STEPHENS (1972)
	Chick (embryogenesis)	Light microscopy	Timetable of normal development	KNOWLTON (1967)
	Frog (metamorphosis)	Light microscopy	Changes during metamorphosis described briefly	WITSCHI (1949)
Ear induction: Mesodermal	*Amblystoma, Rana, Triturus*, chick, *Discoglossus, Triton*	Heterotopic transplants and excision, hybrids	Mesodermal involvement in early phases of vesicle formation	SZEPSENWOL (1933), HARRISON (1933, 1938, 1945), ALBAUM and NESTLER (1937), PASTEELS (1939), ZWILLING (1941), KAWAKAMI (1943a, b, 1949), KOHAN (1944), LEVI-MONTALCINI (1946), MOORE (1946)

Table 1 (continued)

Subject of study	Species (age)	Method(s)	Comments	Relevant references
Ear induction: Neural	*Amblystoma, Triton, Rana, Bombinator, Bufo, Alcipenser,* chicken, rabbit, mouse	Heterotopic transplants, heterostatic transplants, excision, mutants, *in vitro*, X-rays	Neural involvement in ear formation, usually following vesicle invagination	STONE (1931), GUARESCHI (1935), HOLTFRETER (1935), BONNEVIE (1936), ALBAUM and NESTLER (1937), WADDINGTON (1937), SCHMIDT (1938), WATERMAN (1938), WOERDEMANN (1938), KOHAN (1939), YNTEMA (1939, 1950), TRAMPUSCH (1941), HERTWIG (1942), GINSBURG and DETLAFF (1944), BARBASETTI (1948), DETWILER (1948, 1949), DETWILER and VAN DYKE (1950, 1951), GARCIA (1961), DEOL (1966), VAN DER BORDEN (1967), FRIEDMANN (1969), RUBEN (1973), VAN DE WATER and RUBEN (1974)
Regulation of size	Chick, *Amblystoma, Rana*	Extirpation and heteroplastic transplants	Induction of heterotopic ectoderm or heteroplastic environmental factors can regulate ear size	WADDINGTON (1937), RICHARDSON (1932), ZWILLING (1941), YNTEMA (1955)
B. Polarization of the ear	*Amblystoma, Bufo*	Rotation and rotated transplants—homeostatic and heterostatic	Anterior-posterior (AP) axis of otocysts becomes gradually fixed near the end of neurulation. The dorsal-ventral axis is fixed later and more abruptly. At time of fixation, reduplication can occur, around AP axis. Rotations after polarization may realign	STREETER (1907, 1914), SPEMANN (1910), OGAWA (1921. 1926), TOKURA (1924, 1925), HARRISON (1924, 1936a, b, 1945), CHOI (1931), HALL (1937, 1939, 1941), YNTEMA (1939, 1948, 1955), TRAMPUSCH (1941)
C. Proliferation of sensory and supporting cells of the Organ of Corti	Monkey—*Macaca* (25–39 days)	^{3}H-thymidine—short term	Cell division most intense in otocyst at 28–29 days, but continues in cochlear duct through 39th day	WILSON et al., (1975)

	Mice—CBA/J (embryonic-7 days postnatal); mutant mouse (11 days)	^{3}H-thymidine—short- and long-term	Labelings of hair cells, Deiter's cells, Hensen's cells, ganglion cells and pillar cells on embryonic days 12–15; peak proliferation of all elements on day 14. Apical-to-basal proliferation gradient of cells of Organ of Corti; basal-to-apical gradient of ganglion cell proliferation. Cell cycle times	RUBEN (1967, 1973), RUBEN et al., (1971)
D. Differentiation of Organ of Corti	Human (10–25 weeks)	Light microscopy of histological sections, EM, histochemistry	Timetables of histogenesis; basal-to-apical differentiation of inner hair cells prior to outer hair cells; Organ of Corti adult-like by 26 weeks; histochemical and cytochemical changes through 1 mo. postnatal	VAN DER STRICHT (1918, 1919, 1920a, b), ALEXANDER (1926), HELD (1926), KOLMER (1927), BAST and ANSON (1949), OMEROD (1960), WERSALL and FLOCH (1967), NAKAI (1970), MORGAN et al. (1971), YOKOH (1974)
	Human (12–27 weeks)	Surface preparations	Earliest differentiation in middle of basal turn and proceeds in both directions; inner hair cells recognized prior to outer hair cells; quantification of hair cells at 3–6 mo	BREDBERG et al. (1965), BREDBERG (1967, 1968)
	Sheep and goat (20–50 days)	Light microscopy	Timetable of normal development; basal-to-apical differentiation	TITOVA (1970)
	Cat and dog (embryonic day 40-6 weeks postnatal)	Dissections (Retzius); light microscopy, scanning EM, and deaf mutant	Timetable of normal development; presence of kinocilia in newborn hair cells which degenerate later	RETZIUS (1884), PUJOL and MARTY (1968, 1970), PUJOL and HILDING (1973), BOSHER and HALLPIKE (1965), LINDEMAN et al. (1971), MAIR (1973)
	Guinea pig (12–120 mm)	Light microscopy, histochemical, EM, surface preparations	Timetable of histogenesis and changes in PAS reaction, glycogen, RNA, DNA, Golgi bodies, lipids, alkaline phosphatase, acid phosphatase, succinate dehydrogenase, lactic dehydrogenase; basal-to-apical hair cell differentiation	VAN DER STRICHT (1918), CHONDYNICKI (1968), NAKAI and HILDING (1968), HILDING (1969), PUJOL and HILDING (1973)

Tabelle 1 (continued)

Subject of study	Species (age)	Method(s)	Comments	Relevant references
	Rabbit (embryo-36 days postnatal)	Light microscopy, surface preparations, EM, organ culture	Timetable of histogenesis; basal-to-apical differentiation; inner hair cells precede outer hair cells; adult-like at 18 days postnatal	BAGINSKY (1886, 1890), RETZIUS (1884), HELD (1909), WATERMAN (1938), ÄNGGARD (1965), NAKAI and HILDING (1968), HILDING (1969)
	Rat (embryonic to 45 months postnatal)	Light microscopy, surface preparations, histochemical	Timetable of normal development, especially postnatal; quantitative changes in the cellular elements; basal-to-apical differentiation; inner hair cells precede outer hair cells; Na-K-activated ATPase activity increases rapidly during days 12–18; adult-like by 20–50 days postnatal	LANE (1917), VAN DER STRICHT (1918), WADA (1923), LORENTE DE NÓ (1926), BELANGER (1956), FALBE-HANSEN (1967), KUIJPERS (1974)
	Mouse (embryonic day 11-postnatal day 20)	Light microscopy, EM, ^{3}H-nucleic acid, autoradiography	Timetable of normal development; basal-to-apical sequence may begin in middle of basal turn; inner hair cells differentiate earlier than outer hair cells; kinocilia in hair cells at birth disappear by 10 days; decreased RNA production during histogenesis; adult-like by 13–14 days postnatal	TELLO (1931–1932), LORENTE DE NÓ (1933a), OTIS and BRENT (1954), WEIBEL (1957), RAMON Y CAJAL (1960), MIKAELIAN and RUBEN (1965), KIKUCHI and HILDING (1965, 1966), ALFORD and RUBEN (1963), HILDING (1969), RUBEN (1969), SHER (1971)
	Mouse	Organ culture, mutants, irradiation	Histogenesis of sensory epithelium depends on association with rhombencephalon and on innervation by ganglion cells	GRUNEBERG et al. (1940), HERTWIG (1942, 1944), DEOL (1956, 1964a, b, 1968), MIKAELIAN and RUBEN (1964), KIKUCHI and HILNIDG (1967), VAN DER BORDEN (1967), VAN DE WATER and RUBEN (1971, 1974), VAN DE WATER et al. (1973), RUBEN (1973)
	Hamster (embryonic to 20 days postnatal)	Light microscopy	Timetable of normal development, basal-to-apical sequence, adult-like at about 20 days postnatal	STEPHENS (1972), PUJOL et al. (1975)

	Opossum (48–77 days postnatal)	Light microscopy	Middle of basal turn differentiates first; progresses in both directions	Larsell et al. (1935, 1944)
	Chicken (embryogenesis)	Light microscopy, organ culture (light and EM)	Timetables of normal development *in vivo* and *in vitro*; basal-to-apical sequence of differentiation; role of innervation in hair cell differentiation; adult-like at 13–15 days; intracellular junctions during development of sensory elements	Hoadley (1924), Evans (1943), Fell (1928, 1929), Friedmann (1956, 1959, 1969), Proctor and Lawrence (1959), Friedmann and Bird (1961, 1967), McAlpine and Friedmann (1963), Vanzulli and Garcia-Austt (1963), Rebollo and Casas (1964), Vazquez-Nin and Sotelo (1966), Knowlton (1967), Orr (1968, 1975)
	Frog (9 mm to metamorphosis)	Light microscopy	Development of papilla amphibiorium, lagena, and papilla basilaris	Larsell (1934)
E. Origin of ganglion cells				
Proliferation	Mouse	^{3}H-thymidine	Basal-to-apical gradient of proliferation; peak proliferation on day 14	Ruben (1967)
Origin	Human	Light microscopy	Divided opinion on placodal vs. neural crest origin; cells recognizable at 2.8–4 mm stage	Streeter (1906a, 1912), Bartelmez (1922), Cooper (1948), Politzer (1956)
	Rabbit, pig, sheep, cat	Light microscopy	Acoustic ganglia arise exclusively from placodal cells, splitting off from ventromedial wall of vesicle	van Campenhout (1935b), Groth (1939), Halley (1955), Batten (1958)
	Guinea pig	Light microscopy, histochemical sections	Placodal origin, cells migrate from rostral wall of vesicle	Chondynicki (1968)
	Rat	Light microscopy	VII and VIII ganglia initially fused and of neural crest origin	Adelman (1925)
	Mouse	Light microscopy, postnatal mutants	Correlated malformations of neural crest and organ of Corti; support crest contribution to ganglion	Deol (1967, 1970)

Table 1 (continued)

Subject of study	Species (age)	Method(s)	Comments	Relevant references
	Chicken	Light microscopy, organ culture, EM, transplantation, extirpation	Conflicting evidence regarding placodal or crest origin. Histological evidence suggests origin from ventromedial wall of otic cup at approximately 2 days. *In vitro* studies suggest crest origin	FELL (1928–1929), WADDINGTON (1937), PROCTOR and LAWRENCE (1959), REBOLLO and RODRIGUEZ (1964), PROCTOR and PROCTOR (1967), KNOWLTON (1967), VÁZQUEZ-NIN and SOTELO (1968), FRIEDMANN (1969), PARKS and ROBERTSON (1976)
	Amphibian	Light microscopy, extirpation, transplantations	Evidence strongly supports placodal origin	STREETER (1906b, 1907, 1909), BRACHET (1907), STONE (1922), KOSTIR (1924), TOKURA (1925), KNOUFF (1927, 1935), NIESSING (1932), RICHARDSON (1932), VAN CAMPENHOUT (1935a), YNTEMA (1937, 1939), KAAN (1938), ZWILLING (1941), WITSCHI (1949)
	Reptiles	Extirpation	Placodal origin assumed since extirpation of otic capsule eliminated ganglion cells	TOERIEN (1965)
	Fish	Light microscopy	Close proximity of otic cup and ganglion cells supports placodal origin	LANDACRE (1910), VEIT (1924)
F. Innervation of hair cells	Human	EM, light microscopy	At $4^1/_2$ weeks fibers appear to enter vesicle from ganglionic subdivision; at 6 weeks increased fiber outgrowth from ganglion cells; at $7^1/_2$ weeks fibers from ganglion enter epithelium; basal-to-apical sequence of innervation; at 4 mo nerve fibers present in all turns; inner hair cells innervated before outer hair cells	STREETER (1906a), COOPER (1948), BREDBERG (1968), NAKAI (1970), YOKOH (1971, 1974)

Cat	EM, light microscopy	30-day embryo axons seen entering organ of Corti; 58-day embryo afferent and efferent terminals present; innervation of inner hair cells precedes that of outer hair cells	RETZIUS (1884), WINDLE (1932a and b, 1933), PUJOL and HILNIDG (1973)
Rabbit	EM, light microscopy, explantation	Nerve fibers present before hair cells are recognizable; in 27-day embryo, afferent endings visible; 4 days postnatal, afferent and efferent endings abundant	WATERMAN (1938), NAKAI and HILDING (1968), HILDING (1969), PUJOL and HILDING (1973)
Guinea pig	EM and light microscopy	Innervation of hair cells probably begins around embryonic days 43–45; AChE and other enzymatic changes temporally correlated, basal-to-apical sequence. Afferent innervation precedes efferent	ROSSI (1961), CHONDYNICKI (1968), HILDING (1969), PUJOL and HILDING (1973)
Rat	Light microscopy	Nerve fibers present in organ of Corti in early (7–14 mm) embryo when hair cells first identifiable; hair cells innervated in 9-day neonate; innervation temporally correlated with hair cell and ganglion cell growth	LANE (1917), WADA (1923), LORENTE DE NO (1926, 1937)
Mouse	EM and light microscopy on normal and *in vitro* tissue, AChE histochemistry	Nerve fibers present during early otocyst differentiation (12-day embryo) and may influence hair cell differentiation; inner spiral bundle, last fiber system formed; synaptic contacts not present until 5–10 days after birth; basal-to-apical innervation; innervation of inner hair cells prior to outer hair cells; efferent endings present after afferents, just prior to function; afferent innervation correlates with formation of perilymphatic spaces; different AChE endings identifiable first at 7 days postnatal	HELD (1926), TELLO (1931–1932), LORENTE DE NÓ (1933a), RAMON Y CAJAL (1960), KIKUCHI and HILDING (1965), RUBEN (1967), DILLARD (1968), HILDING (1969), SHER (1971), VAN DE WATER and RUBEN (1971), VAN DE WATER et al. (1973), PUJOL and HILDING (1973), SOBKOWICZ et al. (1975), SCOTT (1967)

Table 1 (continued)

Subject of study	Species (age)	Method(s)	Comments	Relevant references
	Hamster	EM and light microscopy	At birth, afferent and efferent nerve fibers present; 2–3 days postnatal, terminals on hair cells; basal-to-apical sequence of innervation	STEPHENS (1972), PUJOL et al. (1975)
	Chicken	EM and light microscopy	Nerve fibers probably present soon after ganglion cell migration (3–4 days); innervation time not clear (see text); but by 10 days first terminals are apparent on basal hair cells; basal-to-apical sequence	BOK (1915), TELLO (1922), PROCTOR and LAWRENCE (1959), REBOLLO and RODRIGUEZ (1964), KNOWLTON (1967), PROCTOR and PROCTOR (1967), ORR (1968), VÁZQUEZ-NIN and SOTELO (1968), FRIEDMANN (1969)

[a] Where original articles give size (crown-rump length), age in weeks was determined from AREY (1965).
[b] Light microscopy refers to examination of histological sections from normal animals, with standard stains.

early appearance of axons, however, electron microscopic examination reveals that synaptic terminals are not present until just prior to cochlear function. In humans, synapses first appear in fetuses at 6—7 months, and in cats, afferent terminals are seen at about 58 days of gestation. The obvious question posed by this contrast is whether the peripheral processes of ganglion cells play some role in cochlear differentiation prior to synaptogenesis. The classic observation that these fibers may "wander" throughout the presumptive cell types of the developing organ suggested an inductive role, which recent studies have confirmed. That is, tissue culture studies on normal avian and mammalian otocysts, as well as *in vivo* and *in vitro* studies on mutant mice, all lead to the conclusion that some aspects of hair cell cytodifferentiation (e.g. formation of cilia) do not occur in the absence of cochlear nerve fibers. Furthermore, in areas where nerve fibers are absent, the morphology of the organ of Corti remains rudimentary. In this regard it may also be important to note that innervation of the organ of Corti occurs first in the basal turn and spreads predominantly apically, as described above, and that inner hair cells are innervated prior to outer hair cells. Finally, RUBEN (1967) has reported a basal-to-apical gradient of ganglion cell proliferation in the mouse. Taken together, and in relation to the differentiation gradients apparent in the organ of Corti, these observations strongly suggest that the ganglion cells may be providing a powerful regulatory and determinative influence upon cytological and morphological differentiation of the organ of Corti.

In addition to the terminals from the cochlear ganglion, hair cells of the organ of Corti receive efferent innervation *via* the crossed and uncrossed olivocochlear bundles. Early studies that concentrated on the presence or absence of nerve fibers did not discriminate between cochlear nerve and olivocochlear fibers, and recent histochemical attempts have met with the difficulty that early in development the ganglion cells, as well as efferents, stain positively for acetylcholinesterase (e.g. DILLARD, 1968). Extensive electron-microscope (EM) studies on a variety of mammals by HILDING and his colleagues (HILDING, 1969; SUGIURA and HILDING, 1970; PUJOL and HILDING, 1973) have shown that efferent terminals are formed after cochlear nerve terminals, and become apparent only slightly (1–2 days) prior to the onset of cochlear function.

II. Development of Central Auditory Pathways

The literature summarized in Table 2 deals with the ontogeny of neurons and their processes in the auditory pathways of the central nervous system. The table is organized by levels of the neuraxis. Within each level the author has attempted to find information on proliferation and migration, development of the soma, development of axonal and dendritic processes, development of synaptic terminals, ontogeny of interneuronal transmission, and where appropriate, myelination. In some cases large bodies of literature have been intentionally omitted, where the investigations were directed at a more general topic (e.g. cortical or thalamic development) and only studies where a primary concern was that for ontogeny of the auditory pathways have been included. In other cases, omission of the classical literature was due to the difficulty in securing adequate translations. References dealing with central nervous system effects of early pathology or mutations of the inner ear have been largely omitted from this table; the reader is referred to the excellent paper by WEST and HARRISON (1973), which thoroughly summarizes the earlier work in this area, as well as presenting a thorough analysis of brainstem abnormalities in the deaf white cat. Finally, studies on the development of synaptic function which have utilized acoustic stimulation are included in Table 3 and discussed under the heading "Onset of auditory function" (Section B.III.).

The details of Table 2 are relatively self-explanatory. Thus this section only attempts to point out some of the general principles, areas of controversy, and areas in which more work seems especially desirable.

The interrelationships of functionally related neurons during ontogeny have been an area of continuing interest to developmental neurobiologists. However, relatively few studies have taken what might be called a "functional systems approach" toward studying the early ontogeny of the auditory pathways. Rather most investigations concentrate on a single nuclear region or on functionally unrelated parts of the brain (e.g. "ontogeny of the diencephalon"). Thus, the vast majority of investigations provide valuable descriptive information on the ontogeny of a particular region, but shed little light on the interrelationships between developing neurons. An examination of the body of literature cited in Table 2 reveals that for virtually every postproliferative phase of development, functional relationships between afferent fibers and their target cells have been hypothesized or conclusively demonstrated. Most modern investigations suggest that the migration of neurons to their final site is not influenced by their afferent innervation (JACOBSON, 1970). On the other hand, the investigations of LARSELL (1934) suggest that the cochlear nuclei in the frog are influenced by the incoming VIIIth nerve fibers during metamorphosis. In the opinion of this author, the matter is not completely closed. MOREST (1969a), suggests that afferent fibers could influence the final stages of nuclear migration within neurons of the medial trapezoid nucleus (see Section 3.2, below). In addition, we have recently acquired evidence that otocyst removal in the chick embryo will influence the final location of cells in *n. angularis* (PARKS and RUBEL, unpubl.). Unfortunately, current techniques for studying the position of proliferation and migratory routes of cells making up a single nuclear region do not allow adequate resolution of this problem. The development of more refined methods, such as isotope injections limited to a precise area of the neural tube, or refinements of genetic marking techniques, may alleviate this problem. With respect to other ontogenetic events, the results are much more convincing; cellular development, dendritic development, postsynaptic specializations, and myelination are probably influenced by afferent fibers and/or the electrical activity of the developing auditory system.

Probably the most pervasive result found in the literature on the ontogeny of synaptic transmission in central auditory pathways is that the primary central pathways are laid down and functional well in advance of cochlear function. This finding, first hypothesized by RAMON Y CAJAL (1960), has been repeatedly confirmed by anatomical and electrophysiological studies. On the other hand, it should not be hastily concluded that chemical or electrical communication between cells is not involved in the establishment of these pathways; essentially nothing is known about the inherent electrical activity of the auditory system prior to the onset of cochlear function. Furthermore, some observations suggest that the early phases of pathway development may not be entirely independent of intercellular regulation. For example, several authors have noted that the first fibers to enter the medulla are from ganglion cells which will innervate the basal (earliest developing) portion of the cochlea. A similar developmental gradient is observed in both first- and second-order nuclei of the chicken brainstem. That is, the rostral (basal projection) areas of both nuclei precede the caudal parts in both cell death and cell growth (RUBEL et al., 1976). It is conceivable that these parallels between gross changes in the cochlea and brainstem nuclei are coincidental, but the findings again point to a need for more detailed studies of parallel developmental processes within functionally connected cell groups.

There are several topics about which little or no information is available on the ontogeny of auditory pathways. The most important of these have been noted above, namely the origin, proliferation, and migration of cells. Investigations on the proliferation and migration of the cochlear nuclei from the rhombic lip (HIS, 1890; PIERCE, 1967) stand as the exceptions; the cell groups have been carefully studied during their apparent migratory route. In addition, the "birthdate" (time of final mitotic cycle) has been established for several nuclei in a few species. However, it is becoming increasingly probable that cellular interactions at the lumen and during migration may be critical for understanding the ontogeny of neural networks. In lieu of new methods for studying these phenomena, the convergence of short-term ^{3}H-thymidine labeling, electron microscopic and Golgi methods may yield valuable hypotheses about the nature of such interactions. For example, we have previously suggested that cells in the brainstem auditory system may establish or specify their topographic connectivity during migration (RUBEL et al., 1976), and LAUDER and BLOOM (1976) suggest that connectivity may influence neuronal proliferation. Detailed studies of labeled neurons will establish the validity of these speculations.

Another area in which little information is currently available is the time at which cellular connections are established between post-migratory neurons within the auditory system. Such contacts may take two forms. On the one hand, a neuron projecting from, for example, the inferior colliculus to the medial geniculate body (MGB) may establish contact with the MGB cell but not form an excitable synapse. In this case, we have little access to either the time of establishment or the pattern of projections with current methods (but see CROSSLAND et al., 1975). On the other hand, at the time when synaptic connections are made, it is possible to identify the afferent projection by electrophysiological methods, and the concomitant changes in both presynaptic and postsynaptic processes by Golgi impregnation and electron microscopy. No single study has combined these methods, but PYSH (1969) has reported a two-fold increase in synaptic density with a large decrease in neuronal density in the inferior colliculus of rat, at a time when the oxidative metabolism dramatically increases and cochlear function begins. In addition, MOREST (1968, 1969a) utilized Golgi impregnations to give an elegant demonstration of an intimate correspondence between the temporal development of dendrites in the medial nucleus of the

Table 2. Development of central auditory pathways[a]

Subject of study	Species	Comments	Relevant references
A. Cochlear nerve			
Myelination, growth of ganglion cells and their central axons	Human, cat, dog, rat, mouse, chicken	Myelination beginning at: human, 6 months; cat, 3–6 days before birth; dog, 3–8 days postnatal; rat, 5 days postnatal; mouse, 5 days postnatal; chicken, 12-day embryo. Myelination and axon diameter growth begin with fibers innervating basal turn and proceed toward apically innervating fibers. Ganglion cell growth and myelination correspond to nerve myelination. Full myelination prolonged	Bechterew (1885), Wada (1923), Lorente de Nó (1926), Langworthy (1933), Rodriguez (1966), Yakovlev and Lecours (1967), Dillard (1968,) Fox (1968a, b), Pujol and Marty (1970), Windle (1971), Raymond et al. (1976)
Cochlear root present in medulla	Human, cat, pig, hamster, chick	Human, 8-week embryo; cat, 11 mm; hamster, 6 days postnatal; chick, 5-day embryo	Streeter (1906a), Bok (1915), Shaner (1934), Windle and Austin (1936), Ramon y Cajal (1960), Rodriguez and Rebollo (1966), Stephens (1972), Rubel et al. (1976)
B. Rhombencephalon			
Origin and proliferation			
Examination of normal tissue	Human, pig, rat, mouse, chick	Most agree that cochlear nuclei arise from rhombic lip at a middle rostralcaudal level (*per contra*, Harmkark); site of origin of other auditory nuclei may be in ventral brainstem (cf. Cooper)	His (1890), Minot (1892), Blake (1900), Ramon y Cajal (1908a, b), Shaner (1934), Cooper (1948), Harmkark (1954)
^{3}H-thymidine	Mouse, chick	Proliferation of cochlear nuclei overlaps with period when nerve enters brain; no universal sequence regarding proliferation at primary, secondary and tertiary regions	Pierce (1967, 1973), Rubel et al. (1976), Smith and Rubel, unpubl.
Migration	Human, mouse, chick, frog	Some authors conclude that incoming acoustic fibers influence migration of cochlear nuclei; ordered "streaming" of the cells from rhombic lip is emphasized; otocyst removal has little influence on migration	His (1890), Larsell (1934), Shaner (1934), Levi-Montalcini (1949), Knowlton (1967), Milonyeni (1967), Pierce (1968), Rubel et al. (1976)

Differentiation			
Cellular	Human, pig, cat, mouse, chick, frog, salamander	Differentiation of primary and secondary nuclei appears to occur simultaneously. Final cell number influenced by afferents through regulation of cell death. Human brainstem nuclei apparent at about 6—7 weeks	RAMON Y CAJAL (1908a, b), FUSE (1913), BOK (1915), RICHARDSON (1932), LARSELL (1934), SHANER (1934), COOPER (1948), LEVI-MONTALCINI (1949), HALL (1964), RONRIGUEZ and REBOLLO (1966), KNOWLTON (1967), MILONYENI (1967), RUBEL et al. (1976), PARKS and ROBERTSON (1976)
Processes	Rabbit, rat, cat, opossum, chick	Axonal outgrowth precedes dendritic elaboration. Dendrites mature in temporal correspondence with attainment of afferent organization	BOK (1915), RAMON Y CAJAL (1960), MOREST (1968, 1969a, b), RUBEL et al. (1976)
Presynaptic terminals	Cat, rabbit opossum, chick	Reduced silver and Golgi preparations show elaboration of calciform endings by progressive alteration of thin sinuous endings, in conjunction with dendritic and perikaryon maturation. No EM	MOREST (1968, 1969a, b), SCHWARTZ (1972)
Synaptic transmission	Rabbit, cat sheep	Electrical stimulation of cochlea or nerve prior to receptor function produces transmission to cortex in rabbit or cat. In sheep, middle ear reflex elicited prior to receptor function	MARTY and THOMAS (1963), MARTY and SHERRER (1964), MARTY (1967), PUJOL and MARTY (1968), MEYERSON (1967)
Myelination	Human, cat, dog, chick	Myelination in human begins around 6 months of gestation and follows sigmoid function; suggestion (LANGWORTHY) that myelination may be regulated by neuronal activity	TILNEY and CASAMAJOR (1924), LANGWORTHY (1933), RODRIGUEZ and REBOLLO (1967) YAKOVLEV and LECOURS (1967), FOX (1968b), SCHWARTZ (1972)
C. Mesencephalon			
Origin and proliferation	Mice, chick	In mice, proliferation of neurons in inferior colliculus and nuclei of lateral lemniscus overlaps with proliferation of cochlear and olivary nuclei cells	PIERCE (1973), KARTEN, unpubl.
Differentiation			
Cellular	Rat	Cell density in inferior colliculus decreases markedly during days 1–14 postnatal	PYSH (1969)
Processes	Human, rat, rabbit, guinea pig, opossum	Dendrites in inferior colliculus present well before onset of function; dendritic development in central nucleus precedes that in superficial layers. Extracellular space decreases around onset of hearing while monoamine oxidase activity increases	COOPER (1948), WAWRZYNIAK (1966), MOREST (1969b), PYSH (1969)

Table 2 (continued)

Subject of study	Species	Comments	Relevant references
Presynaptic terminals	Rat	Synaptic profiles present at birth in the rat; two-fold increase in density of profiles between 1 and 14 days	PYSH (1969)
Synaptic transmission	Cat, rabbit	Present at birth in both species	MARTY and SCHERRER (1964), MARTY (1967), PUJOL and MARTY (1968)
Myelination	Human, cat, dog	Beginning around 6 months' gestation in human, and birth in cat and dog	TILNEY and CASAMAJOR (1924), LANGWORTHY (1933), YAKOVLEV and LECOURS (1967), FOX (1968b)
D. Diencephalon (Medial geniculate)			
Origin, proliferation, and migration	Human, rabbit, rat, mouse, chicken	Alar plate origin as part of dorsal thalamic region; follows migration of ventral thalamus. First identification in humans at about 10 weeks, anlagen recognized at $8^1/_2$ weeks. First neurons formed in dorsal thalamus; rat—days 14 and 15; hamster—days 16–18	HIS (1890, 1904), CLARK (1933), MUIRA (1933), GILBERT (1935), KUHLENBECK (1964), ROSE (1942), COOPER (1948), DEKABAN (1954), STRÖER (1956), NIIMI et al. (1962), COGGESHALL (1964), KEYSER (1972), MCALLISTER and DAS (1977)
Differentiation			
Cellular	Human	At six months, cells are small and multipolar	COOPER (1948)
Processes	Cat	Dendritic development completed between birth and end of second week; principal division differentiates prior to dorsal division	MOREST (1964)
Synaptic transmission	Dog, cat, rabbit	Synaptic transmission at birth as shown by potentials at cortex from stimulation of 8th nerve or inferior colliculus	MARTY and SCHERRER (1964), MARTY (1967), MYSLIVECEK (1968), PUJOL and MARTY (1968)
Myelination	Human	Begins during 7th month of gestation; among earliest thalamic fibers to become myelinated	LANGWORTHY (1933), YAKOVLEV and LECOURS (1967)

A. Telencephalon (Auditory cortex) Differentiation			
Cellular	Human, mouse	Decreasing neuron density and increasing size in human between 8-month fetus and 4-year-old child; especially before term. In mice, some changes during first postnatal month	CONEL (1939–1963), RABINOWICZ (1964, (1967), GYLLENSTEN et al. (1966)
Processes	Dog, human	In the dog, Layer-V pyramidal cells have apical and basal dendrites at birth; in smaller pyramidal cells of Layers-II and -III basal dendrites develop postnatally; rapid period of dendritic growth between 14 and 30 days postnatal. In human, massive dendritic expansion after 8th prenatal month	CONEL (1939–1963), RABINOWICZ (1964, 1967), FOX (1968a, b)
Presynaptic terminals	Rat	First synapses apparent on day 16 of gestation; axodendritic on "newly arrived" cells in marginal layer; axosomatic synapses apparent on day 18	KÖNIG et al. (1975)
Synaptic transmission	Cat, rabbit, rat, guinea pig, dog	At birth, evoked potentials recorded in ipsilateral and contralateral primary and se ondary auditory regions, claustrum and contralateral MGB upon stimulation of medial geniculate body; 2–3 times adult latency; following rate restricted and little augmentation	MARTY and SCHERRER (1964), HASSMANNOVA and MYSLIVECEK (1967), MARTY (1967), CHALOUPKA et al. (1968), HASSMANNOVA et al. (1968), MYSLIVECEK (1968), PUJOL and MARTY (1968)
Myelination	Human, dog, cat	Thalamocortical fiber myelination occurs gradually during postnatal period; correlates with evoked potential latency decrease	LANGWORTHY (1933), MARTY and SCHERRER (1964), YAKOVLEV and LECOURS (1967), FOX (1968a, b)

[a] This table does not include studies that utilized acoustic stimulation (see Table 3: Functional development).

trapezoid body and the arrival of afferent axons. Aside from these examples, there are virtually no detailed investigations of the ontogeny of projections onto cells of the auditory pathway. Coupled with the paucity of knowledge on neurochemistry and its development in the central auditory pathway, these omissions make it impossible to state the temporal contingencies that occur during normal auditory system development.

Two final areas that should receive increased attention are: i) the development of centrifugal projections within the brain; and ii) the ontogeny of topographical specificity. The former has received little attention in any sensory system, and probably plays a critical role in the ontogeny of mature function. The topic of specificity, on the other hand, has received a great deal of attention in the visual system (see Chapters by HIRSCH and LEVENTHAL and BATE). Since the concepts stemming from the work on the visual system could be examined and extended by considerations of the tonotopic and binaural representations in the auditory pathways, it is hoped that future research will be directed at this problem.

III. Development of Auditory Function

The functional characteristics of the auditory system can be assessed by physiological and behavioral means; each has advantages and disadvantages. The readily available microelectrode and macroelectrode methods make it possible to analyze the functional capacity of different parts of the system in such a way as to answer such questions as the following: is the system functional, i.e. does it respond to sound? what frequencies elicit responses? are the changes in relative sensitivity over age? and do changes in frequency of amplitude sensitivity relate to morphological changes in the peripheral or central auditory pathways? More recently a number of laboratories have begun to investigate the ontogeny of "coding properties" of cells within the auditory system, presumably in order to relate properties such as tuning curves, phase locking, or response profiles to the ontogeny of behavioral capacities. Unfortunately, in no case is it clear what specific properties of unit response functions are related to any behavioral capacity, and clear examples of units responding as species-specific "feature detectors" have not been established (CAPRANICA, 1977). Therefore, although studies of unit coding characteristics are of interest for understanding nervous system ontogeny, the results should not be considered indicative of an organism's ability to respond selectively to meaningful environmental acoustic differences. Behavioral investigations of the ontogeny of auditory function, on the other hand, can provide information regarding an animal's minimum perceptual abilities. However, behavioral studies rely on a set of assumptions that are often difficult to meet. These include: i) That there is no ontogenetic change in the sensitivity of the response measure or unconditioned S–R chain; ii) That the topography of the response remains stable over time; and iii) That an organism's sensory capacities are somehow tied to the motor system in such a way as to be able to generate differential responses. Since these requirements can almost never be fully assumed when dealing with a developing system, negative results from behavioral studies cannot usually be interpreted. Furthermore, because a stable response measure is difficult to find, large numbers of subjects are usually required and adequate control procedures or populations are essential. Taken together, these factors often make behavioral studies of sensory ontogenesis more difficult to undertake, and more open to criticism, than physiological investigations. Yet, in the final analysis, we are concerned with understanding the ontogeny of a system (audition) with which the animal can make meaningful interpretations regarding its environment, leading to differential behavioral outcomes. Thus, behavioral investigations using carefully chosen response measures, adequate controls, and sufficiently powerful analytical tools, are of the utmost importance.

In Table 3, studies of functional development are subdivided by species and major response measures. Although this table is primarily intended to summarize the state of knowledge on the onset of auditory function, studies of functional development which focus on more complex auditory processing have also been included. Thus, investigations of cellular response parameters, behavioral differentiation of acoustical properties, conditioning, and perception of species-typical vocalizations are also included. Earlier reviews by GOTTLIEB (1968b, 1971b) on the onset of sensory function, BRADLEY and MISTRETTA (1975) on fetal receptor development, and TEES (1976) were helpful in compiling the bibliography. As earlier, there are several areas that are not covered fully in the table. First, most of the literature on human behavioral development with respect to postnatal auditory and language function has not been included, since postnatal development of human auditory function is thoroughly reviewed by EISENBERG (1976), and there are a number of recent texts on language development. Second, the voluminous literature on human cortical evoked potentials has not been fully covered, but the important findings are believed to have been noted and the references cited provide access to the relevant literature. Third, the excellent work of RUBEN and his colleagues on the development of auditory function and its degeneration in mutant mice is not included (MIKAELIAN and RUBEN, 1964, 1965; MIKAELIAN et al., 1974; BROWN and RUBEN, 1969). As above, discussion of Table 3 will be limited to the important principles, controversies, and questions for future research.

1. Onset of Cochlear Function

One primary goal of research on the ontogeny of auditory function, whether physiological or behavioral, is to identify the relationships between structure and function. The general problem of identifying structure-function relationships is central to all life sciences; ontogeny would seem to be an unusually powerful approach, since parallel ontogenetic changes in structure and function can be investigated. Thus a large number of investigators have attempted to determine "the final morphogenetic event" leading to the establishment of cochlear function by studying cochlear morphology in parallel with either physiological or behavioral indices of functional activity. It should be noted that freeing of the tectorial membrane, development of the *stria vascularis* and associated endocochlear potentials, formation of afferent terminals, formation of efferent terminals, and other aspects of cochlear development have all, at one time or another, been cited as *the* final factor initiating cochlear function. Most investigators seem to prefer giving credit to some sort of neural innervation change, but a minority also propose biochemical changes or morphogenetic changes in hair cells or supporting elements of the organ of Corti. In point of fact, the single most common observation throughout this literature is that simultaneously with the onset of cochlear function, the inner spiral sulcus forms by the regression and/or degeneration of pseudostratified epithelium. This and other changes can be clearly seen in the reproduction of RETZIUS's illustration of cochlear development of the cat (Fig. 2). Various investigators (e.g. BAST and ANSON, 1949; ÄNGGÅRD, 1965; PUJOL and HILDING, 1973) have recognized the probable importance of this event for allowing a shearing force to be set up between the tectorial membrane and the hair cell cilia. Without this sulcus (see Fig. 3) movement of the basilar membrane would not be converted to transverse rocking movements of the organ of Corti, which presumably generate the "adequate stimulus" for hair cell ionic fluctuations.

Although formation of the inner spiral sulcus is of great importance for the establishment of cochlear function, it is probable that neither it nor any of the other events noted above is uniquely responsible for the onset of hearing. As part of his study on the ontogeny of the rat cochlea, which stands as the most thorough investigation of cochlear ontogeny to date, WADA (1923) compared the cochleas of rats that "reacted to a shrill whistle or handclap" with those of littermates that did not respond. He concluded that "the inception of hearing does not coincide with detachment of the tectorial membrane from the papilla spiralis, but with the development of *each constituent* of the papilla spiralis and and the membrane tectoria" (my emphasis) (p. 154). Over half a century later this conclusion still holds; contrary to the notion of a single developmental event that initiates functional integrity, there seems to be some sort of autoregulation, such that simultaneous growth and destruction of constituent cell types leads to a cochlea that is electrically, chemically, and mechanically capable of rudimentary function. And then, quite rapidly, the cochlea attains a maturity that cannot be distinguished from adults of the species. It is not clear whether a single causative agent exists, or each element develops independently, or whether there is a form of autoregulation that synchronizes all elements (through their oxidative metabolism or some other process). However, the last idea should not be hastily discarded. It is now known that coinciding with the ontogenetic events leading to and immediately following functional onset of the organ of Corti, there are corresponding changes in functionally related tissues, such as the maturation of Reissner's membrane, rapid maturation of the *stria vascularis* and concomitant development of the endocochlear potential, growth of spiral ganglion cells, and maturation of central nervous system elements. Not only is there synchrony in the time of occurrence of morphological and chemical changes leading to functional integrity, but the ensuing changes with respect to the *pattern of development* in the cochlea and central nervous system structures presents a similar parallel. The basal-to-apical changes in cochlear development are coincident with rostral-to-caudal changes observed in the medullary nuclei of the chick embryo (RUBEL et al., 1976); basal portions of the cochlea are represented in rostral portions of these brain stem nuclei, and apical (or distal) regions project to caudo-lateral sites (see Section 3).

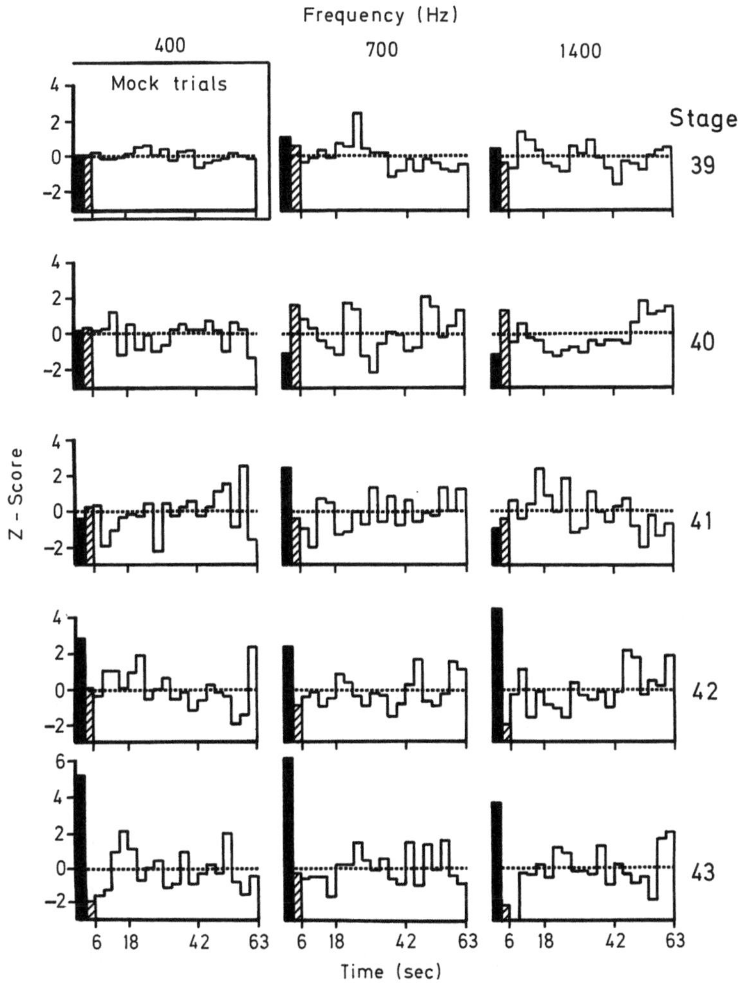

Fig. 6. Each histogram shows normalized activity scores of 12—17 chicken embryos in response to acoustic stimulation. Embryonic stage (Hamburger and Hamilton, 1951) at which subjects were tested are indicated along right margin, frequency of acoustic stimulus along top. Each embryo tested at one age with one frequency. Five trials were presented (5 min apart). In each trial, stimulus was presented (115 dB, SPL) for 3 s (black bar) and activity was electronically recorded for each 3-s period from stimulus onset for succeeding 63 s. Shaded bar indicates immediate poststimulus period. Positive Z-scores indicate increase in activity over baseline, and negative Z-scores indicate inhibition of activity. "Mock" trials are nonstimulus trials; flat histogram indicates no recording artifact. At stage 40 (14—15 days) statistically reliable inhibition of motility was found in response to 700-Hz and 1400-Hz stimulus. At stage 41 (15—16 days) reliable increase in motility was recorded to 700-Hz stimulus. At stages 42 and 43 (16—18 days) all stimuli evoked increased activity. (From Jackson and Rubel, 1978)

Any considerations of the time at which cochlear function begins must take note of the species variability. Whereas Gottlieb (1971b) paid specific attention to the overwhelming similarity in the sequence of sensory system maturation in birds and mammals (i.e. tacile-vestibular-auditory-visual), little is known regarding why particular systems become functional when they do. Obviously it can be asserted that there is some selective advantage brought to bear, or that it is merely coincident with the age and developmental state at which partition or hatching occurs. However, these "explanations" add little to our understanding of what particular selective advantage is accrued. In precocial birds, the auditory system becomes functional, as defined by electrophysiological or behavioral measures, well before hatching. For example, chick embryos will demonstrate a cochlear microphonic or brainstem evoked potential response on day 12 or 13, and as shown in Fig. 6, behavioral responses can be elicited by acoustic stimulation of the intact egg on day 14–15 (Stage 40). While it has been shown in some avian species, and implicated for others, that parent- or sibling-produced sounds have a function in regulating hatching or post-hatching behavior, such claims only appear to concern the last two or three days prior to hatching. Thus, it is not at all clear why the auditory system of the chick or duckling becomes capable of function 6–8 days before hatching, much less why the human auditory system is functional around 30 weeks of gestation. Such early function may be a mere coincidence of developmental stage at parturition. On the other hand, it is important to consider two additional possibilities. First, it is conceivable that in order for the system to be capable of appropriate functioning at the period of development where it will first be utilized for signal analysis, extrinsic input from the external or internal environment is actually being processed near the time functional activity can first be elicited in the laboratory. Second, as suggested by GOTTLIEB (1971a, b), it is also possible that functional activity intrinsic to the auditory system itself is somehow regulating its development. In this case, the fact that function can be elicited in unusual laboratory conditions may be merely an epiphenomenon.

2. Ontogenetic Changes in Stimulus Parameters

In addition to considering the time at which the auditory system first becomes functional, it is important to consider the stimulus parameters which will first elicit a response. In Table 3, these parameters (when available) are briefly noted in the fourth column. While it is clear that responses can be elicited at decreasing stimulus intensities and increasing repetition rates as the animal matures, these parameters will not be discussed. Of more interest are the ontogenetic changes in frequency parameters adequate to evoke a behavioral or physiological response.

Perusal of Table 3 indicates that in virtually every animal in which an adequate range of frequencies has been used, responses can first be elicited by low to low-middle frequency tones. Thus, in the human fetus, the tonal frequencies that first elicit changes in heart rate or motility are in the 500–1500 Hz range. This problem has been most extensively investigated in cat, where MARTY and his colleagues have systematically investigated evoked potentials and cellular responses throughout the auditory system. Thus, information is available regarding onset of responses from the cochlea and from each major nucleus along the central auditory pathways, and behavioral responses to sound. Although the evidence is somewhat conflicting, in that the first cochlear microphonic (CM) and behavioral responses are to frequencies under 1 kHz, while evoked potential and unit responses are to tones in the 0.5 to 2.0 kHz range, all studies agree that responses to frequencies above 2.5 kHz are relatively delayed in comparison with low frequency tones. Similar results (i.e. low or low-to-middle frequencies elicit responses prior to high frequencies) have been found in most mammals and birds where adequate testing has been conducted, and in every reported study the extreme high frequencies for the species in question are the last to reliably elicit a response.

It is now well known that in birds and mammals, apical portions of the basilar membrane respond maximally to low frequencies while basilar portions are most responsive to high frequencies (BEKESY, 1960; DALLOS, 1973). Therefore, if cochlear differentiation and functional development were to correspond, an apical-to-basal differentiation gradient would be predicted for the mammalian organ of Corti and avian basilar papilla. As stressed earlier, however, studies of peripheral differentiation have shown the opposite result. That is, differentiation of the hair cells, innervation of the hair cells, development of the spaces of Nuel, opening of the tunnel of Corti, formation of the inner spiral sulcus, differentiation of the *stria vascularis*, etc., all occur first in the basal or mid-basal turn of the cochlea and last at the apex. Thus there appears to be a dissociation between the ontogeny of functional responses to tonal stimuli and the differentiation of the cochlea. Several suggestions have been put forward to resolve this apparent paradox. FOSS and FLOTTORP (1974) examined the ontogeny of startle responses of cats, rabbits, dogs, and mink to pure tones of 125 Hz to 8 kHz, and consistently found that initial responses were elicited by tones under 1.0 kHz. Based on observations of the locations of the major blood vessels in the *mature* human ear, they concluded that the upper portion of the basal coil and lower portion of the second coil probably receive an adequate blood supply earliest, and that the frequencies represented at those locations are the first to elicit a response. Thus they contend that a dissociation does not exist and that functional ontogeny of the cochlea and corresponding frequency responses are governed by the developing circulatory system. This view, however, ignores the fact that high-frequency responses are the *last* to mature, while in the cochlea, differentiation of the *apical* turn is delayed. Furthermore, their own observation that little is known about the ontogeny of the cochlear vasculature, coupled with the fact that it most probably correlates with the general gradient discussed above, compels this author to accept the dissociation and consider alternative explanations.

SAUNDERS et al. (1973, 1974) observed that the ability to record high-threshold brainstem-evoked potentials to frequencies above 1.3 kHz in chick embryos corresponds to the embryo's entrance into the air space of the egg and the subsequent cavitation of the middle ear. These authors hypothesized that the mesenchymal tissue of the immature middle ear may cause an impedance mismatch leading to selective attenuation of high-frequency stimuli. Careful inspection of their data, however, reveals that cavitation of the middle ear corresponded with a 20–30 dB increase in sensitivity, which was uniform across the 0.1–3.0 kHz range that could be considered. In addition, other indications of the effects of middle ear loading (e.g. WEBSTER, 1962; FOSS and FLOTTORP, 1974) and direct measurements of impedance (ROBERTSON et al., 1968; KEITH, 1975) suggest that attenuation due to middle ear loading would be approximately uniform across the audible frequency range. On the other hand, in the one study that bypassed the middle ear by direct acoustic stimulation of the oval window in neonatal rabbits, ÄNGGÅRD (1965) reported that the earliest cochlear microphonic potentials could be elicited by stimuli in the 2–4 kHz range. Two days later CMs were observed over the entire 0.2–10 kHz range used and there was a uniform threshold reduction of 30–35 dB over the succeeding 20 days. Since no comparable data exist on the ontogeny of CMs in rabbits with intact middle ears, it is not possible to draw conclusions about differential middle ear attenuation during ontogeny. It will be important, however, for future investigators to compare response functions obtained by tonal stimulation of the intact ear with those obtained by direct stimulation of the oval window.

A second explanation of the apparent dissociation between the morphological development of the cochlea and the ontogeny of physiological and behavioral responses has been reiterated (RUBEL et al., 1976). We suggested that the developing cochlea is indeed first responsive to sound in the region which is most mature, i.e. near the middle of the basal turn in mammals and toward the basal one-third in birds. However, the transduction properties of the immature cochlea are probably not identical with those of the adult, and factors such as the greater mass per unit area due to the immaturity of the organ of Corti, basilar membrane, and inner spiral sulcus would cause an inertial load restricting the frequency following of the basilar membrane to low frequencies and tending to damp the displacement of the traveling wave. Thus, we have proposed that in the immature animal the basilar portion of the cochlea is the first to transduce mechanical vibrations into neural signals, but that due to its immaturity (probably less tension on the basilar membrane and/or greater cellular density of the basilar membrane and organ of Corti) its response is primarily to low frequencies. In this case, sound transduction in the cochlea would commence toward the basal region in response to low frequencies and progress first in both directions, and then apically. Concurrently, there would be an increase in the ability to transduce successively higher frequencies at the rapidly maturing basal end of the cochlea, while progressively more apical regions would mature to the stage where responses to low frequencies would be initiated (note that in the mature cochlea the basal region responds to both low and high frequencies, while the apical region responds only to low frequencies). Although our suggestion is slightly more detailed, several previous authors (e.g. LARSELL et al., 1944; VANZULLI and GARCIA-AUSTT, 1963; PUJOL and MARTY, 1970) have suggested that ontogenetic changes in the mechanics of the cochlear partition are responsible for the discrepancy between cochlear ontogeny and functional responses. In addition to its accordance with the morphological observations on cochlear development, this hypothesis gains some support from observations on central nervous system ontogeny. First, we have recently shown (RUBEL et al., 1976) that the differentiation of 2nd- and 3rd-order auditory regions in the chicken brainstem proceeds in a spatial direction which corresponds to the basal-to-apical pattern observed in the cochlea. That is, normal histogenic cell death, volumetric changes, and qualitative signs of differentiation all occur initially in the rostral end of each nucleus, which receives projections from the basal portion of the cochlea and responds to relatively high frequencies in the mature system (RUBEL and PARKS, 1975). The caudal (or apical projection) regions, which respond to low frequencies in the adult, are relatively delayed in

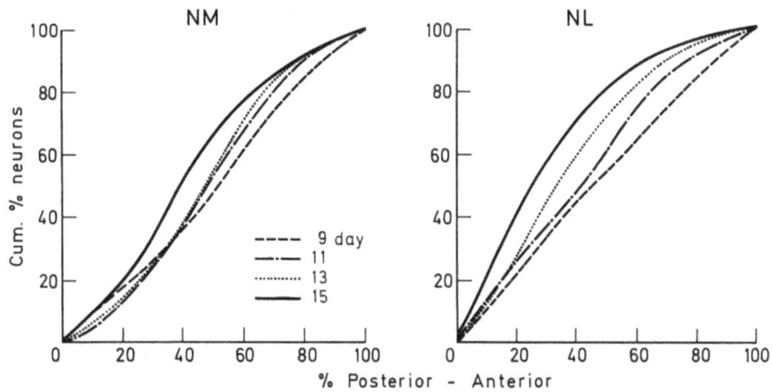

Fig. 7. Spatial distribution of cell numbers in *n. magnocellularis* and *n. laminaris*, across age. To compensate for varying rostrocaudal dimensions of nuclei at different ages, total extent of nuclei is represented in terms of cumulative percentages of distance from posterior to anterior end. Similarly, to compensate for differing numbers of cells at different ages, number of cells is expressed in terms of cumulative percentages of totals at each age. In particular, note increases in initial slope (at posterior end) and decreases in final slope (at anterior end) for both nuclei at successively older ages, indicating systematic shift in relative numbers of cells from anterior portion of each nucleus toward posterior regions. NM, *n. magnocellularis*; NL, *n. laminaris*. (From RUBEL et al., 1976)

their differentiation. This rostral-to-caudal gradient for cell death in *n. magnocellularis* (2nd-order neurons) and *n. laminaris* (3rd-order neurons) is shown in Fig. 7. At day 9 of embryogenesis, the distribution of neurons over the rostral-caudal extent of each nucleus is relatively uniform. By 11 days, this distribution deviates from uniformity at the rostral end, where cell loss has begun. During the following four days (11–15) cell loss occurs at successively more caudal locations, causing the deviation from uniformity (45° line on graphs) to occur at points successively more toward the caudal end. Similar changes occur when nuclear volumes are analyzed and, as noted, for a number of qualitative indices.

If neural activity or some other aspect of afferent input influences these developmental parameters, which it almost certainly does (see Section C. III), this gradient suggests that the neurons in the basal projection regions are the first to receive afferent input, with apical projection regions becoming functional at successively later times. This scheme would also explain the results reported by PUJOL and MARTY (1968). Recording evoked potentials from the primary auditory region (A1) of the feline cerebral cortex 2–3 days after birth, they found responses only to low-to-middle frequencies. More importantly, potentials were localized toward the *rostral portion* of the middle ectosylvian gyrus, which receives projections from the base of the cochlea and which responds maximally to high frequency tones in the adult cat. This finding does not rule out the possibility that high frequencies are selectively attenuated by the immature middle ear. It does suggest, however, that in very young kittens, the basal region of the cochlea is responding to low-frequency stimuli and that the apex is unresponsive to low frequencies, especially in view of the fact that at birth electrical stimulation of the auditory nerve evokes responses throughout the primary auditory projection area.

A final resolution of the factors influencing the development of frequency resolution in the developing cochlea is fundamental for an understanding of normal ontogeny of audition as well as for consideration of the role of experience in the development of the central auditory pathways. In this regard it is interesting to note that low- to middle-frequency sounds predominate in the embryonic environment of mammals and birds. High frequencies are poorly represented endogenously in the intrauterine environment and exogenous high frequencies are attenuated more than lows by the abdominal wall of mammals (see Table 4) and by the eggshell in birds (JACKSON and RUBEL, unpubl.). Since the tonotopic organization of primary auditory regions of the brain have now been described for many mammalian species and quantitative "maps" are available for the brainstem nuclei of the chicken, it should be possible to resolve any ontogenetic changes in frequency representation. One method would be to correlate unit electrophysiological responses to properties with position of the responding neural elements. Another, more exciting, possibility would be to apply recently developed "metabolic" methods such as labeled 2-deoxyglucose autoradiography (KENNEDY et al., 1975). Application of this last-named method to problems of functional development, in general, offers the exciting prospect that stimulation can be applied to an intact organism in its otherwise normal sensory environment.

3. Ontogeny along the Neuraxis

As implied earlier, very little can be said about the particular morphogenetic events responsible for functional changes in the cochlea or central auditory pathways. The cochlear microphonic potential onset correlates with the appearance of a positive increase in amplitude of the endocochlear potential, and a negative summating potential (DC shift in response to stimulation) recorded in the scala media (ÄNGGÅRD, 1965). The summating potential reverses polarity, to become positive, at about 15 days postpartum in the rabbit, which suggests that it may be related to the later innervation of the outer, as opposed to inner, hair cells. However, anatomical confirmation that the time of innervation correlates with this polarity shift has not been forthcoming.

In species where both CMs and nerve potentials have been examined, the onset of responses attributable to VIIIth nerve action potentials seems to lag behind onset of the cochlear microphonic by a day or two (see PUJOL and HILDING, 1973). In the mouse it appears that nerve potentials recorded directly from the VIIIth nerve lag behind those recorded from the round window as well. It is not at all clear why these differences should occur. They may represent differences in recording sensitivity, or real dissociations between the functional elements initiating the various potentials. In any case, no morphological changes have been identified that account for them.

Concurrently with the onset of cochlear and nerve potentials, evoked potentials can be recorded throughout the primary auditory pathways. The onset and elaboration of these evoked potentials have been extensively described for most laboratory animals and for humans. Excellent reviews are available on both human (HECOX, 1975; EISENBERG, 1976) and animal studies (ROSE and ELLINGSON, 1970; MYSLIVEČEK, 1970; SCHERRER et al., 1970). In addition to showing the ages at which the evoked potentials are first elicited by a click or pure tone stimulus, these studies can be summarized as follows: i) Evoked potentials increase in amplitude with age, which is probably due to recruitment of more elements responding to a stimulus of a given magnitude; ii) Thresholds decrease with age, probably indicating mechanical or neural changes in the cochlea and peripheral structures, increased ramification of neural processes, and, tentatively, changes in the efficiency of individual synapses; iii) Latencies of responses decrease markedly during the first two weeks of function and more slowly thereafter, which initially is attributable to growth and myelination of peripheral axons, and then decreases in central transmission time; iv) Increased ability to follow repetitive stimuli, which may be attributed to the development of recurrent inhibitory mechanisms, changes in transmitter availability, or changes in the ionic compartmentalization capacities of the cells; iv) Increased recruitment (i.e. augmentation of response) with low frequency stimulation; vi) Increased complexity of evoked potentials (especially cortical), with immature responses being dominated by the negative component, and then the initial positivity as well as later waves maturing.

Little advance has been made in identifying specific neuronal changes responsible for the various aspects of evoked potential maturation in the auditory system. However, some progress has been made in determining the location(s) of such changes. For example, by recording responses in the medial geniculate nucleus and the auditory cortex, ROSE et al. (1957) were able to establish that the latency reductions were primarily of subcortical origin, rather than in the thalamocortical relay. Similarly, a series of studies on human neonates (see HECOX, 1975) has extrapolated from the waveform identification available on adults to determine that the initial latency change seen in human neonates is in VIIIth nerve transmission time. The extensive series of investigations by the Czechoslovakian group (see MYSLIVEČEK, 1970) is a further example; by using regional stimulation and cortical recording, they have been able to delimit the sites of several ontogenetic changes to either peripheral components or central neural networks. By far the most systematic investigation of the ontogeny of auditory evoked responses from the central nervous system has come from the French group (see MARTY, 1962, 1967; PUJOL, 1972). This series of investigations has established that i) Responses are obtained simultaneously 2–3 days postnatally in kittens in all central nuclei of the primary auditory pathways; ii) The latency decreases occur at successively higher levels of the neuraxis; iii) Responses are elicited by tones (at least three sinusoidal periods) before responses are obtained to clicks; iv) The topography of acoustically-driven synaptic connections in the cortex may be altered during the first postnatal week due to the loss of thalamic inputs to superficial cortical layers (KÖNIG et al., 1972); and v) Auditory responses are seen in association areas of the cerebral cortex concurrently (at 2–3 days in the kitten) with primary responses. With respect to the last finding it is interesting to note that responses to click stimuli were not reliably recorded in the kitten cerebellum until *several weeks* after birth (SHOFER et al., 1969).

4. Neuronal Coding of Pure Tones

In contrast to the plethora of information on evoked potential maturation, there is a relative paucity of data on the response contingencies (stimulus coding) of individual neurons during auditory system maturation. Undoubtedly, one reason that little attention has been directed toward this area is that neurons in which the response contingencies are congruent with particular environmental features ("feature detectors") have been less evident in the auditory than in the visual system. As confidence dwindles in the heuristic value of a model asserting the necessity of hierarchically organized feature construction, it may become apparent that the ability to parametrically control the attributes of acoustic signals will be highly advantageous for understanding the ontogeny of information processing in the brain.

The coding properties of neurons can be defined with respect to either the stimulus properties that reliably elicit an extracellularly recorded neuronal response (either increase or decrease in number of action potentials over time) or the response characteristics themselves. One fundamental form of coding is the tonotopic organization of neurons found within each nuclear region: the characteristic frequency (CF), or lowest threshold excitatory frequency, of neurons within a nuclear region is systematically distributed across the nuclear region. The tonotopic organization undoubtedly results from the sequential spatial organization of the cochlea and topographic projection of each neural region onto its target population. Tonotopic organization is apparent as early as neuronal responses to stimulation can be elicited from the central nervous system (PUJOL and MARTY, 1968; KONISHI, 1973; AITKIN and MOORE, 1975). At the present time, it is not possible to make a definite statement as to the extent of the neural region that is occupied by neurons responsive to auditory stimulation at the early ages, and there is reason to believe that the region responsive to a given frequency may shift during development.

Less developmental work has focused on more complex coding parameters, such as two-tone interactions, tuning curves, response areas, phase locking, response types, and binaural interactions. Since these are critically important considerations for future investigations, the ontogenetic information that is available on each will be briefly described.

a) Two-tone interactions, typically where the cellular response to one tone is inhibited by the addition of another frequency, have not been studied developmentally.

b) Tuning curves indicate the excitatory thresholds of a unit at frequencies surrounding the characteristic frequency (CF) for that cell. A tuning curve typically consists of a U- or V-shaped function, which indicates the frequency specificity of the excitatory response. KIANG et al. (1965) introduced the use of a quantitative measure ("Q") of the sharpness of tuning (Q = characteristic frequency divided by the bandwidth 10 db above excitatory threshold). The Q measure is in direct proportion to the sharpness of tuning (slope of the tuning curve). There are several reasons to suspect that tuning curves may become sharper during the early stages of auditory function. First, the general view of embryonic canalization suggests not only that tissues become increasingly determined, but also that physiological and behavioral processes might become increasingly specific during development (WADDINGTON, 1952; RUBEL and ROSENTHAL, 1975). Second, if neuronal specificity with respect to the frequency of an adequate stimulus has anything to do with an organism's ability to differentiate between frequencies (which is by no means certain), ontogenetic behavioral changes (see below) would suggest that tuning curves become steeper with age. Finally, some authors (e.g. KATSUKI, 1961) have suggested that neuronal responses become more specific as one ascends the neuraxis. If one accepts this notion, it is reasonable to believe that the fine adjustments in the local circuitry that determines the parameters of receptive field size may be established consequent to the formation of major projection pathways. Thus, systematic quantitative studies on tuning parameters throughout the developing auditory system are of prime importance for an understanding of the ontogeny of frequency coding in the auditory system as well as for a general understanding of the ontogeny of functional specificity in perceptual systems. Unfortunately, one investigation (AITKIN and MOORE, 1975) has been reported in which the ontogeny of tuning curves was systematically studied. These authors did report the qualitative observation that units in the kitten inferior colliculus became more frequency-specific with age, and presented quantitative data. Average Q values increased with age. However, as the authors point out, Q values normally increase as a function of frequency, and therefore increased responsiveness to high frequencies may be responsible for this change. Had the authors presented quantitative data separately for low frequencies (e.g. 500–2500 Hz), further interpretation might be possible. In any case, the observations of AITKIN and MOORE are important as a preliminary statement. Future researchers should follow their lead, and provide quantitative data on tuning characteristics at brain stem and VIIIth nerve levels, as well as higher levels of the auditory system.

c) The response area of a neuron can be represented by a family of functions showing spike rate as a function of frequency and intensity of a set of stimuli. Ideally, both excitatory and inhibitory zones are shown, by comparison with the spontaneous activity of a cell. Response areas provide a much more complete picture of the stimulus-response contingencies than tuning curves alone, and often yield a very different picture of the complexities of coding than do tuning curves (YOUNG and BROWNELL, 1976). In addition, response areas are dramatically influenced by anesthetics, sleep, etc. Thus, developmental analyses of unit response areas may be of extreme importance; such analyses, in conjunction with morphological studies on the ontogeny of intranuclear and centripetal connections, may yield clues as to the neuronal pathways responsible for the increased coding complexities; and relationships between the behavioral abilities to analyze acoustic information and the ontogeny of neural function may be revealed.

d) Phase locking refers to the extent to which responses of auditory neurons are "locked" to a portion of the stimulating sinusoid, i.e. the degree to which neuronal response intervals tend to be at the period, or some multiple of the period, of the sinusoidal stimulus. There are good reasons to believe that phase locking may be important for low-frequency sound localization, and it may be involved in other acoustic processing capacities. Phase locking is typically found in neurons responsive to low frequencies (<2 kHz) although it has been reported up to frequencies of about 5 kHz. It is most often studied in neurons at relatively low levels of the neuraxis (VIIIth nerve, cochlear nucleus, olivary nuclei, and inferior colliculus). The only published report on the ontogeny of phase locking is an abstract by JAVEL et al. (1975). In kittens under two weeks of age, they observed phase locking of anteroventral cochlear nucleus cells to frequencies below 300 Hz, but no phase locking to tones above 1000 Hz. It is not known what property of peripheral or central processing is responsible for the limited range of phase locking, but the fact that a cochlear microphonic response to higher frequencies can be recorded suggests sites central to the transduction of mechanical to electrical energy.

e) The ontogeny of binaural response properties of central neurons is another area that has received little attention. One preliminary report (AITKIN and REYNOLDS, 1975) indicates that many inferior colliculus units in very young (5- to 20-day-old) kittens are binaurally responsive. Binaurally excited cells (E/E units) displayed maximum firing to the same frequency applied to either ear (GOLDBERG and BROWN, 1968). Unfortunately, these findings are difficult to interpret; since no controls for interaural communication were performed, ipsilateral stimuli had to be considerably louder (20–40 db) than contralateral tones, and very few units showed opposite responses to the two ears (E/I units) as compared to adults. Future investigations of binaural response properties will be of particular importance. The binaural nature of inputs to brainstem auditory neurons provides a unique advantage for investigations of neurospecificity and the role of experience in the ontogeny of a neuronal structure. Furthermore, quantitative studies on the degree to which binaurally responsive neurons in young animals are "tuned" to similar stimuli will provide additional information on how the fine properties of neuronal coding are established.

f) Response type is the final property by which neuronal responses are often classified. This property refers to the temporal characteristics of responses, as exemplified by post-stimulus-time (PST) histograms. PUJOL and his colleagues (see PUJOL, 1972; CARLIER et al., 1975) have found a common ontogenetic sequence of response types throughout the brain stem auditory system of the kitten. In the VIIIth nerve, cochlear nucleus, and inferior colliculus of very young kittens (less than 7 days old) only "ON" type responses are found; that is, a 500-ms stimulus of adequate frequency and intensity typically evokes one or a few spikes within the first 50 ms and no further activity. At around the seventh day a "rhythmic" response becomes apparent, in which the unit shows alternating period of activity and silence during the stimulus. Initially the silent periods are long (100–300 ms), and they gradually decrease until at 20 days most units respond with an initial burst of discharges followed by sustained activity until the tone is turned off. Although these response types may bear some resemblance to the response types described in the cochlear nucleus by KIANG and his colleagues (e.g. GODFREY et al., 1975), it is probably only a superficial similarity; the ontogenetic sequence is found in the VIIIth nerve, where all units in the adult are of the later-developing "primary" type. It seems reasonable to conclude that these ontogenetic changes are occurring at the periphery. Whether they involve changes in centrally originating efferent pathways (as suggested by CARLIER et al., 1975) is doubtful;

changes in the afferent synapse, such as increased transmitter availability, would be an equally parsimonious interpretation. Future studies in which the periphery is isolated from central influences will be of importance for understanding what acoustic information the developing central nervous system has available. Only then will it be feasible to attempt to understand the ontogeny of auditory information processing in the central nervous system.

5. Responses to Species-Specific Stimuli

The use of pure tone stimuli for auditory neurophysiological investigations has been widely criticized by students of animal behavior, because species-specific acoustic signals usually contain frequency and/or intensity modulations as well as stereotyped temporal modulations between utterances. These critics maintain that neurons selective to the stereotypy ("feature detectors") should be sought, and a number of laboratories are now devoting considerable resources to such investigations. Although some promising results have been reported (e.g. NEWMAN and WOLLBERG, 1973) no convincing examples of cell groups that respond reliably only to the species-specific vocalizations have been found. On the other hand, a number of studies on adult animals have revealed neurons that are selectively responsive to relatively complex parameters such as frequency modulation (e.g. WHITFIELD and EVANS, 1965).

Similar criticisms can obviously be made of the research summarized above on the ontogeny of neuronal responses. While most behavioral research has emphasized that early behavioral responses, especially discriminative responses, are elicited by species-typical complex acoustic stimuli, the neurophysiology of auditory system development has been studied with the aid of pure tones or clicks. One can argue that frequency is a fundamental property of acoustic signals, and therefore merits priority of analysis, but it is also plausible that the auditory system has evolved to give priority to biologically relevant stimuli. In the case of young animals especially, energy and survival considerations would tend to favor the relative disregard for less relevant stimuli. In this respect, it is interesting to note that responses from young organisms can be evoked more readily with tones than clicks, and with long duration tones (500 ms) than short tones (50 ms). Thus, neurophysiological research on young animals, using biologically significant stimulus arrays or synthesized variations thereof, may significantly alter our conceptions of the ontogeny of central auditory processing networks.

6. Behavioral Studies of Sensory Coding

The only way to determine perceptual capacities at any developmental stage is through behavioral studies in which organisms are required to respond differentially to stimuli that differ along some acoustic dimension(s). Only by showing that all stimuli are perceived and that differential responses can be elicited, is it possible to show that a dimension is encoded. The literature on behavioral ontogeny of sensory coding is included in Table 3 and recent reviews of studies in human

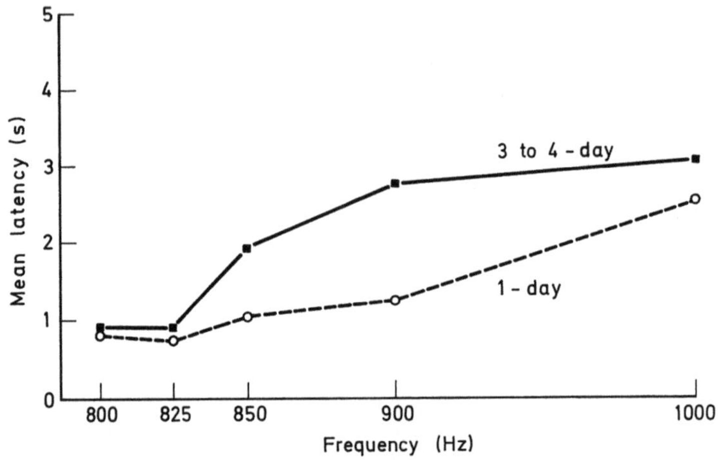

Fig. 8. Generalization gradients as function of stimulus frequency for 1-day-old and 3- to 4-day-old hatchling chickens. An orienting response [duration of distress call inhibition-*Latency* to make distress call following 90 dB (SPL) tone stimulus] was habituated over 12 trials using an 800-Hz stimulus. Separate groups (N = 10) then tested at 800 Hz (control group) 825, 850, 900, or 1000 Hz. One-day-old chicks show differential orientation only to 1000-Hz stimulus, while 3- to 4-day-olds show differential response to 850-Hz stimulus. (From KERR and RUBEL, 1977)

(EISENBERG, 1976; EIMAS, 1975) and in other species (TEES, 1976) should be consulted for more detail.

Most studies have used complex stimuli in order to determine when language-relevant or species-specific stimuli are differentially encoded. For example, several investigators have shown that phonemic information is encoded by human neonates and there is evidence that categorization may be centered around phonemic boundaries quite early in development. These conclusions are reinforced by the findings of both ear and hemispheric differences in neonatal audition. Equally interesting is the finding that the ability of children to extract relevant acoustic information from background noise continues to improve over the first 10 years.

While there is little information on the development of perceptual encoding by other mammals, studies on avian species with tape-recorded signals have indicated that recognition of species-specific or parental vocalizations emerges after an initial period of undifferentiated responses to acoustic signals. Although these investigations do reveal the time at which biologically relevant discriminations can be made, the complex nature of the stimuli makes it difficult to extract information relevant to understanding of the ontogeny of perceptual encoding along any single acoustic dimension. The use of synthetically altered stimuli, in which each dimension can be individually manipulated, is beginning to alleviate this shortcoming, and promises to reveal important principles (see GOTTLIEB, 1974, 1976b).

Our own studies on the ontogeny of perceptual coding have concentrated on the encoding of a unitary acoustic dimension, frequency, in order to determine if perceptual sharpening occurs with respect to this fundamental property. In two studies (RUBEL and ROSENTHAL, 1975; KERR and RUBEL, 1977) using different responses, it has been shown that the encoding of frequency information becomes

Table 3. Functional Development of the auditory system

Species	Dependent measure	Age[a] (E = Embryonic, P = postnatal)	Stimulus conditions[b, c]	Comments	Relevant references
Human	Brainstem evoked potentials	Premature: 34 weeks (E)	Clicks, 10–20/s; 30–55 dB above threshold	No habituation; latency decreases with age; peripheral transmission latency matures earlier (by 6 weeks postnatal) than central transmission, which is mature at about 1 yr postnatal	LEIBERMAN et al. (1973), HECOX and GALAMBOS (1974), BUDA et al. (1975), SALAMY et al. (1975), SCHULMAN-GALAMBOS and GALAMBOS (1975), SALAMY and MCKEAN (1976)
	Cortical evoked potentials	Premature: 23 weeks (E)	Clicks, 0.25/s; 115 dB	Large latency decreases during weeks 23–33 (E); gradual latency decreases postnatally to 1–2 yr; lateral asymmetry to verbal stimuli present in first weeks postnatal	ELLINGSON (1964), WEITZMAN et al. (1965), WEITZMAN and GRAZIANI (1968), AKIYAMA et al. (1969), ENGEL and YOUNG (1969), MONOD and GURMA (1971), SCIBETTA et al. (1971), GRAZIANI and WEITZMAN, (1972), GRAZIANI et al. (1974), MOLFESE et al. (1975)
		In utero: 32–35 weeks	Tones of 1000 Hz applied to maternal abdomen	Recordings from abdomen above vortex suggest cortical EP with initial latency of 100–150 ms	BARDEN et al. (1968), SAKABE et al. (1969)
	Heart rate	*In utero* (26–28 weeks)	Tones at 500–1500 Hz; 100 dB at abdomen; 87–95 dB *in utero*	Stimulus elicits heart rate acceleration; response to stimuli above 3000 Hz seems to develop later (after 30 weeks); similar results from prematurely delivered neonates; at birth response to white noise more reliable than to tonal stimuli; habituation to tones, dishabituation to frequency change recorded in newborns	SONTAG and WALLACE (1936), BERNARD and SONTAG (1947), BARTOSHUK (1962a, b, 1964), MURPHY and SMYTH (1962), DWERNICKA et al. (1964), JOHANSSON et al. (1964), WEDENBERG (1965), BENCH and MITTLER (1967), SMYTH and BENCH (1967), TANAKA and ARAYAMA (1969), GRIMWADE et al. (1971), JOHANSSON (1972, pers. comm.), GOODLIN and SCHMIDT (1972), TURKEWITZ et al. (1972), BERKSON et al. (1974)

Table 3 (continued)

Species	Dependent measure	Age[a] (E=Embryonic, P=postnatal)	Stimulus conditions[b, c]	Comments	Relevant references
	Movement	29–32 weeks. (E); *in utero* or premature	Clicks, complex noises, or tones at 500–2000 Hz	Complex stimuli probably more effective than simple tones; initially 500 and 1000 Hz elicit more reliable response than 2000 Hz or higher; habituation of movement response, which recovers with stimulus rate change; buzzer 20 dB above background does not influence nonnutritive sucking in 35–40 weeks prematures, but inhibition of sucking postnatally	PEIPER (1925), FORBES and FORBES (1927), RAY (1932), SPELT (1938), FLEISCHER (1955), HUTT et al. (1968), LIPSITT (1969), TANAKA and ARAYAMA (1969), KATONA and BERENYI (1974), MILLER (1975)
	Conditioning	32–40 weeks (E); *in utero*	CS "loud clapper"	*In utero*, conditioned movement after 100 trials; postnatally, conditioned sucking or head turning at 2–4 days; number of trials to criterion and frequency discrimination improve with age	KASATKIN and LEVIKOVA (1935), SPELT (1938), LIPSITT (1969), KRULISOVA (1974)
	Directional eye movements	3–10 min (P)	"Clicker"	Auditory localization apparent at birth; at 2–4 days turn away from loud and toward softer sounds; possibly systematic right-left ear threshold differences. Sound influences visual scanning at 2 days (P)	WERTHEIMER (1961), TURKEWITZ et al. (1966), MENDELSON and HAITH (1976)
	Selective attention	4–11 yr (P)	Verbal stimuli, white noise background	Ability to extract verbal stimuli from background improves with increasing age	FISHER and RUBENSTEIN (1975), HOWLAND et al. (1975)
Sheep	Cortical evoked potential	90–100 days (E)	Hand clap or click	First responses correlated with appearance of spontaneous activity in EEG	BERNHARD et al. (1959), MEYERSON (1967) (see MARTY (1967))

Cat	Endocochlear potential	Birth	Resting	Potential increased from 8.8 mV at birth to 80.4 mV in adult; largest increase during days 9–20 (3.5 mV/day); development of *stria vascularis* correlated with EP and progresses from basal-to-apical end	FERNANDEZ and HINOJOSA (1974)
	Cochlear microphonic	61 days (E)	Click, tones at 300–400 Hz, above 80 dB	Internal spiral sulcus forming at basal end; tectorial membrane attachment to supporting cells becoming tenuous; tunnels of Corti and Nuel formed in basal turn. High-frequency response matures postnatally. Adult-like by 3 weeks (P)	PUJOL and MARTY (1970), ROMAND et al. (1970), ROMAND (1971), PUJOL and HILDING (1973)
	8th nerve action potential (unit or compound)	Birth	Click, tones at 800–2000 Hz, 100 dB	Latency threshold and fatigability decrease and frequency range broadens; adult-like by 1 month (P). Unit responses initially of "on" type; by 20 days become "sustained"	ROSE et al. (1957), ROMAND (1971), JEWETT and ROMANO (1972), PUJOL and HILDING (1973), CARLIER et al. (1975)
	Cochlear nucleus unit activity	3 days (P)	Tones at 500–2000 Hz, 100 dB	Latency decrease during first month; responses initially "on" type; at 10 days sustained responses predominate; spontaneous activity increases over first month and inhibitory post-stimulus period decreases in length. Phase locking initially to 300 Hz only; units initially broadly tuned	PUJOL et al. (1969), PUJOL (1972), ROMAND and MARTY (1975), JAVEL et al. (1975)
	Cerebellar potentials	5–6 weeks (P)	Click	When reliably recorded, adult-like in latency and waveform	SHOFER et al. (1969)
	Superior olivary nucleus potentials	2–3 days (P)	Tones at 500–2500 Hz, 100 dB	Afferents from cochlear nucleus to superior olive myelated at birth; 2–3 ms transmission time at age first recorded	ROMAND et al. (1973)

Table 3 (continued)

Species	Dependent measure	Age[a] (E=Embryonic, P=postnatal)	Stimulus conditions[b,c]	Comments	Relevant references
	Inferior colliculus potentials	2–3 days (P)	Tones at 500–2500 Hz, 105 dB	Latency initially long; adult-like by 20 days (P). As frequency range broadens and thresholds drop, units become increasingly frequency-specific and Q_{10}'s increase; tonotopic organization apparent at 6–8 days (P). Middle-ear cavitation hypothesized responsible for increased frequency range	PUJOL and GRANIER (1968), PUJOL (1972), AITKIN and MOORE (1975)
	Medial geniculate potentials	2–3 days (P)	Click, tones at 500–2500 Hz; 105 dB	Response initially seen in pars medialis, Latency adult-like by about 25 days (P); unit responses initially "on" type	ROSE et al. (1957), PUJOL et al. (1970), PUJOL (1972)
	Auditory cerebral cortex (Al) potentials	2–3 days (P)	Tones at 500–2500 Hz; 90 dB	Response to clicks matures later than that to tones (6 days). Evoked potential latency, following rate, and amplitude adult-like by 5th week; latency decrease primarily prior to thalamus; potential initially dominated by negative wave with initial positive wave maturation delayed. First unit responses recorded rostral (basal) area of Al; correlated with cochlear maturation in basal turn. Units responsive up to 15 kHz by 8 days; only upper part of apical coil yet to mature. Units show "on" type response initially, followed by more complex coding (at about 3 weeks)	GROSSMAN (1955), ROSE et al. (1957), ELLINGSON and WILCOTT (1960), MARTY (1962), 1967), ELLINGSON (1964), MARTY and SCHERRER (1964), CHALOUPKA et al. (1968), PUJOL and MARTY (1968), VOLOKHOV (1968), SHOFER et al. (1969), KÖNIG et al. (1971, 1972), PUJOL (1972)

	Cerebral cortex, association area potentials	2–3 days (P)	Loud click, tones	Evoked potentials in orbital, insular, suprasylvian and pericruciate areas at 2–3 days; percent age of units in suprasylvian area responding to click increases from 33% at 8 days to 96% at 1 mo	MAYERS et al. (1971, 1976), PUJOL (1972)
	Behavioral responses	6–7 days (P)	Tones and mixed stimuli 108–120 dB	Startle and eye opening response first to low frequency tone or noise band (125–750 Hz); progressive maturation of higher frequency response; directional orienting matures later, approximately 2–4 weeks	TILNEY and CASAMAJOR (1924), FOSS and FLOTTORP (1974), VOLOKHOV (1968)
Dog	Cochlear microphonic	4–5 days (P)	"Intense" click	Detachment of tectorial membrane with supporting cells and inner spiral sulcus forming and tunnel of Corti formed in basal turn	PUJOL and HILDING (1973)
	8th nerve compound action potential	7–8 days (P)	"Intense" click	Lengthening of outer hair cells and further formation of inner spiral sulcus; adult-like with respect to threshold and frequency spectrum at 15–20 days (P)	PUJOL and HILDING (1973)
	Auditory cortex evoked potentials	6–14 days (P)	click	EP initially positive-negative: latency decreases to adult-like by 5–6 weeks; rapid development of EP amplitude and following characteristics (14–30 days) correlated with basal dendritic growth and myelination	ZAHLAVA et al. (1966), CHALOUPKA et al. (1968), FOX (1968a, b)
	Behavioral responses	9–15 days (P)	Tones and noise at 500–750 Hz, 108–120 dB	Initial startle response to 500–750 Hz; higher frequencies (2–18 kHz) first elicit responses on days 18–22. Initial directional orientation about 16–24 days (P)	VOLOKHOV (1968), FOSS and FLOTTORP (1974)

Table 3 (continued)

Species	Dependent measure	Age[a] (E=Embryonic, P=postnatal)	Stimulus conditions[b, c]	Comments	Relevant references
Mink	Behavioral responses	30 days (P)	Tones and noise at 500–750 Hz, 108–120 dB	Startle responses to higher frequencies mature later	FOSS and FLOTTORP (1974)
Rabbit	Cochlear microphonic (CM) and summating potential (SP)	5 days (P)	Tones at 1–7 kHz, 120 dB	Direct stimulation of oval window. CM threshold and frequency response adult-like by 15–20 days. SP polarity reversal at 10–15 days attributed to differentiation of outer hair cells. Onset correlated with formation of inner spiral sulcus, tunnel of Corti, and spaces of Nuel; occur first in basal turn	ÄNGGÅRD (1965)
	Endocochlear potential	5 days (P)	Tones at 1–7 kHz, 120 dB	Develops homogeneously throughout cochlea during 5–15 days	ÄNGGÅRD (1965)
	Brainstem evoked potentials	5 days (P)	Tones at 3 kHz, 125 dB	Response to tones prior to click; latency decreased between days 5–15	ÄNGGÅRD (1965)
	Auditory cortex evoked potentials	4–6 days (P)	Click	Latency and amplitude characteristics develop over first two months; initial responses appear partially from superficial layers, which correlates with synaptology	MARTY (1962), MARTY and SCHERRER (1964), KLYAVINA and OBRASTOVA (1968), KÖNIG and MARTY (1974)
	Behavioral responses	6–7 days (P)	Tones and noise at 108–120 dB	Initial startle response to 250–750 Hz; directional orientation first seen on day 12–15; conditioning possible on days 10–11	VOLOKHOV (1968), FOSS and FLOTTORP (1974)

Guinea pig	Cochlear microphonic (CM)	48–52 days (P)	Click, tones of 600–2000 Hz, at 100 dB	CM to higher and lower frequencies delayed; amplitude adult-like by birth or a few days after; onset correlated with freeing of tectorial membrane, formation of inner spiral sulcus, development of fluid spaces, and rapid increases in cochlear enzymatic activity	Rawdon-Smith et al. (1938), Chondynicki (1968), Romand et al. (1971), Romand (1971), Pujol and Hilding (1973)
	8th nerve potential	50–53 days (E)	Click, tones of 1–3 kHz, at 110 dB	Between onset and 2 days (P), latency, waveform and amplitude become adult-like	Romand et al. (1971), Romand (1971), Pujol and Hilding (1973)
	Auditory cortex evoked potential (EP)	52–53 days (E)	Click	Latency decreases, amplitude and following rate increase; adult-like soon after birth. At onset negative wave predominates; transient latency increase and decreases anoxia resistance during perinatal period. Responses recorded *in utero* to external stimulus 40 hrs prenatal	Chaloupka et al. (1968), Scibetta and Rosen (1969), Rosen (1971), Sedlacek (1974, 1976)
	Behavioral responses	52–60 days (E)	Tones and noise ("loud")	In prematures, movement response as early as 52 days (E); Preyer by 61 days (E) to "clang"; at birth (65 days) directional orientation and conditioning observable	Avery (1928), Rawdon-Smith et al. (1938), Volokhov (1958), Scibetta and Rosen (1969)
Rat	Endocochlear potential	5 days (P)	—	Low at 8–11 days; rapidly increasing potential during 11–16 days (P); correlates with maturation of *stria vascularis*, morphology, cochlear microphonic, Na-K-activated ATPase activity, and exploration from the nest site	Schmidt and Fernandez (1963), Bosher and Warren (1971), Kuijpers (1974)
	Cochlear microphonic potential	8–9 days (P)	Tones at 200–2000 Hz, 80 dB (re 1 dyne/cm^2)	By 11 days, frequency range up to 10 kHz; 16–20 days (P) adult-like sensitivity and frequency range (to 70 kHz). Opening of meatus (at 12–13 days) lowers thresholds 20–30 dB	Crowley and Hepp-Raymond (1966)

Table 3 (continued)

Species	Dependent measure	Age[a] (E=Embryonic, P=postnatal)	Stimulus conditions[b,c]	Comments	Relevant references
	Brainstem potentials	13 days (P)	Clicks at 70 dB	Latency maturation at relatively younger ages for earlier waves (i.e. most peripheral)	JEWETT and ROMANO (1972)
	Auditory cortex evoked potentials	8–15 days (P)?	Tones and click	Onset varies widely between studies; negativity predominates at early ages; adult-like in latency, waveform, following rate, and amplitude by 28 days	BURES (1953), CHALOUPKA and MYSLIVECEK (1960), CHALOUPKA et al. (1968), VOLOKHOV (1968), KÖNIG and MARTY (1974)
	Behavioral responses	9–12 days (P)	Click, handclap, whistle	Opening of meatus not responsible for functional onset; correlated with initial maturation of responses are apparent innervation of hair cells (day 9), increase in hair cell volume (days 0–10); prolonged increase in ganglion cell volume (days 0–20), development of tunnel of Corti and spaces of Nuel (days 9–15), formation of inner spiral sulcus, etc., (see Wada). Basal-to-apical cochlear maturation. Directed orientation by day 12	SMALL (1899), LANE (1917), WADA (1923), VOLOKHOV (1968), BRACKBILL (1976)
Mouse	Endocochlear potential	Birth	—	Low at birth, adult-like by 14 days (P)	SCHMIDT and FERNANDEZ (1963)

	Cochlear microphonic (CM) and 8th nerve action potentials (AP)	8 days (P)	Clicks, tones of 600–2000 Hz at 100 dB	Genus *Mus:* Meatus opens about day 8 (P); frequency range extends to 8 kHz by day 9 and 40 kHz by day 11; adult-like in 0.6–40 kHz range by 14 days. Click elicited CM at 10–12 days. Cochlear maturation delayed in apical portion. Last morphological change around onset appears to be formation of efferent connections (day 10). AP development parallels CM, but delayed about 24 hrs.	ALFORD and RUBEN (1963), MIKAELIAN and RUBEN (1965), KIKUCHI and HILDING (1965)
	Inferior colliculus potentials	15 days (P)	Tone of 15 kHz, at 100 dB	Did not test prior to 15 days or with other frequencies; increasing amplitude of evoked potential coincides with period of susceptibility to priming	LIEFF et al. (1975)
	Behavioral responses	11 days (P)	Tones of 1–3 kHz, at 95–110 dB, clicks	*Mus:* Onset of Preyer reflex coincides with onset of 8th nerve response; response initially to 1–3 kHz at 11–12 days, spreads to 0.5–36 kHz by 14 days, adult-like by 16 days *Peromyscus:* Startle response to "loud squeak" varies with species (9–23 days P)	ALFORD and RUBEN (1963), MIKAELIAN and RUBEN (1965), POWERS et al. (1966), LAYNE (1968)
Gerbil	Cochlear microphonic (CM) and 8th nerve action potentials (AP)	CM: 12 days (P). AP: 14	Tones of 200–30000 Hz, 80–100 dB	Initial CM responses to almost entire frequency range; thresholds decrease 40–60 dB during 12–23 days (P). Behavioral response, movement, correlated with onset of AP	FINCK et al. (1972)
Opossum (*Didelphys virginiana*)	Endocochlear potential	25–30 days (P)	—	Rapid development between 36 and 75 days; adult-like by 78 days	SCHMIDT and FERNANDEZ (1963)

Table 3 (continued)

Species	Dependent measure	Age[a] (E = Embryonic, P = postnatal)	Stimulus conditions[b, c]	Comments	Relevant references
	Cochlear microphonic (CM)	48–50 days (P)	Loud tones at 1–7 kHz	Low-frequency CMs mature before high-frequency responses; peak frequency sensitivity shifts upward; adult-like by 77 days; upper part of basal turn matures earliest; apex matures latest. Middle ear ossification also correlates with CMs. *D. marsupialis:* onset about 59 days (P)	McCrady et al. (1937, 1940), McClain (1939), Larsell et al. (1944)
	Behavioral responses	50 days (P)	Handclap, shrill whistle	Whole body movement initially; 51 days, movement to loud tones of 1–1.5 kHz; pinna reflex and larger frequency range mature during next days. First behavioral response in *D. marsupialis* at 59 days (P)	Larsell et al. (1935), McCrady et al. (1937)
Chicken	Cochlear microphonic	13 days (E)	Tones of 100–250 Hz; 90–100 dB	Higher frequencies progressively added; threshold frequency increases during development to 2–3 kHz at hatching; amplitude also increases. Possible CM apparent on 10th day (E)	Vanzulli and Garcia-Austt (1963), Saunders et al. (1973)
	Brainstem evoked potentials	11–12 days (E)	Tones of 100–1300 Hz, at 105–120 dB	Response range sensitivity and following rate increase until adult-like at hatching. Thresholds decrease 20–30 dB between day 18 and hatching, possibly due to middle ear cavitation	Saunders et al. (1973, 1974), Saunders (1974)

	Midbrain unit responses	1–5 days (P)	Tones of 100–3600 Hz	Unit responses sharply tuned; no inhibition; EEG already well developed	PETERS et al. (1960), ROSOWSKI et al. (1976)
	Forebrain evoked potentials	18 days (E)	Clicks	Presumably from field L	SEDLACEK (1974)
	Behavioral responses	14 days (E)	Tones of 700, 1000 and 1400 Hz, at 100–115 dB	Reliable change (decrease) in movement on day 14; some authors suggest earlier response. By day 15–16, rhythmic clicks may evoke "conditioned" movement; reliable conditioning to 3 kHz tone by day 17; species-specific maternal call evokes increased activity at 19–20 days. Post-hatching sharpening of perceptual discrimination	GOS (1935), KUO (1938), HUNT (1949), SEDLACEK (1964), GOTTLIEB (1965a), GRIER et al. (1967), SVIDERSKAYA (1967), RUBEL and ROSENTHAL (1975), VINCE et al. (1976), JACKSON and RUBEL (1977), KERR and RUBEL (1977)
Duck	Brainstem potentials	19–20 days (E)	Tones of 120–500 Hz; 60–90 dB	Units responsive to progressively higher frequencies and lower thresholds added through hatching; binaural units apparent at youngest ages. Evoked potentials in hatchlings show lowest threshold at 1–2 kHz	KONISHI (1973), SAUNDERS et al. (1974), WOOLF and CAPRANICA (1975)
	Behavioral responses	Day 22 (E)	Mallard duck maternal call, 55–66 dB	At age where initial responses observed, ducklings show differential response to species-specific maternal vocalization; response initially inhibitory, becomes excitatory on day 24 (E); at pipping stage, movement-contingent stimulation with maternal call results in more movement during stimulus than randomly presented stimulation	GOTTLIEB (1965b, 1966, 1968a and b, 1971a), IMPEKOVEN (1973)

Table 3 (continued)

Species	Dependent measure	Age[a] (E=Embryonic, P=postnatal)	Stimulus condition[b,c]	Comments	Relevant references
Gulls	Behavioral responses	15 days (E)	Species-specific calls, at 80–85 dB	At 15 days (E) increase activity to "crooning," other calls less effective; on first day after hatching, find recognition of individual parent's vocalizations; discrimination increases for 1–3 weeks. Parental recognition at hatching varies with species and type of call	BEER (1970a, 1972), EVANS (1970a, b), IMPEKOVEN (1973)
Guillemots	Behavioral responses	24–36 h before hatching	Species-specific "luring" call	Embryos can discriminate familiar vs. unfamiliar call from same species; at hatching, recognize call of individual parent	TSCHANZ (1968)

[a] Age when auditory response is first reliably seen.
[b] Stimulus used at age where first auditory response is first seen.
[c] Intensity reported at dB *re* 0.0002 dynes/cm^2 unless otherwise noted.

increasingly finer during the first 3–4 days after hatching in the neonatal chicken. Both experiments used an habituation-generalization paradigm; an orienting response to tone bursts of one frequency was first habituated and then a new set of stimuli, differing by 0, 25, 50, 100 or 200 Hz was presented. Increased orientation during the test trials indicated that the new frequency was being differentially encoded. Results of the study by KERR and RUBEL are shown in Fig. 8. One-day-old and 3- to 4-day-old chicks, while making distress calls, were given 12 trials at approximately 30-second intervals. In each trial, a 90 db, 800-Hz tone was presented for 1.6 s. Initially, the subjects oriented to the tone and ceased emitting distress calls for 5–10 s. With successive trials, the orienting response habituated, and after 12 such trials either the same frequency (800 Hz) or a new stimulus was presented. As seen in Fig. 8, 3- to 4-day-old chicks displayed an orienting response (increased latency to emit distress calls) to an 850-Hz stimulus, while one-day-old chicks did not demonstrate differential responding, except to the 1000 Hz stimulus. These results are identical with those obtained using an eye-opening response (RUBEL and ROSENTHAL, 1975) and indicate that a fundamental acoustical property such as frequency becomes increasingly encoded over the first 3–4 days after hatching. The extent to which similar processes are occurring along other acoustic dimensions and to which this result is generalizable to other species is uncertain. Examination of other ontogenetic changes in perceptual encoding is of importance for understanding neural mechanisms underlying the ontogeny of auditory information processing.

IV. Experiential Influences on Auditory System Development

The influence of experiential events on the ontogeny of neural structure and function within the visual system has been a focus of much recent research. Less attention has been given to this problem by investigators concerned with development of audition. Two major reasons can be identified: i) In comparison with visual features less information is available about the neuronal coding of acoustic features; and ii) It is easier to control gross aspects (light vs. dark) of the visual environment. The relative paucity of literature in this area is unfortunate, since in many ways experimental control of individual parameters of acoustic exposure is easier. Furthermore, in the visual system a large part of the processing is performed at the retinal level, which makes it difficult to analyze the relationships between chronic changes in visual exposure, receptor excitation, and changes at each succeeding neuronal level. In the auditory system, the relationships between hair cell excitation and activity are much clearer, at least at the cochlear nucleus level. As will be seen in Section C. III., the unique segregation of the binaural inputs at brainstem levels affords the opportunity for very detailed cellular analyses (e.g. see BENES et al., 1977).

Table 4 summarizes most of the available literature regarding the influences of early auditory experience on cochlear, central nervous system, and behavior development (see also MISTRETTA and BRADLEY, 1977).

The first item in Table 4 is included to point out that the normal mammalian embryonic environment has potential for a reasonable amount of sound exposure, especially in the low frequency range. Avian species also have the potential for acoustic stimulation during embryogenesis. JACKSON and the present author, using a piezoelectric microphone inserted and sealed into chick eggs, have observed that low-frequency signals (<1500 Hz) are attenuated by 10–20 db, and signals above approximately 1800 Hz are transmitted into the egg with little or no attenuation.

The remainder of Table 4 is devoted to the available examples indicating that early environmental experience effects lasting changes on auditory system development. A thorough coverage of this topic would also include congenital pathologies in humans, known genetic malformations in other animals, and the effects of early acoustic trauma. Human congenital pathologies are beyond the scope of this chapter (see OMEROD, 1960; ALTMANN, 1964; SCHUKNECHT et al., 1965; LINDSAY and MATZ, 1966).

Although the extensive literature on genetically caused malformations of the inner ear has not been included, its possible relevance to the understanding of environmental influences is worth noting. Genetically produced malformations of the inner or middle ear provide naturally occurring manipulations. By causing severe alterations in peripheral input to the central nervous system (CNS), these and other pathological conditions can be used to define a "baseline" for indicating how the CNS responds to gross changes in peripheral input. For this reason, as well as for their clinical relevance, studies defining the etiology, timing, physiological consequences, and behavioral deficits resulting from mutations affecting the inner ear are of great importance. Several "animal models" of congenital deafness have been discovered and a few have been studied extensively (e.g. HUDSON et al., 1962; MIKAELIAN and RUBEN, 1964; MIKAELIAN et al., 1965, 1974; DEOL, 1967, 1968, 1970; BROWN and RUBEN, 1969; RUBEN, 1973; SUGA and HATTLER, 1970; MAIR, 1973). The recent introduction of *in vitro* methods for investigation of the etiology of genetically caused inner ear malformations (VAN DE WATER and RUBEN, 1974) promises significant advances in this field. Relatively little work has been directed toward elucidating the effects of degenerative mutations of the inner or middle ear on central nervous system development. The most significant contribution (WEST and HARRISON, 1973) describes severe reductions in cell size in some areas of the cochlear nucleus and superior olivary complex, with little or no atrophy in other areas of these brainstem auditory nuclei in two adult deaf white cats. Although similar but less severe effects were found in one young mutant cat, the subject population was too small to yield definitive conclusions. Future investigations in this area should use mutant mice or some other preparation in which sufficient subject populations are available to permit thorough developmental analyses. A systematic effort along these lines could yield important principles for understanding the role of peripheral input on central nervous system ontogenesis. One caution bears consideration. While mutant preparations may provide a powerful tool, the primary etiology of the syndrome is seldom, if ever, clear. That is, when the primary effect of the mutation is on the developing neural tube (e.g. VAN DE WATER and RUBEN, 1974), the interpretation of central nervous system changes becomes more difficult; any given difference between neurons in normal and mutant animals may result from either direct effects of the mutation, indirect (transneuronal) influences, or a combination of both.

The effects of acoustic trauma (and high-intensity noise exposure) on the developing cochlea, auditory pathways of the brain, and auditory perception, has been thoroughly reviewed by SAUNDERS and BOCK (1977). Thus Table 4 includes only a few of the available references. Two points merit added emphasis. First, it appears that in young as well as older animals, high-intensity noise exposure induces permanent structural changes in the inner ear; in fact, the available data strongly suggest that the cochlea of young animals may be hypersusceptible to acoustic trauma. Second, SAUNDERS and his colleagues have built up a convincing case for the idea that similar events are occurring with the induction of seizure susceptibility through priming in normally seizure-resistant mice. Their data show that primed mice have increased evoked potential thresholds, coupled with hyperresponsitivity to high intensity stimuli. Although they have likened the latter effect to discuss supersensitivity, direct evidence of changes in central nervous system pathways is not available. In any case, the conclusion that priming serves to "deprive" the central nervous system of normal sensory activation indicates that this preparation may be useful for understanding the morphological and physiological effects of auditory deprivation. However, a thorough analysis of the effects of priming on the developing cochlea itself is a critical step; then it will be possible to ascertain how priming influences the activity impinging on the central nervous system (CNS) and thereby alters normal neuronal input.

Systematic attempts to understand how the ontogeny of neurons or neuronal networks is influenced by auditory experience are lacking. The few studies noted in Table 4 that have found changes in CNS anatomy or physiology have failed to distinguish between alterations of normal ontogeny and degenerative effects (cf. SOLOMON and LESSAC, 1968; GOTTLIEB, 1976a). Following examples provided by studies on the mammalian visual system, the principal paradigm has been to subject one group of animals to acoustic deprivation or alteration of the sound patterns heard, and then make comparisons between these "experimental" subjects and normal animals as adults. Typically, one neuronal locus is arbitrarily examined with respect to cellular responses or evoked potentials. While such studies may demonstrate an *influence* of acoustic experience on neural responses, they provide little more. Failure to examine the cochlea or output thereof means that the site of the "effect" is unclear; failure to examine the normal ontogeny of the process under investigation, means that the developmental nature of the "effect" cannot be interpreted; failure to sample systematically within the arbitrarily chosen neuronal population means that the specificity of the "effect" is obscured; and failure to determine how the manipulation itself influences neuronal activity within the auditory system, the mechanism(s) responsible for the effect remain(s) undeterminable. Although these criticisms may appear unduly harsh, it is the opinion of this author that attempts to ascertain the influence of early experience on sensory system development should envision more than demonstrating "an effect"; systematical control of the acoustic experience and careful analysis of structural and functional changes at successive levels of the auditory system could lead to more complete understanding of general principles.

Behavioral analyses of experiential influences on auditory perception have been more systematic, especially those using avian species. Essentially four methods have been used: i) Auditory deprivation; ii) Alterations of the normal acoustic environment by application of artificial stimuli; iii) Artifical application of normally occurring stimuli or some variation thereof; and iv) Inferential analysis drawn on correlations of early acoustic experience and later behavioral responses.

Broad-band auditory deprivation appears to retard the development of several auditory functions. WOLF'S (1943) early studies indicated that when rats are reared in relative sound isolation, learning to use an acoustic signal as a cue for finding food is retarded. TEES (1967a, b) found that perceptual discriminations of stimulus duration and pattern are impaired. Since neither investigator examined normal ontogeny of auditory perception, the developmental nature of these deficits cannot be assessed. Studies on ducklings and chicks, on the other hand, have addressed development. GOTTLIEB'S investigations of duckling embryos and hatchlings (Section C.I) have shown that the development of differential responsiveness to species-typical call characteristics, including repetition rate, dominant frequency, and frequency subunits, is delayed when the organisms are deprived of hearing their own and their siblings' vocalizations. Our own investigations of the influence of auditory experience have concerned the ontogeny of differential responsiveness to stimulus changes that vary along the frequency dimension. After confirming that 3–4-day-old neonatal chickens make finer frequency distinctions than 1-day-old hatchlings (Fig. 8) KERR and RUBEL (1977) used plastic ear plugs to attenuate auditory input over this period. The ear plugs were evaluated by brainstem evoked potential frequency threshold functions, and found to attenuate acoustic input 38–42 dB across the entire audible range. Figure 9 shows the results of generalization testing on deprived animals 45 min after the ear plugs were removed, compared to results from normal 3- to 4-day-old chicks, and chicks that went through the deprivation procedures except for insertion of the ear plugs (shams). The results clearly show that deprived subjects have flatter generalization functions than either normal or sham subjects, and were comparable to 1-day-olds in Fig. 8. It is of importance that the deprived animals do not differ from the other groups at 800 Hz or 1000 Hz, indicating that the results are not due to differences between the groups in either "general arousal" or sensitivity to acoustic stimuli (i.e. increased thresholds), but are specific to categorization of sounds along the frequency dimension. Taken together, Figs. 8 and 9 clearly indicate that deprivation has retarded the normal perceptual sharpening that takes place during the first 3–4 days after hatching in the chicken. Whether this retardation of perceptual discrimination will be corrected during longer periods of deprivation or with exposure to a normal acoustic environment, and the extent to which the perceptual sharpening is dependent on the specific frequencies available in the organism's environment are problems currently being investigated.

Experiments in which an artifical auditory stimulus is applied during some period of embryogenesis, and later behavior is examined (see Table 4, Item 5) are of interest for at least two reasons. First, as noted by GOTTLIEB (1968a), it should not necessarily be presumed that the perception of a sensory stimulus will lead to a measurable motor response. That is, the formation of sensory-driven reflexes, which require a connection between the receptor-activated neural pathways and effector networks, may evolve under different selection pressures than the receptor systems themselves. For example, if externally or internally generated stimuli are important for the organization of the receptor and neural information processing

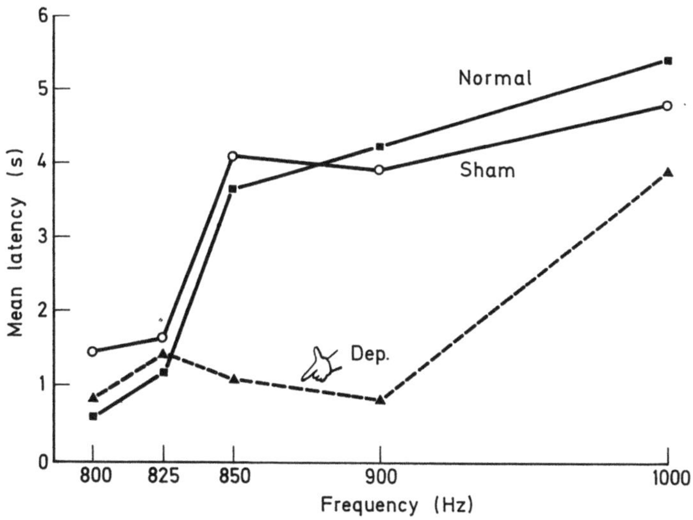

Fig. 9. Same paradigm as Fig. 8. Three groups of 3- to 4-day-old chickens (50/group) were used. Deprived (Dep) subjects had been deprived of 40—45 db acoustic stimulation from embryonic day 18.5 until approximately 1 hr before testing. "Sham" subjects were treated like deprived group except for actual acoustic deprivation. "Normal" subjects were untreated controls. All subjects were habituated (12 trials) to 800-Hz (90 dB, SPL) stimulus and then tested at one of indicated frequencies (each data point represents mean of 10 subjects): 800 Hz (control group for each age), 825 Hz, 850 Hz, 900 Hz, 1000 Hz. "Latency" indicates time to resume distress calling following stimulus, time subject orients to the stimulus. Normal and "sham" subjects show differential orientation at 850 Hz, while subjects deprived of normal acoustic experience do not show differential orientation until 1000 Hz—200 Hz from training stimulus. (From KERR and RUBEL, 1977)

networks, such stimuli may not elicit any observable reflexive response. Thus, one method by which the onset of sensory function can be determined is through brief exposure to a stimulus at one point in development and later assessment of its influence on behavior development. The second and more usual rationale for these experiments is to ascertain how embryonic stimulation may influence later behavioral responses, such as conditioning or following preferences. Of at least equal importance are possible effects of sensory experience on the rate of behavior development, e.g. time at which hatching occurs (see VINCE, 1973).

Experimental manipulations of an organism's normal acoustic environment have been systematically attempted only for avian species[4]. Most such studies are concerned with how specific aspects of an organism's auditory experience come to control species-specific or individual-specific perceptual preference behaviors. By sytematically varying the parameters of acoustic stimuli impinging on the avian embryo, GOTTLIEB has shown that preference behavior can be markedly influenced. Such manipulations may also influence the rate of perceptual development (GOTTLIEB, 1971a) and the rate of behavioral and physiological maturation processes in general (VINCE, 1973). Thus, it is critical that future investigations are

[4] In addition to the studies listed in Table 4, the extensive literatures on the effects of auditory experience on the ontogeny of bird song and human language are relevant, but not considered, as these subjects concern primarily "production" rather than "reception".

Table 4. Experiential influences on auditory system development

Treatment/dependent variable	Species	Comments	Relevant references
1. Sound in prenatal environment Human and goat	Human, goat	Intrauterine ambient noise in low frequency range (20–200 Hz) as high as 90 dB (SPL); attenuation of environmental sounds increases from 20 dB @ 50 Hz to 50 dB @ 2000 Hz	WALKER et al. (1971), BENCH (1968), BENCH et al. (1970)
2. High-intensity sound exposure (cf. SAUNDERS and BOCK (1977) for review)			
Human hearing loss		Available data suggests increased susceptibility to loss of high frequency sensitivity with repeated childhood exposure	WEBER et al. (1967), MILLS (1975), (Review), SAUNDERS and BOCK (Review)
Noise adaptability	Human	Prenatal environment may influence adaptability to postnatal noise exposure	ANDO and HATTORI (1970)
Cochlear changes	Cat, guinea pig, mouse	Immature cochlea shows increased susceptibility to temporary and permanent threshold shifts as well as permanent structural damage	LURIE (1940), PRICE (1972), FALK et al. (1974), SAUNDERS and HIRSCH (1976)
CNS changes	Mouse	"Primed" mice of normally seizure-resistant strains show increase in evoked potential thresholds coupled with increased amplitude of evoked potential at high intensity (loudness recruitment), lowered threshold for Preyer reflex and seizure susceptibility	HENRY (1972), WILLOTT and HENRY (1974), HENRY and HAYTHORN (1975), SAUNDERS and BOCK (1977)
3. Neonatal asphyxiation	Human, kitten	Perinatal asphyxiation can decrease cell number and nuclear volume in cochlear nuclei; changes appear permanent	HALL (1964)
4. Auditory deprivation CNS anatomical changes	Mice	Binaural deprivation during days 3–45 resulted in decreased cell size and neuron density in some brainstem nuclei, and increased packing density in inferior colliculus	WEBSTER and WEBSTER (1976)

CNS physiological changes	Rat	Monaural deprivation during first month resulted in loss of ipsilateral inhibitory influences in inferior colliculus. Deprivation for 8 months resulted in evoked potential threshold increase of 30–60 dB which disappeared following 3 weeks' exposure to normal environment	Batkin et al. (1970), Silverman et al. (1975)
Behavioral changes—mammals	Rat	Early auditory deprivation resulted in deficits in learned response to acoustic signal when compared with visually deprived subjects after 10 days of normal environment. Immediate deficits after deprivation for first 60 days in duration and pattern discrimination learning but no deficits in simple frequency or intensity discrimination learning	Wolf (1943), Gauron and Becker (1959), Tees (1967a and b, 1976)
Behavioral changes—birds	Ducks, Chicks	In duckling deprivation retards development of normal embryonic changes in response to species-typical calls and disrupts posthatching perceptual preferences for normal species-typical calls by interfering with ontogeny and maintenance of high-frequency selectivity. Normal sharpening of frequency generalization gradient fails to occur during deprivation	Gottlieb (1968a, 1971a, 1976b), Kerr and Rubel (1977)
5. Altered acoustic environment			
CNS changes	Rats, Hamsters	Selective exposure of rats to increasing or decreasing frequency modulation during first 4 months tends to depress inferior colliculus unit responses to white noise following *unfamiliar* modulation; little or no effect on number of units responding to either FM stimulus. Global "enrichment" by continuous exposure to radio does not influence latencies. In hamsters continuous exposure to 85 dB white noise during first 3 months increases cortical evoked potential threshold; possible cochlear damage also	Clopton and Winfield (1974, 1976), Mailloux et al. (1974), Pujol et al. (1975)
Lack of CNS change	Cat	No change in inferior colliculus unit responses following increase in daily exposure to 1000 Hz tone	Moore and Aitkin (1975)

Table 4 (continued)

Treatment/dependent variable	Species	Comments	Relevant references
Behavioral changes	Chicks, quail	Exposure to intermittent tone during embryonic days 13–18 increases posthatching "attractiveness" of the familiar frequency; and conditioning to bell during late embryogenesis may influence postnatal response to bell. Brief (2 h) stimulation with broad-band noise may accelerate hatching in quail	HUNT (1949), GRIER et al. (1967), RAJECKI (1974), WOOLF et al. (1976)
6. Artificial variations of normal acoustic environment	Guillemots, gulls, ducklings		
Behavioral influences		Perinatal exposure to parental, siblings' and self-produced calls influences postnatal behavioral response specificity by: i) determining individual characteristics of parental call recognition (guillemots and gulls); or ii) affecting rate of development and maintenance of preference for species-typical characteristics	TSCHANZ (1964, 1968), GOTTLIEB (1966, 1971a, 1976), IMPEKOVEN (1969), IMPEKOVEN and GOLD (1973)
7. Inferential Analysis of Naturally Occurring Prenatal and Perinatal Stimuli			
Postnatal behavior	Gulls	Approach responses of chick initially unselective with respect to parental or unfamiliar call but specific to call-type; over first 2–3 weeks becomes specific to parental vocalizations but nonselective regarding call-type	BEER (1970a, b, 1972), EVANS (1970a, b)
	Humans	Heartbeat sound played to newborn infants resulted in increased weight gain and decreased crying; heartbeat played to 3-year-olds resulted in decrease in latency to fall asleep	SALK (1962)
8. Crossmodal effects			
Cortical development	Mice, rats	Neonatal visual deprivation, eye enucleation or vibrissa removal resulted in altered growth of cells and increased dendritic spines in auditory cortex	GYLLENSTEN et al. (1966), RYUGO et al. (1974)

designed in a way that will allow dissociation of general metabolic from specific perceptual effects.

The last method that has been used to examine the role of auditory experience in perceptual ontogeny attempts to correlate the normal auditory experience of an organism with later perceptual behavior. The research of BEER and EVANS on gulls has convincingly shown that experience of the chick allows it to recognize individual nuances of the parental voice, and to generalize this throughout the range of parental vocalizations. A conceptually similar approach was used in experiments with human neonates and toddlers by SALK (1962). Although the latter publication received much attention in the popular press, it has been largely and unjustly ignored in the scientific literature. The theoretical and practical importance warrants replication and extension of SALK's findings.

Finally, intermodal influences on the ontogeny of the auditory system should be considered. Two investigations suggest hypertrophy of the auditory cortex following alterations in visual and somesthetic stimulation (Table 4, Item 8). While a great deal more research is needed in this area, these studies imply that "increased reliance" upon the auditory system may affect the development of neuronal elements. Thus, an integrated series of behavioral and neurological investigations of the effects of crossmodal manipulations may yield additional understanding of how psychological constructs such as "attention" are transformed into meaningful neuronal attributes contributing to the regulation of behavior.

C. Developmental Neurobiology and Auditory System Ontogeny

This section gives brief descriptions of three research programs in which the vertebrate auditory system is being investigated in order to solve general problems of developmental neurobiology. The purpose is to demonstrate that some of the unique characteristics of the vertebrate auditory system make it especially applicable for illuminating general principles of nervous system and behavior development. In addition to the research program currently under way in the present author's laboratory, GOTTLIEB's investigation of the ontogeny of species-specific perceptual behavior in embryonic and neonatal ducklings, and MOREST's examinations of neuronal migration, synaptic terminal growth, and dendritic elaboration are described. Of the many excellent research programs involving the elaboration of auditory processes, these two have been selected principally because of their contributions toward an understanding of general problems of developmental neurobiology, their systematic nature, and the present author's own familiarity with them. The summaries below are not intended to provide a thorough review of the authors' work, but rather to provide an overview, and to exemplify how study of the developing auditory system may increase understanding of general principles of nervous system and behavior development.

I. The Ontogeny of Species-Specific Perceptual Behavior

While other investigators interested in perceptual development, especially preference behavior, were stressing the importance of imprinting for determining the stimulus characteristics guiding responsiveness to environmental stimuli, GOTTLIEB (1965a) discovered that when given a choice between a parental call of their own species and that of another species, naive ducklings and chicks overwhelmingly approach and follow the species-typical stimulus. Furthermore, even after extensive exposure or active "imprinting training" with the atypical stimuli, a clear preference for the mallard maternal call over that of the chicken or wood duck is shown by Peking (domesticated mallard) ducklings. In addition to calling for a re-evaluation of the interpretations commonly associated with imprinting, this series of studies posed the important question: During the normal course of ontogeny, what factors are responsible for the development of species-typical ("innate") behaviors? GOTTLIEB (1976b) and others have reviewed the various theoretical arguments around this question. Basically it comes down to the following distinction. Do the neural networks responsible for species-typical behaviors arise without relevance to an organism's sensory environment or behavior, or is the stereotype of the final phenotype partially due to commonalities of the developing organisms' interactions with their environment? GOTTLIEB'S original discovery of a marked preference for the species-typical maternal call among naive chicks and ducks was taken by some as an indication that perceptual preference behavior developed without regard to an organism's sensory environment. However, strongly influenced by the teachings of KUO (see KUO, 1967), GOTTLIEB has utilized this preparation to show that an organism's sensory environment, an important part of which is self-generated, plays an important role in the development of species-specific perceptual distinctions.

The specific details of these studies can be found in the original reports and have already been summarized (see GOTTLIEB, 1971a, 1976b). Briefly, in one series of investigations, the embryonic development of the ducklings' ability to differentially respond to the species-typical maternal call was evaluated. It was found that 4–5 days prior to hatching (days 22 and 23) the mallard call causes a reliable decrease in bill movement, while on day 24, when the animals normally enter the air space and pulmonary respiration begins, the mallard call begins to elicit an increase in bill movements and vocalizations while other calls do not elicit reliable changes in these activities. Following collection of these baseline data on normal, communally incubated embryos, GOTTLIEB demonstrated that by incubating eggs in sound-isolated compartments or exposing eggs to increased auditory stimulation (sibling calls played over a loudspeaker), the time at which the bill-clapping response to the mallard maternal call shifted from inhibitory to excitatory could be retarded or accelerated, respectively. While it is not possible to ascertain from these results whether the acoustic manipulations caused a generalized change in developmental rate (e.g. VINCE, 1973) or specifically influenced maturation of the auditory system, these findings do suggest that the kind of acoustic stimulation to which embryos are normally exposed influences the development of differential responding to species-typical vocalizations. The fact that 22-day-old embryos respond selectively to the mallard maternal call indicates that the ontogeny of the inhibitory response is independent of auditory experience.

In the second series of studies, GOTTLIEB directly tested the idea that an organism's own behavior can provide stimulation that influences the development of perceptual preference behavior. The development of a procedure for devocalizing duckling embryos (GOTTLIEB and VANDENBERG, 1968) allowed complete control over the acoustic environment. When devocalized, isolated ducklings were given a choice between the mallard maternal call and that of the wood duck or chicken, they were found to be deficient. Normal ducklings 24 hrs after hatching show an overwhelmingly preferential approach toward the mallard call; ducklings deprived of hearing their own and siblings' vocalizations during the perihatching period did not show a reliable preference between mallard and wood duck calls until later than normal hatchlings, and did not preferentially choose the mallard over the chicken call at 48 or 60 hrs after hatching (GOTTLIEB, 1971a). Upon further analysis of these deficits (see GOTTLIEB, 1976b, for summary), it has been found that both the repetition rate of notes in the maternal call and frequency parameters are bases on which normal ducklings choose the species-typical call. When deprived of hearing their own vocalizations or those of their siblings, deficiencies in the development and the maintenance of the discriminative behavior result, and when provided with supplementary stimulation that contains the appropriate parameters, the preferential responding is regained.

In summary, GOTTLIEB has capitalized on the auditory perceptual behavior of avian embryos and hatchlings to demonstrate that the neural substrate responsible for the ontogeny of species-typical or "innate" behaviors may be powerfully influenced by normally occurring environmental events, some of which are provided by the developing organism itself. GOTTLIEB himself has put it more fully in the following passage from his 1971 monograph:

> Because of the demonstrated contribution of normally occurring stimulation in regulating the embryo's and hatchling's usual responses to the maternal call of their species, it can be said that the epigenesis of species-specific perception is a probabilistic phenomenon, and does not represent merely an unfolding of a fixed or predetermined organic substrate independent of normally occurring sensory stimulation. According to the present results, natural selection, insofar as it has played a role in the evolution and maintenance of species-specific perception, would seem to have involved a selection for the entire developmental manifold, including both the organic and normally occurring stimulative features of ontogeny (pp. 148—149).

II. Post-Mitotic Migration and Differentiation of Neurons in the Central Nervous System

In what is deservedly becoming a classic series of reports (MOREST 1968, 1969a, b, 1970a, b), MOREST has made several important contributions to our understanding of the processes involved in neuronal differentiation. The problem addressed by MOREST can be stated as follows. Given that neurons are assembled in definitive locations, and that at a given location many of the architectural characteristics of both presynaptic and postsynaptic processes are highly stereotyped, how are these factors regulated? More specifically, what causes the soma of a neuron to migrate to a specific location and by what route does it get there? What factors lead to the stereotyped form of terminal arborizations in a given location, especially when collaterals to another location have a completely different form? And by what means do particular dendrites of a neuron acquire their highly stereotyped form, when other dendrites of the same neuron have very different architecture? While MOREST's studies do not, by any means, answer these complex questions, they do provide important insights toward eventual clarification.

In contrast with the experimental investigations described above, MOREST's studies used tissue from normal animals. A second important difference is that while GOTTLIEB's research program has a single overriding theoretical premise, the present studies are descriptive in nature, attempting to define the normal sequence of events occurring in the development of a neuron and then inferring possible causal relationships. Finally, GOTTLIEB'S empirical work is restricted to the ontogeny of perceptual behavior in chicks and ducklings, and the generality of his findings remains to be determined; whereas MOREST made his initial observations on auditory centers of the brainstem, principally the medial trapezoid nucleus (MOREST, 1968, 1969a), but has made the additional important step of assessing their generality in other parts of the brain (MOREST 1969b, 1970a, b).

MOREST has primarily used a modification of the rapid Golgi method (MOREST and MOREST, 1966) and supplemented this with tissue impregnated by Golgi-Cox and Golgi-Kopsch methods. The ontogeny of the principal neurons of the medial nucleus of the trapezoid body formed the basis of the original observations. This nucleus, which receives a principal input in the form of large calyces of Held from the contralateral ventral cochlear nucleus, was chosen because of several unique characteristics. First, the principal cells are large neurons, characteristic in shape, that can be recognized throughout most of mammalian embryogenesis. Second, their dendritic arborizations are sufficiently distinctive that throughout development they can be differentiated from dendrites of the stellate and elongate cells of the same nucleus. Third, each principal neuron perikaryon receives synaptic input from one and only one axonal calyx of Held. Finally, the calyces of Held are among the largest synaptic endings in the mammalian brain; they are architecturally homogeneous, and they are only seen on the cell bodies of the principal neurons; because of these characteristics they can be unambiguously recognized throughout their development. These characteristics, therefore, yield a preparation uniquely suited to analyses of the normal ontogeny of post-mitotic neuronal development, especially with respect to temporal characteristics of changes in presynaptic and postsynaptic elements occurring during the period of synaptogenesis.

Three aspects of neuronal differentiation were clarified in MOREST's papers: i) The final stages of neuronal migration; ii) The elaboration of presynaptic elements; and iii) The elaboration of dendrites. Most previous descriptions of the migratory action of neurons were made from Nissl-stained material and pictured undifferentiated, round cells from the mantle layer moving by ameboid action through adjacent tissue to their final resting place. After arriving at the final destination, axons and dendrites were presumed to extend from the soma, and concurrently or soon thereafter afferent axons formed synapses on the cell body and dendritic processes. By following the presynaptic and postsynaptic elements of the principal neurons of the medial trapezoid nucleus from closely timed litters of cats, rabbits, and opossums, MOREST proposed quite a different view. When first recognized, the neogenic neurons lie near the outer limiting membrane, to which they are attached by a process ending on the membrane. Another process extends toward the lumen. Both processes are characterized by large growth cones and numerous filopodia. As the afferent axons grow into the area that will later become the medial trapezoid nucleus, they initially contact the internally extending primitive process. The axonal extension of the principal neuron apparently arises from this internal protoplasmic process at a site within what will become the nuclear boundary. The nucleus of the neuron moves through this internal primitive process, to a position adjacent to the origin of its axon and where the presynaptic endings are differentiating.

The incipient calyces of Held grow in the trapezoid body through the ipsilateral nucleus, apparently without forming connections with the principal neurons or their processes. Continuing in the decussation of the trapezoid body, they enter the position of the medial nucleus as large growth cones with numerous filopodia radiating through the tissue. When they approach and contact a principal neuron or its internal process, many thin sinuous branches entwine the postsynaptic elements. Gradually, as the neuronal nucleus moves into the vicinity of the afferent ending, the branches either coalesce or are replaced by the broad branches forming definitive calyciform endings. After forming its definitive calyx on one principal neuron, the afferent fiber issues collateral growth cones, which eventually ramify extensively in the surrounding areas of the nucleus.

Dendritic differentiation appears to follow a similar course. Dendrites arise from the neogenic principal neurons as growth cones with abundant filopodia extending from the dendritic ramifications and cell body alike. Gradually, the dendrites establish their stereotyped branching pattern and spine configuration while losing the transitory filopodia. Concurrently with the elaboration of dendritic stereotypy, afferent axons form their characteristic presynaptic configuration. That is, "the development of the dendritic branches accompanies the elaboration of the particular type of axonal plexus that will become synaptically related" (MOREST, 1969a, p. 271).

These descriptions of the events surrounding synaptogenesis and dendritic elaboration are normative. They provide a catalog of several important morphogenetic events, and a timetable of their occurrence. By close examination of simultaneously and successively occurring processes in functionally related neuronal elements, MOREST was also able to make a number of provocative suggestions regarding the importance of temporal contiguities and mutual interactions of incipient presynaptic and postsynaptic processes for sculpting the final functional elements. Although most of the hypotheses put forth regarding the inductive roles of presynaptic and postsynaptic elements remain to be adequately tested, the heuristic value of these contributions for understanding the elaboration of neuronal networks is evident. Furthermore, thorough descriptions of this kind are necessary before definitive experiments can be designed or their results interpreted.

III. Peripheral Influences on the Ontogeny of Neural Structure and Function

The research program presently under way in the present author's own laboratory is in many respects an attempt to integrate the approaches and findings of GOTTLIEB and MOREST. Simply stated, the overall objective is to understand the ways in which the development of neural structure and function are influenced by stimuli impinging upon an organism from its environment. While this is not a unique problem, most previous approaches have utilized the mammalian or amphibian visual system (see Chapter 7). Although these preparations have yielded a number of interesting discoveries, general principles by which experiential events affect the ontogeny of neuronal networks have not been forthcoming. The characteristics of structural and functional organization found in the avian brainstem auditory pathways offer a system uniquely suited to investigations relating neural ontogeny to afferent input.

Two considerations have guided our work on this preparation. First, the assumption is made that an organism's sensory experience influences neural ontogeny through chronic changes in the amount or pattern of afferent activity impinging on the nervous system. Thus, to understand the principles of experiential influences it is necessary to determine the effects of known changes in the quality or quantity of afferent input on the functional and structural ontogeny of neural networks. Viewed in this way, "experiential influences" are considered as one type of tissue interaction that may play a role in the ontogeny of neural elements and the relative influences of both drastic and subtle changes in afferent input at each stage of development are of importance. While total elimination of afferents to a neuron almost certainly interrupts more cellular processes than subtle changes in the pattern of synaptic activation, at least a portion of the syndrome produced by denervation is due to the elimination of synaptic activity. Thus, the examination of developmental changes produced by total removal of a class of afferents should yield a catalog of events, each of which can be systematically evaluated upon progressively more subtle manipulation of the afferent supply.

The second consideration refers to the necessity of using a developmental approach to the processes under examination. That is, for a full appreciation of the principles by which afferent input can influence the development of structure within a neural network, it is necessary to understand the normal sequence and timing of developmental events. Only then is it possible to ascertain the developmental nature of changes induced by perturbations of the neuronal environment (SOLOMON and LESSAC, 1968; GOTTLIEB, 1976a). Furthermore, as exemplified by the work of MOREST, a thorough normative analysis can often provide valuable insights into the possible tissue interactions responsible for the ontogeny of normal neural structure and function.

These and other considerations led the present author to select the avian brainstem auditory pathways for investigation. More specifically, *nucleus magnocellularis* and *nucleus laminaris*, which are comprised of 2nd- and 3rd-order auditory neurons respectively, appeared to be uniquely suited to investigations of the role of afferents in neuronal development and maintenance. Furthermore, behavioral studies on the ontogeny of auditory information processing provided an interesting behavioral parallel in which the role of acoustic experience could be investigated. Our initial investigations (RUBEL and PARKS, 1975; PARKS and RUBEL, 1975) were primarily intended to provide detailed structural and functional information on the organization of this neuronal system, in order to determine its applicability to this line of investigation. The normal morphological ontogeny of *n. magnocellularis* and *n. laminaris* at the light microscope level was then described (RUBEL et al., 1976), and these observations are currently being extended by neurophysiological and electron microscopic analyses of normal development. Finally, the present author's group has begun to describe how modifications of afferent input can alter the development and maintenance of the normal neuronal elements in this system (PARKS and ROBERTSON, 1976; JACKSON and RUBEL, 1976; BENES et al., 1977; PARKS and RUBEL, 1977). The following account briefly describes the results of these investigations. The emphasis will not be a close examination of the results obtained in these studies, but to show how the unique characteristics of this particular neuronal system make it ideally suited to a thorough analysis of the various roles that afferent information may play in the ontogeny of neural structure and function.

The primary auditory pathways to the medulla of the chicken embryo and hatchling are schematically shown in Fig. 10. Axons of the first-order neurons, situated in the cochlear ganglion, enter the dorsolateral quadrant of the medulla and bifurcate into a lateral and a medial bundle. The lateral branches terminate in a cochleotopic manner in nucleus angularis (NA), which is a triangular nucleus composed of mixed cell types lying in the dorsolateral angle of the medulla. The medial branches of the VIIIth nerve course along the floor of the fourth ventricle (IV) to terminate in *n. magnocellularis* (NM). N. magnocellularis is a well-defined cluster of large round or ovoid cell bodies, which are either devoid of dendrites or have short bushy rudimentary dendrites. The cells are organized in a columnar manner (dorsal-ventral or dorsolateral-ventromedial) with VIIIth nerve afferents coursing between the columns and terminating principally on the cell bodies as large endbulbs. As also shown in Fig. 10 *n. laminaris* (NL) in chickens is composed of a discrete monolayer of cell bodies with bipolar dendritic orientations. Cells of *n. laminaris* receive binaural, spatially segregated innervation from the magnocellular nuclei; axons from NM pass in the uncrossed dorsal cochlear tract to innervate the dorsal dendrites and cell bodies of the ipsilateral *n. laminaris*, and decussate in the crossed dorsal cochlear tract to innervate the ventral dendrites and the cell bodies of the contralateral laminar nucleus. This relatively simple structural organization, the accessibility of avian embryos for neurophysiological studies or peripheral manipulations, and the possibilities for precise control over the acoustic environments of embryos and hatchlings, warranted further investigation of the functional organization of *n. magnocellularis* and *n. laminaris.*

Neurophysiological examination of these nuclei and a detailed analysis of the projections from *n. magnocellularis* to *n. laminaris* revealed the following properties. *N. magnocellularis* cells respond only to a narrow range of frequencies played to the ipsilateral ear. The cells display sharp excitatory tuning curves and primary-type response histograms, and always have a definable excitatory characteristic

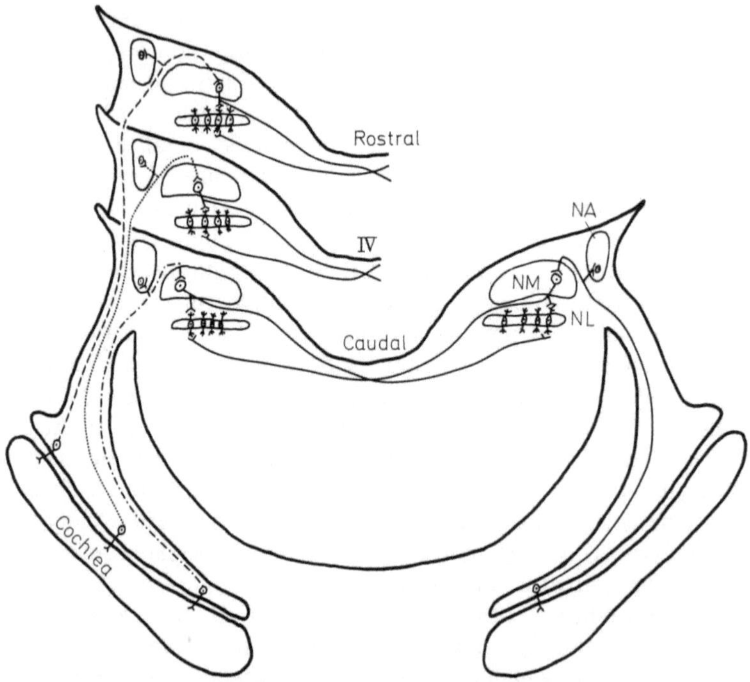

Fig. 10. Schematic drawing showing projections from cochlea *via* cochlear ganglion to *n. angularis* (NA) and *n. magnocellularis* (NM); and spatially segregated, bilateral projections from *n. magnocellularis* to *n. laminaris* (NL). IV—4th ventricle

frequency. The cells are organized in isofrequency columns, in which all cells respond to a similar characteristic frequency, and the physiologically-defined columnar organization corresponds to the anatomically observed cell columns. *N. laminaris* cells are binaurally activated by acoustic stimuli and usually show similar characteristic frequencies and thresholds to stimulation of each ear. That is, NL neurons are maximally activated by the same frequency range applied to either ear. In other respects their responses are similar to neurons of NM. The organization of neurons of both nuclei is characterized by a highly stereotyped and similar tonotopic organization; cells maximally responsive to low frequencies are situated in the caudolateral aspect of each nucleus and higher frequencies activate cells at progressively rostromedial positions. This tonotopic organization is sufficiently stereotyped that quantitative analyses of the relationships between the position of a cell in each nucleus and its characteristic frequency allows accurate prediction of this functional characteristic (within 200–400 Hz) from positional information alone. The utility of this finding is that it allows independent prediction of the characteristic frequency of a neuron, which can be compared to results obtained as a function of manipulations of an organism's acoustic environment.

Small lesions made in *n. magnocellularis* through tungsten microelectrodes after recording the characteristic frequency of cells in the lesioned area allowed further delineation of the organization of projections from *n. magnocellularis* on one side of the brain to both laminar nuclei. Analysis of this tissue confirmed the spatial segregation of binaural projections onto each laminar nucleus (see Fig. 10). In addition, it was found that a lesion placed in a restricted portion of *magnocellularis* resulted in terminal degeneration in corresponding rostrocaudal and mediolateral positions in the two laminar nuclei. That is, there is a precise tonotopic and topographic projection from each region of *n. magnocellularis* to the dorsal side of ipsilateral *n. laminaris* cells and the ventral aspect of neurons in the corresponding position on the other side of the medulla. Furthermore, observations of Golgi-stained tissue have revealed that the dendrites of *n. laminaris* cells are confined to the glia-free margin surrounding the perikaryon lamina (see Fig. 13). These results allow precise delineation of the receptive cell surfaces of *n. laminaris* neurons as well as unusually precise specification of the source and functional properties of input to each dendritic surface of the neurons.

The morphogenesis of *n. magnocellularis* and *n. laminaris* was then investigated in several closely timed series of chicken embryos (RUBEL et al., 1976). The primary purpose of this investigation was to provide qualitative and quantitative baseline data describing the timing and extent of major morphogenetic events in the ontogeny of this neuronal network. Qualitative changes in nuclear structure were studied in Nissl- and silver-stained tissue from embryos between five days of incubation and hatching. The times at which *n. magnocellularis* and *n. laminaris* neurons pass through their final cycle of DNA synthesis ("cell birth-dates") were determined by ^{3}H-thymidine labeling. Finally, quantitative analyses of cell number and nuclear volume were conducted for each nucleus over the embryonic period from nine days of incubation through hatching.

The first major morphogenetic event to be considered in the analysis of any neural region is the developmental stage and duration of time during which cellular proliferation occurs. Cumulative labeling with ^{3}H-thymidine can be used to determine the time at which cellular proliferation has ceased and the occurrence of heavy labeling is a good indicator that the time of isotope injection was during one of the final mitotic cycles. Thus a thorough analysis by means of this technique allows one to estimate the period of development during which the cells of a given nuclear region pass through their final proliferative phase. Our analysis indicated that there is no overlap in the final period of DNA synthesis of NM and NL. NM cells consistently showed heavy labeling when injections were made at 60 hrs (2.5 days) of incubation and there was no indication of reliable thymidine uptake at 72 hrs of incubation. NL neurons, on the other hand, did not show heavy labeling unless the isotope was injected at 84 or 96 hrs of incubation. Figures 11 and 12 demonstrate these relationships in two animals, which were incubated together and treated identically except for the time of a single application of tritiated thymidine. Both animals were sacrificed at 17 days of incubation. Administration of ^{3}H-thymidine at 60 hrs of incubation (Fig. 11) produced heavy labeling of NM cells and relatively fewer grains over NL neurons, while application at 84 hrs (Fig. 12) resulted in heavy labeling of NL alone.

Summarizing the labeling experiments, it appears that the majority of NM neurons leave the mitotic cycle between 48 and 72 hrs of incubation, while the final production of NL cells is delayed by approximately one day. Both cell groups are produced in the region of the rhombic lip, and when first recognized in the embryo

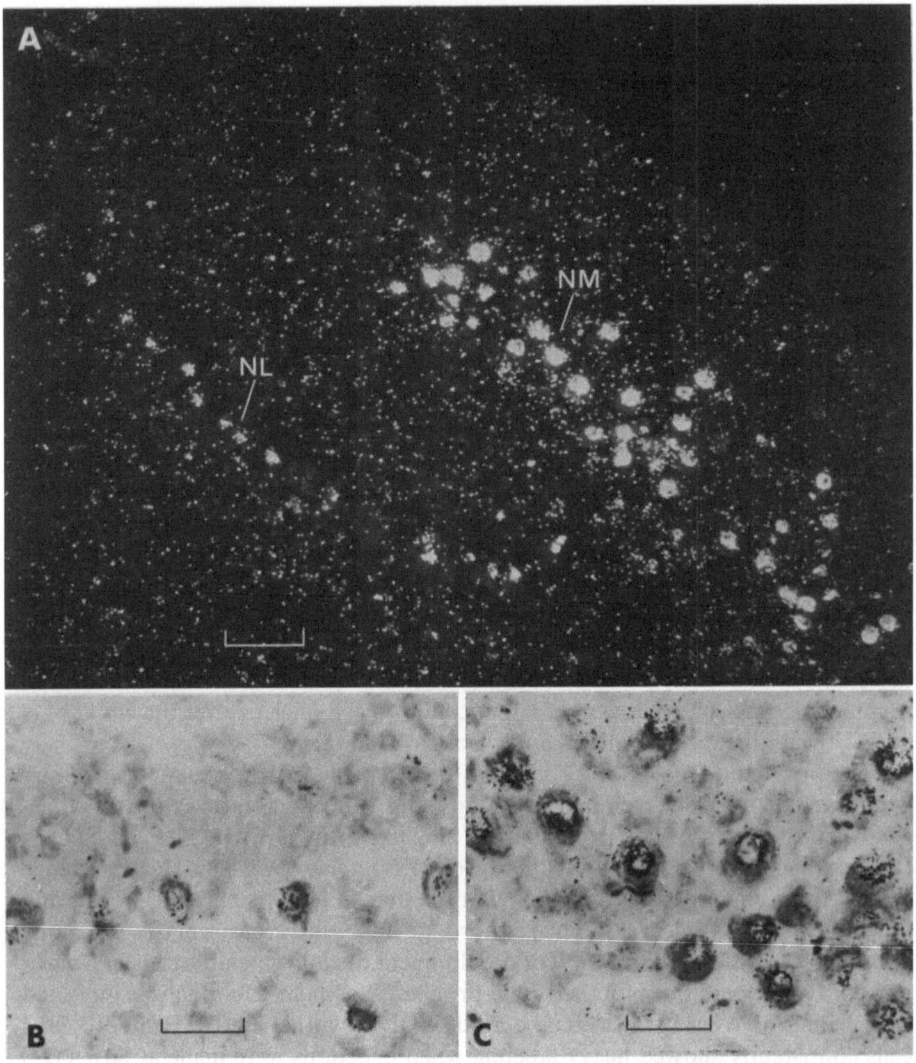

Fig. 11. (A) Autoradiograph of *n. magnocellularis* and *n. laminaris* as seen under darkfield optics. Tritiated thymidine (25 μCi) was injected at 60 hrs of incubation, and animal was sacrificed at 17 days of incubation. *N. magnocellularis* cells are seen to be heavily labeled and *n. laminaris* cells are lightly labeled. Bar = 50 μm. (B) Light-field photomicrograph of Nissl-stained *n. laminaris* cells, taken from same section as above. Silver grains are seen over nuclei of nearly all cells. Bar = 20 μm. (C) Lightfield photomicrograph of Nissl-stained *n. magnocellularis* cells from same section, showing dense silver grains over the nuclei of cells. Bar = 20 μm. (From RUBEL et al., 1976)

(about six days) they overlap throughout most of their rostrocaudal and mediolateral extent. It is therefore reasonable to hypothesize that both cell groups are produced by a single progenitor population, with NM cells being formed first and migrating away from the mother cells, and then, in a second wave of mitotic activity, the final population of NL cells being formed around 84–100 hrs of incubation. This temporal sequence of cellular proliferation, coupled with the fact

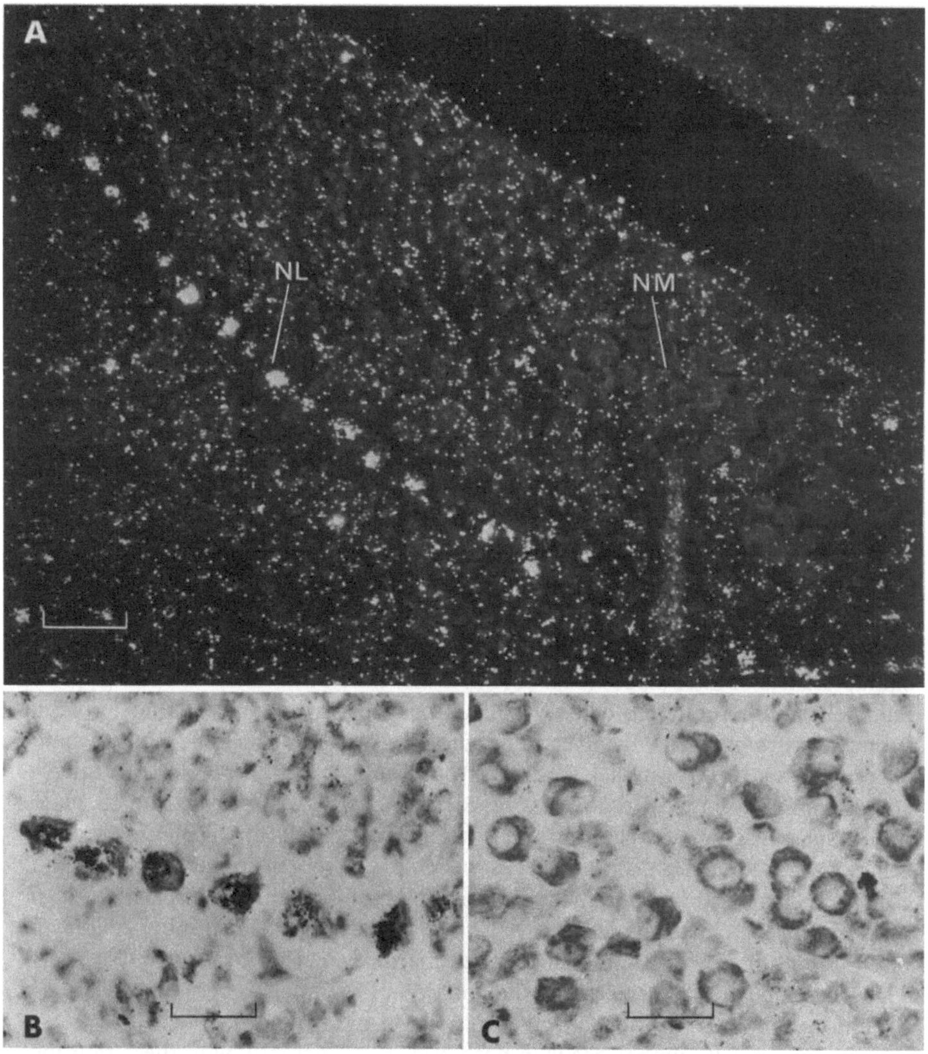

Fig. 12. (A) Darkfield photomicrograph of *n. laminaris* and *n. magnocellularis* showing result of tritiated-thymidine labeling at 84 hrs. Animal was from same group as that shown in Fig. 11, and was handled identically except for time of thymidine injections. *N. laminaris* cells can be seen to be heavily labeled, while *n. magnocellularis* neurons are not labeled. Bar = 50 µm; (B) Lightfield photomicrograph of Nissl-stained *n. laminaris* cells taken from same section as above. Nuclei of cells show dense accumulations of silver grains. Bar = 20 µm; (C) Lightfield photomicrograph of *n. magnocellularis* cells taken from same section. Few grains are seen over any cells, and labeling over cells of nucleus was no higher than background. Bar = 20 µm. (From Rubel et al., 1976)

that NL cells come to lie directly ventral to NM, suggests that the cells may interact during their proliferative or migratory phases. The hypothesis that presynaptic cells may influence proliferation or the early stages of differentiation of their postsynaptic target population has often been considered (e.g. see Lauder and Bloom, 1974). However, there is little evidence to support this conjecture and no uniformity in the temporal sequence of proliferation of synaptically related cells. A more common finding is that cell groups that will later be in synaptic contact pass through one another during the migratory phase of development (ANGEVINE and SIDMAN, 1961; MIALE and SIDMAN, 1961; HANAWAY, 1967; COWAN, 1971; LANGMAN et al., 1971). This finding has led to the suggestion that cellular interactions during nerve cell migration may influence the pattern of synaptic connections that will later be manifest, or, as discussed earlier, may influence the migratory route of the nucleus through primitive protoplasmic processes. Little direct evidence for this hypothesis is available for central nervous system tissue, but we have observed in embryos around days 5–7 of incubation that cells labeled at the same time as NL cells can be observed streaming through the magnocellular nucleus. In any case, the occurrence and possible functions of such interactions might be particularly investigable in this system because it is known that there is a precise and relatively punctate topographic projection between the nuclei, which can be investigated morphologically or physiologically, and in addition, the two nuclei have nonoverlapping times of origin, which allows independent radiolabeling of either nuclear group.

Following proliferation and migration, these cell groups come to attain their stereotyped cytoarchitectural characteristics over the period from 9 to 15 days of incubation. By 9 days of incubation (Stage 34) both cell groups can be easily recognized although neither has assumed the cytoarchitectural or positional characteristics of later stages (Fig. 13A). At this time both nuclei are composed of relatively undifferentiated, densely packed cell bodies and there is no apparent subnuclear organization. In silver-impregnated tissue, it is apparent that the major afferent axons to both nuclei are present, although preliminary electron microscopic examination revealed no synapses at this time. Between days 9 and 13, major morphogenetic changes occur in both nuclei. The NM cells are displaced medially, cell density diminishes, and the cells become aligned into their characteristic columnar organization. The most striking changes occur in NL, where from the undifferentiated cluster of cells seen at day 9 a precisely defined monocellular or bicellular lamina, with a uniform glia-free margin, emerges by day 13 (Fig. 13B). This change begins in the rostromedial portion of the nucleus between days 9 and 11 and is completed in the caudolateral portion by around day 15.

Changes in cell number occur in both nuclei concurrently with these morphogenetic events (Fig. 14). Whereas NM loses only about 18% of its cellular population and all cell death appears to occur between embryonic days 11 and 13, cell death in NL is much more extensive (about 80%) and takes place over a considerably longer period (days 9–15). These changes in cell number are of interest for a number of reasons. First, while there are large differences in the amount and duration of cell death, the period of maximal cell loss in both nuclei is between 11 and 13 days of incubation. Since proliferation of NM and NL is separated by about 24–36 hrs, it appears that whatever factors regulate cell death may serve to bring the two nuclei into ontogenetic synchrony. Two such factors are the establishment of connections with an appropriate target population (COWAN, 1973; HAMBURGER, 1975) and the reception of afferent innervation (LEVI-MONTALCINI, 1949). Unfortunately, most experimental investigators of embryonic cell death have not been able to differentiate between these factors, because the

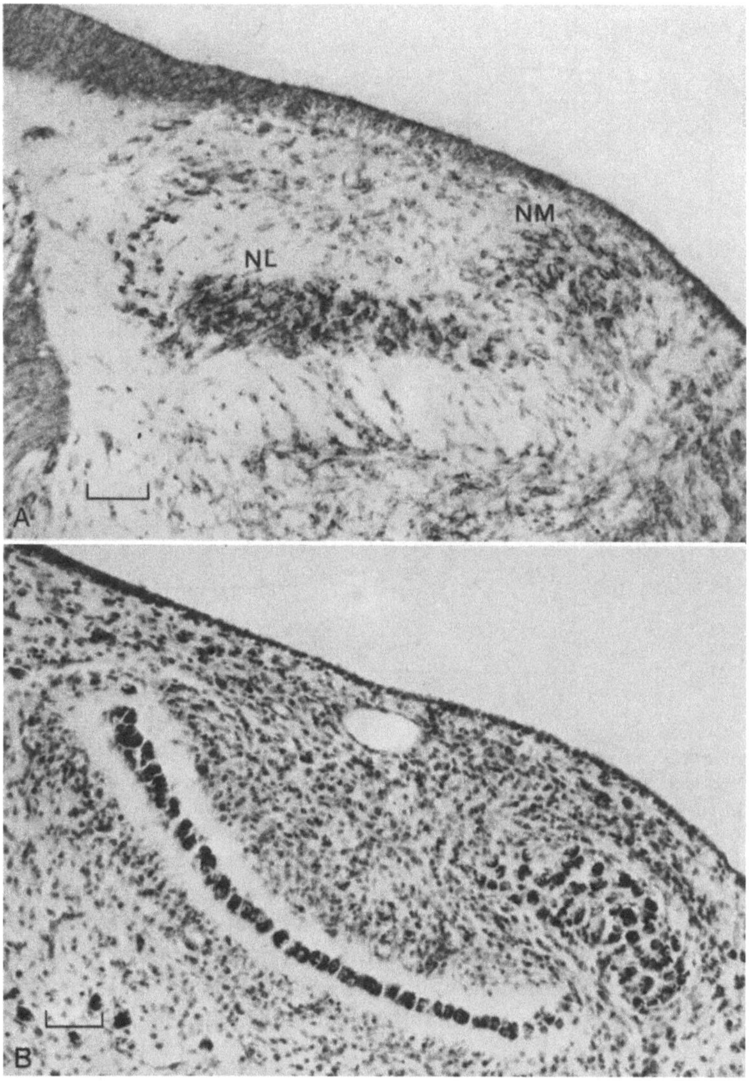

Fig. 13. (A) Nissl-stained section through *n. magnocellularis* and *n. laminaris* of 90-day chick. *N. laminaris* has not yet taken on its characteristic single-laminar appearance at this time; however, glia-free margin that surrounds it at later ages is already becoming apparent. Bar = 50 µm; (B) Nissl-stained section of *n. magnocellularis* and *n. laminaris* at 13 days of incubation. *N. laminaris* can be seen surrounded by glia-free neuropil, with lateral division (left) 2—3 cells deep, and medial division a single-cell layer. Bar = 50 µm. (From RUBEL et al., 1976)

manipulations performed altered both the target population and the afferent input, the latter often transneuronally. In this respect, several factors suggest that these avian auditory nuclei may be particularly useful, namely: both nuclei are relatively small and well-defined; the periods of cell death and cell proliferation are widely separated in time; both afferent and efferent axons have formed before the time of cell death; and neurophysiological investigations are relatively routine in

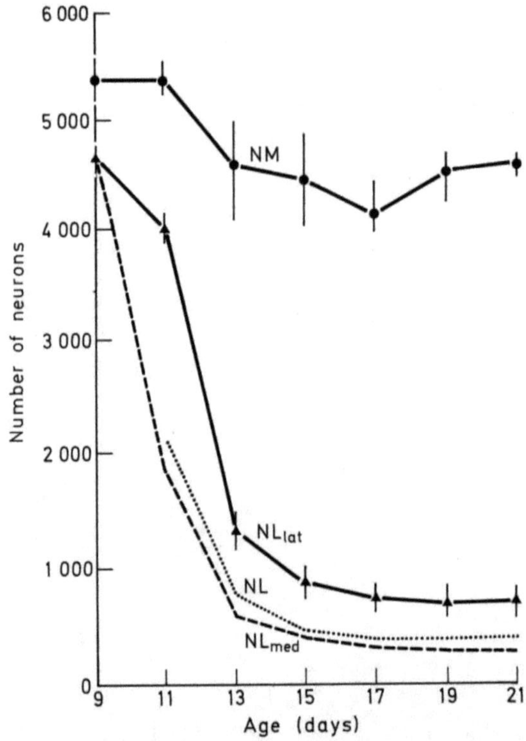

Fig. 14. Cell counts in *n. magnocellularis* and *n. laminaris* from 9 days to hatching. Points are means for each age, and vertical bars show ranges. Lines showing counts for divisions of *n. laminaris* are mean values. NM, *n. magnocellularis*; NL, *n. laminaris* (total); NL lat., lateral division of *n. laminaris*; NL med., medial division of *n. laminaris*

avian embryos, such that peripheral manipulations removing the major source of afferents to NM can be routinely accomplished without interfering with its efferent target, and the tectal target cells of NL can be removed embryonically without damaging their afferent innervation directly or transsynaptically. COWAN and his colleagues have made great contributions toward our understanding of embryonic cell death and have cautioned that for "further elucidation of the hypothetical relationships between cell death and the establishment of either afferent or efferent functional connections, it is necessary to know exactly when synapses are made, when functional activity emerges, and to be able to vary these factors in a systematic way" (ROGERS and COWAN, 1973). NM and NL of the chicken embryo appear to be the most suitable neuronal system in the vertebrate brain for accomplishing these goals.

The volumetric analyses of NM and NL (Fig. 15) also indicate a surprising amount of developmental synchrony in the enlargement of cell bodies and neuropil areas which occurs after the period of cell death. Since it is quite likely that functional connections are established by day 15, it will be important to understand the role that afferent activity may play in the regulation of this parameter as well. It is therefore important that the period of rapid cellular and neuropil expansions can be temporally separated from that of cell death.

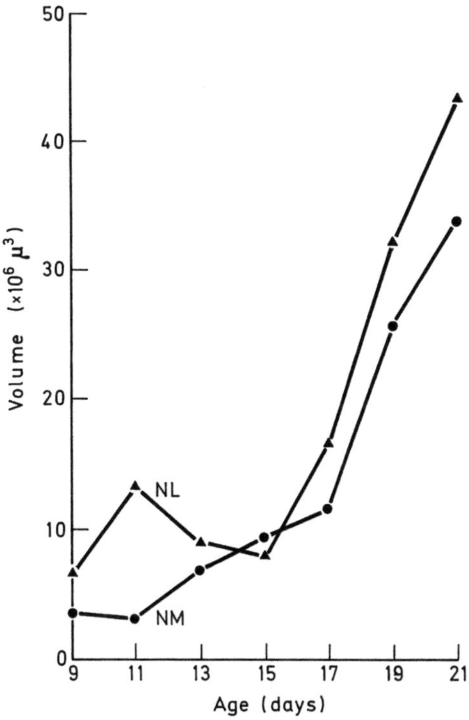

Fig. 15. Total volumes of *n. magnocellularis* and *n. laminaris* as function of age. NM, *n. magnocellularis*; NL, *n. laminaris*

Summarizing the ontogenetic information, *n. magnocellularis* (NM) and *n. laminaris* (NL) display the following important characteristics: i) The neurons of NM and NL go through their final phase of DNA synthesis over restricted and temporally separated time periods, which allows independent labeling of either neuronal population; ii) The period of cell death does not overlap with the proliferative period; iii) The alignment of NL cells is temporally correlated with, and possibly results from, cell death; iv) Cell loss occurs over a defined time period, has a definite spatial gradient (see Fig. 7) and, in the case of NL, is quite extensive; and v) Volumetric changes in cell bodies and neuropil regions occur in a period when functional connections are established and occur after the principal time of cell loss. These factors, in conjunction with the information on normal structural and functional connections, the accessibility of the avian embryo for surgery, neurophysiology, and pharmacology, and the potential for controlling embryonic and hatchling acoustic environment further suggest that these brainstem auditory pathways will serve as an excellent preparation for investigations of cellular interactions in the developing nervous system.

Our first attempts at manipulating the cellular environment have used direct deafferentation. In one series of investigations (PARKS and ROBERTSON, 1976; JACKSON and RUBEL, 1976) the effects of removing the VIIIth nerve afferents either prior to synaptogenesis (otocyst removal) or well after (at hatching, three months, or three years) have been studied in *n. magnocellularis*. In support of LEVI-

Montalcini (1949) it was found that removal of the otocyst had little or no effect on magnocellularis cell number until the period of normal cell death, after which time both the amount and period of cell death were greatly enhanced. By 19 days of incubation, the deafferented n. magnocellularis had 40–50% fewer cells than the normal population. Surprisingly, this effect was not age-dependent; cochlea removal in hatchling, three-month-old, or adult chickens also resulted in a similar amount of transneuronal cell loss in *n. magnocellularis* within 45 days after surgery.

A second series of investigations (Benes et al., 1977) examined the degree to which transneuronal changes are specific to the postsynaptic membrane surface to which the input has been removed. By deafferenting one side of the NL cells it was possible to compare the deafferented dendritic surface with the other dendritic region of the same neurons, which had their normal input intact. An EM morphometric analysis indicated rapid and complete degeneration of the deafferented dendrites, while the opposite dendrites of these neurons retained their normal qualitative and quantitative characteristics.

While the above studies by no means answer the question of how both structural and functional characteristics of afferent input regulate neuronal ontogeny, it is hoped that progressively more subtle manipulations of afferent activity and increasingly sensitive measures of neuronal structure and function in this system will yield a more thorough understanding of how an organism's external environment regulates the ontogeny of neural networks.

D. Acknowledgement

I wish to thank Dr. Deborah McLinn for assistance in typing and editing the manuscript plus numerous excursions to the library for checking reference citations; Messrs Thomas N. Parks and Daniel J. Smith for their suggestions on the manuscript; and Mr. Harold Stanislaw for assistance with photographic reproductions. The author's research reported herein is funded by Grant No. BNS 76-03006 from the National Science Foundation and funds from the Deafness Research Foundation.

References

Adelman, H. B.: The development of the neural folds and cranial ganglion of the rat. J. comp. Neurol. **39**, 19—123 (1925).

Ades, H. W., Engstrom, H.: Inner ear studies. Acta otolaryngol. (Stockholm) Suppl. **302**, 1—126 (1972).

Aitkin, L. M., Moore, D. R.: Inferior colliculus. II. Development of tuning characteristics and tonotopic organization in central nucleus of the neonatal cat. J. Neurophysiol. **38**, 1208—1216 (1975).

Aitkin, L. M., Reynolds, A.: Development of binaural responses in the kitten inferior colliculus. Neurosci. Letters **1**, 315—319 (1975).

Akiyama, Y., Schultz, F. J., Schultz, M. A., Parmalee, A. H., Jr.: Acoustically evoked responses in premature and full-term newborn infants. Electroencephalogr. clin. Neurophysiol. **26**, 371—380 (1969).

Albaum, H. G., Nestler, H. A.: Xenoplastic ear induction between *Rana pipiens* and *Amblystoma punctatum*. J. exp. Zool. **75**, 1—10 (1937).

Alexander, G.: Entwicklungsgeschichte, Anthropologie, Varietäten. In: Handbuch der Hals-, Nasen- u. Ohren-Heilkunde. Berlin: Springer 1926.

Alford, B. R., Ruben, R. J.: Physiological, behavioral, and anatomical correlates of the development of hearing in the mouse. Ann. Otol. Rhinol. Laryngol. **72**, 237—247 (1963).

Altmann, E.: Normal development of the ear and its mechanics. Arch. Otolaryngol. **52**, 725—766 (1950).

Altmann, F.: The inner ear in genetically determined deafness. Acta otolaryngol. (Stockholm) Suppl. **187** (1964).

Anderson, H., Matthiessen, M. E., Jorgensen, B. M.: Growth of the otic cavities in the human fetus. Acta otolaryngol. (Stockholm) **68**, 243—249 (1969).

Ando, Y., Hattori, H.: Effects of intense noise during fetal life upon postnatal adaptability (statistical study of the reactions of babies to aircraft noise). J. Acoust. Soc. Amer. **47**, 1128—1130 (1970).

Angevine, J. B., Jr., Sidman, R. L.: Autoradiographic study of cell migration during histogenesis of cerebral cortex in the mouse. Nature (Lond.) **192**, 766—768 (1961).

Änggård, L.: An electrophysiological study of the development of cochlear function in the rabbit. Acta otolaryngol. (Stockholm) Suppl. **203**, 1—64 (1965).

Anson, B. J.: The early development of the membranous labyrinth in mammalian embryos, with special reference to the endolymphatic duct and the utriculo-endolymphatic duct. Anat. Rec. **59**, Suppl. **1**, 15—25 (1934).

Arey, L. B.: Developmental anatomy. Philadelphia: Saunders 1965.

Avery, G. T.: Responses of fetal guinea pigs prematurely delivered. Genet. Psychol. Monogr. **3**, 248—331 (1928).

Baginsky, B.: Zur Entwicklung der Gehörschnecke. Arch. micr. Anat. **28**, 14—37 (1886).

Baginsky, B.: Über den Ursprung und den zentralen Verlauf des Nervus acusticus des Kaninchens und der Katze. Virchow's Arch. path. Anat. **119**, 81 (1890).

Barbasetti, M. A.: Sulle relazioni causali di sviluppo "otocista-cresta neurale" in trapianti embrionaly negli Anfibi. Rend. Acad. Naz. Lincei. **4**, Ser. **8**, 489—493 (1948).

Barden, T. P., Peltzman, P., Graham, J. T.: Human fetal electroencephalographic response to intrauterine acoustic signals. Amer. J. Obstet. Gynecol. **100**, 1128—1134 (1968).

Bartelmez, G. W.: The origin of the otic and optic primordia in man. J. comp. Neurol. **34**, 201—232 (1922).

Bartoshuk, A. K.: Response decrement with repeated elicitation of human neonatal cardiac acceleration to sound. J. comp. Physiol. Psychol. **55**, 9—13 (1962a).

Bartoshuk, A. K.: Human neonatal cardiac acceleration to sound: habituation and dishabituation. Percept. Motor Skills **15**, 15—27 (1962b).

Bartoshuk, A. K.: Human neonatal cardiac response to sound: a power function. Psychonom. Sci. **1**, 151—152 (1964).

Bast, T. H., Anson, B. J.: The Temporal Bone and the Ear. Springfield, Ill.: Thomas 1949.

Batkin, S., Groth, H., Watson, J. R., Ansberry, M.: Effects of auditory deprivation on the development of auditory sensitivity in albino rats. Electroencephalogr. clin. Neurophysiol. **28**, 351—359 (1970).

Batten, E. H.: The origin of the acoustic ganglion in the sheep. J. Embryol. exp. Morph. **6**, 597—615 (1958).

Bechterew, W.: Über die innere Abteilung des Strichkörpers und des achten Hirnnerven. Neurol. Centralblatt **3**, 145 (1885).

Beer, C. G.: On the responses of laughing gull chicks to the calls of adults. I. Recognition of the voices of the parents. Anim. Behav. **18**, 652—660 (1970a).

Beer, C. G.: On the responses of laughing gull chicks (*Larus atricilla*) to the calls of adults. II. Age changes and responses to different types of calls. Anim. Behav. **18**, 661—667 (1970b).

Beer, C. G.: Individual recognition of voice and its development in birds. Proceedings of the 15th International Ornithological Congress, 1972.

Békésy, G. V.: Experiments in Hearing. New York: McGraw 1960.

Belanger, L. F.: Observations in the development, structure, and composition of the cochlea of the rat. Ann. Otol. Rhinol. Laryngol. **65**, 1060—1073 (1956).

Bench, J.: Sound transmission to the human fetus through the maternal abdominal wall. J. Genet. Psychol. **113**, 85—87 (1968).
Bench, R.J., Anderson, J.H., Hoare, M.: Measurement system for fetal audiometry. J. Acoust. Soc. Amer. **47**, 1602—1606 (1970).
Bench, R.J., Mittler, P.J.: Changes in heart rate in response to auditory stimulation in the human fetus. (Abstract). Bull. Brit. Psychol. Soc. **20**, 14A (1967).
Benes, F.M., Parks, T.N., Rubel, E.W.: Rapid dendritic atrophy following deafferentation: an EM morphometric analysis. Brain Res. **122**, 1—13 (1977).
Berkson, G., Wasserman, G.A., Behrman, R.E.: Heart rate response to auditory stimulus in premature infants. Psychophysiology **11**, 244—246 (1974).
Bernard, J., Sontag, L.W.: Reactivity to tonal stimulation; preliminary report. J. Genet. Psychol. **70**, 205—210 (1947).
Bernhard, C.G., Kaiser, I.H., Kolmodin, G.M.: On the development of cortical activity in fetal sheep. Acta physiol. scand. **47**, 333—349 (1959).
Blake, J.A.: The roof and lateral recesses of the fourth ventricle, considered morphologically and embryologically. J. comp. Neurol. **10**, 79—108 (1900).
Boettcher, A.: Über Entwicklung und Bau des Gehörlabyrinthes: Nachuntersuchungen an Säugetieren. Verhandl. d. K. Leop. Carol. Deutsche Akad. Naturforsch. **35**, 293 (1869).
Bok, S.T.: Die Entwicklung der Hirnnerven und ihrer zentralen Bahnen. Die stimulogene Fibrillation. Fol. neurobiol. **9**, 475—565 (1915).
Bonnevie, K.: Abortive differentiation of ear vesicles following hereditary brain anomaly in "short-tailed waltzing mice". Genetica **18**, 105—125 (1936).
Bosher, S.K., Hallpike, C.S.: Observations on the histological features, development, and pathogenesis of the inner ear degeneration of the deaf white cat. Proc. roy. Soc. B**162**, 147—170 (1965).
Bosher, S.K., Warren, R.L.: A study of the electrochemistry and osmotic relationships of the cochlear fluids in the neonatal rat at the time of the development of the endocochlear potential. J. Physiol. (Lond.) **212**, 739—671 (1971).
Brachet, A.: Recherches sur l'ontogenèse de la tête chez les Amphibiens. Arch. Biol. (Liège, Paris) **23**, 165—258 (1907).
Brackbill, Y.: The development of response to continuous auditory stimulation in rats treated neonatally with 6-hydroxydopamine. Dev. Psychobiol. **9**, 5—15 (1976).
Bradley, R.M., Mistretta, C.M.: Fetal sensory receptors. Physiol. Rev. **55**, 352—382 (1975).
Bredberg, G.: The human cochlea during development and aging. J. Laryngol. Otol. **81**, 739—358 (1967).
Bredberg, G.: Cellular pattern and nerve supply of the human organ of Corti. Acta otolaryngol. (Stockholm) Suppl. **236**, 1—135 (1968).
Bredberg, G., Engstrom, H., Ades, H.W.: Cellular pattern and nerve supply of the human organ of Corti: a preliminary report. Arch. Otolaryngol. **82**, 462—469 (1965).
Bross, M., Zubek, J.P.: Effects of auditory deprivation on visual resolving power. Canad. J. Psychol. **29**, 340—347 (1975).
Brown, J.S.: The clinical applications of the early development of the human ear. Laryngoscope **74**, 1172—1182 (1964).
Brown, P.G., Ruben, R.J.: The endocochlear potential in the shaker-1 (sh-1/sh-1) mouse. Acta otolaryngol. (Stockholm) **68**, 14—20 (1969).
Buda, F., McLean, C., Salamy, A.: Human brainstem potentials: maturational changes in auditory transmission. (Abstract). Electroencephalog. clin. Neurophysiol. **39**, 434—434 (1975).
Bureš, J.: [Convulsions: readiness to reflexogenic palsy in rats and mice during ontogeny.] (In Czech.) Česk. Fysiol. **2**, 265—273 (1953).
Burr, H.S.: An electrodynamic theory of development suggested by studies of the proliferation rates in the brain of *Amblystoma*. J. comp. Neurol. **56**, 347—371 (1932).
Candiollo, L., Levi, A.C.: Studies on the morphogenesis of the middle ear muscles in man. Arch. klin. exp. Ohr.-, Nas.-, Kehlk.-Heilk. **195**, 55—67 (1969).
Capranica, R.: Auditory processes in sound communication. Symposium presented at the 61st Annual Meeting of the Federation of American Societies for Experimental Biology. Chicago, Illinois, April, 1977.
Carlier, E., Abonnenc, M., Pujol, R.: Maturation des réponses unitaires à la stimulation tonale dans le nerf cochléaire du caton. J. Physiol. (Paris) **70**, 129—138 (1975).

Chaloupka,Z., Myslivеček,J.: [A study of the ontogenetical development of the auditory analyzer.] (In Czech). Česk Fysiol. **9**, 423—429 (1960).
Chaloupka,Z., Záhlava,J., Myslivеček,J., Hassmannova,J.: Development of functional abilities of the auditory system. In: Ontogenesis of the Brain. Prague: Charles University Press 1968.
Choi,M.: Determination of the ear and side-specificity of the ear region ectoderm in amphibian embryos. Fol. anat. japon. **9**, 315—352 (1931).
Chondynicki,S.: Embryogenesis of the auditory part of the inner ear in the guinea pig. Acta theriol. **13**, 219—260 (1968).
Clark,W.E.L.: Medial geniculate body and nucleus isthmi. J. Anat. (Lond.) **67**, 536—548 (1933).
Clopton,B.M., Winfield,J.A.: Early auditory experience modifies unit responsiveness to temporal auditory patterns in the inferior colliculus. (Abstract). J. Acoust. Soc. Amer. **55**, 468 (1974).
Clopton,B.M., Winfield,J.A.: Effect of early exposure to patterned sound on unit activity in rat inferior colliculus. J. Neurophysiol. **39**, 1081—1089 (1976).
Coggeshall,R.E.: A study of diencephalic development in the albino rat. J. comp. Neurol. **122**, 241—269 (1964).
Conel,J. Leroy: The postnatal development of the human cerebral cortex, 1–8. Cambridge, Mass.: Harvard University Press 1939—1963.
Cooper,E.R.A.: The development of the human auditory pathway from the cochlear ganglion to the medial geniculate body. Acta Anat. **5**, 99—122 (1948).
Cowan,W.M.: Studies on the development of the avian visual system. In: Cellular Aspects of Neural Growth and Differentiation. Los Angeles: University of California Press 1971.
Cowan,W.M.: Neuronal death as a regulative mechanism in the control of cell number in the nervous system. In: Development and Aging in the Nervous System. New York: Academic 1973.
Cowdry,E.V.: The development of the cytoplasmic constituents of the nerve cells in the chick. Amer. J. Anat. **15**, 389—427 (1914).
Crossland,W.J., Cowan,W.M., Rogers,L.A.: Studies on the development of the chick optic tectum. IV. An autoradiographic study of development of retino-tectal connections. Brain Res. **91**, 1—23 (1975).
Crowley,D.E., Hepp-Raymond,M.-C.: Development of cochlear function in the ear of the infant rat. J. comp. Physiol. Psychol. **62**, 427—432 (1966).
Dallos,P.: The auditory periphery. New York: Academic 1973.
Dekaban,A.: Human thalamus. An anatomical, developmental, and pathological study. II. Development of the human thalamic nuclei. J. comp. Neurol. **100**, 63—97 (1954).
Deol,M.S.: Anatomy and development of mutants pirouette, shaker-1 and waltzer in mouse. Proc. roy. Soc. B**145**, 206—213 (1956).
Deol,M.S.: The abnormalities of the inner ear in kreisler mice. J. Embryol. exp. Morph. **12**, 475—490 (1964a).
Deol,M.S.: The origin of the abnormalities of the inner ear in dreher mice. J. Embryol. exp. Morph. **12**, 727—733 (1964b).
Deol,M.S.: Influence of the neural tube on the differentiation of the inner ear in the mammalian embryo. Nature **209**, 219—220 (1966).
Deol,M.S.: The neural crest and the acoustic ganglion. J. Embryol. exp. Morph. **17**, 533—541 (1967).
Deol,M.S.: Inherited diseases of the inner ear in man in the light of studies on the mouse. J. med. Genet. **5**, 137—158 (1968).
Deol,M.S.: The origin of the acoustic ganglion and effects of the genedominant spotting (MV) in the mouse. J. exp. Morph. **23**, 773—784 (1970).
Detwiler,S.: Further quantitative studies on locomotor capacities of larval *Amblystoma* following surgical procedures upon the embryonic brain. J. exp. Zool. **108**, 45—73 (1948).
Detwiler,S.R.: The responses of *Amblystoma* larvae with midbrain replaced by a supernumerary medulla. J. exp. Zool. **110**, 321—325 (1949).
Detwiler,S.R., van Dyke,R.J.: The role of the medulla in the differentiation of the otic vesicle. J. exp. Zool. **113**, 179—200 (1950).
Detwiler,S.R., van Dyke,R.J.: Recent experiments on the differentiation of the labyrinth in *Amblystoma*. J. exp. Zool. **118**, 389—406 (1951).
Dillard,R.G.: Myelination and acetylcholinesterase activity in the development of the eighth nerve. M.D. thesis, Yale University, New Haven, Connecticut 1968.

Dwernicka, B., Jasienska, A., Smolarz, W., Wawryk, R.: Attempt at determining the fetal reaction to acoustic stimulation. Acta otolaryngol. (Stockholm) **57**, 571—574 (1964).
Eimas, P. D.: Speech perception in early infancy. In: Infant Perception: from Sensation to Cognition, 2. New York: Academic 1975.
Eisenberg, R. B.: Auditory competence in early life. Baltimore, Md.: University Park Press 1976.
Ellingson, R. J.: Cerebral electrical responses to auditory and visual stimuli in the infant/human and subhuman studies. In: Neurological and electroencephalographic correlative studies. London: Grune 1964.
Ellingson, R. J., Wilcott, R. C.: Development of evoked responses in visual and auditory cortices of kittens. J. Neurophysiol. **23**, 363—375 (1960).
Engel, R., Young, N. B.: Calibrated pure tone audiograms in normal neonates based on evoked electroencephalographic responses. Neuropediatrie **1**, 149—160 (1969).
Evans, H. J.: The independent differentiation of the sensory areas of the avian inner ear. Biol. Bull. **84**, 252—262 (1943).
Evans, R. M.: Parental recognition of the "mew call" in black-billed gulls (*Larus dulieri*). Auk **87**, 503—513 (1970a).
Evans, R. M.: Imprinting and mobility in young ring-billed gulls (*Larus delawarensis*). Anim. Behav. Monogr. **3**, 195—248 (1970b).
Falbe-Hansen, J.: On glycogen in the cochlear duct of fetuses and young of albino rats. Acta otolaryngol. (Stockholm) **63**, 340—346 (1967).
Falk, S. A., Cook, R. V., Haseman, J. D., Sanders, G. M.: Noise-induced inner ear damage in newborn and adult guinea pigs. Laryngoscope **84**, 444—453 (1974).
Fell, H. B.: The development *in vitro* of the isolated otocyst of the embryonic fowl. Arch. exp. Zellforsch. **7**, 69—81 (1928—1929).
Fernández, C., Hinojosa, R.: Postnatal development of endocochlear potential and stria vascularis in the cat. Acta otolaryngol. (Stockholm) **78**, 173—186 (1974).
Finck, F., Schneck, C. D., Hartman, E. F.: Development of cochlear function in the neonate Mongolian gerbil (*Meriones unguiculatus*). J. comp. Physiol. Psychol. **78**, 375—380 (1972).
Fisher, L., Rubenstein, G.: Developmental aspects of auditory and visual scanning. Percept. Motor Skills **40**, 583—586 (1975).
Fleischer, K.: Untersuchungen zur Entwicklung der Innenohrfunktion (Intrauterine Kindsbewegungen nach Schallreizen). Z. Laryngol. Rhinol. Otol. **34**, 733—740 (1955).
Forbes, H. S., Forbes, H. B.: Fetal sense reaction: hearing. J. comp. Physiol. Psychol. **7**, 353—355 (1927).
Foss, I., Flottorp, G.: A comparative study of the development of hearing and vision in various species commonly used in experiments. Acta otolaryngol. (Stockholm) **77**, 202—214 (1974).
Fox, M. W.: Neuronal development and ontogeny of evoked potentials in auditory and visual cortex of the dog. Electroencephalogr. clin. Neurophysiol. **24**, 213—226 (1968a).
Fox, M. W.: Multidisciplinary studies of the postnatal development of the CNS and behavior of the dog. In: Ontogenesis of the Brain. Prague: Charles University Press 1968b.
Friedmann, I.: *In vitro* culture of the isolated otocyst of the embryonic fowl. Ann. Otol. Rhinol. Laryngol. **65**, 98—107 (1956).
Friedmann, I.: Electron microscope observations on *in vitro* cultures of the isolated fowl embryo otocyst. J. Biophys. Biochem. Cytol. **5**, 263—268 (1959).
Friedmann, I.: The innervation of the developing fowl embryo otocyst *in vivo* and *in vitro*. Acta otolaryngol. (Stockholm) **67**, 224—238 (1969).
Friedmann, I., Bird, E. S.: The effect of ototoxic antibiotics and of penicillin on the sensory areas of isolated fowl embryo otocyst in organ cultures: An electron-microscope study. J. path. Bact. **81**, 89—90 (1961).
Friedmann, I., Bird, E. S.: Electron microscopic studies of the isolated fowl embryo otocyst in tissue culture. J. ultrastruct. Res. **20**, 356—365 (1967).
Fuse, G.: Das Ganglion ventrale und das Tuberculum acusticum bei einigen Säugern und beim Menschen. Arb. Hirnanat. Inst. Zürich **7**, 1—210 (1913).
Garcia, F. M.: X. reunion de la Sociedad Española de Otorrinolaringologica **59**, 230 (1961). (Noted by van de Water and Ruben, 1976).
Gauron, E. G., Becker, W. C.: The effects of early sensory deprivation on adult rat behavior under competition stress: an attempt at replication of a study by Alexander Wolf. J. comp. Physiol. Psychol. **52**, 689—693 (1959).

Gilbert, M.S.: The early development of the human diencephalon. J. comp. Neurol. **62**, 81—107 (1935).

Ginsberg, A.S., Dettlaff, T.: Experiments on transplantation and removal of organ rudiments in embryos of *Acipenser stellatus* in early developmental stages. Trans. Acad. Sci. U.R.S.S. **44**, 209—212 (1944).

Godfrey, D.A., Kiang, Y.S., Norris, B.E.: Single unit activity in the posteroventral cochlear nucleus of the cat. J. comp. Neurol. **162**, 247—268 (1975).

Goldberg, J.M., Brown, P.B.: Functional organization of the dog superior olivary complex: an anatomical and electrophysiological study. J. Neurophysiol. **31**, 649—656 (1968).

Goodlin, R.C., Schmidt, W.: Human fetal arousal levels as indicated by heart rate recordings. Amer. J. Obstet. Gynecol. **114**, 613—621 (1972).

Gos, M.: Les réflexes conditionnels chez l'embryon d'oiseau. Bull. Soc. roy. Sci. (Liège) **4**, 194—199, 246—250 (1935).

Gottlieb, G.: Imprinting in relation to parental and species identification by avian neonates. J. comp. Physiol. Psychol. **59**, 345—356 (1965a).

Gottlieb, G.: Prenatal auditory sensitivity in chickens and ducks. Science **147**, 1596—1958 (1965b).

Gottlieb, G.: Species identification by avian neonates: Contributory effect of perinatal auditory stimulation. Anim. Behav. **14**, 282—290 (1966).

Gottlieb, G.: Species-specific perception in duck embryos. In: Ontogenesis of the Brain. Prague: Charles University Press 1968a.

Gottlieb, G.: Prenatal behavior of birds. Quart. Rev. Biol. **43**, 148—174 (1968b).

Gottlieb, G.: Development of Species Identification in Birds: an inquiry into the prenatal determinants of perception. Chicago: University of Chicago Press 1971a.

Gottlieb, G.: Ontogenesis of sensory function in birds and mammals. In: The Biopsychology of Development. New York: Academic 1971b.

Gottlieb, G.: On the acoustic basis of species identification in wood ducklings (*Aix sponsa*). J. comp. Physiol. Psychol. **87**, 1038—1048 (1974).

Gottlieb, G.: The roles of experience in the development of behavior and the nervous system. In: Studies on the Development of Behavior and the Nervous System, Vol. 3. New York: Academic 1976a.

Gottlieb, G.: Early development of species-specific auditory perception in birds. In: Studies on the Development of Behavior and the Nervous System, Vol. 3. New York: Academic 1976b.

Graziani, L.J., Katz, L., Cracco, R.Q., Cracco, J.B., Weitzman, E.D.: The maturation and interrelationship of EEG patterns and auditory evoked response in premature infants. Electroencephalogr. clin. Neurophysiol. **36**, 367—375 (1974).

Graziani, L.J., Weitzman, E.D.: Sensory evoked responses in the neonatal period and their clinical application. In: International Handbook of Electroencephalography and Clinical Neurophysiology. Amsterdam: Elsevier 1972.

Grier, J.B., Counter, S.A., Shearer, W.M.: Prenatal auditory imprinting in chickens. Science **155**, 1692—1693 (1967).

Grimwade, J.C., Walker, D.W., Bartlett, M., Gordon, S., Wood, C.: Human fetal heart rate change and movement in response to sound and vibration. Amer. J. Obstet. Gynecol. **109**, 86—90 (1971).

Grossman, C.: Electro-ontogenesis of cerebral activity. Arch. Neurol. Psychiat. (Chicago) **74**, 186—204 (1955).

Groth, W.: Der Ursprung der Labyrinthplacode und des Ganglion statoacusticum in Vergleich zur Genese des Riechorgans beim Kaninchen: Zugleich ein Beitrag zur Homologisierung der Genese von Gehör-, Riech- und Sehorgan beim Säuger. Z. mikroskop. anat. Forsch. **45**, 393 (1939).

Gruneberg, H., Hallpike, D.S., Ledoux, A.: Observations on the structure, development, and electrical reactions of the internal ear of the shaker-1 mouse. Proc. roy. Soc. B**129**, 154—173 (1940).

Guareschi, C.: Studi sulla determinazione dell'orecchio interno degli anfibi anuri. Arch. Ital. Anat. Embriol. **35**, 97—129 (1935).

Gyllensten, L., Malmfors, T., Norrlin, M.L.: Growth alteration in the auditory cortex of visually deprived mice. J. comp. Neurol. **126**, 463—470 (1966).

Hall, E.K.: The determination of the axes of the embryonic ear: An experimental study by the method of 90° rotations. J. exp. Zool. **75**, 11—39 (1937).

Hall, E.K.: On the duration of the polarization process in the ear primordium of embryos of *Amblystoma punctatum* (Linn.). J. exp. Zool. **82**, 173—192 (1939).

Hall, E. K.: Reversal of polarization in the ear primordium of *Amblystoma punctatum*. J. exp. Zool. **86**, 141—151 (1941).
Hall, J. G.: The cochlea and cochlear nuclei in neonatal asphyxia. Acta otolaryngol. (Stockholm) Suppl. **194**, 1—93 (1964).
Halley, G.: The placodal relations of the neural crest in the domestic cat. J. Anat. **89**, 133—152 (1955).
Hamburger, V.: Experimental analysis of the dual origin of the trigeminal ganglion in the chick embryo. J. exp. Zool. **148**, 91—123 (1961).
Hamburger, V.: Cell death in the development of the lateral motor column of the chick embryo. J. comp. Neurol. **160**, 535—546 (1975).
Hamburger, V., Hamilton, H. L.: A series of normal stages in the development of the chick embryo. J. Morph. **88**, 49—92 (1951).
Hanaway, J.: Formation and differentiation of the external granule layer of the chick cerebellum. J. comp. Neurol. **131**, 1—14 (1967).
Hardy, J. B.: Congenital hearing defects. In: The Nervous System, Vol. 3. New York: Raven 1975.
Harmkark, W.: The rhombic lip and its derivatives in relation to the theory of neurobiotaxis. In: Aspects of Cerebellar Anatomy. Oslo: Johan Grundt Tanum 1954.
Harrison, R. G.: Experiments on the development of the internal ear (Abstract.). Science **59**, 448 (1924).
Harrison, R. G.: Factors concerned in the development of the ear in *Amblystoma punctatum*. Anat. Rec. **64**, Suppl. 1, 38—39 (1933).
Harrison, R. G.: Relations of symmetry in the developing ear of *Amblystoma punctatum*. Proc. nat. Acad. Sci. (Wash.) **22**, 238—247 (1936a).
Harrison, R. G.: Relations of symmetry in the developing embryo. Coll. Net. **11**, 217—226 (1936b).
Harrison, R. G.: Further investigation of the factors concerned in the development of the ear (Abstract). Anat. Rec. **70**, Suppl. 3, 35 (1938).
Harrison, R. G.: Relations of symmetry in the developing embryo. Trans. Conn. Acad. Arts Sci. **36**, 277—330 (1945).
Hassmannova, J., Myslivećek, J.: Maturation of the primary cortical response to stimulation of medial geniculate body. Electroencephalogr. clin. Neurol. **22**, 547—555 (1967).
Hassmannova, J., Rokyta, R., Záhlava, J., Myslivećek, J.: Intercentral relationships in the auditory system during ontogeny. In: Ontogenesis of the Brain. Prague: Charles University Press 1968.
Hecox, K.: Electrophysiological correlates of human auditory development. In: Infant Perception: from Sensation to Cognition, Vol. 2. New York: Academic 1975.
Hecox, K., Galambos, R.: Brainstem auditory evoked responses in human infants. Arch. otolaryngol. (Stockholm) **99**, 30—33 (1974).
Held, H.: Untersuchungen über den feineren Bau des Ohrlabyrinthes der Wirbeltiere. II. Zur Entwicklungsgeschichte des Cortischen Organs und der Macula acustica bei Säugetieren und Vögeln. Abhandl. Math.-Phys. C. K. Sachs. Ges. Wissensch. **13**, 193 (1909).
Held, H.: Die Cochlea der Säuger und der Vögel, ihre Entwicklung und ihr Bau. In: Handbuch der Normalen und Pathologischen Physiologie. Berlin: Julius Springer (1926).
Henry, K. R.: Pinna reflex thresholds and audiogenic seizures: Developmental changes after acoustic priming. J. comp. Physiol. Psychol. **79**, 77—81 (1972).
Henry, K. R., Haythorn, M. M.: Auditory similarities associated with genetic and experimental acoustic deprivation. J. comp. Physiol. Psychol. **89**, 213—218 (1975).
Hertwig, P.: Neue Mutationen und Kopplungsgruppen bei der Hausmaus. Z. Indukt. Abstamm. Vererb. **80**, 220—247 (1942).
Hertwig, P.: Die Genese der Hirn- und Gehörgangsmißbildungen bei Röntgen-mutierten Kreisler-Mäusen. Z. menschl. Vererb.-Konstitutionslehre **28**, 327—353 (1944).
Hilding, D. A.: Electron microscopy of the developing hearing organ. Laryngoscope **79**, 1691—1704 (1969).
His, W.: Die Entwicklung des menschlichen Rautenhirns vom Ende des ersten bis zum Beginn des dritten Monats. Abhandl. Math.-Phys. K. K. Sachs. Ges. Wissensch. **29**, 1—72 (1890).
His, W.: Die Entwicklung des menschlichen Gehirns während der ersten Monate. Leipzig: S. Hirzel 1904.
Hoadley, L.: The independent differentiation of isolated chick primordia in chorio-allantoic grafts. I. The eye, nasal region, and mesencephalon. Biol. Bull. **46**, 281—315 (1924).
Holtfreter, J.: Über das Verhalten von Anurenektoderm in Urodelenkeim. Roux' Arch. Entw.-Mech. **133**, 427—494 (1935).

Howland, R., Rasbury, W., Heilman, K. M., Hammer, L.: Development of auditory figure-ground discrimination and ear asymmetry under monaural stimulus presentation. Dev. Med. Child Neurol. **17**, 325—332 (1975).

Hoyte, D. A. N.: The postnatal growth of the ear capsule in the rabbit. Amer. J. Anat. **108**, 1—16 (1961).

Hudson, W. R., Durham, N. C., Ruben, R.: Hereditary deafness in the Dalmatian dog. Arch. Otolaryng. **75**, 213—219 (1962).

Hunt, E. L.: Establishment of conditioned responses in chick embryos. J. comp. Psychol. **42**, 107—117 (1949).

Huschke, E.: Erste Bildungsgeschichte des Auges und Ohres. (Versammlung Naturforscher u. Ärzte zu Hamburg.) Isis von Oken 1831. As cited in Bast and Anson 1949.

Hutt, S. J., Hutt, C., Lenard, H. G., Bernuth, H. V., Montjewerff, W. J.: Auditory responsivity in the human neonate. Nature **218**, 888—890 (1968).

Impekoven, M.: Auditory stimuli for pecking in the gull chick *Larus atricilla*. Effect of prenatal experience on postnatal behavior. Paper presented at XIth International Ethology Conference, Rennes, 1969.

Impekoven, M.: Response-contingent prenatal experience of maternal calls in the Peking duck *(Anas platyrhynches)*. Anim. Behav. **21**, 164—168 (1973).

Impekoven, M., Gold, P. S.: Prenatal origins of parent-young interactions in birds: a naturalistic approach. In: Studies on the Development of Behavior and the Nervous System, Vol. 1. New York: Academic 1973.

Jackson, J. R. H., Rubel, E. W.: Rapid transneuronal degeneration following cochlea removal in chickens. Anat. Rec. **184**, 434—435 (1976).

Jackson, J. R. H., Rubel, E. W.: Ontogeny of behavioral responsiveness to sound in the chick embryo *as indicated by electrical recordings of motility*. J. Comp. Physiol. Psychol. (in press) (1977).

Jacobson, M.: Developmental neurobiology. New York: Holt 1970.

Javel, E., Brugge, J. E., Kitzes, L. M.: Response properties of single neurons in anteroventral (AVCN) of newborn cat: frequency selectivity and temporal order of discharges. Paper presented at 90th Meeting of Acoustical Society of America, San Francisco, 1975.

Jewett, D. L., Romano, M. N.: Neonatal development of the auditory system: evoked potentials averaged from the scalp of rat and cat. Brain Res. **36**, 101—115 (1972).

Johansson, R., Wedenberg, E., Westin, B.: Measurement of tone response by the human fetus. A preliminary report. Acta otolaryngol. (Stockholm) **57**, 188—192 (1964).

Kaan, H. W.: Further studies on the auditory vesicle and cartilaginous capsule of *Amblystoma punctatum*. J. exp. Zool. **78**, 159—183 (1938).

Kasatkin, N. I., Levikova, A. M.: On the development of early conditioned reflexes and differentiation of auditory stimuli in infants. J. exp. Psychol. **18**, 1—19 (1935).

Katona, F., Berenyi, M.: Differential reactions and habituations to acoustical and visual stimulation in neonates. Activ. Nerv. **16**, 305 (1974).

Katsuki, Y.: Neural mechanism of auditory sensation in the cat. In: Sensory Communication. Cambridge, Mass.: MIT Press 1961.

Kawakami, I.: Inductions of the cerebral sensory organs. II. Inductive effect of the archenteron roof. Kagaku **11**, 399—402 (1943a).

Kawakami, I.: Inductions of the cerebral sensory organs, III. Inductive effects of the uninvaginated portion of the dorsal blastopore lip. Bot. Zool. **11**, 859—862 (1943b).

Kawakami, I.: Inductive effects of the heated archenteron roof and invaginated portion of the blastophoric lip. Siebetu **4**, 41—45 (1949).

Keith, R. W.: Middle ear function in neonates. Arch. Otolaryngol. **101**, 376—369 (1975).

Kennedy, C., des Rosiers, M., Revich, M., Sharp, F., Sokoloff, L.: Mapping of functional pathways by autoradiographic survey of local metabolic rate with ^{14}C-deoxyglucose. Science **187**, 850—853 (1975).

Kerr, L. M., Rubel, E. W.: Effects of auditory deprivation on perceptual sharpening in the chick. Paper presented at the Annual Meeting of the Animal Behavior Society, June, 1977.

Keyser, A.: Development of the diencephalon of the Chinese hamster. Acta Anat. **83**, Suppl. **59**, 1—181 (1972).

Kiang, N. Y.-S., Watanabe, T., Thomas, E. C., Clark, L. F.: Discharge Patterns of Single Fibers in the Cat's Auditory Nerve. Cambridge, Mass.: MIT Press 1965.

Kikuchi,K., Hilding,D.: The development of the organ of Corti in the mouse. Acta otolaryngol. (Stockholm) **60**, 207—222 (1965).

Kikuchi,K., Hilding,D.: The development of the stria vascularis in the mouse. Acta otolaryng. (Stockholm) **60**, 277—291 (1966).

Klyavina,M., Obraztsova,G.A.: The influence of the sound stimulus intensity on time and amplitude characteristics of the cerebral cortex evoked potentials in early postnatal ontogenesis of the rabbit. In: Ontogenesis of the Brain. Prague: Charles University Press 1968.

Knauff,R.A.: The origin of the cranial ganglia of Rana. J. Comp. Neurol. **44**, 259—361 (1927).

Knouff,R.A.: The development of ectodermal placodes in *Rana pipiens*. J. comp. Neurol. **62**, 17—65 (1935).

Knowlton,V.Y.: Correlation of the development of membranous and bony labyrinthes, sensory ganglia, nerves, and brain centers of the chick embryo. J. Morph. **121**, 179—208 (1967).

Kölliker,R.A.: Entwicklungsgeschichte des Menschen und der höheren Tiere. Leipzig: W.Englemann 1861.

König,N., Marty,R.: On the functions and structure of deep layers of immature auditory cortex. J. Physiol. (Paris) **68**, 145—155 (1974).

König,N., Pujol,R., Marty,R.: Potentials évoqués et réponses unitaires du cortex auditif du chat dans la période postnatale. J. Physiol. (Paris) **63** (Suppl.), 1—2 (1971).

König,N., Pujol,R., Marty,R.: A laminar study of evoked potentials and unit responses in the auditory cortex of the postnatal cat. Brain Res. **36**, 469—473 (1972).

König,N., Roch,G., Marty,R.: The onset of synaptogenesis in rat temporal cortex. Anat. Embryol. **148**, 73—87 (1975).

Kohan,R.E.: Inductive action of medulla oblongata on the body epithelium of amphibia. Trans. Acad. Sci. U.R.S.S. **23**, 307—310 (1939).

Kohan,R.: The chordamesoderm as an inductor of the ear vesicle. Trans. Acad. Sci. U.R.S.S. **45**, 39—41 (1944).

Kolmer,W.: Gehörorgan. In: Handbuch der mikroskopischen Anatomie des Menschen, Bd. 2. Berlin: Springer 1927.

Konishi,M.: Development of auditory neuronal responses in avian embryos. Proc. nat. Acad. Sci. (Wash.) **70**, 1795—1798 (1973).

Kostir,W.J.: An analysis of the cranial ganglia of an embryo salamander, *Amblystoma Jeffersonianum*. Ohio J. Sci. **24**, 230—263 (1924).

Krulisova,H.: Age differences in alimentary conditioning in infants in the first months of life. Activ. Nerv. Sup. (Praha). **16**, 113—114 (1974).

Kuhlenbeck,H.: The ontogenetic development of the diencephalic centers in a bird's brain (chick) and comparison with reptilian and mammalian diencephalon. J. comp. Neurol. **122**, 241—269 (1964).

Kuijpers,W.: Na-K-ATPase activity in the cochlea of the rat during development. Acta Otolaryngol. **78**, 341—344 (1974).

Kuo,Z.-Y.: Ontogeny of embryonic behavior in aves: XII. Stages in the development of physiological activities in the chick embryo. Amer. J. Psychol. **51**, 361—368 (1938).

Kuo,Z.-Y.: The dynamics of behavior development: An epigenetic view. New York: Random House 1967.

Landacre,F.L.: The origin of the cranial ganglia in *Ameiurus*. J. comp. Neurol. **20**, 309—411 (1910).

Lane,H.H.: The correlation between structure and function in the development of the special senses of the white rat. Univ. Okla. Bull. New Ser. No. 140, Univ. Study No. 8, 1—88 (1917).

Langman,J., Shimada,M., Haden,C.: Formation and migration of neuroblasts. In: Cellular Aspects of Neural Growth and Differentiation. Los Angeles: University of California Press 1971.

Langworthy,O.R.: Development of behavior patterns and myelinization of the nervous system in the human fetus and infant. Contrib. Embryol. **24**, 1—143 (1933).

Larsell,O.: The differentiation of the peripheral and central acoustic apparatus of the frog. J. comp. Neurol. **60**, 473—527 (1934).

Larsell,O., McCrady,E., Jr, Larsell,J.F.: Development of the organ of Corti in relation to the inception of hearing. Arch. Otolaryngol. **40**, 233—248 (1944).

Larsell,O., McCrady,E., Jr, Zimmerman,A.A.: Morphological and functional development of the membranous labyrinth in the opossum. J. comp. Neurol. **62**, 95—118 (1935).

Lauder,J.M., Bloom,F.E.: Ontogeny of monoamine neurons in the locus coeruleus, raphe nuclei, and substantia nigra of the rat. I. Cell differentiation. J. comp. Neurol. **155**, 469—482 (1974).

Layne, J. M.: Ontogeny. In: Biology of *Peromyscus (Rodentia)*. Amer. Soc. Mammologists Spec. Publ. No. 2, 148—253 (1968).

Leiberman, F., Sohner, H., Szabo, G.: Cochlear audiometry (electrocochleography) during the neonatal period. Dev. Med. Child Neurol. **15**, 8—13 (1973).

Levi-Montalcini, R.: Richerche sperimentalli sulla determinazione del placode otico nell'embrione di pollo. C.R. Acad. Lincei Ser. 8, **1**, 443—448 (1946).

Levi-Montalcini, R.: The development of the acousticovestibular centers in the chick embryo in the absence of the afferent root fibers and of descending fiber tracts. J. comp. Neurol. **91**, 209—241 (1949).

Lieff, B. D., Permut, A., Schlesinger, R., Sharpless, S. K.: Developmental changes in auditory evoked potentials in the inferior colliculi of mice during periods of susceptibility to priming. Exp. Neurol. **46**, 534—541 (1975).

Lindeman, H. H., Ades, H. W., Bredberg, G., Engstrom, H.: The sensory hairs and the tectorial membrane in the development of the cat's organ of Corti. Acta otolaryngol. (Stockholm) **72**, 229—242 (1971).

Lindsay, J. R., Matz, G. R.: The differentiation of acquired congenital from genetically determined inner ear deafness. Ann. Otol. **75**, 830—843 (1966).

Lipsitt, L. P.: Learning capacities of the human infant. In: Brain and Early Behavior. New York: Academic 1969.

Lorente de Nó, R.: Etudes sur l'anatomie et la physiologie du labyrinthe de l'oreille et du VIIIe nerf. Trab. Inst. Cajal Invest. Biol. **24**, 53—54 (1926).

Lorente de Nó, R.: Anatomy of the eighth nerve. Central projection of the nerve endings of the internal ear. Laryngoscope **43**, 1—38 (1933a).

Lorente de Nó, R.: Anatomy of the eighth nerve. III. General plan of structure of the primary cochlear nuclei. Laryngoscope **43**, 327—350 (1933b).

Lorente de Nó, R.: The neural mechanism of hearing. Anatomy and physiology. The sensory ending in the cochlea. Laryngoscope **47**, 373—377 (1937).

Lurie, M. H.: Studies of acquired and inherited deafness in animals. J. Acoust. Soc. Amer. **11**, 420—426 (1940).

McAllister, J. P., Das, G. D.: Neurogenesis in the epithalamus, dorsal thalamus, and ventral thalamus of the rat: An autoradiographic and cytological study. J. comp. Neurol. **172**, 647—686 (1977).

McAlpine, J. C., Friedmann, I.: A histochemical study of the effects of ototoxic antibodies on the isolated embryonic otocysts of the fowl *(Gallus domesticus)*. J. path. Bact. **86**, 477—486 (1963).

McClain, J. A.: The development of the auditory ossicles of the opossum *(Didelphys virginiana)*. J. Morph. **64**, 211—265 (1939).

McCrady, E., Jr, Wever, E. G., Bray, C. W.: The development of hearing in the opossum. J. exp. Zool. **75**, 503—517 (1937).

McCrady, E., Jr, Wever, E. G., Bray, C. W.: A further investigation of the development of hearing in the opossum. J. comp. Psychol. **30**, 17—21 (1940).

Mailloux, J. G., Edwards, H. P., Barry, W. F., Rawsell, H. C., Achorn, E. G.: Effects of differential rearing on cortical evoked potentials of the albino rat. J. comp. Physiol. Psychol. **87**, 475—480 (1974).

Mair, I. W. S.: Hereditary deafness in the white cat. Acta otolaryngol. (Stockholm) Suppl. **314**, 1—48 (1968).

Marty, R.: Développement postnatal des réponses sensorielles du cortex cérébral chez le chat et le lapin. Arch. Anat. Microscop. Morphol. exp. **51**, 129—264 (1962).

Marty, R.: Maturation postnatale du système auditif. In: Regional Development of the Brain in Early Life. Philadelphia: Davis 1967.

Marty, R., Scherrer, J.: Critères de maturation des systèmes afférents corticaux. In: Growth and Maturation of the Brain. Progress in Brain Research, Vol. 4. New York: Elsevier 1964.

Marty, R., Thomas, J.: Réponse électro-corticale à la stimulation du nerf cochléaire chez le chat nouveau-né. J. Physiol. (Paris) **55**, 165—166 (1963).

Mayers, K. S., Robertson, R. T., Rubel, E. W., Thompson, R. F.: Development of polysensory responses in association cortex of the kitten. Science **171**, 1037—1038 (1971).

Mayers, K. S., Robertson, R. T., Teyler, T. J., Thompson, R. F.: Rèsponse properties of single units in an association area of the kitten neocortex. Physiol. Behav. **16**, 151—161 (1976).

Mendelson, M. J., Haith, M. M.: The relation between audition and vision in the human newborn. Soc. Res. Child Dev. Monogr. 167 (1976).

Meyerson, B.: Comments to paper by R. Marty. In: Regional Development of the Brain in Early Life. Philadelphia: Davis 1967.
Miale, I. L., Sidman, R. L.: An autoradiographic analysis of histogenesis in the rat cerebellum. Exp. Neurol. **4**, 277—298 (1961).
Mikaelian, D., Alford, B. R., Ruben, R. J.: Cochlear potentials and VIII nerve action potentials in normal and genetically deaf mice. Ann. Oto, Rhino, Laryngol. **74**, 146—158 (1965).
Mikaelian, D. O., Ruben, R. J.: Hearing degeneration in shaker-1 mice. Correlation of physiological observations with behavioral responses and with cochlear anatomy. Arch. Otolaryng. **80**, 418—430 (1964).
Mikaelian, D., Ruben, R. J.: Development of hearing in the normal CBA-J mouse: Correlation of physiological observations with behavioral responses and with cochlear anatomy. Ann. Oto-Rhino-Laryngol. **59**, 451—461 (1965).
Mikaelian, D. O., Warfield, D., Norris, O.: Genetic progressive hearing loss in the C57/b16 mouse. Acta otolaryng. (Stockholm) **77**, 327—334 (1974).
Miller, L. K.: Effects of auditory stimulation upon non-nutritive sucking by premature infants. Percept. Motor Skills **40**, 879—885 (1975).
Mills, J. H.: Noise and children: A review of literature. J. Acoust. Soc. Amer. **58**, 767—779 (1975).
Milonyeni, M.: The late stages of the development of the primary cochlear nuclei. Brain Res. **4**, 334—344 (1967).
Minot, C. S.: Human Embryology. New York: Wood 1892.
Mistretta, C. M., Bradley, R. M.: Effects of early sensory experience on brain and behavioral development. In: Studies on the Development of Behavior and the Nervous System, Vol. 4. New York: Academic (in press) (1977).
Molfese, D. L., Freeman, R. B., Palmero, D. S.: The ontogeny of brain lateralization for speech and nonspeech stimuli. Brain Lang. **2**, 356—368 (1975).
Monod, N., Garma, L.: Auditory responsitivity in the human premature. Biol. Neonat. **17**, 292—316 (1971).
Moore, J. A.: Studies in the development of frog hybrids. I. Embryonic development in the cross *Rana pipiens* × *Rana sylvacia*. J. exp. Zool. **101**, 173—219 (1946).
Moore, D. R., Aitkin, L. M.: Rearing in an acoustically unusual environment: effects on neural auditory activity. Neurosci. Letters **1**, 29—34 (1975).
Morest, D. K.: The neuronal architecture of the medial geniculate body of the cat. J. Anat. (Lond.) **98**, 611—630 (1964).
Morest, D. K.: The growth of synaptic endings in the mammalian brain: a study of the calyces of the trapezoid body. Z. Anat. Entwickl.-Gesch. **127**, 201—220 (1968).
Morest, D. K.: The differentiation of cerebral dendrites: A study of the post-migratory neuroblast in the medial nucleus of the trapezoid body. Z. Anat. Entwickl.-Gesch. **128**, 271—289 (1969a).
Morest, D. K.: The growth of dendrites in the mammalian brain. Z. Anat. Entwickl.-Gesch. **128**, 290—316 (1969b).
Morest, D. K.: A study of neurogenesis of the forebrain of opossum pouch young. Z. Anat. Entwickl.-Gesch. **130**, 265—305 (1970a).
Morest, D. K.: The pattern of neurogenesis in the retina of the rat. Z. Anat. Entwickl.-Gesch. **131**, 45—67 (1970b).
Morest, D. K., Morest, R. R.: Perfusion-fixation of the brain with chrome-osmium solutions for the rapid Golgi method. Amer. J. Anat. **118**, 811—832 (1966).
Morgan, A., Charachon, D., Bringuier, N.: Disorders of the auditory apparatus caused by embryopathy or fetopathy: prophylaxis and treatment. Acta otolaryngol. (Stockholm) Suppl. **291**, 1—27 (1971).
Muira, R.: Über die Differenzierung der Grundbestandteile im Zwischenhirn des Kaninchens. Anat. Anz. **61**, 433—475 (1933).
Murphy, K. P., Smyth, C. N.: Response of the fetus to auditory stimulation. Lancet **5**, 972—973 (1962).
Mysliveček, J.: Subcortical auditory and visual relays and specific cortical relays. In: Ontogenesis of the Brain. Prague: Charles University 1968.
Mysliveček, J.: Electrophysiology of the developing brain: central and Eastern European contributions. In: Developmental Neurobiology. Springfield, Illinois: Thomas 1970.
Nakai, Y.: An electron microscopic study of the human fetus. Pract. Oto- Rhino-Laryngol. **32**, 257—267 (1970).

Nakai, Y., Hilding, D.: Cochlear development. Some electron microscopic observations of maturation of hair cells, spiral ganglion, and Reissner's membrane. Acta otolaryngol. (Stockholm) **66**, 369—385 (1968).
Newman, J. D., Wollberg, Z.: Multiple coding of species-specific vocalizations in auditory cortex of squirrel monkeys. Brain Res. **54**, 287—304 (1973).
Niessing, C.: Die Entwicklung der cranialen Ganglien bei Amphibien. Morph. Jb. **70**, 472—530 (1932).
Niimi, K., Hazada, K., Kusaka, Y., Kishi, J.: The ontogenetic development of the diencephalon of the mouse. Tokushima J. exp. Med. **8**, 203—238 (1962).
Ogawa, C.: Experiments on the orientation of the ear vesicle in amphibian larvae. J. exp. Zool. **34**, 17—43 (1921).
Ogawa, C.: Einige Experimente zur Entwicklungsmechanik der Amphibien-Hörbläschen. Fol. anat. japon. **4**, 413—431 (1926).
Omerod, F. C.: The pathology of congenital deafness in the child. In: The Modern Educational Treatment of Deafness. Manchester: Manchester University Press 1960.
O'Rahilly, R.: The early development of the otic vesicle in staged human embryos. J. Embryol. exp. Morph. **11**, 741—755 (1963).
Orr, M. F.: Histogenesis of sensory epithelium in reaggregates of dissociated embryonic chick otocysts. Dev. Biol. **17**, 39—54 (1968).
Orr, M. F.: Light and electron microscopic study of the embryonic chick otocyst: the effects of trypsin and Ca- and Mg-free salt solution. Dev. Biol. **47**, 325—340 (1975).
Otis, E., Brent, R.: Equivalent ages in mouse and human embryos. Anat. Rec. **120**, 33—53 (1954).
Parks, T. N., Robertson, J.: The effects of otocyst removal on the development of chick brain stem auditory nuclei. Anat. Rec. **184**, 497—498 (1976).
Parks, T. N., Rubel, E. W.: Organization and development of brain stem auditory nuclei of the chicken: organization of projections from *n. magnocellularis* to *n. laminaris*. J. comp. Neurol. **164**, 435—448 (1975).
Parks, T. N., Rubel, E. W.: Embryonic deafferentiation produces a delayed, secondary migration of *n. angularis* to an ectopic position. Paper presented at the annual meeting of the Society for Neuroscience, Anaheim, Calif., November, 1977.
Pasteels, J.: Les effets de la centrifugation axiale de l'oeuf féconde et insegmenté chez les amphibiens anoures. Bull. Acad. Belg. Cl. Sci. **25**, 334—345 (1939).
Peiper, A.: Sinnesempfindungen des Kindes vor seiner Geburt. Monatsschr. Kinderheilk. **29**, 236—241 (1925).
Peters, J. J., Vonderahe, A. R., Huesman, A. A.: Chronological development of electrical activity in the optic lobes, cerebellum, and cerebrum of the chick embryo. Physiol. Zool. **33**, 225—231 (1960).
Pierce, E. T.: Histogenesis of the dorsal and ventral cochlear nuclei in the mouse. An autoradiographic study. J. comp. Neurol. **131**, 27—54 (1967).
Pierce, E. T.: Time of origin of neurons in the brain stem of the mouse. In: Progress in Brain Research, Vol. 40. New York: Elsevier 1973.
Politzer, G.: Die Entstehung des Ganglions acusticum beim Menschen. Acta Anat. **26**, 1—13 (1956).
Powers, B. M., Warfield, D., Ruben, R. J.: Development of the Pryer response to different frequencies in normal CBA-J mice. J. Aud. Res. **6**, 426—435 (1966).
Price, G. R.: Loss of cochlear microphonic sensitivity in young cat ears exposed to intense sound (Abstract). J. Acoust. Soc. Amer. **51**, 104 (1972).
Proctor, C. A., Lawrence, M.: The development of sensory cell innervation in the inner ear. Univ. of Mich. Med. Bull. **25**, 37—48 (1959).
Proctor, C. A., Proctor, B.: Understanding hereditary nerve deafness. Arch. Otolaryngol. **85**, 23—40 (1967).
Pujol, R.: Development of tone-burst responses along the auditory pathway in the cat. Acta otolaryngol. (Stockholm) **74**, 383—391 (1972).
Pujol, R., Abennenc, M., Rebillard, J.: Development and plasticity in the auditory system: methodological approaches and first results. In: Aspects of Plasticity. Paris: INSERM 1975.
Pujol, R., Granier, M. R.: Maturation des réponses auditives dans le colliculus inferior chez le chat. J. Physiol. (Paris). **60** (Suppl. 2), 525 (1968).
Pujol, R., Hilding, D.: Anatomy and physiology of the onset of auditory function. Acta otolaryngol. (Stockholm) **76**, 1—10 (1973).
Pujol, R., König, N., Romand, R.: Etude neurophysiologique de la maturation du corps genouillé median. J. Physiol. (Paris) **62** (Suppl. 2), 305 (1970).

Pujol, R., Marty, R.: Structural and physiological relationships of the maturing auditory system. In: Ontogenesis of the Brain. Prague: Charles University Press 1968.
Pujol, R., Marty, R.: Postnatal maturation of the cochlea of the cat. J. comp. Neurol. **139**, 115—125 (1970).
Pujol, R., Romand, R., Granier, M. R., Marty, R.: Maturation des réponses à la stimulation tonale dans les noyaux cochléaires. J. Physiol. (Paris) **61** (Suppl. 2), 380 (1969).
Pysh, J. J.: The development of the extracellular space in neonatal rat inferior colliculus: An electron microscopic study. Amer. J. Anat. **124**, 411—430 (1969).
Rabinowicz, T.: The cerebral cortex of the premature infant of the 8th month. Progr. Brain Res. **4**, 39—86 (1964).
Rabinowicz, T.: Quantitative appraisal of the cerebral cortex of the premature infant of 8 months. In: Regional Development of the Brain in Early Life. Philadelphia: Davis 1967.
Rajecki, D. W.: Effects of prenatal exposure to auditory or visual stimulation of postnatal distress vocalizations in chicks. Behav. Biol. **11**, 525—536 (1974).
Ramon y Cajal, S.: Terminación periférica del nervio acústico de las aves. Trab. Lab. Invest. Biol. Univ. Madrid **6**, 161—176 (1908a).
Ramon y Cajal, S.: Les ganglions terminaux du nerf acoustique des oiseaux. Trav. Lab. Rech. Biol. (Madrid) **6**, 195—224 (1908b).
Ramon y Cajal, S.: Studies on vertebrate neurogenesis. New York: Thomas 1960 (Translated by Lloyd Guth).
Rawdon-Smith, A. F., Charmichael, L., Wellman, B.: Electrical responses from the cochlea of the fetal guinea pig. J. exp. Psychol. **23**, 531—535 (1938).
Ray, W. S.: A preliminary report on a study of fetal conditioning. Child Dev. **3**, 175—177 (1932).
Raymond, R., Sans, A., Romand, M. R., Marty, R.: The structural maturation of the stato-acoustic nerve in the cat. J. comp. Neurol. **170**, 1—16 (1976).
Rebollo, M. A., Casas, M.: Desarollo del organo de Corti y de la vía auditiva en el embrión de pollo. I. Histogénesis del organo de Corti. An. Fac. Med. Montevideo **49**, 35—41 (1964).
Rebollo, M. A., Rodriguez, M. M.: Desarollo del organo de Corti y de la vía auditiva en el embrión de pollo. II. Histogénesis de ganglio de Corti. An. Fac. Med. Montevideo **49**, 42—46 (1964).
Retzius, G.: Das Gehörorgan der Wirbeltiere. II. Das Gehörorgan der Reptilien, der Vögel und Säugetiere. Stockholm: Samson and Wallin 1884.
Richardson, D.: Some effects of the heteroplastic transplantation of the ear vesicle in *Amblystoma*. J. exp. Zool. **63**, 413—445 (1932).
Robertson, E., Peterson, J., Lamb, L.: Relative impedance measurements in young children. Arch. Otolaryng. **88**, 162—188 (1968).
Rodriguez, M. M., Rebollo, M. A.: Desarollo del organo de Corti y de la vía auditiva en el embrión de pollo. III. Nervio coclear. An. Fac. Med. Montevideo **51**, 40—44 (1966).
Rogers, L. A., Cowan, W.: The development of the mesencephalic nucleus of the trigeminal nerve in the chick. J. comp. Neurol. **147**, 291—320 (1973).
Romand, R.: Maturation des potentiels cochléaires dans la période périnatale chez le chat et chez le cobaye. J. Physiol. (Paris) **63**, 763—782 (1971).
Romand, R., Granier, M. R., Marty, R.: Développement postnatal de l'activité provoqûee dans l'olive superieure laterale chez le chat par la stimulation sonore. J. Physiol. (Paris) **66**, 303—315 (1973).
Romand, R., Granier, M. R., Pujol, R.: Potentiels cochléaires chez le cobaye dans la periode prénatale. J. Physiol. (Paris) **63**, 280A—281A (1971).
Romand, R., Marty, R.: Postnatal maturation of the cochlear nuclei in the cat: A neurophysiological study. Brain Res. **83**, 225—233 (1975).
Romand, R., Pujol, R., König, N., Marty, R.: Maturation électrophysiologique de la cochlée. C. R. Ser. 1970, D**270**, 2476—2479.
Rose, G. H., Ellingson, R. J.: Ontogenesis of evoked potentials. In: Developmental Neurobiology. Springfield, Ill.: Thomas 1970.
Rose, J. E.: The ontogenetic development of the rabbit's diencephalon. J. comp. Neurol. **77**, 61—129 (1942).
Rose, J. E., Adrian, H., Santibanez, G.: Electrical signs of maturation in the auditory system of the kitten. Acta neurol. Latinamer. **3**, 133—143 (1957).
Rose, J. E., Sobkowicz, H. M., Bereman, B.: Growth in culture of peripheral axons of the spiral neurons in response to displacement of the receptors. J. Neurocytol. **6**, 49—70 (1977).

Rosen,M.: Developmental fetal EEG studies in the guinea pig, with comments on the sheep, and human fetus. In: Brain Development and Behavior. New York: Academic 1971.
Rosowski,J.J., Bock,G.R., Saunders,J.C.: Frequency selectivity in single neurons in the mesencephalicus lateralis pars dorsalis (MLD) of the neonatal chick. J. Acoust. Soc. Amer. **59**, Supp. 1, S47 (A) (1976).
Rossi,C.: L'acétylcholinesterase au cours du développement de l'oreille interne du cobaye. Acta Otolaryngol. Suppl. **170**, 1—91 (1961).
Rubel,E.W., Parks,T.N.: Organization and development of brain stem auditory nuclei of the chicken: tonotopic organization of *n. magnocellularis* and *n. laminaris*. J. comp. Neurol. **164**, 411—434 (1975).
Rubel,E.W., Rosenthal,M.H.: The ontogeny of auditory frequency generalization in the chicken. J. exp. Psychol.: Anim. Behav. Proc. **1**, 287—297 (1975).
Rubel,E.W., Smith,D.J., Miller,L.C.: Organization and development of brain stem auditory nuclei of the chicken: ontogeny of *n. magnocellularis* and *n. laminaris*. J. comp. Neurol. **166**, 469—490 (1976).
Ruben,R.J.: Development of the inner ear of the mouse: a radioautographic study of terminal mitoses. Acta otolaryngol. (Stockholm) Suppl. **220**, 1—44 (1967).
Ruben,R.J.: The synthesis of DNA and RNA in the developing inner ear. Laryngoscope **79**, 1546—1556 (1969).
Ruben,R.J.: Development and cell kinetis of the Kreisler (kr/kr) mouse. Laryngoscope **83**, 1440—1468 (1973).
Ruben,R.J., van de Water,T., Polesky,A.: Cell kinetics of the 11- and 12-day mouse otocysts. Laryngoscope **81**, 1708—1718 (1971).
Ryugo,D.K., Ryugo,R., Glabus,A., Killackey,H.P.: Increased spine density in auditory cortex following visual or somatic deafferentation. Brain Res. **90**, 143—146 (1974).
Sakabe,N., Arayama,T., Suzuki,T.: Human fetal evoked responses to acoustic stimulation. Acta otolaryngol. (Stockholm) Suppl. **252**, 29—36 (1969).
Salamy,A., McKean,C.M., Buda,R.B.: Maturational changes in auditory transmission as reflected in human brainstem potentials. Brain Res. **96**, 361—366 (1975).
Salamy,A., McKean,C.M.: Postnatal development of human brainstem potentials during the first year of life. Electroencephalog. Clin. Neurophysiol. **40**, 418—426 (1976).
Salk,L.: Mothers' heartbeat as an imprinting stimulus. Trans. N. Y. Acad. Sci. **24**, 753—763 (1962).
Saunders,J.C.: The development of auditory evoked responses in the chick embryo. Minerva Otorhinolaryngol. **24**, 221—229 (1974).
Saunders,J.C., Bock,G.R.: Influences of early auditory trauma on auditory development. In: Studies on the Development of Behavior and the Nervous System, Vol. 4. New York: Academic (in press) 1977.
Saunders,J.C., Coles,R.B., Gates,G.R.: The development of auditory evoked responses on the cochlea and cochlear nuclei of the chick. Brain Res. **63**, 59—74 (1973).
Saunders,J.C., Gates,G.R., Coles,R.B.: Brainstem evoked responses as an index of hearing thresholds in one-day-old chicks and ducklings. J. comp. Physiol. Psychol. **86**, 426—431 (1974).
Saunders,J.C., Hirsch,K.A.: Changes in cochlear microphonic sensitivity after priming C57BL/6J mice at various ages for audiogenic seizures. J. comp. Physiol. Psychol. **90**, 212—220 (1976).
Scherrer,J., Verley,R., Garma,L.: A review of French studies in the ontogenetical field. In: Developmental Neurobiology. Springfield, Ill.: Thomas 1970.
Schmidt,G.A.: Die morphogenetische Bedeutung des Nervensystems. I. Die Korrelation in der Entwicklung des Gehörorgans. Russk. Arkh. Anat. **18**, 298—344, 475—477 (1938).
Schmidt,R.S., Fernández,C.: Development of mammalian endocochlear potential. J. exp. Zool. **153**, 227—236 (1963).
Schuknecht,H.F., Igarishi,M., Gacek,R.R.: The pathological types of cochleo-saccular degeneration. Acta otolaryngol. (Stockholm) **59**, 154—167 (1965).
Schulman-Galambos,C., Galambos,R.: Brainstem auditory-evoked responses in premature infants. J. Speech Hear. Res. **18**, 456—465 (1975).
Schwartz,I.R.: The development of terminals in the cat medial superior olive (Abstract). Anat. Rec. **172**, 401 (1972).
Scibetta,J.J., Rosen,M.B.: Responses evoked by sound in the fetal guinea pig. Obstet. Gynecol. **33**, 830—836 (1969).
Scibetta,J.J., Rosen,M.B., Hochberg,C.J., Chik,L.: Human fetal brain response to sound during labor. Amer. J. Obstet. Gynecol. **109**, 82—85 (1971).

Scott, G. L., Sobkowicz, H. G., Bereman, B., Rose, J. E.: Ultrastructure of synaptic profiles in the developing organ of Corti in culture (Abstract). Neurosci. Abs. **2**, 25 (1976).
Sedlaček, J.: Further findings on the conditions of formation of the temporary connection in chick embryos. Physiol. Bohemoslav. **13**, 411—420 (1964).
Sedlaček, J.: Proceedings: The prenatal development of sensory systems in precocial mammals and birds. Act. Nerv. Super. (Praha) **16**, 303—305 (1974).
Sedlaček, J.: Fetal and neonatal development of evoked responses in guinea pig auditory cortex. Physiol. Bohemoslav. **25**, 13—21 (1976).
Shaner, R. F.: The development of the nuclei and tracts related to the acoustic nerve in the pig. J. comp. Neurol. **60**, 5—19 (1934).
Sher, A. E.: The embryonic and postnatal development of the inner ear of the mouse. Acta otolaryngol. (Stockholm) Suppl. **285**, 1—77 (1971).
Shofer, R. J., Sax, D. S., Strom, M. G.: Analysis of auditory and cerebrocortically evoked activity in the immature and adult cat cerebellum. In: Neurobiology of Cerebellar Evolution and Development. Chicago: A. M. A. 1969.
Silverman, M. S., Clopton, B. M., Flammino, F.: Development of single cell binaural interaction in the inferior colliculus. Paper presented at the 89th Meeting of the Acoust. Society of America. Austin, Texas, 1975.
Small, W. S.: Notes on the psychic development of the young white rat. Amer. J. Psychol. **11**, 80—100 (1899).
Smyth, C. M., Bench, R. J.: Fetal response to audiogenic stimulation as an indication of neurological function. In: Digest of 7th International Conference on Medicine and Biomedical Engineering, Stockholm, 1967.
Sobkowicz, H. M., Bereman, B., Rose, J. E.: Organotypic development of the organ of Corti in culture. J. Neurocytol. **4**, 543—572 (1975).
Solomon, R., Lessac, M. S.: A control group design for experimental studies of developmental processes. Psychol. Bull. **70**, 145—209 (1968).
Sontag, L. W., Wallace, R. F.: Changes in the rate of the human fetal heart in response to vibratory stimuli. Amer. J. Dis. Childh. **51**, 583—589 (1936).
Spelt, D. K.: Conditioned responses in the human fetus in utero. Psychol. Bull. **35**, 712—713 (1938).
Spemann, H.: Die Entwicklung des invertierten Hörgrübchens zum Labyrinth. Ein kritischer Beitrag zur Strukturlehre der Organanlagen. Roux' Arch. Entw.-Mech. **30**, 437—458 (1910).
Spoendlin, H.: The organization of the cochlear receptor. Adv. Oto-Rhino-Laryngol. **13**, 1—227 (1966).
Spoendlin, H.: Ultrastructure and peripheral innervation pattern of the receptor in relation to the first coding of the acoustic message. In: Hearing in Vertebrates: A CIBA Foundation Symposium. Boston: Little, Brown 1968.
Spoendlin, H.: Structural basis of peripheral frequency analysis. In: Frequency Analysis and Periodicity Detection in Hearing. Leiden: Sijthoff 1970.
Spoendlin, H.: Neuroanatomy of the cochlea. In: Psychophysical Methods and Physiological Facts in Audition. Berlin: Springer 1974.
Stephens, C. B.: Development of the middle and inner ear in the golden hamster (*Mesocricetus auratus*). Acta otolaryngol. (Stockholm) Suppl. **296**, 1—51 (1972).
Stone, L. S.: Experiments on the development of the cranial ganglia and the lateral line sense organs in *Amblystoma punctatum*. J. exp. Zool. **35**, 421—496 (1922).
Stone, L. S.: Induction of the ear by the medulla and its relation to experiments on the lateralis system in amphibians (Abstract). Science **74**, 577 (1931).
Streeter, G. L.: On the development of the membranous labyrinth and the acoustic and facial nerves in the human embryo. Amer. J. Anat. **6**, 139—165 (1906a).
Streeter, G. L.: Some experiments on the developing ear vesicle of the tadpole with relation to equilibrium. J. exp. Zool. **3**, 543—558 (1906b).
Streeter, G. L.: Some factors in the development of the amphibian ear vesicle, and further experiments on equilibrium. J. exp. Zool. **4**, 431—445 (1907).
Streeter, G. L.: Experimental observations on the development of the amphibian ear vesicle. Anat. Rec. **3**, 199—201 (1909).
Streeter, G. L.: The development of the nervous system. In: Manual of Human Embryology, Vol. 2. Philadelphia: Lippincott 1912.

Streeter, G. L.: Experimental evidence concerning the determination of posture of the membranous labyrinth in amphibian embryos. J. exp. Zool. **16**, 149—176 (1914).
Streeter, G. L.: The developmental alterations in the vascular system of the brain of the human embryo. Contrib. Embryol. **8**, 5—54 (1918).
Ströer, W. F. H.: Studies on the diencephalon. I. The embryology of the diencephalon of the rat. J. comp. Neurol. **105**, 1—24 (1956).
Suga, F., Hattler, K. W.: Physiological and histopathological correlates of hereditary deafness in animals. Laryngoscope **80**, 1—104 (1970).
Sugiura, A., Hilding, D. A.: Cochleo-saccular degeneration in the Hedlund white mink. Acta Otolaryngol. (Stockholm) **69**, 126—137 (1970).
Sviderskaya, G. E.: Effects of sound on the motor activity of chick embryos. Bull. exp. Biol. Med. (U.S.S.R.) **63**, 24—28 (1967). (Translated from the Russian).
Szepsenwol, J.: Recherches sur les centres organisateurs des vésicules auditives chez des embryons de poulets omphalocéphales obtenus expérimentalement. Arch. Anat. Microscop. **29**, 5—94 (1933).
Tanaka, Y., Arayama, T.: Fetal responses to acoustic stimuli. Pract. Oto-Rhino-Laryngol. **31**, 269—273 (1969).
Tees, R. C.: The effects of early auditory restriction in the rat on adult duration discrimination. J. Aud. Res. **7**, 195—207 (1967a).
Tees, R. C.: Effects of early auditory restriction in the rat on adult pattern discrimination. J. comp. Physiol. Psychol. **63**, 389—392 (1967b).
Tees, R. C.: Perceptual development in mammals. In: Neural and Behavioral Specificity. New York: Academic 1976.
Tello, J. F.: Les différenciations neuronales dans l'embryon du poulet pendant les premiers jours de l'incubation. Trab. Lab. Invest. Biol. Univ. Madrid **21**, 1—93 (1922).
Tello, J. F.: Le réticule des cellules ciliées du labyrinthe chez la souris et son indépendance des terminaisons nerveuses de la VIIIe paire. Trab. Lab. Invest. Biol. Univ. Madrid **27**, 167—186 (1931–1932).
Tilney, F., Casamajor, L.: Myelinogeny as applied to the study of behavior. A. M. A. Arch. Neurol. Psychiat. **12**, 1—66 (1924).
Titova, L. K.: Development of receptor structures in the inner ear of vertebrates. Washington, D. C.: National Aeronautical Administration Technical Transactions NASA TT F-165 1970.
Toerien, M. J.: An experimental approach to the development of the ear capsule in the turtle *Chelydra serpentina*. J. Embryol. exp. Morph. **13**, 141—149 (1965).
Tokura, R.: Zur Frage der Hörbläscheninversion. Fol. anat. japon. **2**, 97—118 (1924).
Tokura, R.: Entwicklungsmechanische Untersuchungen über das Hörbläschen und das akustische, sowie faciale Ganglion bei den Anuren. Fol. anat. japon. **3**, 173—208 (1925).
Trampusch, H. A. L.: On ear-induction. Acta Neerland. Morphol. **4**, 195—213 (1941).
Tschanz, B.: Beobachtungen und Experimente zur Entstehung der "persönlichen" Beziehung zwischen Jungvögel und Eltern bei Trottellummen. In: Verhandl. Schweiz. Naturforsch. Ges., Zürich: 1964.
Tschanz, B.: Trottellummen: die Entstehung der persönlichen Beziehung zwischen Jungvogel und Eltern. Z. Tierpsychol. **4** (Suppl.), 1—103 (1968).
Turkewitz, G., Birch, H. G., Moreau, T., Levy, L., Cornwell, A. C.: Effects of intensity of auditory stimulation on directional eye movements in the human neonate. Anim. Behav. **14**, 93—101 (1966).
Turkewitz, G., Birch, H. G., Cooper, K. K.: Responsiveness to simple and complex auditory stimuli in the human newborn. Dev. Psychobiol. **5**, 7—19 (1972).
van Campenhout, E.: Origine du ganglion acoustique chez le porc. Arch. biol. (Paris) **46**, 273—286 (1935a).
van Campenhout, E.: Experimental researches on the origin of the acoustic ganglion in amphibian embryos. J. exp. Zool. **72**, 175—193 (1935b).
van der Borden, J.: Experimental synotia dorsalis in mouse embryos. Ann. Otol. **76**, 129—148 (1967).
Van der Stricht, O.: The genesis and structure of the membrane tectoria and the crista spiralis of the cochlea. Contrib. Embryol. **7**, 55—86 (1918).
Van der Stricht, O.: The development of the pillar cells, tunnel space, and Nuel's spaces in the organ of Corti. J. comp. Neurol. **30**, 283—326 (1919).
Van der Stricht, O.: The arrangement and structure of sustentacular cells and hair cells in the developing organ of Corti. Contrib. Embryol. **9**, 109—142 (1920a).

Van der Stricht, O.: Les membranes tectrices des crêtes et des taches acoustiques. Arch. biol. (Paris) **31**, 299—321 (1920b).

Van de Water, T. R., Hewood, P., Ruben, R. J.: Development of sensory structures in organ cultures of the 12th and 13th gestation day mouse embryo inner ears. Ann. Otol. **82** Suppl. 4, 1—18 (1973).

Van de Water, T. R., Ruben, R. J.: Organ culture of the mammalian inner ear. Acta otolaryngol. (Stockholm) **71**, 303—312 (1971).

Van de Water, T. R., Ruben, R. J.: Quantification of the *in vitro* development of the mouse embryo inner ear. Ann. Otol. **82** Suppl. 4, 19—21 (1973).

Van de Water, T. R., Ruben, R. J.: Organ culture of the mammalian inner ear: a tool to study inner ear deafness. Laryngoscope **84**, 738—749 (1974).

Van de Water, T. R., Ruben, R. J.: Organogenesis of the ear. In: Scientific Foundations of Otolaryngology. London: Heinemann Medical 1976.

Vanzulli, A., Garcia-Austt, E.: Development of cochlear microphonic potentials in the chick embryo. Acta neurophysiol. Latinoamer. **9**, 19—23 (1963).

Vázquez-Nin, G. H., Sotelo, J. R.: Neurofibrillar differentiation during embryonary growth. J. comp. Neurol. **128**, 313—331 (1966).

Vázquez-Nin, G. H., Sotelo, J. R.: Electron microscope study of the developing nerve terminals in the acoustic organs of the chick embryo. Z. Zellforsch. mikr. Anat. **92**, 325—338 (1968).

Veit, O.: Beiträge zur Kenntnis des Kopfes der Wirbeltiere. II. Frühstadien der Entwicklung des Kopfes von Lepidosteus osseus und ihre prinzipielle Bedeutung für die Kephalogenese der Wirbeltiere. Morph. Jb. **53**, 319—390 (1924).

Vince, M. A.: Some environmental effects on the activity and development of the avian embryo. In: Studies on the Development of Behavior and the Nervous System, 1. Behavioral Embryology. New York: Academic 1973.

Vince, M., Reader, M., Tolhurst, B.: Effects of stimulation on embryonic activity in the chick. J. comp. Physiol. Psychol. **90**, 221—230 (1976).

Volokhov, A. A.: Comparative studies of the functional development of analyzer system in animals in the process of ontogenesis. Prog. Brain Res. **22**, 527—540 (1968).

Wada, T.: Anatomical and physiological studies on the growth of the inner ear of the albino rat. Amer. Anat. Mem. **10**, 1—74 (1923).

Waddington, C. H.: The determination of the auditory placode in the chick. J. exp. Biol. **14**, 232—239 (1937).

Waddington, C. H.: The epigenetics of birds. London: Cambridge University Press 1952.

Walker, D., Grimwade, J., Wood, C.: Intrauterine noise: a component of the fetal environment. Amer. J. Obstet. Gynecol. **109**, 91—95 (1971).

Waterman, A. J.: The development of the inner ear rudiment of the rabbit embryo in a foreign environment. Amer. J. Anat. **63**, 161—217 (1938).

Wawrzyniak, M.: Histochemical activity of some enzymes in the mesencephalon during otogenic development of the rabbit and guinea pig. IV. Fol. histochem. cytochem. **4**, 373—388 (1966).

Weber, H. J., McGovern, F. J., Zink, D.: An evaluation of 1000 children with hearing loss. J. Speech Hear. Disab. **32**, 343—354 (1967).

Webster, D. B.: A function of the enlarged middle-ear cavities of the kangaroo rat, *Dipodomys*. Physiol. Zool. **35**, 248—255 (1962).

Webster, D. B., Webster, M.: Morphological effects of acoustic deprivation on the brainstem auditory system of CBA/J mice. Paper presented to the 89th American Association of Anatomists. Louisville, Kentucky, April, 1976.

Wedenberg, E.: Prenatal test of hearing. Acta otolaryngol. (Stockholm) Suppl. **206**, 27—32 (1965).

Weibel, E. R.: Zur Kenntnis der Differenzierungsvorgänge im Epithel des Ductus cochlearis. Acta Anat. **29**, 53—90 (1957).

Weitzman, E. D., Fishbein, W., Graziani, L.: Auditory evoked responses obtained from the scalp electroencephalogram of the full term human neonate during sleep. Pediatrics **35**, 458—462 (1965).

Weitzman, E. D., Graziani, L. J.: Maturation and topography of the auditory evoked response of the prematurely born infant. Dev. Psychobiol. **1**, 79—89 (1968).

Wersall, J., Flock, A.: Morphological aspects of cochlear hair cell physiology. In: Henry Ford Hospital International Symposium. Boston: Little, Brown 1967.

Wertheimer, R. G. E.: Psychomotor coordination of auditory and visual space at birth. Science **134**, 1692 (1961).

West, C. D., Harrison, J. M.: Transneuronal cell atrophy in the congenitally deaf white cat. J. comp. Neurol. **151**, 377—398 (1973).

Weston, J. A.: Neural crest cell migration and differentiation. In: Cellular Aspects of Neural Growth and Differentiation. Los Angeles: University of California Press 1971.

Wever, E. G.: Origin and evolution of the ear of vertebrates. In: Evolution of Brain and Behavior in Vertebrates. Hillsdale, N. J.: L. Erlbaum Association 1976.

Whitfield, I. C., Evans, E. F.: Responses of auditory cortical neurons to stimuli of changing frequency. J. Neurophysiol. **28**, 655—672 (1965).

Willott, J. F., Henry, K. R.: Auditory evoked potentials: developmental changes of threshold and amplitude following early acoustic trauma. J. comp. Physiol. Psychol. **86**, 1—7 (1974).

Wilson, D. B., Sawyer, R. H., Hendrickx, A. G.: Proliferation gradients in the inner ear of the monkey (*Macaca mulatta*) embryo. J. comp. Neurol. **164**, 23—29 (1975).

Wilson, J. T.: Observations upon young human embryos. J. Anat. Physiol. **48**, 315 (1914).

Windle, W. F.: The neurofibrillar structure of the 5.5 mm cat embryo. J. comp. Neurol. **55**, 315—331 (1932a).

Windle, W. F.: Neurofibrillar structure of the 7 mm cat embryo. J. comp. Neurol. **55**, 99—138 (1932b).

Windle, W. F.: Neurofibrillar development in the central nervous system of cat embryos between 8 and 12 mm long. J. comp. Neurol. **58**, 643—723 (1933).

Windle, W. F.: Origin and early development of neural elements in the human brain. In: The Biopsychology of Development. New York: Academic 1971.

Windle, W. F., Austin, H. F.: Neurofibrillar development in the central nervous system of chick embryo up to 5 days incubation. J. comp. Neurol. **63**, 431—463 (1936).

Witschi, R.: The larval ear of the frog and its transformation during metamorphosis. Z. Naturforsch. **46**, 230—242 (1949).

Woerdeman, M. W.: Inducing capacity of the embryonic eye. Proc. Konink. Nederl. Akad. Wetens. **41**, 336—343 (1938).

Wolf, A.: The dynamics of the selective inhibition of specific functions in neurosis. A preliminary report. Psychosom. Med. **5**, 27—38 (1943).

Woolf, N. K., Bixby, J. L., Capranica, R. R.: Prenatal experience and avian development: brief auditory stimulation accelerates the hatching of Japanese quail. Science **194**, 959—960 (1976).

Yakovlev, P. I., Lecours, A.-R.: The myelogenetic cycles of the regional maturation of the brain. In: Regional Development of the Brain in Early Life. Philadelphia: Davis 1967.

Yntema, C. L.: An experimental study of the origin of the cells which constitute the VIIth and VIIIth ganglia and nerves in the embryo of *Amblystoma punctatum*. J. exp. Zool. **75**, 75—101 (1937).

Yntema, C. L.: Self-determination of heterotopic ear ectoderm in the embryo of *Amblystoma punctatum*. J. exp. Zool. **80**, 1—18 (1939).

Yntema, C. L.: The symmetry of ears induced from disharmonic ectoderm. Anat. Rec. **100** (Suppl. 1), 95 (1948).

Yntema, C. L.: An analysis of induction of the ear from foreign ectoderm in the salamander embryo. J. exp. Zool. **113**, 211—244 (1950).

Yntema, C. L.: Ear and nose. In: Analysis of Development. Philadelphia: Saunders 1955.

Yokoh, Y.: Early formation of nerve fibers in the human otocyst. Acta Anat. **80**, 99—106 (1971).

Yokoh, Y.: On sensory cells in the human otocyst. Acta Anat. **87**, 72—76 (1974).

Young, E. D., Brownell, W. E.: Response to tones and noise of single cells in dorsal cochlear nucleus of unanesthetized cats. J. Neurophysiol. **39**, 282—300 (1976).

Záhlava, J., Chaloupka, Z., Myslivećek, J.: The development of primary cortical auditory responses at various stimulation frequencies in dogs. Activ. Nerv. Sup. (Praha) **8**, 175—176 (1966).

Zwilling, E. J.: The determination of the otic vesicle in *Rana pipiens*. J. exp. Zool. **86**, 333—342 (1941).

Chapter 6

The Development of Somatosensory Thalamus in Mammals

MADGE E. SCHEIBEL* and ARNOLD B. SCHEIBEL, Los Angeles, California, USA

With 21 Figures

Contents

* deceased.

A. Introduction

Consideration of the ontogenetic development of sensory thalamus and its interrelations with cerebral cortex is fraught with numerous problems, both technical and conceptual. The histological matrix is one of the most complex in the entire nervous system, comprising as it does a variety of highly specialized pre- and postsynaptic elements, linked together in part, by intricate synaptic ensembles or glomeruli (FAMIGLIETTI, 1970; HARDING, 1971; JONES and POWELL, 1969; RALSTON, 1971; RALSTON and HERMANN, 1968; SCHEIBEL et al., 1972; SPACEK and LIEBERMAN, 1974; SZENTAGOTHAI, 1963; SZANTAGOTHAI et al., 1966; TÖMBÖL, 1966). The physiological characteristics of this area are equally varied and have resulted in the development of a number of models, none of which is completely satisfying (ANDERSEN and ANDERSSON, 1968; ANDERSEN et al., 1964; PURPURA, 1969; SCHEIBEL and SCHEIBEL, 1966, 1967; SCHEIBEL et al., 1973; SCHLAG and WASZAK, 1971; STERIADE and WYZINSKI, 1972; WASZAK, 1974). Knowledge of thalamic synaptic chemistry and pharmacology is still limited, although substantial advances have been made in the past decade (ANDERSEN and CURTIS, 1964; GEREBTZOFF, 1970; MCCANCE et al., 1968; MCLENNAN et al., 1968). Finally, the functional role and clinical significance of thalamic activity remain enigmatic despite an abundant literature and the apparent success of a limited repertoire of surgical intervention (COOPER, 1969; COOPER et al., 1971; HASSLER, 1959; JASPER and BERTRAND, 1966; SPIEGEL and WYCIS, 1952; VAN BUREN and BORK, 1972).

Quite clearly, the development of data on fine structure and function in the thalamus is still in an early phase, marked as much by promise as by significant disclosure. Nonetheless, the density of existing data is attested in the literature of the last decade, and is underlined by a recent series of thalamic symposia, which have significantly focused existing information and proposed early-order models (FRIGYESI et al., 1972; PURPURA and YAHR, 1966; RISS et al., 1972).

Very much less information is available on the morphogenesis and functional maturation of these systems. Accordingly, an examination of developments in this field at the present juncture may be both premature—and appropriate. It is certainly premature insofar as we can pretend to substantial understanding of the sequence of morphological development, especially at high levels of resolution, and its relation to functional displays and developing behavior. It is nonetheless appropriate in that the modest body of data that does exist can still be marshalled within a narrow compass and presented with considerable economy. These are, at best, mixed blessings, and can be expected to be remedied with time.

For the purpose of this review, we shall depend significantly on our own structural and functional studies of developing somatosensory thalamus (the ventrobasal complex, VB) and thalamocortical interactions, adding to our presentation data from any source that may enrich, modify, or explain. The greatest present lack turns out to be a comprehensive electron-microscopic analysis of developing thalamic tissue, with special reference to maturation of the synaptic bag and its contents, and the ripening of the thalamic glomerular complex. One useful exception is KARLSSON's (1967) electron-microscopic study of synaptic development in the lateral geniculate body, which may reveal developmental principles applicable to neighboring sensory centers.

Most of our own structural studies are based on analysis of Golgi impregnations of rats, mice, and kittens up to six weeks of age. Several litters of each species were obtained at three to seven days before birth, to provide some idea of the extent of antenatal development. Such sequences can then be compared with neurally mature or adult tissue. Control material derived from routine neurohistological studies support and supplement the Golgi sections, of which we have produced 10,000 over a period of almost 10 years.

Very young rats and mice are particularly useful in the study of patterning of afferent fibers and synaptic terminals. In these forms, the upper brain stem structures at issue are so small that they can frequently be studied, together with associated portions of basal ganglia and cortex, in individual sections. Furthermore, by using very thick sections (100—200 μm) of lightly stained material, a single tissue specimen may, in some cases, represent an appreciable fraction of the total depth of the nuclear complex. Fortuitous impregnations may then literally supply a map of individual fibers, often revealing probable areas of origin as well as course and pattern of termination, in some detail.

Kitten material, on the other hand, is more useful for the study of postsynaptic elements, particularly the differentiation of dendritic domains. There appear to be broader and more obvious differences between cell types in carnivores than in the smaller and more generalized rodent brain. The distinction between short- and long-axoned cells, in particular, is very much easier to make. In fact, there is reason to believe that short-axoned elements are minimal in mature rodent thalamus, and may be totally invisible in the newborn for some reason. While these factors make structural analysis of several species appropriate in a study of this sort, it is still necessary to remember that species differences, both temporal and topographic, must also be considered. For instance, the rodent appears to possess a large N. *parafascicularis* but no N. *centremedianum* at all, in contradistinction to the cat and primate (MEHLER, 1966). Additionally, the level of neural maturation in the 15-day-old rat is obviously quite different from that of a similarly aged kitten. If such variables are appropriately accounted for, however, the advantages of multi-species analysis, as indicated above, can outweigh the drawbacks.

The functional studies depend entirely on data derived from kittens maintained with chronically implanted electrodes. Well over 100 animals have received implants and recordings made with the animals in the unrestrained or partly restrained state over periods of two to five weeks. Surgical procedures have been performed as early as six hours after birth to assure a fairly sensitive estimate of thalamocortical functional states in the perinatal period.

B. Morphogenesis of Presynaptic Systems

I. Ascending Afferents as seen with the Light Microscope

A number of presynaptic fiber systems converge on the ventrobasal (VB) complex, and each terminal appears to develop a unique and idiomatic architecture. Previous studies have outlined the general characteristics of these components and given some indication of how their individual morphology may relate to their functional roles (CAJAL, 1900, 1911; SCHEIBEL and SCHEIBEL, 1966; SZENTÁGOTHAI, 1967; TÖMBÖL, 1966). We have not been able to impregnate

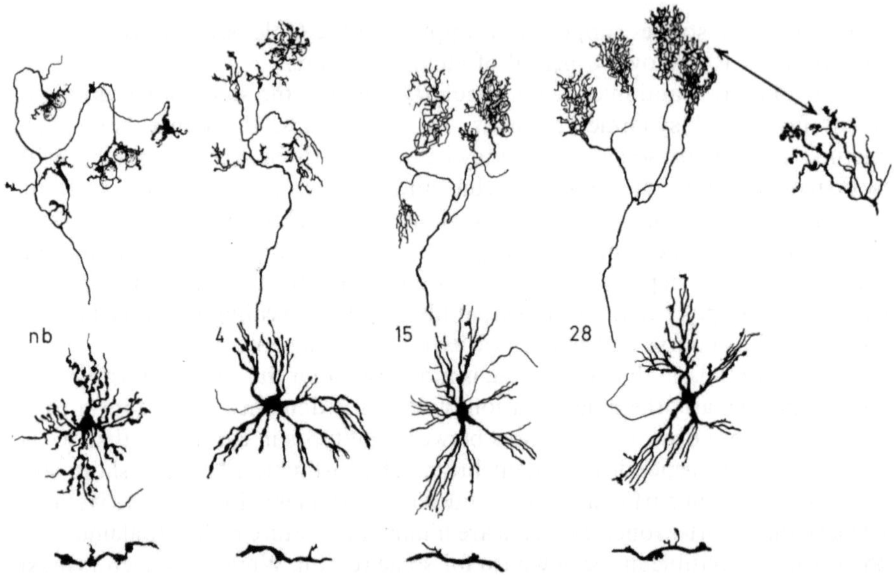

Fig. 1. Maturation of certain elements of ventrobasal nucleus in newborn (nb) and 4-, 15-, and 28-day-old rat. Upper series shows development of lemniscal afferent starting with little more than an axonal growth cone with tiny terminal extensions. Progressive extension and division of these terminals produces recognizable bushy arbors by end of second week. Later, these terminals' axonal branchlets develop small clusters of nodules, which become most important presynaptic element of arbor. These are seen in partly formed state at 28th day (arrow). Middle series shows that maturation of thalamocortical relay cells consists largely in elaboration of dendrite domain, extension of secondary dendrite branches, loss of typical dendritic nodules, and gradual appearance of dendritic surface specializations such as long, nodulated spine-like structures, excrescences, etc. Lower series of structures shows this sequence at higher power. Drawings based on modified Golgi preparations

material at a sufficiently immature period to mark the initial appearance of such fibers at their thalamic terminal sites, but examination of late prenatal mouse and rat material has allowed reconstruction of many significant effects.

1. Ascending Lemniscal Afferents

Ascending lemniscal afferents are already present in the late postnatal period (5—7 days before term), bearing exaggerated growth cones or small calyces from which wisp-like extensions are beginning to develop (SCHEIBEL and SCHEIBEL, 1966). There is rapid development of this system in the days immediately following birth, with the appearance of a number of short collaterals from each extension. These form multiple rudimentary nests which appear to fit around one or more thalamic cell bodies (Fig. 1, nb). These rudimentary bushy terminals gradually increase in size and complexity, apparently keeping pace with the developing extent and complexity of the postsynaptic dendritic domain. This *pari-passu* growth relationship between pre- and postsynaptic elements is reminiscent of the ascent of climbing fibers through the cerebellar molecular layer, time-locked to the growth of the receptive Purkinje dendrite arbor (CAJAL, 1911). Note that this sequence of perinatal arbor development in rat and mouse is somewhat less precocious than that in the newborn cat, whose sensory arbors are already quite dense at birth. The nodulated clusters that appear to constitute the effective synaptic terminals of the mature arbor (SZENTAGOTHAI, 1967) are not seen in any of our preparations even as late as day 10 or day 14 of life, and must be assumed to develop later. Their appearance has been documented by both light and electron microscopists, and they undoubtedly contribute significantly to the formation of thalamic glomeruli (JONES and POWELL, 1969; SZENTÁGOTHAI, 1967; TÖMBÖL, 1966).

We have previously shown that terminal arbors generated by the lemniscal afferent system are arranged in a distributive sequence from the hilar entry zone, producing an onion-skin arrangement of concentric representational fields that can be seen particularly well in the sagittal and frontal planes. It seemed interesting to attempt to find out whether this spatial relationship of afferent and efferent systems might be originally specified in the pre- or the postsynaptic systems involved. While no definitive answer to this question has been found it is worth recording that in a succession of Nissl control stains of the thalamic sensory fields in late pre- and early postnatal rats, no hint of nerve cell grouping is visible. However, Golgi impregnations performed on littermates at the same times show highly immature lemniscal terminal arbors already arranged in successive laminae. From these data we conclude that order is impressed upon the postsynaptic cell ensemble by the presynaptic afferents.

2. Presynaptic Terminals of Spinothalamic and "Nonspecific" Origin

These two types of terminal are generally difficult to distinguish from each other. Degeneration studies indicate that the former are concentrated in the posterior portion of the ventrobasal field (BOWSHER, 1957; GETZ, 1952; MEHLER, 1957), while nonspecific afferents have a more generalized distribution (NAUTA and KUYPERS, 1958; SCHEIBEL and SCHEIBEL, 1958, 1967). Both appear as rather small-caliber fibers undergoing a limited degree of rather diffuse, nonpatterned branching. Their terminal axons are widely dispersed geographically and show considerable degrees of overlap. Individual axon terminals bear very small growth cones and, subsequently, terminal boutons. Axons originating in the nonspecific portions of the brain stem (reticular formation) and thalamus (intralaminar system) appear to enter the ventrobasal fields from all sides and spread widely throughout.

II. Descending Afferents as seen with the Light Microscope

Descending afferents to somatosensory thalamus have been described from cerebral cortex and from nucleus reticularis thalami (CAJAL, 1900, 1911;

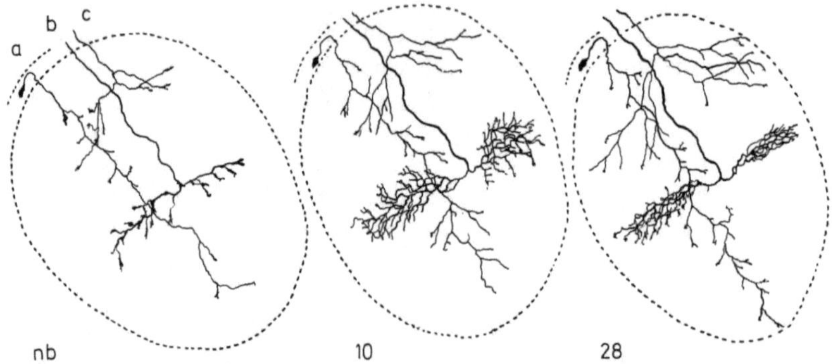

Fig. 2. Maturation of axonal terminal neuropil from 3 descending thalamopetal afferents in newborn (nb), 10-day-, and 28-day-old rat. Fiber "a" comes from nucleus *reticularis thalami*, fibers "b" and "c" from cerebral cortex. Note that all 3 are present in recognizable, if rudimentary fashion at birth, and that neuropil terminal pattern of fiber "b" is actually more extensive at 10 days than it is at 28 days. Drawings based on Golgi preparations

MINDERHOUD, 1971; RINVIK, 1972; SCHEIBEL and SCHEIBEL, 1966a, b; SCHEIBEL et al., 1972a). At least two patterns seem to be associated with the former, although their precise cortical source has not been determined. It seems fairly clear that both S_1 and S_2 project upon the ventrobasal complex (RINVIK, 1972), but we do not know whether the two fiber types relate to the two cortical sources, respectively.

1. Corticifugal Fibers

As suggested by CAJAL's studies (1900, 1911) and developed more fully in the present authors' (SCHEIBEL and SCHEIBEL, 1966b, 1967, 1971a; SCHEIBEL et al., 1971a), thalamopetal fibers of presumed cortical origin are characterized by both compact and diffuse terminal patterns (Fig. 2). The large-caliber axons generating dense, discoid terminal plexuses perpendicular to the long axis of the ventrobasal nucleus are among the first to be recognizable. The abrupt right-angle turn, which characterizes them even at the earliest perinatal epoch recorded (Fig. 2), is easily seen in satisfactory Golgi impregnations. The secondary branching from this transversely oriented preterminal is initially rudimentary (Fig. 2, nb) but with further development becomes increasingly dense, forming fairly well-developed transverse discoid plexuses by the 10th to 15th postnatal day (Fig. 2 [10]). Comparison of immature and mature neuropil segments indicates not only active growth through multiple-order rebranching, but also a dynamic molding and reshaping process, which must entail selective axonal loss in achievement of the final morphological pattern. As an example, note the obvious thinning of the transverse neuropil segment or disc in Fig. 2 between the 10th and 28th days.

The second type of fiber generates a diffuse and spreading plexus covering appreciable amounts of thalamic sensory matrix. It appears to be finer in caliber and is sometimes more difficult to identify, especially in the immediate pre- and

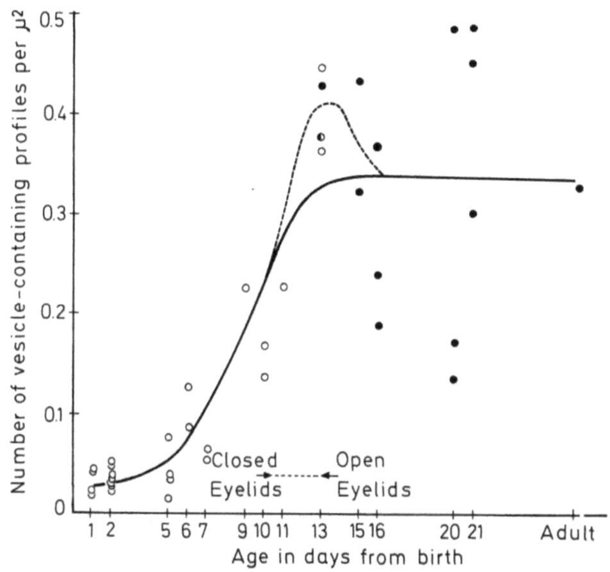

Fig. 3. Diagram showing frequency of vesicle-containing profiles in lateral geniculate of 38 rats ranging in age from 1 to 21 days. Open circles represent unmyelinated specimens, half-filled circles represent specimens with areas of suspected myelinization, and filled circles those where myelinization has been initiated. Vesicle-containing profiles are present on first day, and density increases between days 5 and 10. After day 10 considerable variation is seen. Typical value for adult specimen also represented at extreme right. General 10-fold increase in density of vesicle-containing profiles is found in this material. Myelinization appears to be closely related to opening of eyes, which occurs at about the 12th day in these litters. Dashed line above curve of average values represents average value for day 13; continuation shows for average values from days 10 to 16. From KARLSSON (1967); with permission of author and publisher

postnatal periods. The rather formless terminal system of each element gradually expands during the initial 10 to 14 days of extrauterine life and soon shows evidence of the tri-dimensional conical field and the fore-to-aft stacking previously described (SCHEIBEL and SCHEIBEL, 1966b, 1971a).

2. Fibers from N. *reticularis thalami*

The caudal-projecting axons of neurons in N. *reticularis thalami* are identifiable at birth by their long, relatively straight trajectories and the occasional short collaterals generated throughout the thalamic ventral complex of nuclei. Judging by the length of the collateral branches and the presence of small growth cones on many of them, it seems unlikely that they have established synaptic connections in significant numbers with the postsynaptic ensemble at this time. In the 12- to 15-day-old kitten, and in the 7- to 10-day-old rat, these fibers appear to be a good deal more mature, usually terminating in individual or small clusters of terminal boutons, presumably upon the receptive postsynaptic neurons, which may be the thalamocortical projection cells (SCHEIBEL et al., 1972). Their possible relation to the development of thalamic slow waves will be discussed.

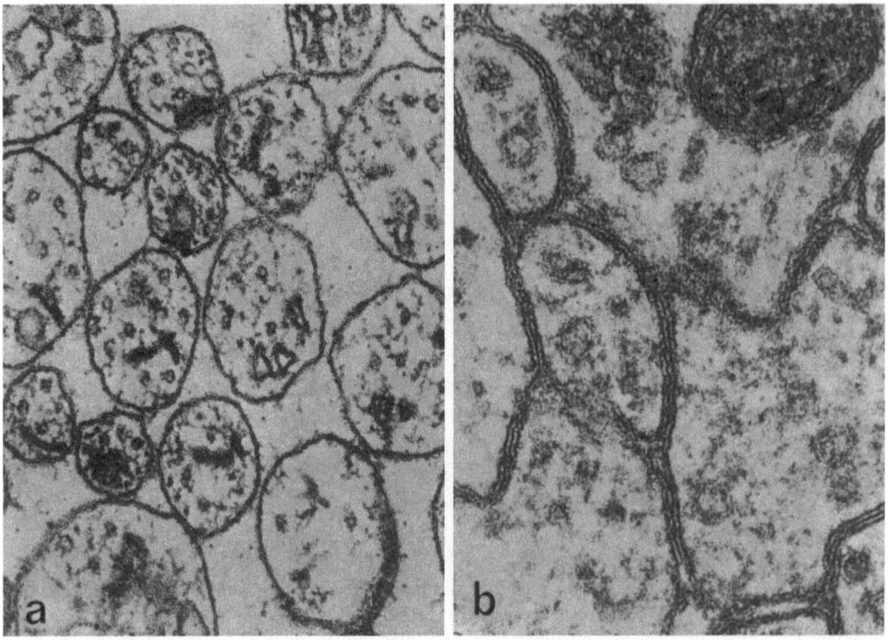

Fig. 4a and b. Comparison between lateral geniculate neuropil of 2 day and 10 day old rat showing difference in packing density of small axons and amount of extracellular space. In the latter, virtually all membranes are in almost continuous contact with each other. Original electron microscopic magnifications; (a) 100000; (b) 290000. From KARLSSON (1967); with permission of author and publisher

III. Afferent Systems as seen with the Electron Microscope

There have been very few electron-microscopic studies of developing thalamus thus far. One of the few relevant studies of this sort (KARLSSON, 1967) has approached the problem of perinatal maturation in rat lateral geniculate. It seems likely that many of the observations reported in this paper are applicable to the somesthetic thalamus. Between the 1st and 21st postnatal days, there is a marked increase in the number of synaptic profiles, a 10-fold increase often being recorded in the first two weeks (Fig. 3). Further enhancement in total number of synaptic structures continues well beyond 15 days, but the rate falls off steadily. Conversely, the volume of extracellular space is markedly decreased over this same period while the amount of neuropil, i.e. identifiable profiles, increases sharply (Fig. 4). Although a review of the organizational patterns in mature thalamic neuropil goes beyond the scope of this article, it is worth indicating that at least three distinct classes of axon terminals have been identified. Their appearance, distribution on thalamocortical relay neurons, and possible source are summarized in Fig. 5 which accompany descriptive legend [from studies by SPACEK and LIEBERMAN (1973)].

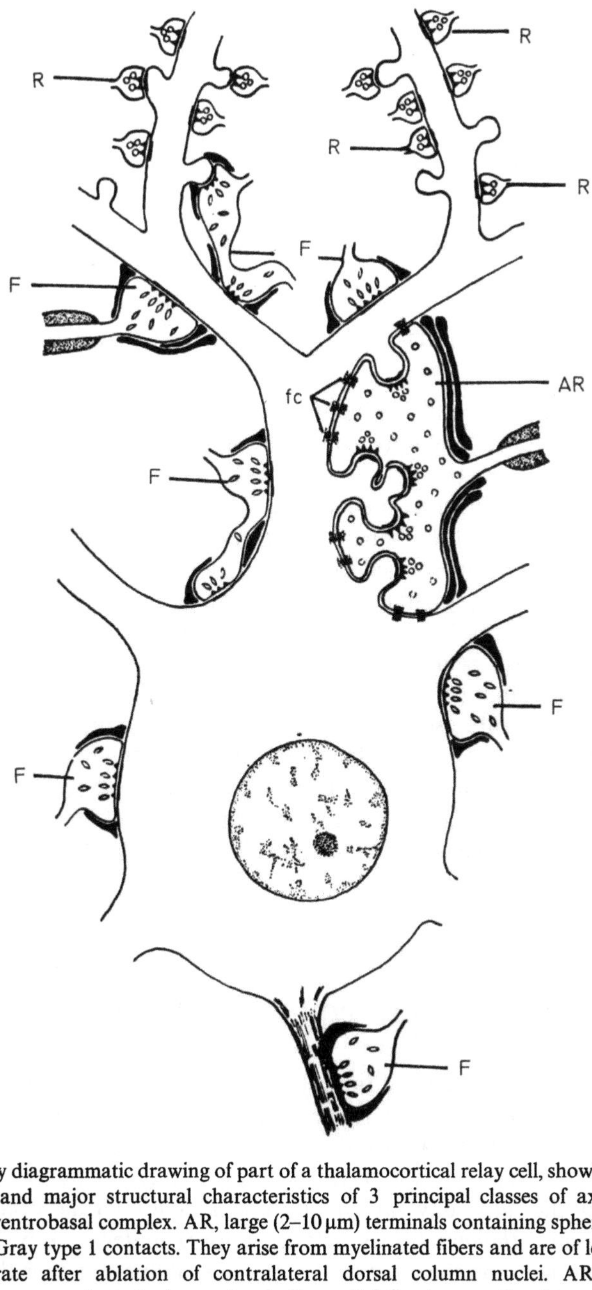

Fig. 5. Slightly diagrammatic drawing of part of a thalamocortical relay cell, showing approximate size, distribution, and major structural characteristics of 3 principal classes of axon terminals so far identified in ventrobasal complex. AR, large (2–10 µm) terminals containing spherical synaptic vesicles and making Gray type 1 contacts. They arise from myelinated fibers and are of lemniscal origin, since they degenerate after ablation of contralateral dorsal column nuclei. AR boutons terminate predominantly or exclusively in anatomically well-defined synaptic glomeruli, associated with dendrites of 1.5–3.0 µm diameter bearing excrescences, rarely also with cell bodies. Synapses are made with dendritic (or somal) excrescences, filamentous contacts with dendritic shaft. Medium-sized, irregularly shaped terminals 1–4 µm in diameter are called F boutons and make synaptic contacts similar to Gray type 2 endings. They have flattened vesicles and terminate on the shafts of primary and secondary dendrites. They also appear to be the only terminals found on the axonal initial segment. Very small (1 µm) terminals containing spherical vesicles and making Gray type I synapses are called R boutons. Many undoubtedly originate in cerebral cortex, since they degenerate massively after lesions of primary sensory cortex. They predominate on small dendrites and on their simple spine-like excrescences. From SPACEK and LIEBERMAN (1973); with permission of author and publisher

C. Morphogenesis of Postsynaptic Systems

I. Classification of Thalamic Neurons

The classic studies of CAJAL identified two major cell types in somesthetic thalamus; the long-axoned thalamocortical projection elements named by him "...Buschzellen ou cellule en buisson" (CAJAL, 1911) because of the characteristic dendrite bushes; and the short-axoned or local circuit element (TÖMBÖL, 1967; SCHEIBEL et al., 1972). Recently a third type, constituting about 5% of the total neuron population has been described in the ventrobasal complex. This has been called the "integrator cell" (SCHEIBEL and SCHEIBEL, 1966b), on the assumption that its diffuse, reticular cell-like dendritic domain and its typically bifurcating rostro-caudal axon system bearing a number of collaterals might fit it for some type of integrative or summative role.

A fourth type of element in sensory (and adjacent intralaminar) thalamic fields has also been reported, a very small cell (5–7 μm soma) bearing several richly tufted dendrites, and without a demonstrable axon (SCHEIBEL et al., 1972b). If this is, indeed, an axonless neuron (Fig. 10b), like retinal amacrine cells and olfactory bulb granules, it may serve some as yet unspecified function further enriching the thalamic structural substrate.

II. The Time of Origin of Thalamic Cells

ANGEVINE (1970) has used autoradiography for experimental determination of the time of origin of thalamic cells. These results are summarized in Fig. 6 and show several interesting features. In the mouse, cells of the ventral thalamic complex appear to develop at the 11th gestational day, followed by massive proliferation until the 14th day, at which time activity virtually ceases. The adjacent N. *reticularis thalami* shows activity approximately 1 day before the ventral nuclear complex, and the most medial portion of this system is the latest to reflect the proliferative phase.

Autoradiographic analysis of the olfactory bulb indicates that large neurons generally develop earlier than small ones (HINDS, 1968). This appears to agree with observations in perinatal cerebral cortex (PURPURA, 1961; SCHEIBEL and SCHEIBEL, 1963, 1964; STENSAAS, 1968) and in cerebellum. This is probably also true in thalamic sensory fields although in all Golgi-stained kitten material that we have studied both cell populations are recognizable. It has proven very much more difficult to make this differentiation in rat and mouse, perhaps because of the more limited number of short-axoned cells in such forms.

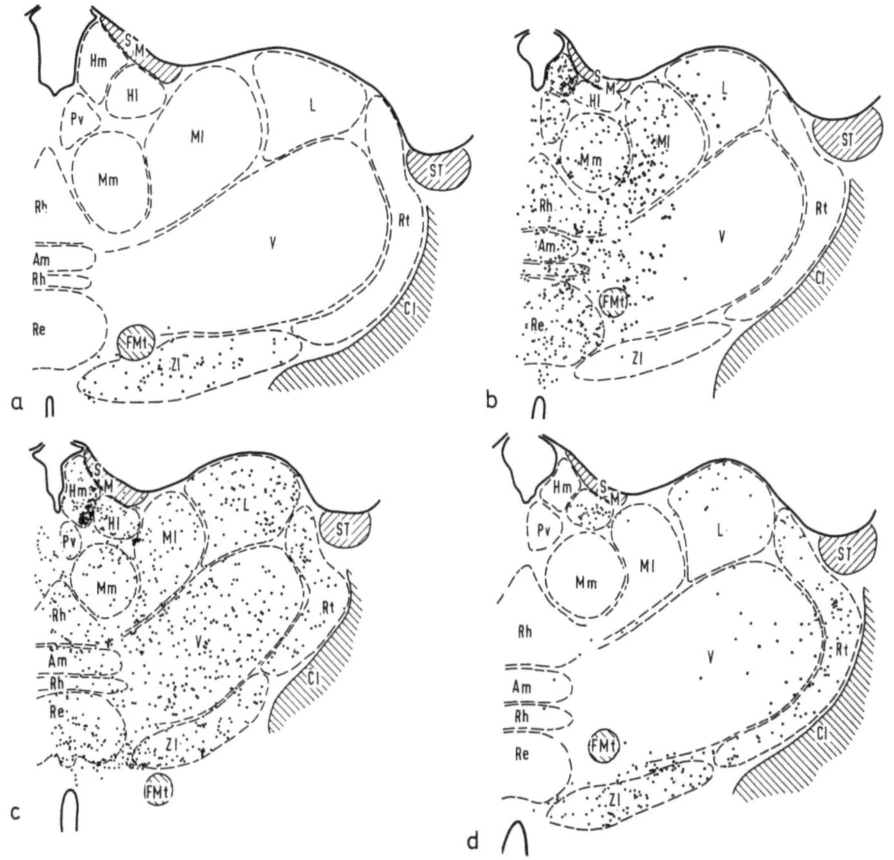

Fig. 6a–d. Results of autoradiographic study of time of origin of neurons at middle level of mouse diencephalon. Positions of heavily radioactive neurons alone are charted; (a) single injection of ^{3}H-thymidine on day 10 of gestation, as seen in autoradiogram from animal killed on day 75 after birth; (b) as (a) but with injection of ^{3}H-thymidine on day 11 of gestation; (c) as (a) but with injection of ^{3}H-thymidine on day 12 of gestation; (d) as (a) but with injection of ^{3}H-thymidine on day 13 of gestation. Thalamic neurons of somatosensory (ventrobasal) thalamus appear to originate between late on the 10th and early on 13th gestational days, with more medially positioned cells leading more lateral elements in time of development. Abbreviations: Hm, medial *habenula*; Hl, lateral *habenula*; SM, *stria medullaris thalami*; Pv, paraventricular; Rh, rhomboidal; Am, anteromedial; Re, *reuniens*; Mm, medial part of medial thalamic n.; Ml, lateral part of medial thalamic nucleus; L, lateral; V, ventral complex; Rt, *reticularis thalami*; CI, internal capsule; ST, *stria terminalis*; ZI, *zona incerta*; FMt, mamillothalamic *fasciculus*. From ANGEVINE (1970); with permission of author and publisher

III. Morphological Changes in Thalamic Neurons during the Postnatal Period

Figures 1, 7, and 8 summarize the morphological changes in the postnatal period. Bushy thalamocortical neurons already show the short primary dendrite stems and early secondary bush-like branching as compared with the simpler, more extended dendrite systems of "integrator cells" and the somewhat smaller,

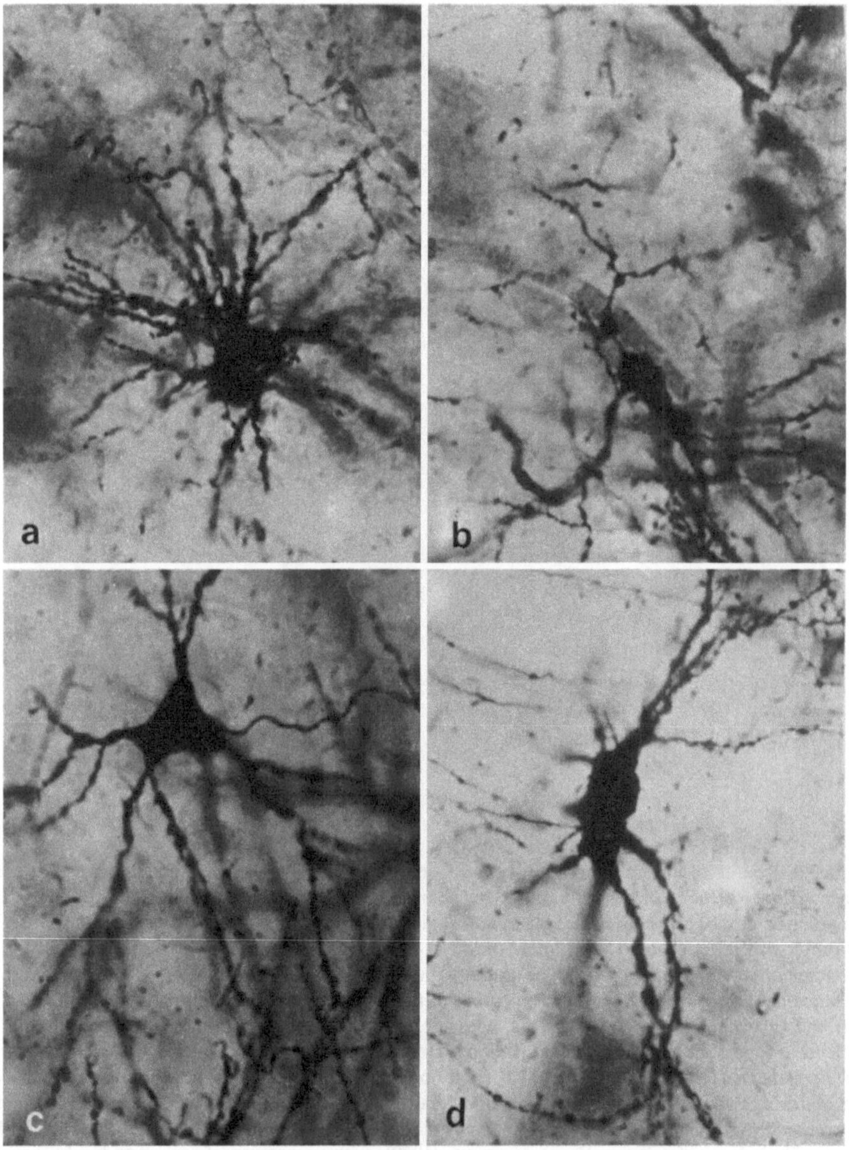

Fig. 7a–d. Neurons of ventrobasal thalamus in newborn kitten: (a) very immature thalamocortical relay cell showing poorly differentiated dendrite system with multiple nodulated dendrite stalks; (b) local circuit cell showing initial bifurcations of axon; (c) somewhat more mature thalamocortical relay neuron showing smaller number of dendrites better differentiated into primary and secondary stalks, and initial portion of axon; (d) thalamic neuron with simpler dendritic structure, possibly an "integrator" cell. Modified Golgi method; 440

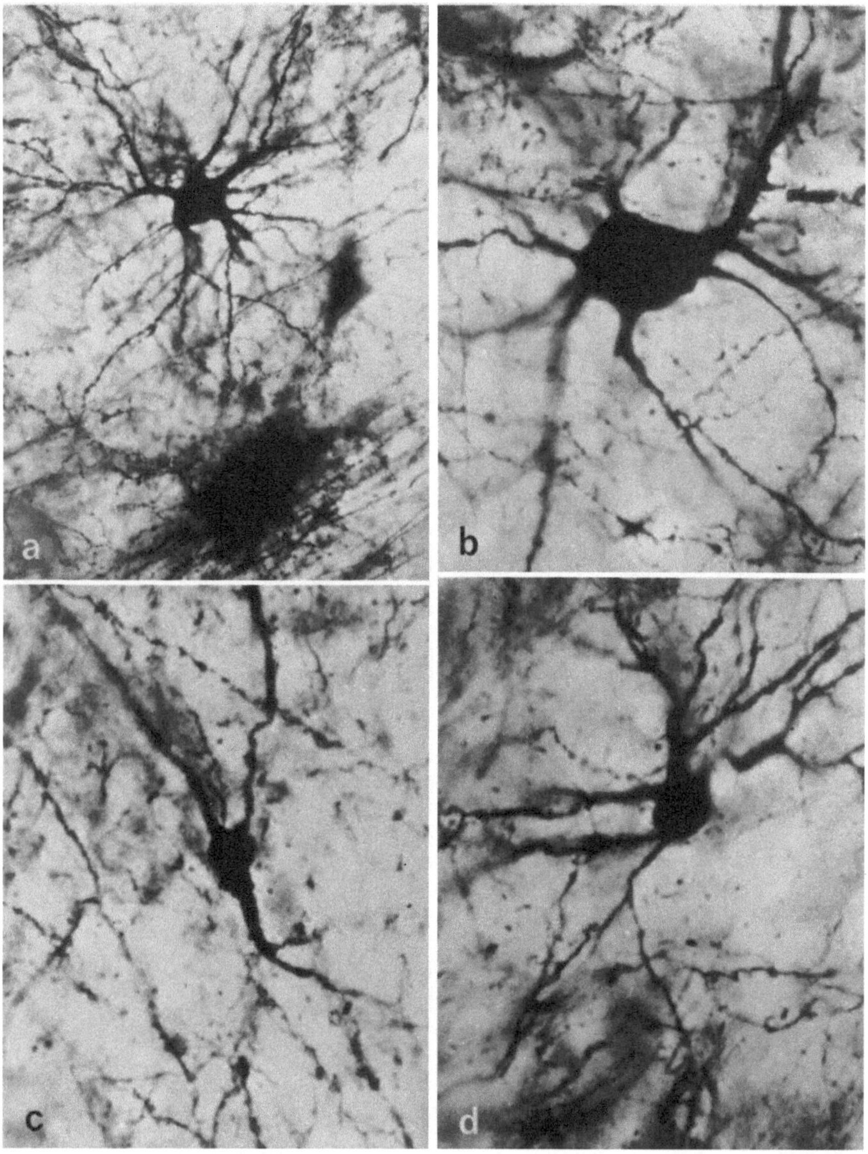

Fig. 8a–d. Neurons of ventrobasal thalamus in 15-day-old kitten. (a) Thalamocortical relay neuron showing many characteristics of mature cell, although dendritic system is not fully developed; (b) thalamocortical neuron showing *soma*, initial segment, and first portion of axon. No recurrent collaterals visible. (c) neuron with simple dendritic pattern, probably an "integrator" cell; (d) local circuit neuron showing initial bifurcations of axon. Modified Golgi method; (a) and (c) 220 × ; (b) and (d) 440 ×

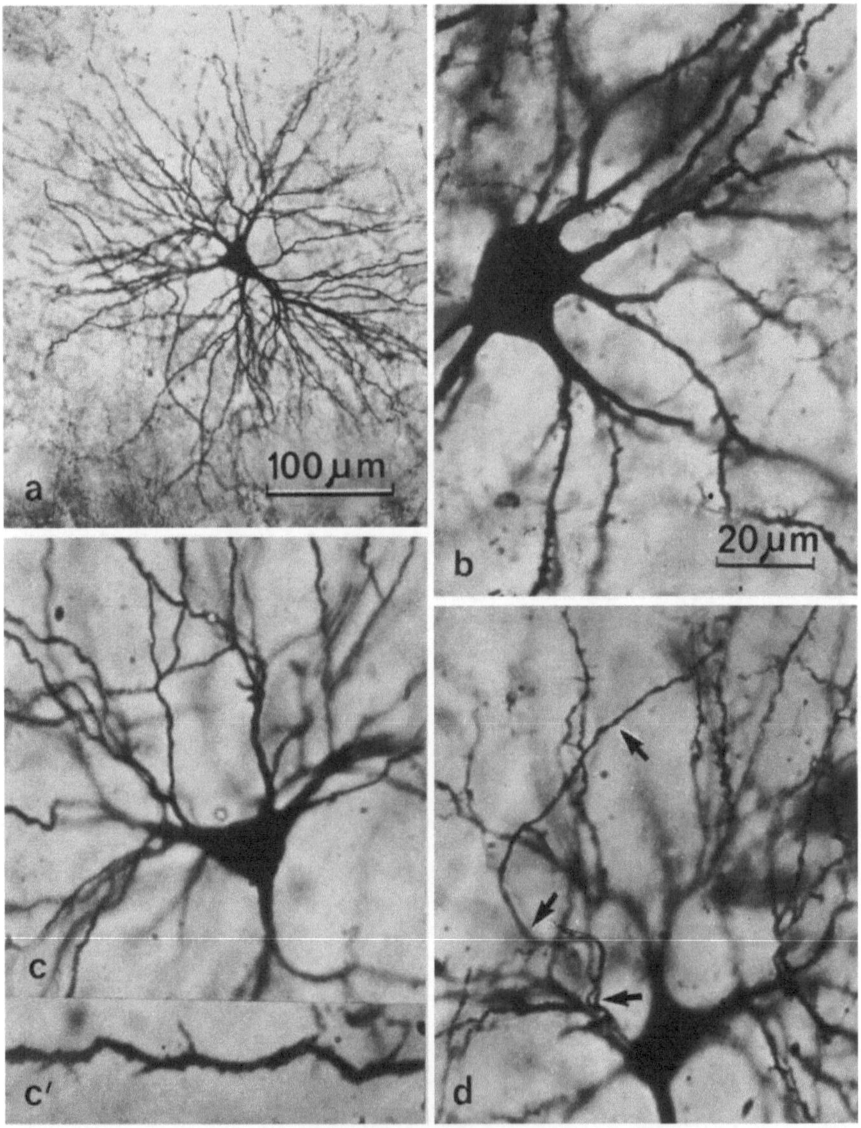

Fig. 9a–d. Neurons of ventrobasal thalamus in neurally mature cat. (a) Fully developed thalamocortical relay neuron showing size and complexity of the dendritic domain; (b) similar cell photographed at higher power to show details of primary and secondary dendrites and a few spine-like excrescences; (c) thalamocortical relay neuron with dendritic domain made up of finer-caliber, smoother dendrite shafts; (c′) portion of a relay cell dendrite showing several spine-like excrescences; (d) relay cell showing initial portion of axon (arrows). Two fine collaterals seen on this axon but do not show in this photograph. Modified Golgi method; (a) 110 ×; (b)–(d) 440 ×; (c′) 660 ×

short-axoned (local circuit) elements. The dendrites of all neurons are richly nodulated—a presumed concomitant of focal zones of enlarged cisternae in the endoplasmic reticulum. These probably represent dendritic areas of active protein synthesis (PETERS, 1971). No spines are present, although numerous hair-like side branchlets of 3–10 µm, both with and without terminal nodules, can be seen.

In the 15-day-old cat (Fig. 8), dendrite systems show the same general characteristics, although they are generally longer and the dendrite domains more nearly resemble those of the adult preparation, especially in the case of thalamocortical projection cells. Dendrite surfaces are still nodulated, although the constrictions between adjacent nodules are not so deep (Figs. 1, 7, and 8). Occasional nodulated excrescences of the type seen on mature relay and local circuit cells are now seen. Structural characteristics of the mature thalamocortical neuron are summarized in Fig. 9.

IV. Axons of Thalamic Cells

Axons of thalamocortical relay and local circuit cells have been seen in preparations of rodent, carnivore, and primate. Myelinization in the ventral complex is still in its early stages in the 15-day-old kitten and, as a result, axons can sometimes be followed for considerable distances. While the initial portions of the multiply branching short axoned cells are easily identified (Figs. 7b and 8d), as previously noted (SCHEIBEL et al., 1972b, c), neither the number nor the caliber of collateral branches from thalamocortical relay cells is striking. Recurrent collaterals can be seen on a limited number of relay cells in 15-day material and also in 2- to 3-month-old cat thalamus. The present authors estimate the incidence of such collaterals as occurring in about 15–20% of cells which is appreciably less than some investigators have reported. This matter is of more than academic interest, since the recurrent collateral has been invoked as an important part of a Renshaw-type feedback circuit considered by some (ANDERSEN et al., 1964; ANDERSEN and ANDERSSON, 1968) as substrate to the Adrian effect (rhythmic wave trains following a thalamopetal volley) (ADRIAN, 1951). The present authors' own assessment of the relatively minor role played by recurrent collaterals in the ventrobasal complex is in agreement with studies on the lateral geniculate nucleus by FAMIGLIETTI (1970). Note also that TÖMBÖL (1974) has recently concluded that recurrent collaterals of thalamocortical neurons are too few and too fine to exert a significant recurrent effect on thalamic nuclear activity.

V. Dendro-Dendritic Communication

The significance of dendro-dendritic communication in thalamic sensory neuropil has been considered during the past 5 years (FAMIGLIETTI, 1970; HARDING, 1971; RALSTON, 1971; RALSTON and HERMAN, 1965; SCHEIBEL et al., 1972a, b, 1973; SPACEK and LIEBERMAN, 1974; SZENTAGOTHAI, 1963, 1967) in a

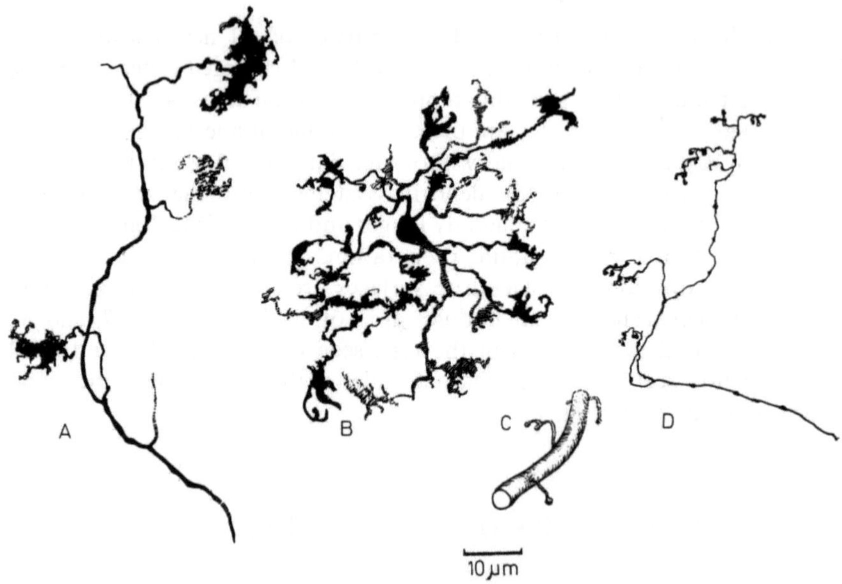

Fig. 10A–D. Details of pre- and postsynaptic structure in mature ventrobasal nucleus. (A) terminal dendritic tufts on local circuit-cell processes. (B) small axonless cell showing many terminal structures. (C) spine-like structures on dendrite of thalamocortical relay cell. (D) bouton clusters on axon terminals of sensory bushy arbor. Drawings based on material stained with modified Golgi methods

series of communications that have also stressed the profusion of specific dendritic structures (projections, excrescences, etc.) presumed to be involved in these interactions. Our studies indicate that these structures are, at best, only rudimentary in the 15-day-old kitten and begin to appear mature some time after the fourth week of life. This correlates well with the maturation of thalamic slow-wave trains as will be seen in the next section.

In general, Golgi analysis of both thalamic sensory and adjacent medial intralaminar nuclei reveals two forms of dendritic specialization. Mossy tuft-like structures (Fig. 10A) are found on the dendrite tips of most local circuit interneurons, and in smaller numbers on some dendritic termini of thalamocortical relay cells. Long spine-like structures or nodulated appendages of varying size are present in limited numbers along the dendrite shafts of most local circuit neurons (Figs. 10C and D and 11b, d, e, and g). The evidence suggests that the mossy tuft bears a presynaptic relation to the dendritic appendage, thereby creating a pattern of dendrodendritic synaptic links throughout dorsal thalamus. Thalamocortical cells also appear to be synaptically engaged by this system through their own tuft apparatus.

The dendritic appendages are almost certainly postsynaptic in functional role. Several electron-microscopic studies identify them as the central receptive core of synaptic complexes or glomeruli upon which a number of profiles characterized as presynaptic appear to terminate (JONES and POWELL, 1969; RALSTON, 1971; RALSTON and HERMANN, 1969). This postulated postsynaptic function is completely consistent with the apparent role of "true" spines found along the

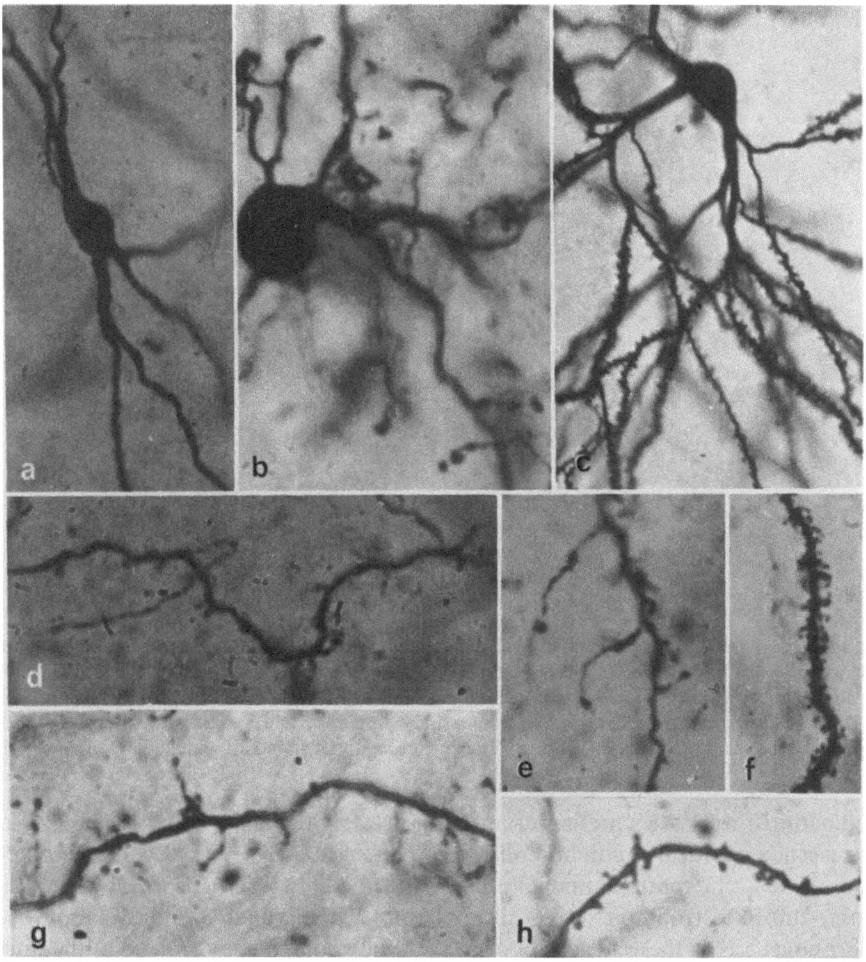

Fig. 11a–h. Structural details of local circuit neurons in ventrobasal thalamus of adult cat. (a) General configuration of local circuit neuron showing relatively smooth proximal dendrites with small numbers of spine-like excrescences. (b) higher-power view of another such cell, showing elaborate nodulated excrescences; (c) striatal neuron from same section, shown for comparison. Note dense concentrations of spines and their short spike-like appearance. (d), (e), (g), and (h) different types of dendritic spine-like and nodulated excrescences, showing variation and appreciable length of some of structures. (f) Dendrite from cortical pyramidal neuron on same section, shown for comparison. Note density of spines, short length, and nodulated appearance. Modified Golgi methods; (a) 350 × ; (b) and (c) 440 × ; (d)–(h), 660 ×

dendrites of neurons in cortical systems. In the case of the thalamic appendages, however, the heterogeneity of terminating afferent elements and the surrounding glial capsule all contribute to forming a synaptic complex or glomerulus that appears larger and more complex than any cortical structure so far described.

It has not yet proved possible to determine the time course of development for each of these structures individually. As indicated, they can be recognized in the 15-day-old kitten, though they are clearly immature morphologically. By the end

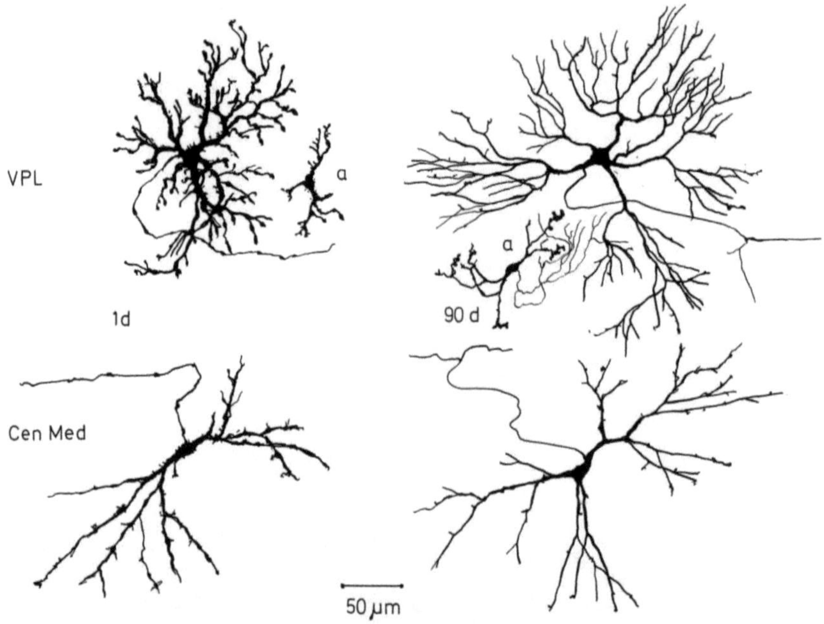

Fig. 12. Maturative patterns in lateral and medial thalamic neurons. Thalamocortical neuron and local circuit cell, a, of 1-day-old ventral posterolateral, VPL, are very immature, but have reached virtually the adult state at 90 days. Neurons of the central medial nucleus, Cen Med, are relatively more mature at 1 day although dendritic tree may not be entirely developed and surface structures are still of primitive type. At 90 days, cell is mature. Drawn from material stained by modified Golgi method

of the fourth week they are larger and more clearly delineated, and the spine-like excrescences on local circuit and relay cells can no longer be visually differentiated from those in the mature animal, at least with the light microscope. However, the mossy tuft-like structures have still not attained their final size and complexity. Morphogenesis of these structures appears finally complete by the end of the third month (85th to 90th postnatal day) (Fig. 12). Assuming that a complex dendrodendritic system of synaptic interactions is based on these structures and that these relationships, in turn, underlie thalamic synchrony, it might be predicted that the synchronization of thalamic slow-wave sequences could not reasonably be expected to appear before this time. Incomplete evidence from the present authors' own studies (see below) suggests that this is true.

VI. The Cortical Termination of Thalamocortical Relay Cells

The cortical termination was initially recognized by Cajal (1891), and was named "fibers of S. Ramon" by Kölliker (1896). The characteristic cone-shaped arbor generated by these fibers is already fairly well-developed at the time of birth, although the individual arbors are more pleomorphic, less intensively branched, and show less evidence of overlap. With maturation, there is a certain regularization of pattern in each arbor, and the "partially shifted overlap" that LORENTE DE NO (1949) described is more fully developed.

In view of the late development of local circuit (stellate, Golgi type II) elements of cerebral cortex, it seems probable that the thalamocortical relay fibers initially terminate on the apical shafts of fifth-layer pyramids and possibly on the basilar shafts of third-layer pyramids. The dendrite and soma membrane of pyramids of perinatal animals is provided with at least some postsynaptic coupling sites in the irregulary distributed heteromorphic pseudospines (SCHEIBEL and SCHEIBEL, 1971b) that have been described on these cells. That these attachments are maintained and possibly increased in number with maturation is suggested by Golgi studies in primary visual cortex after visual deprivation (GLOBUS and SCHEIBEL, 1967; VALVERDE, 1967) and apparently confirmed by electron microscopy (JONES and POWELL, 1970; GAREY, 1970; GAREY and POWELL, 1971). However it also seems likely that as fourth-layer stellate cells mature, they begin to share in the presynaptic influx from sensory thalamus. It has also been suggested that they are the exclusive postsynaptic receptive elements and that spine changes on pyramids represent a transsynaptic phenomenon (SZENTÁGOTHAI, 1973).

It is worth noting that thalamocortical projections of the midline intralaminar system terminate in a simpler system of fibers that appears to branch several times in the subgriseal white matter, then ascends toward the cortical surface as a parallel system of axons. As pointed out previously, these undoubtedly represent the pluri-areal afferents of LORENTE DE NO (1949) and we have been able to trace some under direct visual control from anterior midline thalamus to orbitofrontal cortex (SCHEIBEL and SCHEIBEL, 1967). These axons are recognizable at birth although we have not been able to identify axodendritic contacts in the kitten until 7 to 10 days later. This has led us to make the suggestion that these axodendritic articulations may not be established until the second week of postnatal life, perhaps in part by some degree of lateral movement of the ascending intracortical portion of the axons accompanied by sprouting of dendritic spines and presynaptic terminal branchlets (SCHEIBEL and SCHEIBEL, 1963, 1964). The great difficulty in obtaining cortical desynchronization by high frequency stimulation of brain stem or midline thalamus, and the rapid fatiguability of the response when obtained in the perinatal kitten are undoubted reflections of the unfinished state of this synaptic system (SCHEIBEL and SCHEIBEL, 1963, 1964; PURPURA, 1961).

D. Development of Electrical Activity

In the data that follow, we shall consider several aspects of the functional development of somatosensory thalamus, presenting parallel data from mesencephalic tegmentum and midline thalamus where necessary for purposes of comparison. These data will include: 1) developmental changes in single unit activity in midbrain and thalamus during wakefulness (A), slow-wave sleep (SWS) or its equivalent, and paradoxical or fast-wave sleep (FWS); 2) the ontogenesis and structure of bursts of unit activity that eventually characterize thalamic discharge patterns during SWS; and 3) the relationship of these bursts to trains of slow waves or spindles, which are equally characteristic of thalamocortical operations.

I. Development of Single Unit Activity

It has proved difficult to record discreet single unit activity in the unrestrained kitten with chronically implanted electrodes before the third postnatal day. At about this time, spike activity can first be distinguished from background noise in midline thalamic and mesencephalic tegmental areas. Laterally situated thalamic zones, comprising mainly those nuclei concerned with specific sensory systems, do not begin to show discreet unit responses until the seventh to tenth postnatal day.

At the third postnatal day, state changes are visible in the mesencephalic tegmentum and pretectum (Fig. 13). These are characterized by a brief, approximately 30-s, decrease in unit activity as the kitten puts its head down and enters a sleep-like state, followed by an increase during the immediately following twitching or fast-wave phase of sleep (Fig. 13).

By the fourth or fifth day of life there is considerable change in unit behavior. Both mesencephalon and midline thalamic nuclei show unit activity and all of these units show state changes. The change is most prominent in mesencephalic tegmental units, which are now more consistently active during FWS than during either of the other two states (Fig. 14). The relatively precocious development of these state-related changes in tegmental units suggests an earlier maturation of circuit and/or synaptic mediator systems here, followed by somewhat later maturation in midline thalamus and lateral thalamic systems. Anatomical studies of neurons in these three areas also indicate a differential rate in maturation patterns. Brain-stem reticular (tegmental) cells show a relatively advanced level of development at birth. Dendrite envelopes of medial thalamic neurons are almost as precocious, while lateral thalamic neurons (i.e., VPL, VL, etc.) are appreciably more immature in the perinatal period (Fig. 12).

The development of state-specific changes in unit firing behavior continues through the third week of life. However, two diverging trends become apparent at this stage. Discharge frequencies of lateral thalamic units begin to match or exceed those of the previously more active midline units during FWS (Fig. 14). In addition, what we interpret as medium- and large-sized units (on the basis of apparent spike amplitude and stability) begin to show evidence of more extensive state changes than small units. At the 18th postnatal day, this trend is even more pronounced with some large- and medium-sized thalamic units showing response ranges during FWS that may be two to three times that seen in the A state. On the other hand, most large- and medium-sized thalamic units in medial thalamus begin to show higher discharge rates in the A state, a pattern characteristic of adult

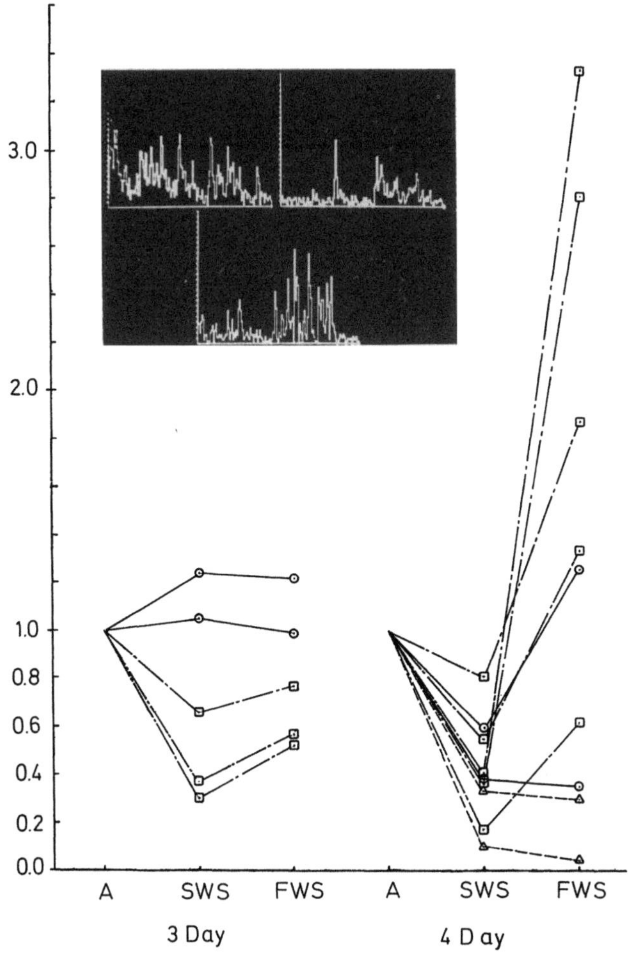

Fig. 13. Patterns of unit firing activity and state-dependent stages during first 4 postnatal days. Kitten implanted on first day, and without significant unit activity on second day. Key: squares, mesencephalic tegmental units; open circles midline thalamic units; triangles, lateral thalamic units. State changes become visible on fourth postnatal day and then most obviously in tegmental units. In general, higher-amplitude units show greater degrees of change than lower-amplitude elements. A, awake; SWS, slow-wave sleep; FWS, fast wave sleep. Inset: spike frequency histogram showing differences in unit activity of large tegmental neuron at 4th postnatal day in waking state, slow-wave, and fast-wave sleep phases. Each period is a 60-s epoch

activity. By the 40th postnatal day, this pattern of state changes will be consistent in both thalamic areas.

Spike interval histograms summarize the state-dependent differences in spike discharge patterns in the three major states during development. As Fig. 15 shows, there is progressive differentiation of spike patterning, especially during SWS and FWS, with higher-density bunching of spikes (i.e. bursts) becoming apparent in SWS during the interval between the 16th and 44th postnatal days, while firing intervals show increasing dispersion during FSW in the same period.

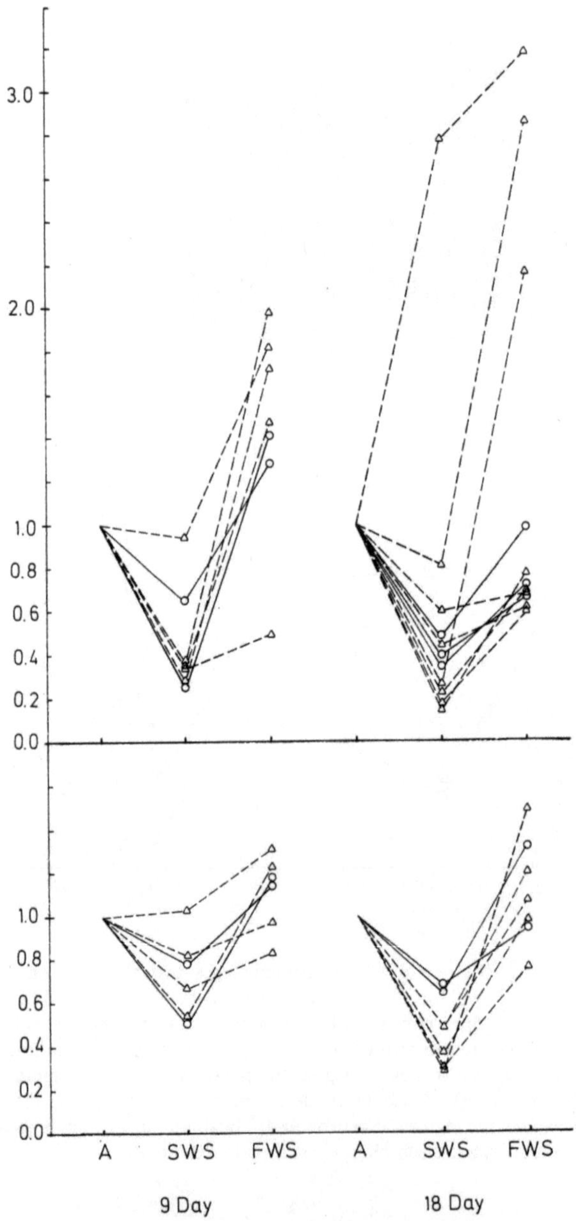

Fig. 14. State-dependent changes of medial (circle) and lateral (triangle) kitten thalamic cells, with large neurons above, small neurons below. Note progressive state-dependent differentiation of unit activity, with large neurons of lateral thalamus showing most obvious trend toward higher firing rates during FWS

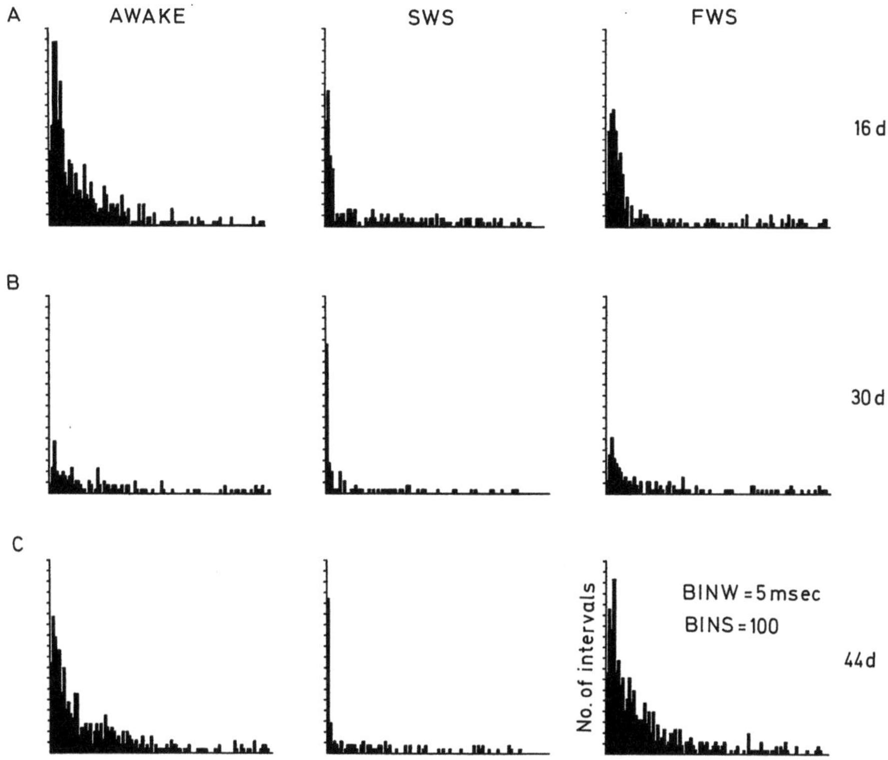

Fig. 15. Spike interval histograms comparing activity of lateral thalamic units in 3 states at 3 different ages. Note progressive tightening of unit firing patterns during slow-wave sleep from approximately 20 msec at 16 days to 5 msec or less at 44 days. As this burst configuration develops during S.W.S., there is further dispersal of firing patterns during fast-wave sleep. Binwidth, 5 msec

II. Unit Bursts

Short repetitive bursts of thalamic unit activity represent another characteristic aspect of neuronal behavior during SWS to develop some time after birth. No bursting can be seen during the first week, and at the 10th postnatal day bursting is still an occasional and isolated phenomenon recorded from only 1 or 2 electrodes in a cluster (Fig. 16). By the 40th day, bursting activity is apparent in most leads of each microelectrode cluster, in both medial and lateral thalamus. However, bursts still involve only a small proportion of the active units, and the rate at which bursts are produced is still rather low (Fig. 17). Furthermore, synchronization of burst discharges (i.e. among the several active leads in a microelectrode cluster) is still not seen at this time. Such synchronization will become a more regular feature of discharge patterns in the neurally mature thalamus, and is seen particularly well in the barbiturate-anesthetized cat (Fig. 18).

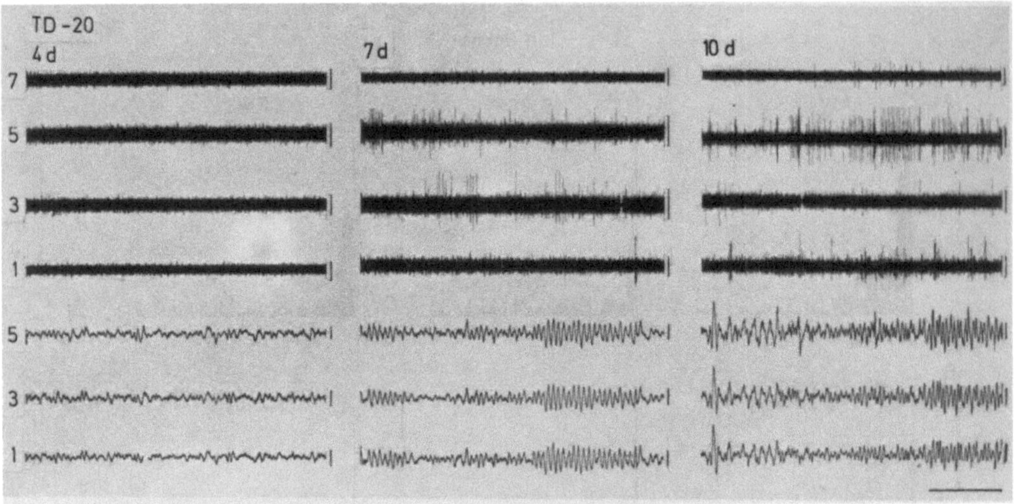

Fig. 16. Development of single unit firing patterns and slow waves in lateral (ventrobasal) thalamus. Neither spike discharges nor clear-cut patterning of thalamically derived slow waves are seen at 4th postnatal day. Spike discharges are visible at 7th day, as are runs of rhythmic slow waves similar to spindles. Both types of activity are more obvious at day 10, as are scattered spike bursts, especially in leads 1 and 5. All recording stations were part of one microelectrode cluster, and slow wave trains were recorded from microelectrodes via multiplexing. Time line: 1 s

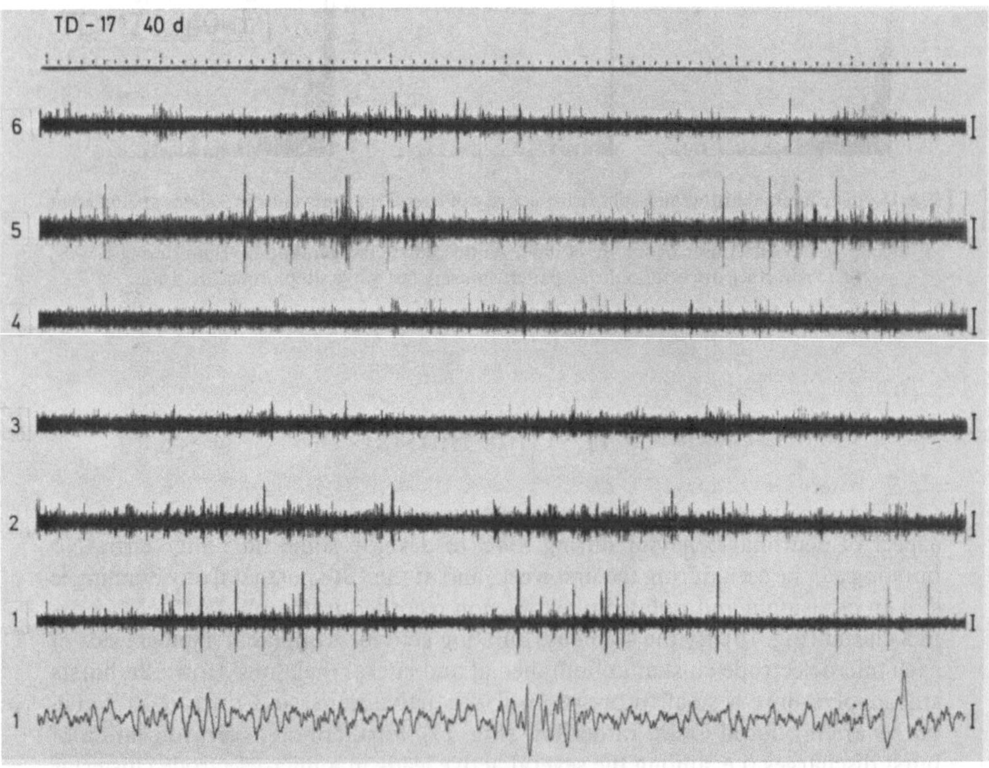

Fig. 17. Spike and slow-wave activity led from elements of two closely spaced microelectrode clusters in ventrobasal complex of 40-day-old kitten. Spike bursts scattered throughout all leads, but little or no correlation among them or with short runs of synchronous slow waves. Time marker: small divisions, 1/10 s; large divisions, 1 s

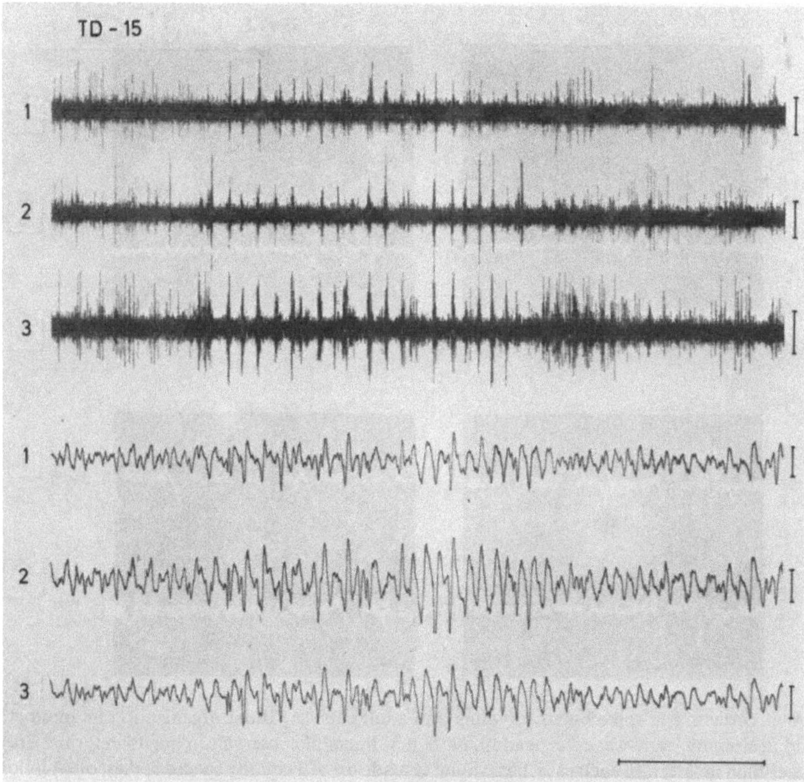

Fig. 18. Synchronization among spike bursts and between spike bursts and slow-wave trains in lightly barbiturized young adult cat. Derivations from single microelectrode cluster in ventrobasal complex, with multiplexed leadoffs to show simultaneous spike and slow-wave phenomena at closely adjacent thalamic points. Recording tips of 3 leads are approximately 1.2 mm apart. Time line: 1 s

III. Burst Structure

When spike bursts are studied at sufficiently rapid sweep speeds to resolve their component unitary discharges, several features can be noted. Each burst tends to be composed of three spikes followed by a space approximately equal to two interspike intervals, and this is followed, in turn, by one or two additional spike discharges (Fig. 19). The total length of each burst varies with the interspike intervals and the total number of spikes. In general, interspike intervals are longest in young kittens and become progressively briefer with maturation. For example, at 16 days of age (Fig. 19B), total burst length may equal 20 msec, while at 30 days this has fallen to 10 msec or less. Small additional decrements in burst length can be noted until day 44–45 day of life, when the temporal dimension of the burst approximates that of the adult cat.

Spike bursts have not been seen during FWS. However, during these periods, individual spikes are frequently succeeded by miniature or abortive spikes occurring at intervals closely approximating those of the secondary spikes of a

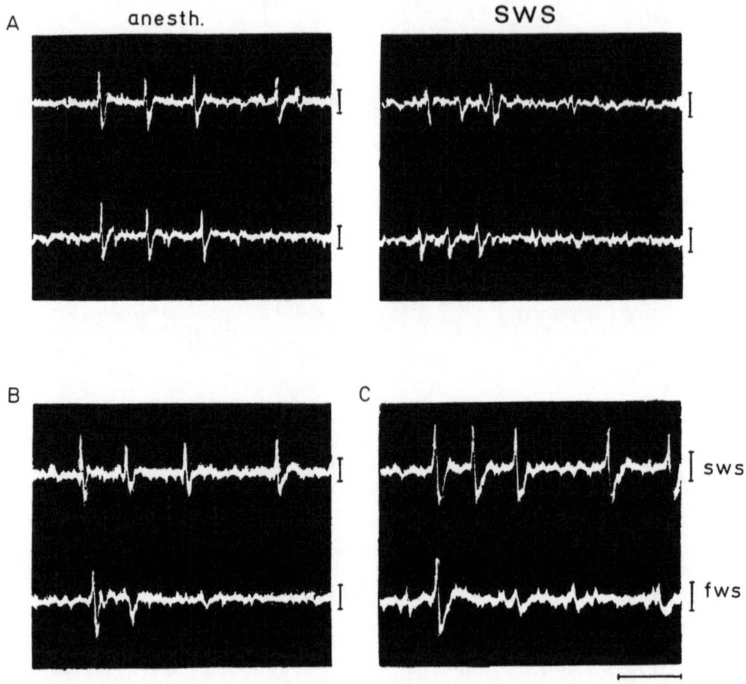

Fig. 19A–C. Structure of spike bursts in young and adult cats. (A) Burst organization in anesthetized adult and in sleeping (slow-wave sleep) adult. Note that interspike intervals in bursts in former are 50–60% longer than in latter. In each case, basic burst is made up of 3 equally spaced spikes, often followed by one or more smaller or abortive spikes after double interval. (B) 16-day-kitten; (C) 40-day-kitten. Upper traces show similar burst structure. Interspike intervals become shorter with age. Lower traces suggest that during fast-wave sleep, while bursts are not seen, individual spikes may be followed by small or abortive spikes following at approximately same intervals as those of fully fledged bursts seen in SWS. See text. Vertical scale, 100 mV; horizontal scale, 10 msec

burst recorded during SWS (Fig. 19B and C, lower traces). This would seem to indicate that a tendency toward bursting behavior under conditions other than SWS is somehow suppressed, perhaps by ascending influences from the lower brain stem.

IV. Slow Waves

Visual examination of thalamic wave patterns indicates the initial appearance of rhythmic activity at the sixth to eighth postnatal day (Fig. 16). This correlates quite well with the previous description of the development of cortical spindle bursts between the 5th and 10th postnatal day (Scheibel and Scheibel, 1963, 1964). However, the total fraction of rhythmic activity remains low for several weeks. This is probably why fast Fourier analysis of randomly selected half-second epochs during wakeful and slow-wave-sleep states, with an average 25 to the

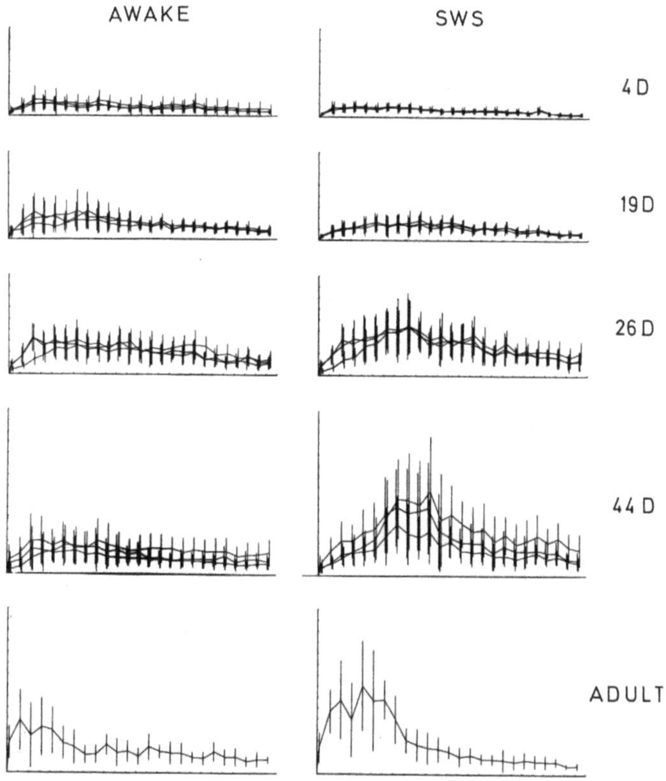

Fig. 20. Fast Fourier analysis of half-sec epochs, showing development of slow rhythmic waves in sensory thalamus. Although some slow-wave trains can be identified at end of first week of life, Fourier analysis shows no evidence of frequency peaking until fourth week (16–20 Hz), with eventual decrease to 10–14 Hz in young adult state

group, revealed essentially flat spectra for the first three weeks (Fig. 20). At 25–26 days, frequency peaking at 16–20 Hz becomes obvious in SWS records. Similar analyses of adult records reveal peaking at 8–10 Hz.

Intracellular analyses by ANDERSEN et al. (1964) have emphasized the relationship of thalamic slow waves to cellular postsynaptic potentials and spike bursts. ANDERSEN (1966) has pointed out that the length of the intracellular IPSP acts as a timing device for the thalamic slow wave episode (spindle) while the postanodal exaltation (p.a.e.) postulated to follow each IPSP is believed to function as the trigger for the next sequence of neuronal discharges and related slow waves. In the first 30–40 days, recordable spike bursts involve only a small proportion of thalamic units and the rate of spike burst production is very low. Figure 17 indicates that even at 40 days of age, a thalamic multiple-electrode cluster reveals only a scattering of bursts with no obvious cross-channel synchrony, and equally little evidence of time-locking to thalamic spindle waves. If a relationship between spindle activity and unit bursts does exist in the kitten, it would appear to involve the combined activity of many units in a neuronal ensemble, some of which may be active during one cycle of the wave and others at another. The net result may be a volley of bursts synchronized with the slow waves. The individual bursts produced by a given unit are much more widely spaced. It is only in the mature animal that thalamic units develop the capacity to generate repetitive bursting activity at the same rate as the rhythmic wave activity, and in synchrony with many other surrounding units.

E. Discussion and Overview

Under optimal conditions, the reviewer looks toward his Discussion as an opportunity for effecting synthesis of disparate data from many sources. Here he should be able to produce a model with some degree of thematic continuity and sufficient probity to provide others with a stepping stone for their own endeavors. It is not required that models be completely true; only that they be useful. If, in the generalizations that are presented, there is some operational principle which, explicitly or implicitly, can be appropriately applied within the general framework of neural function, a valid purpose has been served.

The data we have just considered would appear to yield such opportunities only sparingly, due to limitations in quantity and to uncertainty as to the meaning of those data that are available. Clearly, we need a more closely grained evaluation of developmental morphology in somatosensory thalamus, with particular emphasis upon the step-by-step maturation of synaptic systems seen at the levels of resolution provided by electron microscopy. At the functional level, we have only scratched the surface. We are still deficient in knowledge of immature thalamic response to afferent volleys at both gross and individual cellular levels. Nor is there significant intracellular data that can be related to the electrogenesis of membrane activity and the early development of nonpropagating and propagating processes.

Within the limited compass of present information, we can offer only suggestions as to trend and status, with the expectation that in the near future, in our laboratory and in others, more thorough studies at higher resolution will add to our understanding of the structures and mechanisms that are involved.

I. Thalamic Structural Maturation

Thalamic developmental studies are consistent insofar as they indicate a general degree of immaturity of sensory thalamic structure at birth. Several growth gradients can be described, however. The presumably less evolved, more generalized neurons of the thalamic nonspecific core are more nearly mature than the cells in the more laterally situated somatosensory zones. We have not been able to describe differences in the rate of maturation between thalamocortical relay cells and the short-axoned (Golgi type II) local circuit elements, at least in young postnatal kittens. However, it should also be pointed out that in newborn thalamic tissue we have been unable to find any evidence of the small axonless elements (Fig. 10B) recently described in the adult (SCHEIBEL et al., 1972a, b). While there are several putative interpretations of this, ANGEVINE'S (1970) radioautographic studies suggest that their initial development can occur no later than that of other thalamic neurons. We must, for the moment, assume that they are present in some unrecognizable form in the perinatal thalamus, are not easily stained by Golgi techniques at this epoch, or migrate into the area from a distant source later in postnatal development.

Note that all thalamic afferents, both ascending and descending, are present at birth, in recognizable form. This is particularly noteworthy in rats and mice where cerebral and cerebellar cortices are particularly immature.

In contrast, we have the distinct impression that structural specializations on the surfaces of thalamic cell dendrites (i.e. nodes, excrescences, spine-like processes, etc.) are among the last elements to mature. This is of special interest because of the likelihood that these structures are involved in the dendrodendritic linkages that characterize somatosensory thalamus, and the probable relation of this system, in turn, with synchronization of thalamic slow-wave activity (SCHEIBEL et al., 1973) (see below).

II. Thalamic Functional Maturation

The functional maturation of somatosensory thalamus remains a particularly challenging problem. As indicated in the previous section, thalamic electrogenesis comprises a number of phenomena, including spikes and slow waves, individually and together, independent of one another and apparently time-locked or in synchrony. We have no clear notion of how each of these neural activities contributes to thalamic operations, since the essential nature of thalamic processing is still unknown. In fact, we are still essentially limited to the suggestion made over a half century ago (HEAD, 1920) that the thalamus is that station where sensation becomes conscious. At the phenomenological level at least, there is persuasive evidence implicating rhythmic thalamic activity as pacemaker to cortical rhythmus (ANDERSEN and ANDERSSON, 1968; JASPER and BERTRAND, 1966). The trains of thalamic spindles that are believed to trip rhythmic synchronous cortical wave ensembles are relevant to thalamic ontogeny as a possible index of maturation, especially if the structural elements that form the substrate to the functional output can be identified. This, in itself, poses a difficult problem, since there is no consensus on the source of thalamic synchronous wave trains. At least four hypothetical substrate systems have been proposed and since one or more of them undoubtedly includes the relevant thalamic elements, a brief review is indicated at this point.

III. Historical Review of Concepts of Thalamic Functional Activity

A number of significant observations on the nature of rhythmic cortical waves and speculations on their possible causal mechanisms were published in the early decades of the twentieth century. PRAWDICZ-NEMINSKI (1925) interpreted them as essentially local phenomena, namely the phased beat of groups of cortical neurons. BERGER's (1929 et seq.) studies of the alpha rhythm and its susceptibility to blockade during orienting activity led him to postulate a "work center" affecting all cortical activity, which he thought was probably extracortical in origin. However, he points out that his suggestion regarding thalamus as a source of cortical slow-wave "inhibition" goes back to even earlier work. TRAVIS and DORSEY (1932) showed that "synchronism" of homologous cortical areas of the right and left side in the rat brain persisted following section of the *corpus callosum.* Similarly, BERZE (1911) considered that the thalamus regulated the "tonus" of the entire "psychocerebral" apparatus, turning the cerebral cortex on or off.

In the late nineteen-thirties, two alternative models for cortical rhythmic activity were given serious consideration. The first grew out of BREMER's (1938b) concept of cortical slow waves as expressions of local nonpropagating potentials. A logical consequence of this concept of cellular autorhythmicity was the interaction of cortical cell ensembles and the algebraic summation of their potentials. Thus alpha-blockade could be attributed to increasing cortical asynchrony without postulating a change in modulative control from a central (subcortical) agency.

In opposition to this model was the reverberating-circuit concept of CAMPION and SMITH (1934), which received apparent support a few years later from the studies of DUSSER DE BARENNE and MCCULLOCH (1938) and CHANG (1950). Although the concept of a continuous circular movement between thalamus and cortex was especially attractive in the light of the reverberating circuit model of LORENTE DE NO (1936), several critical contributions by BREMER (1938a) and especially by ADRIAN (1941, 1951) showed it to be untenable. While BREMER demonstrated that the waves making up the cortical EEG did not have the characteristics that might be expected as a consequence of a reverberating phenomenon and, further, that cortical oscillations decreased following deafferentation, ADRIAN showed that thalamic rhythmic activity persisted after total interruption of thalamocortical links. Thus cortical rhythms appeared to depend on thalamic activity, while thalamic synchrony was not similarly dependent on cortex. The classic description of BREMER's "cerveau isolé" preparation (1935) provided evidence, by implication, that brain-stem tissue caudal to thalamus was necessary neither for the maintenance of slow rhythmic cortical potentials nor for thalamic synchrony.

The experiments by MORISON and DEMPSEY (1942, 1943) and their colleagues in the late thirties and early forties focused attention on midline and intralaminar systems as essential to triggering and maintenance of rhythmic bursts on the cortical surface. This concept was further documented and enriched by JASPER (1949) and his colleagues, who demonstrated a number of obligatory intrathalamic components (i.e. ventral anterior nucleus and N. *reticularis thalami*) involved in the phenomenon. Although the controlling effects of midline thalamus seemed well documented, several investigators also showed that associated synchronous thalamic waves were recordable in specific relay nuclei as well as in medial nonspecific fields.

The rhythmic thalamic discharge demonstrated by ADRIAN in response to an afferent volley was reevaluated by ANDERSEN et al. (1964) and their colleagues with the aid of intracellular techniques. Iterative sequences of short EPSPs alternating with longer IPSPs appeared to represent the reflection at the single-cell level of the rhythmic waves of ADRIAN. In explanation of these phenomena, ANDERSEN and ANDERSSON (1968) postulated the activation of a powerful recurrent inhibitory process mediated through Renshaw-like interneurons blocking the EPSP and associated spike burst. The ensuing IPSP was in turn conceived of as being terminated by a rebound mechanism peculiar to thalamic cell membranes, which they called "postanodal exaltation". Since all portions of medial and lateral thalamus were believed to share this capacity for recursive inhibition and active rebound, no one pace-making site was envisaged. They preferred the notion of a "facultative pacemaker", whereby any thalamic site might be able to initiate a sequence of rhythmic thalamic activity with consequent synchronous spindle bursting of the appropriate portion of cortex.

An alternative interpretation of the nature of thalamic synchrony has been proposed by PURPURA (1969) without recourse to the concept of postanodal membrane rebound. His model is based on studies which have demonstrated the frequency-specific nature of medial (nonspecific) thalamic control over relay nuclei as well as powerful elements of reciprocal control exerted by specific over nonspecific fields. This hypothesis stresses the importance of the matrix of excitatory and inhibitory interneurons operating in concert to initiate EPSP-IPSP sequences. The actual nature of the circuit(s) in question has not yet been described.

Although both of these models have appealing features, neither has won general acceptance, and other mechanisms have been proposed to account for thalamic synchrony and pace-making activity. Following our own description of the caudal projection of N. *reticularis thalami* axons and our suggestion that these might act as an inhibitory feedback control over thalamic activity (SCHEIBEL and SCHEIBEL, 1966, 1967), several investigators have attempted to specify the relationship between single unit bursts in N. *reticularis* and activity in relay nuclei such as VL and VB (SCHLAG and WASZAK, 1970; LAMARRE et al., 1971; MASSION, 1968; WASZAK, 1974). The interaction is not a simple one since neither the burst period nor the interburst intervals are sufficiently regular to provide convincing evidence (STERIADE and WYZINSKI, 1972) for direct causal relations between episodic high-frequency N. *reticularis* activity and recurrent inhibitory sequencing of dorsal thalamic cells in a manner similar to the Renshaw type of recurrent inhibition model. However, SCHLAG and WASZAK (1970) and WASZAK (1974) have suggested that N. *reticularis* activity may generate a descending inhibitory tonus operating on dorsal thalamic elements, modulated in turn by other interneurons. STERIADE and WYZINSKI (1972) conceive of both N. *reticularis* and VL elements as under phasic control of a population of as yet unspecified interneurons. The studies of LAMARRE and coworkers (FILION et al., 1971; LAMARRE et al., 1971) also question the efficacy of the reticularis feedback system as a direct modulator of dorsal thalamic cell activity, although their experimental conditions differ somewhat from the others.

IV. A Personal Concept of the Thalamic Matrix

Our own concept of the thalamic neuronal matrix within which synchronogenic processes arise is based in part on our Golgi data, some of which we have reviewed in earlier sections, in conjunction with findings from electron microscopy. The most commanding feature of such a model is the continuous structural interaction of local circuit cells synaptically linked via dendrodendritic junctions. This veritable dendritic meshwork extends across both specific and nonspecific thalamic fields, one of the main differences between the two fields being the presence of thalamocortical relay cells in the latter. These large projection cells are structurally quite different from the local circuit cells, but have a small number of rosettes or tufts on dendritic branches or termini similar to the latter. While the majority of these tufts are probably postsynaptic to the bouton clusters borne by terminating lemniscal afferent fibers, we believe they are also presynaptic to the dendritic spines or appendages of adjacent local circuit cells. The thalamocortical relay cells can thus be considered "plugged in" to the interneuronal stroma already described.

This complex system of interrelated neurons provides a rich substrate, not only for the rapid spread of activation from an initially excited focus but also for the rhythmic waxing and waning of activity via the many two-way synapses linking so many elements of the endless dendritic chain. Such a path would be independent of the connections developed by axon systems of local circuit and relay cells. The much-discussed recurrent collaterals of thalamocortical axons were found in only 15 to 20% of all the cells that we have studied in the adult cat thalamus and would seem to provide (as previously indicated) an inadequate substrate for the powerful recursive effects ascribed to them (ANDERSEN et al., 1964; ANDERSEN, 1966; ANDERSSON et al., 1971a, b).

The substrate of dendritic linkages that we postulate (SCHEIBEL et al., 1973) depends basically on a double functional role for the local circuit cells. They are conceived of as being postsynaptic along the proximal portion of their dendrites and presynaptic at their tuft-bearing terminals. Such an arrangement suggests that each dendrite of a local circuit element may have the capacity to function independently as an input-output element, an assumption that cannot yet be fully supported by physiological evidence. However, it is worthy of note that RALSTON (1971) has made a similar suggestion about the possible mode of operation of thalamic neurons with presynaptic dendrites. Although the details of his model differ from ours, he also suggests that possibility that each dendritic branch may independently influence postsynaptic elements.

While a dendritically linked substrate of this kind provides the capability for rhythmic oscillation at certain resonant frequencies, the tuning and fractionation

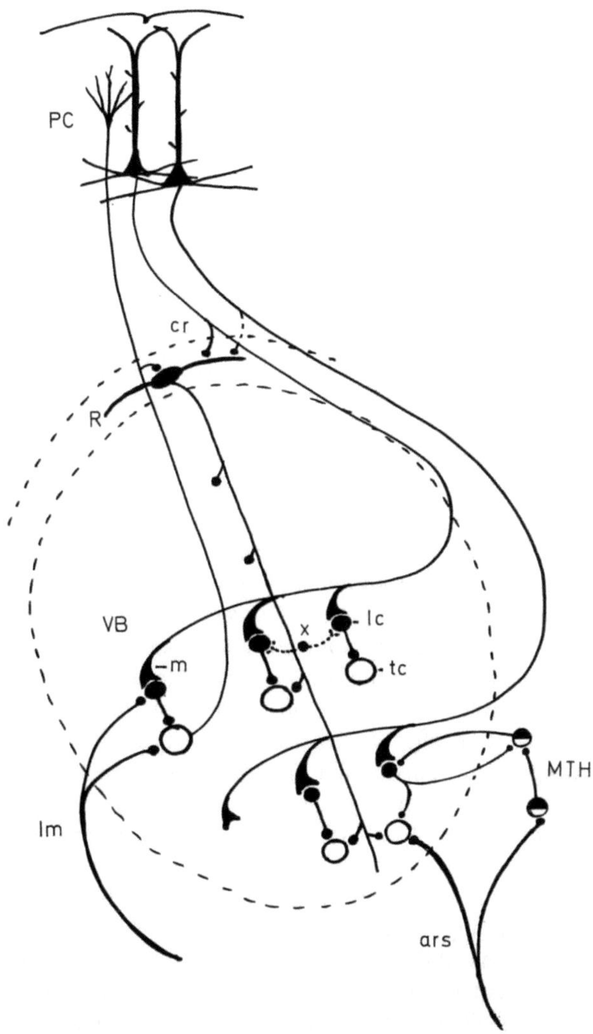

Fig. 21. Highly schematic summary of some extra nuclear systems converging on neuronal ensembles of the somatosensory (VB) thalamus. These systems are believed to mature somewhat earlier than complex and poorly understood intranuclear matrix, as described in text. MTH, midline (nonspecific) thalamus; PC, pericruciate cortex; R, *n. reticularis thalami*; ars, ascending reticular system; lm, medial lemniscus; lc, local circuit neuron; m, corticothalamic terminals (horizontal plexus type); tc, thalamocortical relay neuron; cr; cortico-n. *reticularis* branches; x, possible position of small axonless cells

of this process undoubtedly depends on related systems. Candidates for this role are several intrathalamic complexes, notably the medial nonspecific portion of the thalamic ensemble and the nucleus *reticularis thalami*. In addition, portions of the brain stem reticular core almost certainly exert powerful shaping effects on the system, probably *via* mediation of both of these intrathalamic zones (PURPURA, 1969; SCHEIBEL and SCHEIBEL, 1966b; Fig. 21).

Several bodies of data point to midline thalamus as capable of leading and controlling rhythmic activity throughout the remainder of the thalamus (ANDERSSON et al., 1971a and b), possibly through frequency-specific control mechanisms (PURPURA, 1969, 1970). It has been suggested that a gating mechanism may operate "*via* a balance between the activity in inhibitory and excitatory connections" (ANDERSSON et al., 1971a). This system may also act as the final common path for modulatory effects exerted by portions of the brain-stem reticular core (PURPURA, 1969, 1970; SCHEIBEL and SCHEIBEL, 1966a and b).

The functional relationship of the N. *reticularis thalami* to specific and nonspecific thalamic fields has not yet been fully worked out. Anatomic studies have shown that cells of this nucleus send their axons back into the thalamus, as already described (SCHEIBEL and SCHEIBEL, 1966a; MINDERHOUD, 1971), rather than rostrally toward the cortex, as previously believed. Furthermore, there appears to be a highly organized relationship between sectors of N. *reticularis thalami* on the one hand with portions of thalamus and cerebral cortex on the other (SCHEIBEL and SCHEIBEL, 1966a, 1972; MINDERHOUD, 1971).

Recent studies have appeared to lend support to our suggestion that the n. reticularis thalami may exert negative feedback control over thalamic elements, although the effect is by no means simple and apparently varies with the phase of the sleep-wakefulness cycle of the organism (FILION et al., 1971). As yet, no single definitive experiment has specified the source(s) and mode(s) of control of rhythmic thalamic activity. Nevertheless, on the basis of available structural and functional studies, it seems reasonable to conceive of a model characterized by elaborate dendrite interrelations that offer a substrate for rhythmic activity based on circulating resonating waves. This oscillatory process may be thought of as being under the competing control of 2 or more modulatory systems, the medial nonspecific thalamus and the N. *reticularis thalami*. The effects of these systems vary, depending on the nature and amount of ascending drive from the major sensory inputs, the various subnuclei of the brain stem reticular core, and to a smaller extent, on descending corticofugal impulses. We suggest that, in addition to these tuning operations, the demonstrated specificity of linkages between n. reticularis thalami with cortical sectors rostrally and with thalamic sites caudally, enable fractionation of synchronous processes among the thalamocortical fields. Documentation already exists for some degree of uniqueness in each of these vertically oriented subsystems (ANDERSEN et al., 1964; MINDERHOUD, 1971; SCHEIBEL and SCHEIBEL, 1972; RINVIK, 1972), although further evidence with a finer grain of functional resolution will be necessary for more complete understanding of rhythmic thalamic and cortical processes. Figure 21 summarizes some of these patterns of interconnections and their relative precocity compared to the development of intranuclear connections.

F. Conclusion and Synthesis

Analysis of the morphogenesis of somatosensory thalamus seems to suggest that extranuclear systems are ahead of intranuclear (i.e. within the ventrobasal complex) systems in development by an appreciable time; perhaps by as much as one to three weeks in the young cat. Midline (nonspecific) thalamic neurons, and the central tegmental reticular core are seen to be closer to maturity than thalamocortical relay and local circuit cells at birth, and this differential remains obvious for at least the first month of postnatal life. Axonal afferents impinging on VB, i.e. from lemniscal, corticothalamic, tegmental, and N. *reticularis thalami* systems, are also in place and show at least the rudiments of terminal neuropil patterns in the immediate perinatal period. In contrast, those postsynaptic specializations that characterize thalamic sensory fields cannot be seen, and only become obvious in the second and third months of life. From this it might reasonably be concluded that mechanisms of extranuclear somatosensory control become functional before intrinsic mechanisms.

Such a conclusion is undoubtedly premature in the absence of adequate electron microscopic analysis of each element of the developing system. Nonetheless, if we accept the general thesis, whatever its inadequacies, certain interesting consequences follow. First, it would appear that the capacity for extranuclear control and modulation of somatosensory operations may become functional before those intranuclear processing operations over which they are to exert control. Second, input-output lines traversing the ventrobasal complex would appear to achieve some degree of operative status before maturation of the intranuclear systems that will eventually condition the signal between input and output. Both of these assumptions point toward a pattern of functional development where nuclear operations are activated in a series of discreet steps. In such a schema, each step represents a more complex functional level and each increment is a consequence of the maturation and coming on line of an element of the structural mosaic.

Some evidence for such a step-wise pattern in functional maturation can be adduced from looking back over the maturation of electrical activity already described. Individual unit discharges can first be distinguished from background noise at the end of the first week, while the earliest evidence of grouping of unit discharges into bursts is not seen until the 10th day. Such spike-burst activity is initially isolated and infrequent, and only becomes recognizable in all leads at approximately the 40th day of postnatal life. At this time there is also nascent evidence of developing synchronization of such bursts with thalamic spindles, but such spindle-unit burst synchrony does not become a feature of all recording channels until full neural maturation in the kitten (from three months of age) has been achieved. The capacity to generate consistent and repetitive spike bursts at the same rate as the rhythmic waves comes even later.

In regard to slow waves, the sequence is somewhat more blurred, but, in the kitten at least, there seems to be little doubt that short runs of rhythmic thalamic slow waves similar to cortical spindles appear for the first time between the sixth and eighth postnatal days, depending on the general degree of maturation of the particular litter, and the individual, at birth. The total amount of rhythmic slow-wave activity remains quite low for some time, showing an appreciable increase only after the first three weeks of life. Even then, frequency peaking of the slow-wave runs at this period is in the range of 16—20 Hz, fully twice what it will be at maturity.

It is even more difficult to reconstruct a sequence of maturation for this system from the point of view of its perceptive-cognitive-behavioral role. We can only assume that the gradual development of the organism's tactile, proprioceptive, kinesthetic self-image is as much a function of the finite rate at which new experience can be collected and interpreted as it is the maturation of the mechanisms that make this processing possible. To borrow a speculative notion from an earlier work (SCHEIBEL and SCHEIBEL, 1971a), perhaps the development and "fleshing out" of the body image is, in the final analysis, the ultimate measure of the process of maturation of the somatosensory thalamus.

G. Acknowledgement

The development of this review and the research upon which it is based were supported by grants from the National Institutes of Health: NS 1063, 10567, 11678, and CHD 00972.

The functional data were gathered in collaboration with Dr. THOMAS L. DAVIES. We thank Dr. ROBERT D. LINDSAY for help and direction in the computer analysis of the data.

References

Adrian, E. D.: Afferent discharges to the cerebral cortex from peripheral sense organs. J. Physiol. (Lond.) **100**, 159—191 (1941).

Adrian, E. D.: Rhythmic discharges from the thalamus. J. Physiol. (Lond.) **113**, 9—10 (1951).

Andersen, P.: Rhythmic 10/s activity in the thalamus. In: The Thalamus (D. Purpura and M. Yahr, Eds.). New York: Columbia University Press 1966, pp. 143—151.

Andersen, P., Andersson, S. A.: Physiological Basis of the Alpha Rhythm. New York: Appleton-Century-Crofts 1968.

Andersen, P., Curtis, D. R.: The pharmacology of the synaptic and acetylcholine-induced excitation of ventrobasal thalamic neurons. Acta physiol. scand. **61**, 100—120 (1964).

Andersen, P., Eccles, J. C., Sears, T. A.: The ventrobasal complex of the thalamus: types of cells, their responses, and functional organization. J. Physiol. (Lond.) **174**, 370—399 (1964).

Andersson, S. A., Holmgren, E., Manson, J. R.: Synchronization and desynchronization in the thalamus of the unanesthetized decorticate cat. Electroenceph. clin. Neurophysiol. **31**, 335—345 (1971a).

Andersson, S. A., Holmgren, E., Manson, J. R.: Localized thalamic rhythmicity induced by spinal and cortical lesions. Electroenceph. clin. Neurophysiol. **31**, 347—356 (1971b).

Angevine, J. B.: Time of neuron origin in the diencephalon of the mouse. An autoradiographic study. J. comp. Neurol. **139**, 129—188 (1970).

Berger, H.: On the electroencephalogram of man (1929 et seq.) (Translated by P. Gloor) Electroenceph. clin. Neurophysiol., Suppl. **28** (1969).

Berze, J.: Bewußtseinstonus. Psychischer Antreib und Hirnstamm. Wien. med. Wschr. (1932) No. 11; cited by Berger (1969: *op. cit.*).

Bowsher, D.: Termination of the central pain pathway in man. The conscious appreciation of pain. Brain **80**, 606—622 (1957).

Bremer, F.: Cerveau isolé et physiologie du sommeil. C.R. Soc. Biol. Paris **118**, 1235—1241 (1935).

Bremer, F.: Effets de la déafferentation complète d'une region de l'écorce cérébrale sur son activité électrique spontanée. C.R. Soc. Biol. Paris **127**, 355—359 (1938a).

Bremer, F.: L'activité électrique de l'écore cérébrale (Actualités scientifiques et industrielles **658**, 46). Paris: Hermann 1938b.

Cajal, R. y S.: Sur la structure de l'écorce cérébrale de quelques mammifères. La Cellule **8**, 1 (1891).

Cajal, R. y S.: Contribución al estudio de la vie sensitiva central y de la estructura del talama optico. Rev. trim. Microgr. **5**, 185—198 (1900).

Cajal, R. y S.: Histologie du Système Nerveux de l'Homme et des Vertébrés, Vol. 2. Paris: Maloine 1911.

Campion, G.C., Smith, G.F.: The Neural Basis of Thought. London: Regan, Paul, Trench Trubner 1934.

Chang, H.T.: The repetitive discharges of corticothalamic reverberating circuits. J. Neurophysiol. **13**, 235—257 (1950).

Cooper, I.S.: Involuntary Movement Disorders. New York: Harper and Row 1969.

Cooper, I.S., Waltz, J.M., Amin, I.: Pulvinectomy—a preliminary report. J. amer. Ger. Soc. **19**, 553—555 (1971).

Dusser de Barenne, J.G., McCulloch, W.S.: The direct functional interrelation of sensory cortex and optic thalamus. J. Neurophysiol. **1**, 176—186 (1938).

Famiglietti, E.V.: Dendro-dendritic synapses in the lateral geniculate nucleus of the cat. Brain Res. **20**, 181—191 (1970).

Filion, M., Lamarre, Y., Cordeau, J.P.: Neuronal discharges of the ventrolateral nucleus of the thalamus during sleep and wakefulness in the cat. II. Evoked activity. Exp. Brain Res. **12**, 499—508 (1971).

Frigyesi, T.L., Rinvik, E., Yahr, M.D.: Corticothalamic Projections and Sensorimotor Activities. New York: Raven 1972.

Garey, L.J.: The termination of thalamo-cortical fibers in the visual cortex of the cat and monkey. J. Physiol. (Lond.) **210**, 15—17P (1970).

Garey, L.J., Powell, T.P.S.: An experimental study of the termination of the lateral geniculo-cortical pathway in the cat and monkey. Proc. roy. Soc. B **179**, 41—63 (1971).

Gerebtzoff, M.A.: Anatomie du thalamus et histoenzymologie. Acta neurol. belg. **70**, 683—686 (1970).

Getz, B.: The termination of spinothalamic fibers in the cat as studied by the method of terminal degeneration. Acta anat. **16**, 271—290 (1952).

Globus, A., Scheibel, A.B.: Synaptic loci on visual cortical neurons of the rabbit: The specific afferent radiation. Exp. Neurol. **18**, 116—131 (1967).

Harding, R.N.: Dendro-dendritic synapses, including reciprocal synapses in the ventro-lateral nucleus of the monkey thalamus. Brain Res. **34**, 181—185 (1971).

Hassler, R.: Sterotactic brain sugery for extrapyramidal motor disturbances. In: Introduction to Stereotaxis with an Atlas of the Human Brain (G. Schaltenbrand and P. Bailey, Eds.). New York: Grune & Stratton 1959, pp. 472—488.

Head, H.: Studies in Neurology (2 vols.). London: H. Frowde; Hodder and Stoughton, Ltd., 1920

Hinds, J.W.: Autoradiographic study of histogenesis in the mouse olfactory bulb. I. Time of origin of neurons and neuroglia. J. comp. Neurol. **134**, 287—304 (1968).

Jasper, H.H.: Diffuse projection systems: the integrative action of the thalamic reticular system. Electroenceph. clin. Neurophysiol. **1**, 405—420 (1949).

Jasper, H.H., Bertrand, G.: Thalamic units involved in somatic sensation and voluntary and involuntary movements in man. In: The Thalamus (D.P. Purpura and M.D. Yahr, Eds.). New York: Columbia University Press 1966.

Jones, E.G., Powell, T.P.S.: Electron microscopy of synaptic glomeruli in the thalamic relay nuclei of the cat. Proc. roy. Soc. Lond. (Biol.) **172**, 153—171 (1969).

Jones, E.G., Powell, T.P.S.: An electron microscopic study of the laminar pattern and mode of termination of extrinsic afferent fibers in the somatic sensory cortex of the cat. Phil. Trans. roy. Soc. B **257**, 45—62 (1970).

Karlsson, U.: Observations on the postnatal development of neuronal structures in the lateral geniculate nucleus of the rat by electron microscopy. J. ultrastruc. Res. **17**, 158—175 (1967).

Kölliker, A.: Handbuch der Gewebelehre des Menschen, 6th ed., Vol. 2 (cited by Ramon y Cajal: Histologie du système nerveux de l'homme et des vertébrés (2 vols.). Paris, A. Malvine 1909—1911).

Lamarre, Y., Filion, M., Cordeau, J.P.: Neuronal discharges of the ventrolateral nucleus of the thalamus during sleep and wakefulness in the cat. I. Spontaneous activity. Exp. Brain Res. **12**, 480—498 (1971).

Lorente de No, R.: Cortical excitatory state and synchronism in the control of bioelectric autonomous rhythms (Discussion). Cold Spr. Harb. Symp. quant. Biol. **4**, 336—337 (1936).

Lorente de No, R.: Cerebral cortex: architecture, intracortical connections, motor projections. In: Physiology of the Nervous System (J.F.Fulton, Ed.), 3rd ed. London: Oxford University Press 1949, pp. 288—330.

Massion, J.: Etude d'une structure motrice thalamique, le noyau ventrolatéral et de sa régulation pas les afférences sensorielles. Thesis, Faculté des Sciences de Paris, 1968.

McCance, I., Phillips, J.W., Westerman, R.A.: Acetylcholine sensitivity of thalamic neurons: its relationship to synaptic transmission. Brit. J. Pharmacol. **32**, 635—658 (1968).

McLennan, H., Hoffman, R.D., Marshall, K.C.: Patterns of excitation of thalamic neurons by amino acids and by acetylcholine. Nature **219**, 387—388 (1968).

Mehler, W.R.: The mammalian "pain tract" in phylogeny. Anat. Rec. **127**, 332 (1957).

Mehler, W.R.: Further notes on the center median nucleus of Luys. In: The Thalamus (D.P.Purpura and M.D.Yahr, Eds.). New York: Columbia University Press 1966, pp. 109—122.

Minderhoud, J.M.: An anatomical study of the efferent connections of the thalamic reticular nucleus. Exp. Brain Res. **12**, 435—446 (1971).

Morison, R.S., Dempsey, E.W.: A study of thalamocortical relations. Amer. J. Physiol. **135**, 281—292 (1942).

Morison, R.S., Dempsey, E.W.: Mechanism of thalamocortical augmentation and repetition. Amer. J. Physiol. **138**, 297—308 (1943).

Nauta, W.J.H., Kuypers, H.G.J.M.: Some ascending pathways in the brain stem reticular formation. In: Reticular Formation of the Brain (H.H.Jasper et al., Eds.). Boston: Little Brown 1958, pp. 3—30.

Peters, A.: Stellate cells of the rat parietal cortex. J. comp. Neurol. **141**, 345—373 (1971).

Práwdicz-Neminski, W.W.: Zur Kenntnis der elektrischen und der Innervations-Vorgänge in den funktionellen Elementen und Geweben des tierischen Organismus. Elektrocerebrogramm der Säugetiere. Pflügers Arch. **209**, 362—382 (1925).

Purpura, D.P.: Analysis of axodendritic synaptic organization in immature cerebral cortex. Ann. N.Y. Acad. Sci. **94**, 604—654 (1961).

Purpura, D.P.: Interneuronal mechanisms in thalamically induced synchronizing and desynchronizing activities. In: The Interneuron (M.A.Brazier, Ed.). Berkely: University of California Press 1969, pp. 467—496.

Purpura, D.P.: Operations and processes in thalamic and synaptically related neural subsystems. In: The Neurosciences, Second Study Program (F.Schmitt, Ed.). New York: Rockefeller University Press 1970, pp. 458—470.

Purpura, D.P., Yahr, M.: The Thalamus. New York: Columbia University Press 1966.

Ralston, H.J.: Evidence for presynaptic dendrites and a proposal for their mechanism of action. Nature **230**, 585—587 (1971).

Ralston, H.J., Hermann, M.M.: The fine structure of neurons and synapses in the ventrobasal thalamus of the cat. Brain Res. **14**, 77—97 (1968).

Rinvik, E.: Organization of thalamic connections from motor and somatosensory cortical areas in the cat. In: Corticothalamic Projections and Sensorimotor Activities (T.Frigyesi et al., Eds.). New York: Raven 1972, pp. 57—88.

Riss, W., Koizumi, K., Brooks, C.M.: Basic thalamic structure and function. Brain Behav. Evol. **6**, 1—560 (1972).

Scheibel, A.B., Scheibel, M.E.: Specialized organizational patterns within the nucleus reticularis thalami of the cat. Exp. Neurol. **34**, 316—322 (1972).

Scheibel, M.E., Scheibel, A.B.: Structural substrates for integrative patterns in the brain stem reticular core. In: Reticular Formation of the Brain (H.H.Jasper et al., Eds.). Boston: Little Brown 1958, pp. 31—55.

Scheibel, M.E., Scheibel, A.B.: Some structuro-functional correlates of maturation in young cats. Electroenceph. clin. Neurophysiol., Suppl. **24**, 235—246 (1963).

Scheibel, M.E., Scheibel, A.B.: Some structural and functional substrates of development in young cats. In: The Developing Brain (H.Himwich and W.Himwich, Eds.) (Progress in Brain Research, Vol. 9). Amsterdam: Elsevier 1964, pp. 6—25.

Scheibel, M.E., Scheibel, A.B.: The organization of the nucleus reticularis thalami; a Golgi study. Brain Res. **1**, 43—62 (1966a).

Scheibel, M. E., Scheibel, A. B.: Patterns of organization in specific and nonspecific thalamic fields. In: The Thalamus (D.P.Purpura and M.D.Yahr, Eds.). New York: Columbia University Press 1966b, pp. 13—46.

Scheibel, M. E., Scheibel, A. B.: Structural organization of nonspecific thalamic nuclei and their projection toward cortex. Brain Res. **6**, 60—94 (1967).

Scheibel, M. E., Scheibel, A. B.: Thalamus and body image—a model. Biol. Psychiat. **3**, 71—76 (1971a).

Scheibel, M. E., Scheibel, A. B.: Selected structural-functional correlation in postnatal brain. In: Brain Development and Behavior (M.B.Sterman et al., Eds.). New York: Academic 1971b, pp. 1—21).

Scheibel, M. E., Davies, T. L., Scheibel, A. B.: On dendrodendritic relations in the dorsal thalamus of the adult cat. Exp. Neurol. **36**, 519—529 (1972a).

Scheibel, M. E., Davies, T. L., Scheibel, A. B.: An unusual axonless cell in the thalamus of the adult cat. Exp. Neurol. **36**, 512—518 (1972b).

Scheibel, M. E., Davies, T. L., Scheibel, A. B.: On thalamic substrates of cortical synchrony. Neurology **23**, 300—304 (1973).

Scheibel, M. E., Scheibel, A. B., Davies, T. L.: Some substrates for centrifugal control over thalamic cell ensembles. In: Corticothalamic Projections and Sensorimotor Activities (T. L. Frigyesi et al., Eds.). New York: Raven 1972c, pp. 131—156).

Schlag, J., Waszak, M.: Characteristics of unit responses in nucleus reticularis thalami. Brain Res. **21**, 286—288 (1971).

Špaček, J., Lieberman, A. R.: Distribution of synaptic types on thalamocortical projection neurons of rat ventrobasal thalamus. J. Anat. (Lond.) **117**, 212—213 (1973).

Špaček, J., Lieberman, A. R.: Ultrastructure and three-dimensional organization of synaptic glomeruli in rat somatosensory thalamus. J. Anat. (Lond.) **117**, 487—516 (1974).

Spiegel, E. A., Wycis, H. T.: Stereoencephalotomy (Thalamotomy and Related Procedures). New York: Grune and Stratton 1952.

Stensaas, L. J.: The development of hippocampal and dorsolateral pallial regions of the cerebral hemispheres in fetal rabbits. VI. Ninety millimeter stage, cortical differentiation. J. comp. Neurol. **132**, 93—108 (1968).

Steriade, M., Wyzinski, P.: Cortically elicted activities in thalamic reticularis neurons. Brain Res. **42**, 514—520 (1972).

Szentágothai, J.: The structure of the synapse in the lateral geniculate body. Acta anat. (Basel) **55**, 166—185 (1963).

Szentágothai, J.: The anatomy of complex integrative units in the nervous system. In: Recent Developments in Neurobiology in Hungary, Vol. I (K. Lissák, Ed.). Budapest: Akadémiai Kiadó 1967, pp. 9—45).

Szentágothai, J.: Synaptology of the visual cortex. In: Central Visual Information B (R. Jung, Ed.), (Handbook of Sensory Physiology VII/3). Berlin: Springer 1973.

Szentágothai, J., Hamori, J., Tömböl, T.: Degeneration and electron microscopic analysis of the synaptic glomeruli in the lateral geniculate body, Exp. Brain Res. **2**, 283—301 (1966).

Tömböl, T.: Short neurons and their synaptic relations in the specific thalamic nuclei. Brain Res. **3**, 307—326 (1966).

Tömböl, T.: Collateral axonal arborization. In: Golgi Centennial Symposium (M. Santini, Ed.). New York: Raven 1974, pp. 131—141.

Travis, H., Dorsey, T.: Action current studies of simultaneously active disparate fields of the central nervous system of the rat. Abst. Zbl. Neurol. **65**, 635 (1932).

Valverde, F.: Apical dendritic spines of the visual cortex and light deprivation in the mouse. Exp. Brain Res. **3**, 337—352 (1967).

Van Buren, J. M., Bork, R. C.: Variations and Connections of the Human Thalamus (2 vols.). New York: Springer 1972.

Waszak, M.: Firing patterns of neurons in the rostral and ventral part of nucleus reticularis thalami during EEG spindles. Exp. Neurol. **43**, 38—59 (1974).

Chapter 7

Functional Modification of the Developing Visual System

HELMUT V. B. HIRSCH and AUDIE GENE LEVENTHAL, Albany, New York (USA)

With 31 Figures

Contents

A. Introduction

"... neurons develop in two complementary modes, which together permit all possible neuronal functions to be represented: functions that are innately predetermined as well as those that develop as a result of individual experience. In the first mode of development, neurons can express their functions only in a predetermined way as highly predictable patterns of behavior which are characteristic of each species. In the second mode of development neurons form populations with such diverse potentials that all possible contingencies of function can be realized... From this *plenum formarum* some neuronal functions and not others will be actualized" (JACOBSON, 1974, p. 152).

Some behavior patterns reflect the individual organism's past history, while others are stereotyped and characteristic of all members of the species. Developmental processes must thus provide for the formation of a nervous system capable of guiding behavior with a high degree of constancy from one member of a species to the next, without sacrificing the flexibility of response that makes survival in an unpredictable environment possible.

Over the past two decades, a great deal of information has been gathered about the function, organization, and development of neurons within the vertebrate visual system. Much of this evidence suggests that certain neurons within the mammalian visual system do not require specific sensory input for maintenance or for development of their response properties, while other cells do not achieve or maintain normal function without such an input (SHERMAN et al., 1972; LEVENTHAL and HIRSCH, 1975; BLAKEMORE and VAN SLUYTERS, 1975; HIRSCH and LEVENTHAL, 1976). The visual system, therefore, provides us with the opportunity to compare and contrast the *intrinsic* genetic controls of neuronal development with the *extrinsic*, highly variable controls imposed by the individual organism's environment and experience.

We have not attempted to be thorough and exhaustive in reviewing the many studies that have dealt with the role of sensory stimulation in the development of the mammalian visual system. Rather, we have been somewhat arbitrary in the hope of generating an integrated overview that can link a lot of the data about the development of the visual system with a few common ideas. No doubt we have omitted some studies and interpreted others in a manner that reflects our own theoretical bias. Certainly the thesis we present will prove to be an oversimplification, and thus will eventually be shown to be incorrect. It is our hope, however, that by presenting it we can bring some order into what is one of the most chaotic and rapidly expanding fields of neurobiological research.

B. The Organization of the Mammalian Visual System

I. X-, Y-, and W-Cells in the Retina

Three classes of ganglion cells have been identified in the retina of cat: (1) X-cells (ENROTH-CUGELL and ROBSON, 1966; STONE and FUKUDA, 1974) or sustained cells (IKEDA and WRIGHT, 1972; CLELAND et al., 1973) or Type II cells (FUKADA, 1971); (2) Y-cells (ENROTH-CUGELL and ROBSON, 1966; STONE and FUKUDA, 1974) or transient cells (CLELAND et al., 1971, 1973; IKEDA and WRIGHT, 1972) or Type I cells (FUKADA, 1971); (3) W-cells (STONE and HOFFMANN, 1972; STONE and FUKUDA, 1974).

X-type ganglion cells have axons that conduct action potentials slowly, sum the excitation of different parts of their receptive fields linearly, and display a sustained response to a stationary spot situated properly within their response

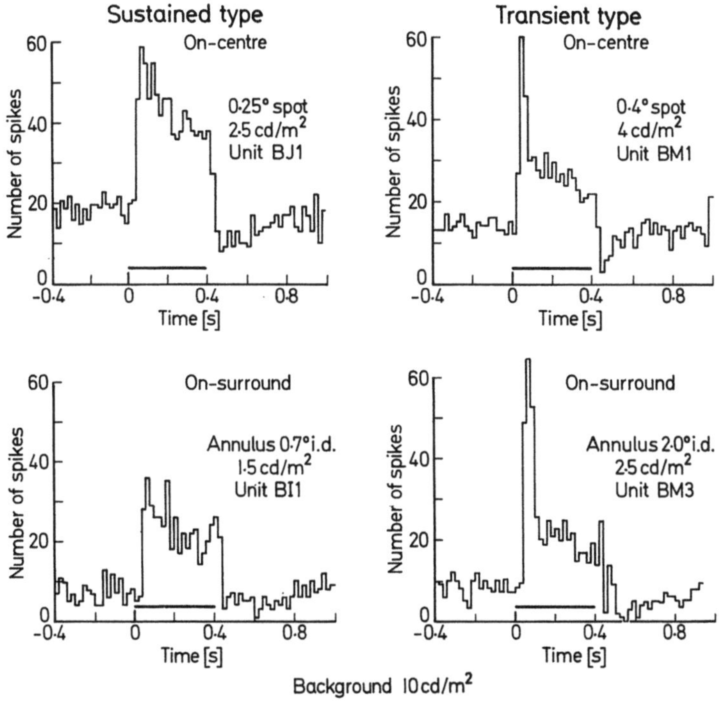

Fig. 1. Post-stimulus time-histograms showing the common characteristics of responses from an on-center and an on-surround sustained unit, and similarly for transient units. Note the considerable differences in the initial transient of sustained and transient units regardless of center or surround origin. Responses were to stimuli flashed on for 0.4 s (shown by dark bar) every 3.4 s and the luminance was adjusted to about four times threshold. The bin width was 20 ms, and 20 responses were summed. Multiply ordinate values by 2.5 to convert to impulses/s. [Reproduced with permission from CLELAND et al. (1973), Fig. 7]

field. In addition, X-cells have relatively small receptive fields and prefer low stimulus velocities. Finally, X-cells exhibit a high maintained discharge rate and this rate is modulated about its mean level when a grating pattern of high spatial frequency crosses the unit's receptive field (Figs. 1–3).

Y-type ganglion cells, on the other hand, possess axons that conduct action potentials rapidly, sum excitation of various parts of their receptive fields nonlinearly, and respond in a transient fashion to stimulation of appropriate regions of their response fields. Furthermore, Y-cells have larger receptive fields and respond to a wider range of stimulus velocities than do X-cells. Finally, Y-cells, unlike X-cells, display a relatively low mean maintained discharge rate and this rate increases transiently when a high-frequency grating pattern is moved across the unit's receptive field.

The group of ganglion cells identified most recently, W-cells, have axons that conduct action potentials more slowly than either X-cells or Y-cells, and many W-cells lack the antagonistic center-surround organization typical of other ganglion cells (KUFFLER, 1953; STONE and HOFFMANN, 1972; HOFFMANN, 1973; STONE and FUKUDA, 1974). This class is quite heterogeneous and contains cells with varied

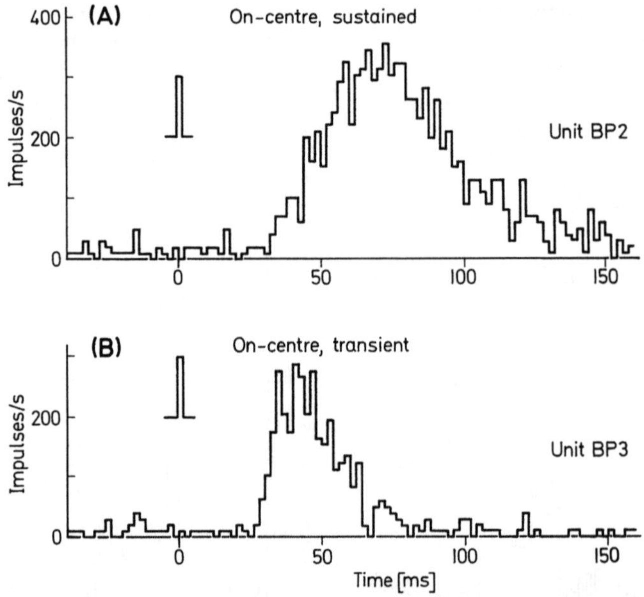

Fig. 2. Impulse-responses. Each graph is post-stimulus time-histogram (2 ms channel width) of summed responses to 50 presentations of brief stimulus flash (timing shown on the left) applied to center of receptive field against background illumination of 0.34 cd/m². A, on-center sustained ganglion cell, 0.25° spot stimulus at 36 cd/m² for 2 ms. B, on-center transient cell, 0.25° spot stimulus at 180 cd/m² for 2 ms. In each case, stimulus was about 4—8 times threshold. [Reproduced with permission from CLELAND et al. (1973), Fig. 9]

receptive field characteristics (STONE and HOFFMANN, 1972; STONE and FUKUDA, 1974). Tonic W-cells display a sustained response to a stimulus that is positioned appropriately and have high maintained rates of activity, while phasic W-cells respond transiently to appropriate stimulation and exhibit lower maintained discharge rates. In addition, some W-cells respond selectively to stimuli moving in a particular direction (direction-selective cells); other W-cells' maintained response rate is suppressed by both light and dark stimuli positioned over the receptive field center (suppressed-by-contrast cells). Finally, the receptive fields of W-cells are similar in size to those of Y-cells, and many W-cells respond to stimuli moving rapidly, although this response is not as marked as with Y-cells.

The three classes of cells described above are distributed differentially throughout the retina and differ in cell size (BOYCOTT and WÄSSLE, 1974). W-cells have small cell bodies and are concentrated within the central retina and the visual streak (ROWE and STONE, 1976). X-cells are also encountered most frequently within the central retina and their somas are of moderate size. Finally, the relative frequency of Y-cells, the class of ganglion cells in which the cell bodies are largest, is low in the central retina and increases with eccentricity (ENROTH-CUGELL and ROBSON, 1966; CLELAND et al., 1973; FUKUDA and STONE, 1974, 1975).

The visual systems of a number of species are reported to contain classes of retinal ganglion cells analogous to those that have been described in the cat's visual system. In particular, similar classes of cells have been described in the

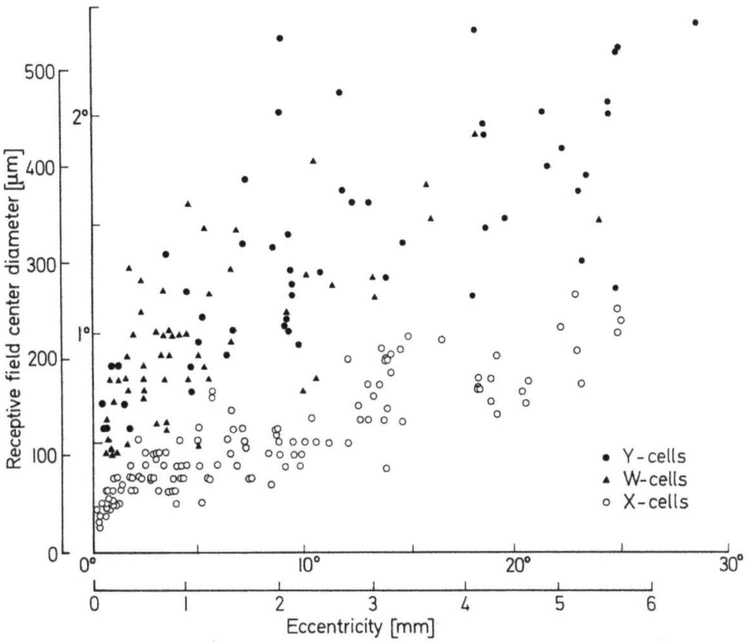

Fig. 3. Receptive field center diameter plotted as a function of eccentricity for Y-, X-, and W-cells. [Reproduced with permission from STONE and FUKUDA (1974), Fig. 11]

primate retina: tonic ganglion cells and phasic ganglion cells exhibit responses and distributions similar to those of X-cells and of Y-cells respectively (GOURAS, 1968, 1969; de MONASTERIO and GOURAS, 1975). Furthermore, X-cells, Y-cells, and W-cells have also been observed in the rabbit retina (CALDWELL and DAW, 1976). The presence of these three classes of neurons in a number of species suggests that they play an important role in the function and in the organization of the mammalian visual system in general.

II. Central Projections of X-, Y-, and W-Cells

The central projections of these three groups of ganglion cells differ markedly. In particular, the very slow conducting axons of W-cells (3–15 m/s) project mainly to the midbrain, with many terminating in the superior colliculus (HOFFMANN, 1973; FUKUDA and STONE, 1974). A limited number of W-cells also appear to project to the visual cortex via the C-laminae of the lateral geniculate nucleus (LGN) (WILSON and STONE, 1975). In contrast, the axons of X-cells, which conduct at intermediate rates (20–25 m/s), project mainly to the forebrain, with many terminating in the lateral geniculate nucleus (CLELAND et al., 1971; HOFFMANN et al., 1972; FUKUDA and STONE, 1974). Relatively few X-cells project to the midbrain, and virtually none project to the superior colliculus (HOFFMANN, 1973;

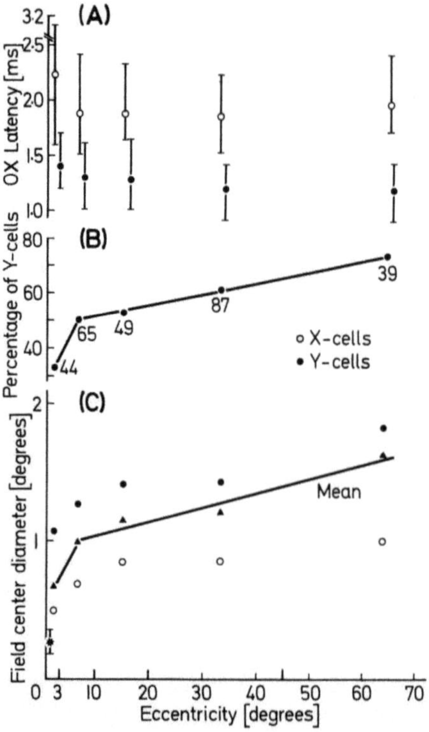

Fig. 4A–C. In this figure, LGN cells are divided into five groups on basis of retinal eccentricity. (A) for each eccentricity group, mean latency to electrical stimulation of optic chiasm of Y-cells (closed circles) and of X-cells (open circles) is plotted, and for each mean vertical bars represent range of observed values. (B) percentage of Y-cells (number of Y-cells as percentage of total of X- and Y-cells) is plotted for each eccentricity group. Number of cells in each eccentricity group is shown. (C) for each eccentricity group, mean diameter of center regions of receptive fields is plotted with closed circle for Y-cells and open circle for X-cells. Mean diameter of all fields in a group is plotted with a △ symbol. ⋆ represents mean center diameter of 12 X-fields recorded along two electrode tracks, which were smallest in the 0–3° eccentricity group. Vertical bar represents range of their field center diameters about this mean. [Reproduced with permission from HOFFMANN et al. (1972), Fig. 4]

FUKUDA and STONE, 1974). Finally, Y-cells have the fastest conducting axons (35–45 m/s), and many of these terminate directly in the lateral geniculate nucleus as well as in the superior colliculus (CLELAND et al., 1971; HOFFMANN et al., 1972; FUKUDA and STONE, 1974). Retinal Y-cells may also project indirectly to the superior colliculus *via* connections with Y-cells in the LGN and complex cells in the cortex (HOFFMANN, 1973; PALMER and ROSENQUIST, 1974; HOFFMANN and SHERMAN, 1974, 1975).

Many neurons in the geniculo-cortical pathways of the cat display response properties which reflect the nature of the ganglion cell afferents they receive (Fig. 4). In particular, relay cells in the lateral geniculate nucleus have been shown to receive direct monosynaptic inputs from X-cells, Y-cells, or W-cells in the retina (CLELAND et al., 1971; HOFFMANN and STONE, 1971; HOFFMANN et al., 1972; MASON, 1975; WILSON and STONE, 1975). These neurons display response characteristics and distributions similar to those of the retinal X-cells, Y-cells, and W-cells which provide their afferent input and have been termed X-, Y-, and W-type relay cells (WILSON et al., 1976).

The presence of both X-cells and Y-cells has also been suggested in the LGN of the rat, the tree shrew, and the monkey (FUKUDA and SUGITANI, 1974; SHERMAN et al., 1976; DREHER et al., 1976). An anatomical segregation of these two classes is evident in the LGN of both old- and new-world monkeys: X-cells were found only in the parvocellular laminae and Y-cells were observed only in the magnocellular laminae (DREHER et al., 1976 [*M. nemestrina*]; SHERMAN et al., 1976 [*Aotus*]).

Neurons in the visual cortex of the cat are also thought to receive direct monosynaptic inputs from X-cells, Y-cells, and W-cells in the LGN (HOFFMANN and STONE, 1971; STONE and DREHER, 1973; SINGER et al., 1975; WILSON and STONE, 1975)[1]. In certain respects, the receptive field characteristics of these cells reflect the response properties of the LGN neurons that terminate upon them. Specifically, cortical cells with predominantly X-type afferents (SA cells) have small receptive fields and respond well only to stimuli moving slowly, while neurons in the visual cortex with mainly Y-cell afferents (FA cells and IFA cells) have large receptive fields and respond reliably to a wider range of stimulus velocities (HOFFMANN and STONE, 1971; STONE and DREHER, 1973; DREHER and SANDERSON, 1973; SINGER et al., 1975) (Figs. 5 and 6). In addition, SA cortical cells are very selective for stimulus orientation and many have discrete, parallel excitatory and inhibitory regions, while FA cortical cells are less selective for stimulus orientation and have "on-off" type receptive fields (STONE and DREHER, 1973). Thus, it has been suggested that simple cortical cells (HUBEL and WIESEL, 1962; PETTIGREW et al., 1968) receive LGN X-cell afferents, and complex cortical cells (HUBEL and WIESEL, 1962; PETTIGREW et al., 1968) receive LGN Y-cell afferents (HOFFMANN and STONE, 1971)[2]. Consistent with this, the velocity tuning and the distribution of cortical complex cells has been observed to correspond with the velocity tuning and the distribution of Y-cells in the LGN (MOVSHON, 1975; WILSON and SHERMAN, 1976). However, convergence of X-cell and of Y-cell afferents upon both simple and complex cortical cells has also been reported (SINGER et al., 1975).

Finally, it has been suggested that a specific class of neurons in the cat's visual cortex displaying responses indicative of an LGN X-cell afferent input (these cells

---

[1] LGN X-cells are thought to project predominantly to area 17, while LGN Y-cells project to area 18 as well as to area 17 (STONE and DREHER, 1973; FUKUDA and SUGITANI, 1974).

[2] Cells in the mammalian striate cortex have been termed simple if their receptive fields can be subdivided into distinct, mutually antagonistic excitatory (on) and inhibitory (off) summating regions. The responses of a simple cell to patterned visual stimulation can be inferred from the orientation and the position of these excitatory and inhibitory regions. The response of a complex cell to patterned visual stimulation cannot be inferred easily from the orientation and distribution of the excitatory and inhibitory regions that comprise its response field. Most of these cells do not have distinct excitatory and inhibitory regions within their response field; they typically display an on-off response to flashing spots of light (HUBEL and WIESEL, 1959, 1962). Corollary, response properties of simple cells are: small response fields, preferences for slow-moving stimuli, low maintained discharge rates, spatially offset on and off excitatory centers, and narrow orientation tuning curves. Corollary properties of complex cells are: larger receptive fields, preferences for rapid stimulus motion, high maintained discharge rates, and broader orientation tuning curves (HUBEL and WIESEL, 1962; PETTIGREW et al., 1968; BISHOP et al., 1971a–c, 1973; HENRY et al., 1974; WATKINS and BERKELEY, 1974; MOVSHON, 1975; SCHILLER et al., 1976a–d; WILSON and SHERMAN, 1976).

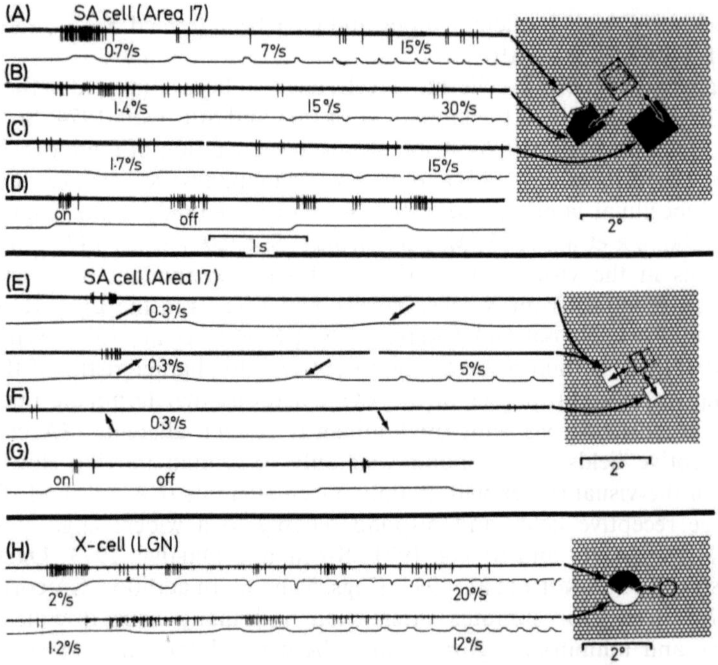

Fig. 5A–H. Responses of two area 17 SA cells and of an LGN X-cell to visual stimuli. Trace below each spike trace is output of a device monitoring luminance changes approximately at center of receptive field. Traces goes up as luminance increases, down as luminance decreases. Duration of deflection is thus inversely related to speed of a moving object. Stimulus situations drawn schematically on right. Continuous outlines represent receptive fields. Dotted outlines represent size and position of stationary flashing stimuli. Receptive fields and stimuli drawn to scale. (A)–(D): Responses of an SA cell. Cell responded to appropriately oriented, moving light (A) and dark (B) bars, but only at low speeds. Cell was unresponsive to a wrongly oriented bar (C). It gave on-off responses to a stationary flashing rectangle (D). Stimulus drawing shows only half of each of bar stimuli used to elicit responses in (A) and (B). (E)–(G): Responses of another SA cell. Cell responded to an appropriately oriented, moving light bar (E), but only at low speeds and in one direction. Cell was thus direction-selective. It was unresponsive to an inappropriately oriented, moving bar (F). It gave on-responses to a stationary, flashing bar (G). (H): Responses of an off-center LGN X-cell to moving light and dark spots. Like SA cells, X-cell is responsive only to slowly moving stimuli. Hatched circle represents center region of receptive field, as plotted with stationary flashing spots. Stimulus drawing shows only half of each stimulus spot. [Reproduced with permission from STONE and DREHER (1973), Fig. 9]

have very small receptive fields and respond only to stimuli moving slowly) respond preferentially to either horizontal or vertical lines, while cortical neurons with response characteristics suggestive of a Y-cell afferent input (these cells have larger receptive fields and respond to high stimulus velocities) show no bias in the distribution of their orientation preferences (LEVENTHAL and HIRSCH, 1977) (Fig. 7). The cortical distribution of these two groups of cells is similar to the distribution of X-cells and of Y-cells in the lateral geniculate nucleus.

Neurons in the cat's midbrain also exhibit responses that appear to be determined by the nature of their retinal ganglion cell inputs. In particular, neurons in the cat's superior colliculus that receive afferents directly from retinal W-cells (W-direct pathway) exhibit velocity selectivities similar to those of W-type

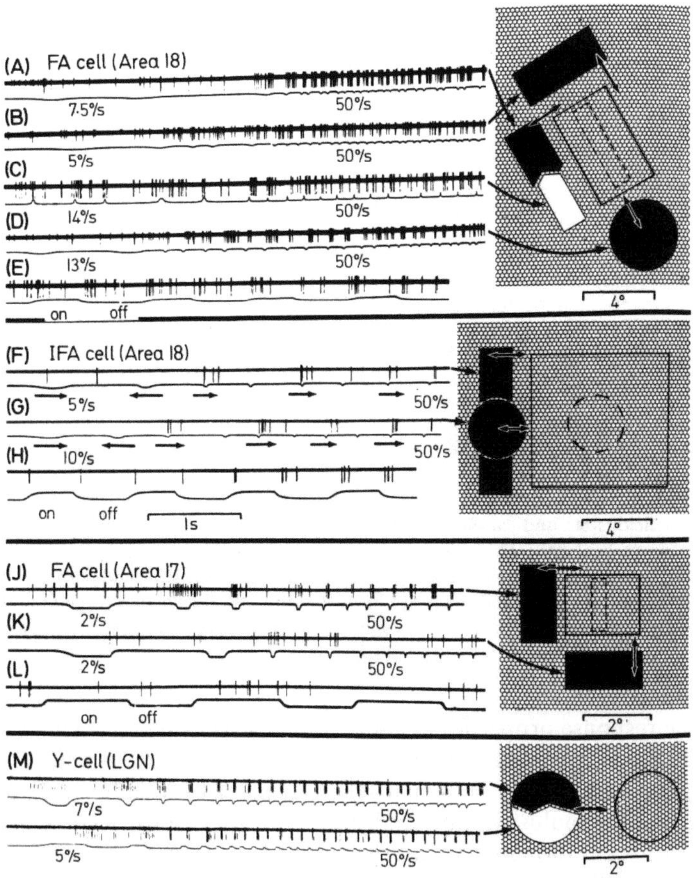

Fig. 6A–M. Responses of cortical FA and IFA cells and of an LGN Y-cell to visual stimuli, represented as in Fig. 5. (A)–(E): Responses of area 18 FA cell. Cell responded to a 2° wide black bar, most clearly at higher speeds [50°/s, (A)]. It responded less strongly to a similarly oriented 1° wide light bar (C) and to a 2° black bar moved at right angles (B). It also responded well to a moving 4° black disc (D). In each case, response was most clear at higher speeds. Cell gave on-off responses to stationary flashing stimulus (E). Stimulus drawing shows only half of each of the bar stimuli used to elicit responses in (A) and (C). (F)–(H): Responses of an area 18 IFA cell. Cell responded to a moving 2° wide bar, most clearly at higher speeds (F). It was direction-selective, responding only as stimulus moved left to right, except at highest speed tested (50°/s), when weak response was evoked by movement in opposite direction. Cell was similarly responsive to a 3° diameter black disc (G). It gave on-off responses to a stationary flashing spot (H). (J)–(L): Responses of an area 17 FA cell. Cell responded to a vertically oriented, moving, 2° wide black bar (J). Response was clear to slow and fast movements. Cell was unresponsive to the same stimulus moving at right angles (K). It gave a weak on-off response to a stationary flashing stimulus (L). (M): Responses of an off-center LGN Y-cell to moving light and dark spots. Response is clear at all speeds tested. Stimulus drawing shows only half of each stimulus spot. [Reproduced with permission from STONE and DREHER (1973), Fig. 10]

ganglion cells (HOFFMANN, 1973; HOFFMANN and SHERMAN, 1974). In addition, as one might expect, many collicular neurons that receive W-cell inputs are direction-selective; presumably, this reflects the direction-selectivity of certain retinal W-cells. In addition, binocular collicular cells that receive W-cell afferents and exhibit responses similar to those of "suppressed-by-contrast" retinal W-cells have been

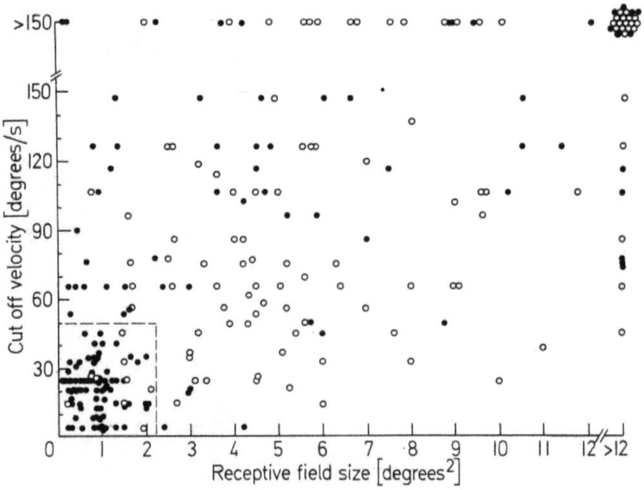

Fig. 7. Receptive field sizes and cutoff velocities of cortical neurons preferring either horizontal ($0 \pm 22.5°$) lines or vertical ($90 \pm 22.5°$) lines (solid circles) and either 45° ($45 \pm 22.5°$) lines or 135° ($135 \pm 22.5°$) lines (open circles). Preponderance of solid circles in lower left-hand corner of this figure indicates that most neurons with small receptive fields and low cutoff velocities respond preferentially either to horizontal or to vertical lines. [Reproduced with permission from LEVENTHAL and HIRSCH (1977), Fig. 1]

reported. The response properties of some collicular cells reflect an afferent input from Y-cells. These collicular cells respond to very high stimulus velocities, as do Y-type ganglion cells. Finally, collicular cells influenced by Y-indirect afferents also respond to rapid stimulus motion and many are direction-selective, as are the complex cortical cells thought to terminate upon them (HOFFMANN, 1973; PALMER and ROSENQUIST, 1974; SINGER et al., 1975).

C. The Role of Sensory Stimulation in the Development of the Visual System

I. Introduction

Environmental influences play a vital but poorly understood role in the development of the nervous system. While there is ample evidence that the environment affects the developing visual system (reviewed by BLAKEMORE, 1974; GUILLERY, 1974; PETTIGREW, 1974a; HIRSCH and JACOBSON, 1975; GROBSTEIN and CHOW, 1975, 1976; DANIELS and PETTIGREW, 1976), the mechanisms involved are virtually unknown (JACOBSON, 1970, 1978).

Sensory stimulation can exert its effects on the developing visual system at many points along the visual pathways. Great excitement has been generated in recent years by descriptions of stimulation-induced changes within the visual cortex. All too often, however, it is difficult to verify that the observed cortical effects are not manifestations of alterations at lower levels in the visual pathways. We shall, therefore, first consider the role that sensory stimulation plays in the development of the retina and the lateral geniculate nucleus and then proceed to an examination of the influence of sensory input on tectal and cortical development.

II. The Retina

1. Anatomical Studies

In cat and rat, much of retinal development takes place before birth (WEIDMAN and KUWABARA, 1968; KUWABARA and WEIDMAN, 1974; CRAGG, 1975a). The outer segments of the photoreceptors, however, are not evident until the fifth postnatal day; the disc structures containing the photopigment are the last essential structure to develop (CRAGG, 1975a). In the cat, retinal dark discharge is dependent upon the intactness of the outer segments of the retina (BURKE and HAYHOW, 1968). Neuronal activity thus cannot be the impetus initiating synaptic development in the retina, since synaptogenesis appears to precede retinal development of the photoreceptors and thus of retinal function. However, spontaneous activity in the visual cortex and in the lateral geniculate nucleus is only partially dependent upon retinal activity (ARDUINI and PINNEO, 1963), and it remains possible that neuronal activity could initiate synapse development in these visual areas (CRAGG, 1975a).

Results of studies of the effects of visual stimulation on the development of the retina are contradictory. This difficulty may stem from the fact that many experimenters have used albino rats whose retinae suffer considerable damage following exposure to even moderate levels of illumination. In this species, exposure to continuous illumination produces severe degenerative changes in the outer layers of the retina (outer plexiform and photoreceptor layers) (NOELL et al., 1966; GORN and KUWABARA, 1967; KUWABARA and GORN, 1968; GRIGNOLO et al., 1969; O'STEEN, 1970; HANSSON, 1970; O'STEEN and LYTLE, 1971; BENNETT et al., 1972, 1973a, b; FIFKOVA, 1972a). Similar damage to the rabbit retina has also been reported following exposure to constant illumination (LAWWILL, 1973).

Following visual deprivation produced either by lid suture, which eliminates patterned stimulation and reduces retinal illumination (FIFKOVA, 1972b, 1973), or by a contact occluder, which presumably induces more severe reduction in retinal illumination (SOSULA and GLOW, 1971), increases have been reported in the number of amacrine cell to bipolar cell contacts in the inner plexiform layer of the deprived rat retina. There is also some evidence of increases in the number of amacrine cell to ganglion cell synapses (SOSULA and GLOW, 1971).

Interpretation of the results of these studies is difficult, since their design included changes in overall retinal illumination and in the degree to which the incident light was patterned. It is thus not possible to determine which of these variables was responsible for the observed effects. One recent study, however, has

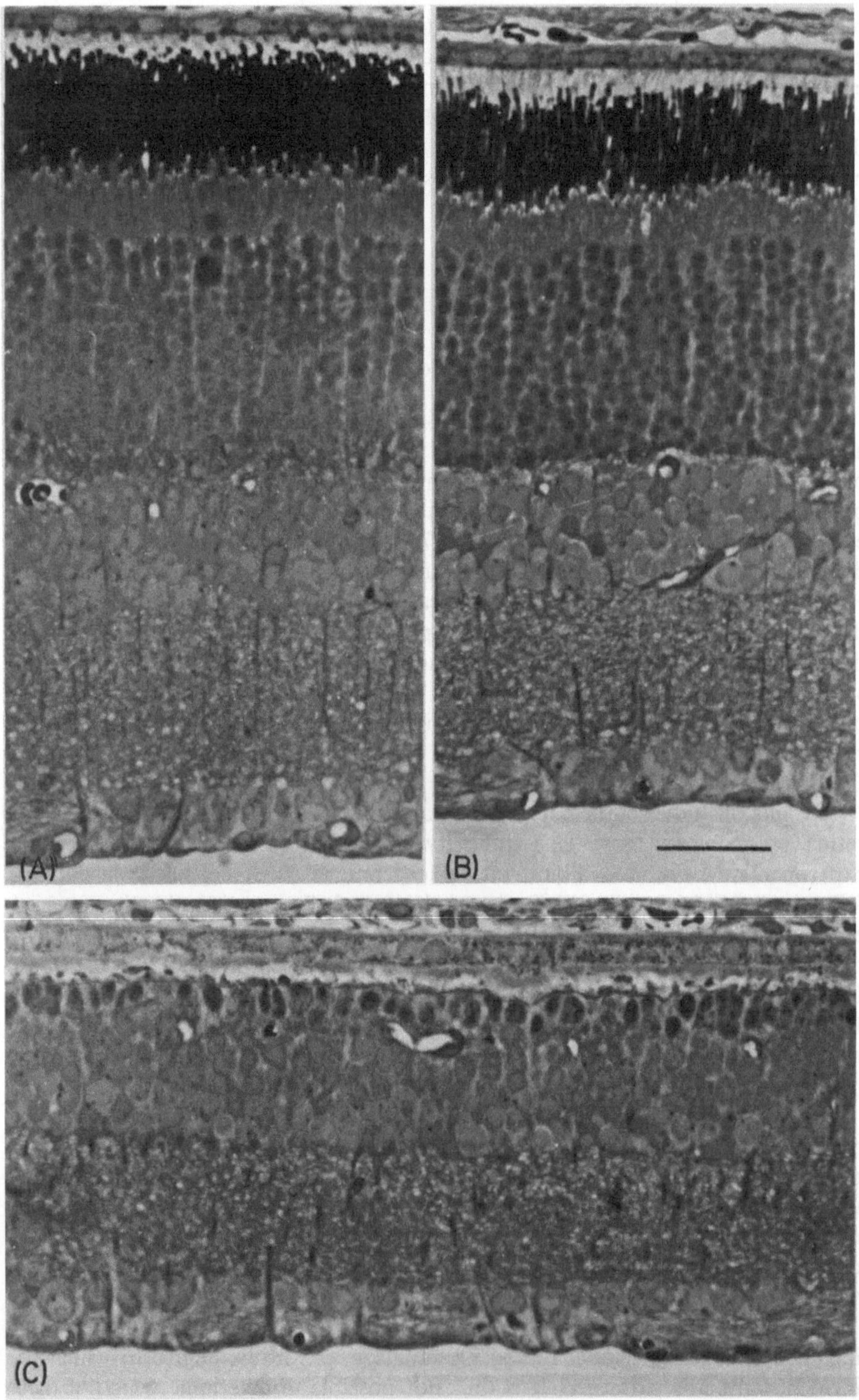
(A)
(B)
(C)

attempted to separate these variables (CHERNENKO and WEST, 1976). The results of this work showed that rats that had been reared in continuous bright light suffered severe damage to the outer segments of the retina, whether or not the animals' eyes had been sutured shut to reduce the incidence of patterned stimulation. In addition, the layers of the retina of rats reared in total darkness were thicker than those of animals reared in dim light (Fig. 8). The only synaptic changes these authors observed were increases in the number of amacrine cell to ganglion cell synapses in the outer third of the inner plexiform layer. Finally, in contrast to earlier reports (SOSULA and GLOW, 1971; FIFKOVA, 1972b), the overall number of amacrine cell and bipolar cell synapses was the same in deprived and control animals in this study. In summary, from this well-controlled study, amacrine to ganglion cell synapses emerge as the primary site of structural changes that occur in the inner layers of the retina. Since a functional amacrine cell input may be necessary for the normal response of transient retinal ganglion cells (WERBLIN and DOWLING, 1969), these results could be interpreted to suggest that visual stimulation must selectively affect these cells.

2. Physiological Studies

Physiological studies of the effects of visual deprivation upon development of the retina indicate that retinal ganglion cells of adult cats continue to discharge in animals placed in total darkness for 2–3 years (BURKE and HAYHOW, 1968). In addition, cats deprived of visual experience from birth to 12–18 months of age by lid-suture of either one eye (monocular deprivation) or of both eyes (binocular deprivation) develop retinae that contain normal proportions of sustained and transient ganglion cells (SHERMAN and STONE, 1973). Finally, it has also been suggested that controlled early visual exposure does not effect the receptive field characteristics of ganglion cells in the rabbit retina (DAW and WYATT, 1974). Visual deprivation thus does not appear to prevent the onset of normal retinal function.

III. The Lateral Geniculate Nucleus

1. Anatomical Studies

Anatomical effects of monocular deprivation have been demonstrated repeatedly in the lateral geniculate nucleus. In particular, abnormalities such as decreases in mean cell size and cell density occur in those layers of the binocular

◁ Fig. 8. Light micrographs of rat retinas exposed to various light intensities for eight weeks (following two weeks of dark-rearing from birth). (A) From a lid-sutured, dark-reared rat. (B) From an unsutured, low light (9.8 lux)-reared rat. (C) From an unsutured, bright light (1960 lux)-reared rat. Total thicknesses of the retinas within each condition were progressively reduced by successively brighter exposures. Unsutured, bright light-reared group (C) showed almost complete degeneration of outer segments and 90% degeneration of receptor nuclei in comparison to lid-sutured, dark-reared group. All micrographs reproduced at same magnification. Scale marker (B) = 30 μm. [Modified and reproduced with permission from CHERNENKO and WEST (1976), Figs. 2–4]

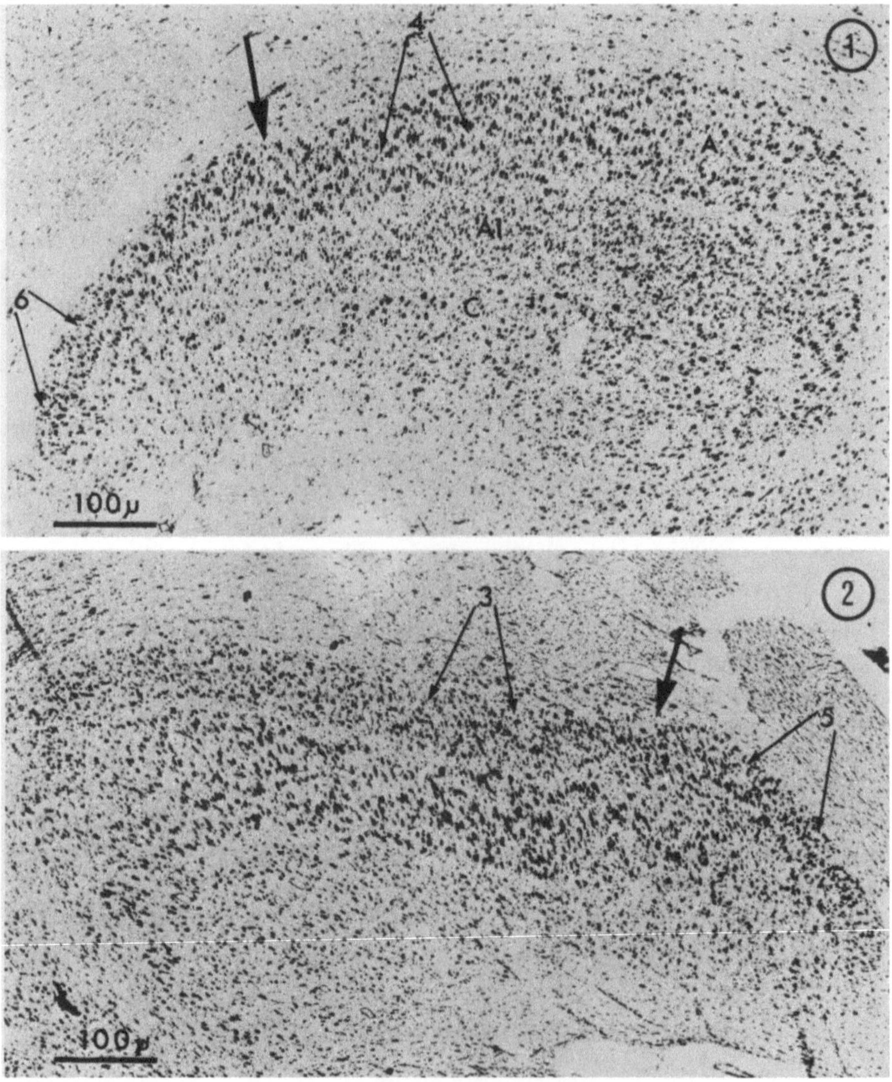

Fig. 9. (1) Frontal section through the right lateral geniculate nucleus of a kitten whose right eye had been sutured for 89 days. Cells in lamina A appear normal and cells in lamina A1 are abnormally small, pale-staining, and closely packed. Areas between small numbered arrows are shown at higher magnification in Fig. 10 (4, 6). Laminae A, A1, and C are labeled, and border between monocular and binocular segments of lamina A is indicated by a thick arrow. (2) Frontal section through left lateral geniculate nucleus of same kitten as above. Cells in binocular segment of lamina A appear abnormal, but cells in monocular segment appear unaffected. Labeling as above. [Reproduced with permission from GUILLERY and STELZNER (1970), Plate 1]

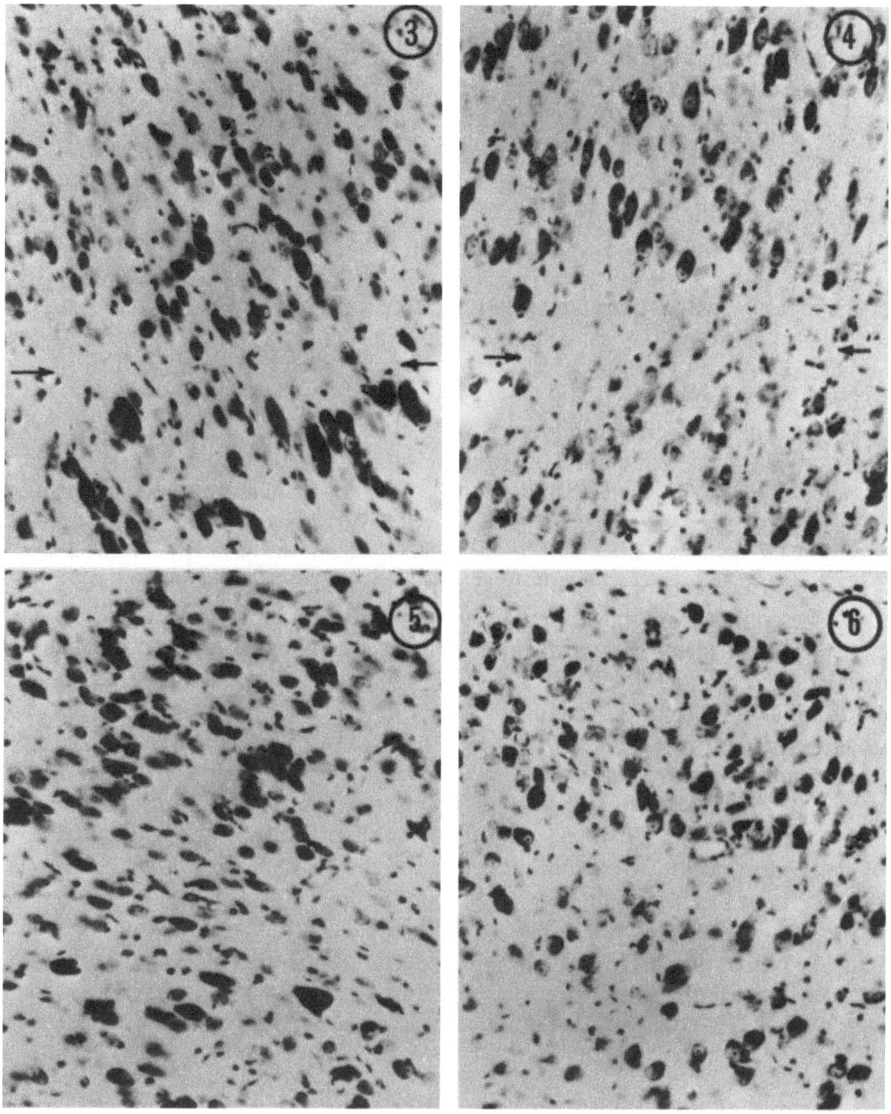

Fig. 10. (3) Higher magnification of binocular segment of laminae A and A1 from contralateral lateral geniculate nucleus in Fig. 9 (2). Small cells of lamina A are in upper part of figure and larger normal cells of lamina A1 in lower part. Arrows indicate interlaminar zone. (4) Higher magnification of binocular segment of laminae A and A1 from ipsilateral lateral geniculate nucleus shown in Fig. 9 (1). Small cells of lamina A1 are in lower part of figure and larger normal cells of lamina A are in upper part. Arrows indicate interlaminar zone. (5) and (6) Higher magnification of monocular segments from sections shown in Fig. 9 (1) and (2) respectively. Note there is no significant cell shrinkage in this segment of nucleus. Some of cells of lamina C are seen in lower parts of figures. [Reproduced with permission from Guillery and Stelzner (1970), Plate 2]

segment of the LGN that are activated by the deprived eye in cat, dog, squirrel, rat, rabbit, monkey, and tree shrew (WIESEL and HUBEL, 1963a; FIFKOVA and HASSLER, 1969; GUILLERY and KAAS, 1974a; GUILLERY, 1974; CHOW and SPEAR, 1974; SHERMAN and WILSON, 1974; HUBEL et al., 1976; CASAGRANDE et al., 1976) (Fig. 9). It has been suggested that these effects are due to a competitive interaction between the two eyes; marked changes were observed in the binocular segment of the geniculate, while cells in the monocular segment representing the deprived eye were affected less severely (GUILLERY and STELZNER, 1970; GUILLERY, 1972, 1973a, b, 1974; SHERMAN and WILSON, 1975) (Fig. 10). Similarly, no shrinkage occurs in an artificial monocular segment (critical segment) produced in the binocular portion of the geniculate in those layers that are innervated by the deprived eye (GUILLERY, 1972)[3].

There is some evidence that the geniculate changes following visual deprivation were caused by a general shrinkage of cells in the deprived laminae (WIESEL and HUBEL, 1963a; GUILLERY, 1973a). However, at least some of the observed effects may have been due to a selective diminution or atrophy of large geniculate cells in the affected layers. Presumably, this resulted from a reduction in the number of LGN Y-cells, the class of geniculate neurons suggested to have the largest somas.

Anatomical effects of binocular deprivation have also been documented in the LGN. These differ from the consequences of monocular deprivation. In particular, where there is a decrease in the mean cell size of geniculate neurons in cats that have been deprived binocularly (WIESEL and HUBEL, 1963a; HICKEY et al., 1977), this decrease is smaller than the one produced by monocular deprivation (WIESEL and HUBEL, 1965a, b). In fact, a number of studies have failed to demonstrate any significant anatomical effect of binocular deprivation (CHOW and STEWART, 1972; GUILLERY, 1973a; HENDRICKSON and BOOTHE, 1976). Furthermore, unlike the localized decrease in cell size produced by monocular deprivation, the effects of binocular deprivation are found throughout the LGN. Closure of both eyes, therefore, may have a slight but "pure" deprivation effect, since no competition between the two eyes should be possible. Finally, early suggestions that a general shrinkage or loss of cells in the LGN had occurred following binocular deprivation have also been challenged by reports suggesting that a loss of large Y-cells from all layers of the LGN could have resulted in the effect observed (SHERMAN et al., 1972).

2. Physiological Studies

Early studies reported no apparent physiological abnormalities in the lateral geniculate nucleus of the cat following visual deprivation. In particular, experimenters found that cells studied in deprived animals exhibit the antagonistic center-surround organization characteristic of adult geniculate neurons (WIESEL and HUBEL, 1963a, 1965a; CHOW and STEWART, 1972). More recently, however,

[3] An artificial monocular segment can be produced if a small lesion is made in the central retina of one of an animal's eyes soon after birth. Degeneration of the layers of the LGN innervated by the damaged retinal region then occurs, and this produces a small region in the geniculate layers innervated by the other eye where no binocular competition should be possible (GUILLERY, 1972).

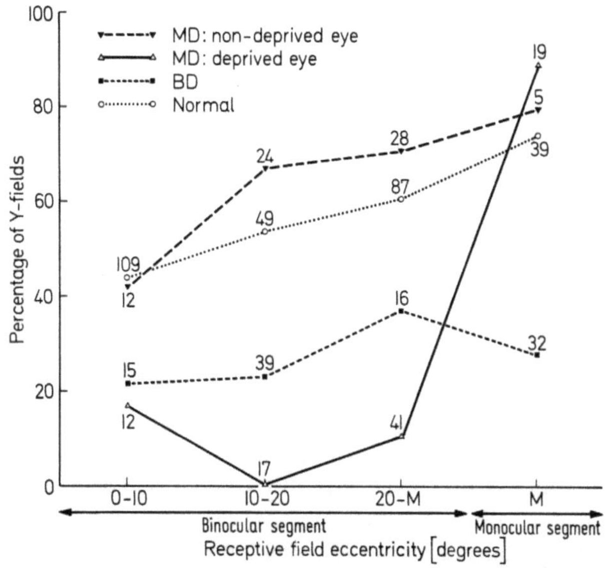

Fig. 11. Variation with eccentricity in visual field of frequency of Y-fields, expressed as percentage of total population (X-fields and Y-fields). This relationship is shown separately, as indicated, for nondeprived-eye receptive fields of monocularly deprived (MD) cats, for deprived-eye receptive fields of MD cats, for receptive fields of binocularly deprived (BD) cats, and for receptive fields of normal cats. Numbers next to each point indicate total number of fields from which percentage of Y-fields was calculated for that point. To minimize random sampling errors in these percentages, each point includes units pooled from at least four different penetrations, using four different electrodes. Only exception to this is that five units representing nondeprived eye fields of MD cats in monocular segment were all encountered along a single electrode penetration. [Reproduced with permission from SHERMAN et al. (1972), Fig. 2]

alterations have been observed in the physiology of the lateral geniculate nucleus following visual deprivation.

A number of reports suggest that a specific class of geniculate neurons is affected severely by both monocular and binocular deprivation. Specifically, both monocular and binocular lid suture result in a reduction in the number of functional Y-cells, but not of X-cells, in those layers of the binocular segment of the lateral geniculate nucleus of the cat or tree shrew that receive projections from the deprived eye (SHERMAN et al., 1972, 1975; HOFFMANN and CYNADER, 1975; NORTON et al., 1976)[4] (Fig. 11). However, apparently normal numbers of Y-cells can be recorded electrophysiologically from the critical segment of monocularly deprived cats (SHERMAN et al., 1972, 1974, 1975; HOFFMANN and CYNADER, 1975).

Other physiological abnormalities have also been observed in the LGN of deprived cats. In particular, binocular inhibitory interactions among the cat's lateral geniculate neurons appear diminished following binocular deprivation (SHERMAN and SANDERSON, 1972). Similarly, inhibitory interactions may also be reduced in the LGN of cats deprived of vision in one eye (EYSEL and GAEDT, 1971). The absence of functional Y-cells may contribute to the decreased inhibitory

[4] Monocular deprivation also reduces the number of functional Y-cells in the medial interlaminar nucleus of the LGN (KRATZ et al., 1976a).

interaction observed among the lateral geniculate neurons of deprived cats. In particular, while there appears to be little convergence of excitatory retinal X-cell and Y-cell afferents upon individual geniculate neurons (CLELAND et al., 1971), it has been suggested that there is considerable convergence of inhibition mediated by Y-cells upon cells receiving excitatory X-cell inputs (HOFFMANN et al., 1972; SINGER and BEDWORTH, 1973; SINGER et al., 1975). The diminution of functional Y-cells in the LGN following deprivation may thus result in a lessening of this inhibition and account for much of the observed effect.

Selective exposure to lines of only one orientation during rearing also affects the response properties of geniculate neurons. In particular, kittens deprived of all visual experience other than daily exposure to a periodic vertical grating are reported to display diminished lateral geniculate mass responses when tested with horizontal, vertical or oblique gratings of the same spatial frequency as the grating the animal viewed during rearing (MAFFEI and FIORENTINI, 1975). In addition, exposure of kittens to a vertically-striped cylinder has been reported to modify the distribution of orientation biases displayed by some geniculate neurons; most of these respond preferentially to lines of the same orientation or orthogonal to the lines the animal viewed during rearing (DANIELS et al., 1975). This effect is thought to be exerted predominantly upon Y-cells in the LGN.

In conclusion, various forms of early visual deprivation result in a lateral geniculate nucleus that is abnormal anatomically as well as physiologically. Furthermore, since apparently normal numbers of geniculate X-cells develop in deprived cats, a selective dependence upon early visual experience for normal function of geniculate Y-cells may mediate these effects in many cases. An innate determination of the response properties of X-cells in the LGN is therefore indicated; the role of visual experience in the development of geniculate Y-cells is unclear at present. It has been demonstrated, however, that few identifiable X-cells or Y-cells are present among the immature neurons in the LGN of young kittens (NORMAN et al., 1975). The absence of functional Y-cells, but not of X-cells, in deprived animals can thus be interpreted to suggest that early visual experience contributes to the development of, rather than simply to the maintenance of, these neurons.

IV. The Superior Colliculus

1. Anatomical Studies

Experience-dependent changes have been observed in the development of synaptic contacts in the superior colliculus. LUND and LUND (1972a, b) have investigated the time course of synaptogenesis in the superior colliculus of the rat. These authors suggest that in this species synaptic development occurs in three phases. The first two of these are independant of early experience, since they occur in both normal and lid-sutured rats. The third phase, however, which occurs from 20 to 40 days after birth, appears to be dependent upon visual stimulation: it is arrested in animals in which both eyes have been sutured shut from birth. The changes that result from binocular deprivation can be reversed if the animal receives patterned visual stimulation before day 30, but not if the deprivation continues beyond days 35–45. There is thus a critical period during which patterned visual stimulation is necessary if synaptogenesis is to be completed in a normal fashion.

Postnatal changes in the size and the density of cells in the superior colliculus of the hamster have also been observed (PTACEK and FAGAN-DUBIN, 1974). In this species the size of collicular cells has been reported to nearly double during the first two weeks of life until a maximum value is reached at approximately day 14 (the time of normal eye-opening). In contrast, the density of collicular cells is highest at birth and decreases during early postnatal life. Since a reduction in cell density may reflect an increase in connectivity (SHARIFF, 1953; HAUG, 1956), this decrease may reflect the synaptogenesis that occurs during this period (LUND and LUND, 1972a, b). Finally, it remains to be determined whether the observed increase in mean cell size reflects a general increase in the size of collicular cells, or, rather, the specific growth of certain large neurons.

2. Physiological Studies

Early visual experience appears to be necessary for normal collicular development. Few neurons in the superior colliculus of very young kittens exhibit response properties typical of cells in the adult animal. Specifically, the proportion of neurons that are direction-selective, motion-sensitive, or activated by both eyes is low in young kittens (STEIN et al., 1973a; NORTON, 1974) but increases during the first few weeks of life (Fig. 12). Visual deprivation retards this development (HOFFMANN and SHERMAN, 1974, 1975), as do lesions of the visual cortex (STEIN and MAGALHAES-CASTRO, 1975). The only collicular neurons to develop normally in visually deprived animals are those with response properties that appear to be mediated by the W-direct pathway.

Monocular deprivation virtually eliminates the Y-indirect pathway from the deprived eye to the binocular segment of the cat's superior colliculus (HOFFMANN and SHERMAN, 1974). Similarly, Y-indirect afferents to the entire superior colliculus cannot be demonstrated in cats deprived of vision binocularly from birth to 10–12 months of age (HOFFMANN and SHERMAN, 1975). The Y-direct pathway also appears reduced in these animals. In addition, collicular cells in cats deprived of visual experience display reduced motion sensitivity, direction selectivity and binocularity (HOFFMANN and SHERMAN, 1974, 1975; WICKELGREN and STERLING, 1969b) (Fig. 13). Presumably, many of these abnormalities result from the absence of the Y-indirect pathway to the superior colliculus of these animals. Supporting this suggestion, collicular neurons receiving Y-indirect inputs via complex cortical cells appear to be direction-selective, binocular, and responsive to rapid stimulus motion (HOFFMANN, 1973; PALMER and ROSENQUIST, 1974; SINGER et al., 1975). Furthermore, the number of collicular cells exhibiting these response properties is reduced following lesions or cooling of the striate cortex (WICKELGREN and STERLING, 1969a, b; ROSENQUIST and PALMER, 1971). Finally, as with decorticate animals, the intact W-direct pathway could mediate much of the limited direction selectivity and

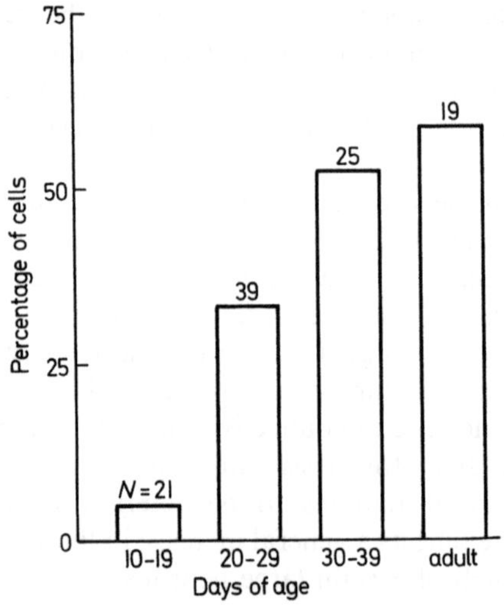

Fig. 12a. Development of direction-selective responses with age. Height of each bar indicates proportion of cells in each age group that gave direction-selective responses. N indicates number of cells examined for this property in each age group. [Reproduced with permission from NORTON (1974), Fig. 2]

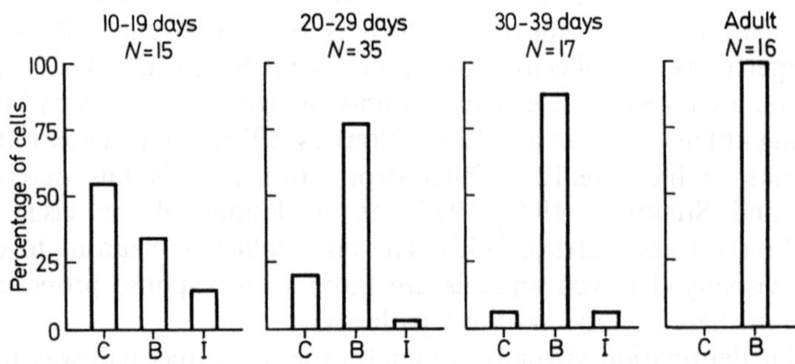

Fig. 12b. Development of ocular dominance pattern in superior colliculus of the cat. Column C shows proportion of cells driven entirely or strongly dominated by contralateral eye (Groups 1 and 2 after HUBEL and WIESEL). Column B shows the proportion of cells driven by both eyes (Groups 3–5 according to HUBEL and WIESEL). Column I shows proportion of cells driven or dominated by ipsilateral eye (Groups 6 and 7). N is number of cells studied in each age group. [Reproduced with permission from NORTON (1974), Fig. 3]

motion sensitivity that is observed among collicular cells in visually deprived cats (HOFFMANN and SHERMAN, 1974, 1975).

The elimination of Y-indirect inputs to the superior colliculus in deprived cats appears to reflect the absence of Y-cells from the LGN of these animals and not a disruption of the projections from the visual cortex to the superior colliculus; electrical stimulation of the cortex indicates that many cortico-tectal connections remain functional (HOFFMANN and SHERMAN, 1975).

There is also indirect evidence that development of the afferent input to the colliculus that is mediated by Y-cells is more dependent upon sensory stimulation than is development of the W-cell afferent input. In particular, the response latency

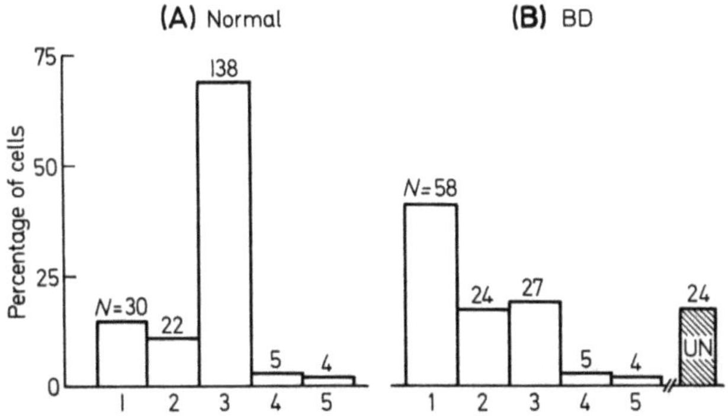

Fig. 13a. Ocular dominance distributions for collicular neurons from normal (left) and binocularly deprived (right) cats. For each, horizontal axis indicates following ocular dominance groups: 1, neurons activated exclusively by contralateral eye; 5, neurons activated exclusively by ipsilateral eye; 2–4, binocularly activated neurons such that 2 represents neurons dominated by the contralateral eye, 3 represents neurons driven almost equally well by each eye, and 4 represents neurons dominated by ipsilateral eye. Vertical scale shows percentage of 1-, 2-, 3-, 4-, or 5-type neurons. Left: 199 neurons from colliculus of normal cats. Right: 142 neurons from colliculus of binocularly deprived cats. Of these 142 neurons, 24 were unresponsive to visual stimuli. [Reproduced with permission from HOFFMANN and SHERMAN (1975), Fig. 3]

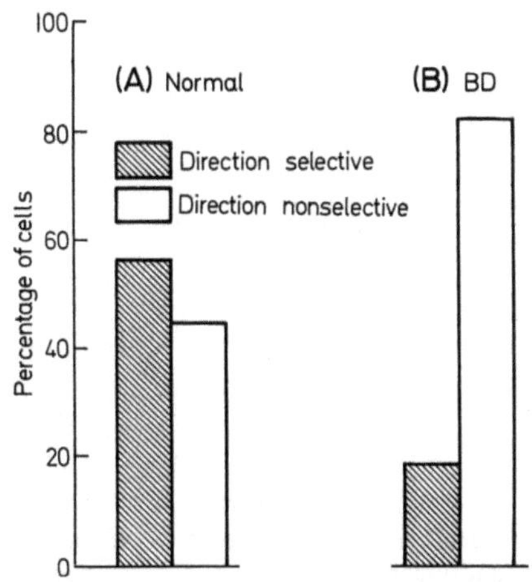

Fig. 13b. Proportion of direction-selective neurons from superior colliculus of normal (left) and binocularly deprived (right) cats. Black bars represent direction-selective cells, open bars represent cells responding almost equally well for all directions of stimulus movement. Vertical scale shows percentage of neurons in each group. Left: 384 collicular cells from normal cats. Right: 147 collicular cells from binocularly deprived cats. [Reproduced with permission from HOFFMANN and SHERMAN (1975), Fig. 4]

of collicular neurons to visual stimulation decreases during the first few weeks of life (STEIN et al., 1973b); this change in latency may reflect the development of a rapidly conducting Y-cell afferent input to the colliculus. In visually deprived adult cats only the slow-conducting W-cell afferent projection to the colliculus can be demonstrated (HOFFMANN and SHERMAN, 1974, 1975).

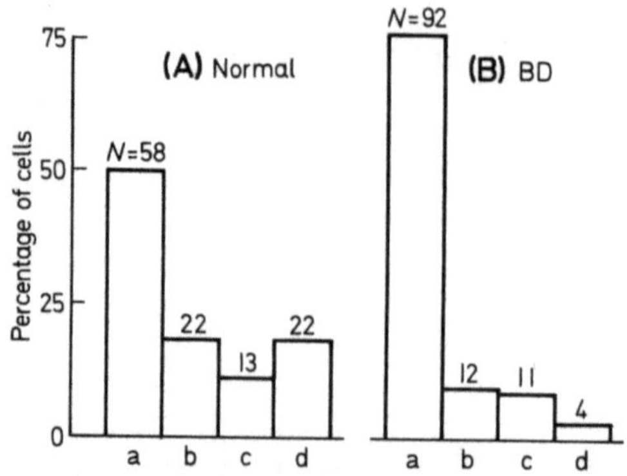

Fig. 13c. Selectivity for speed of stimulus movement measured in superior collicular neurons of normal (left) and binocularly deprived (right) cats. For each, horizontal axis indicates preferred stimulus speed groupings as follows: a, neurons preferring speeds of less than 5°/s (in binocularly deprived cats, these cells mostly responded only to stimulus speeds less than 1°/s); b, neurons preferring speeds of 5–50°/s; c, neurons responding to speeds up to 100°/s; and d, neurons responding to speeds well over 100°/s. Vertical scale gives percentage of a, b, c, or d neurons in each group. Left: 115 collicular neurons from normal cats. Right: 119 collicular neurons from binocularly deprived cats. [Reproduced with permission from HOFFMANN and SHERMAN (1975), Fig. 5]

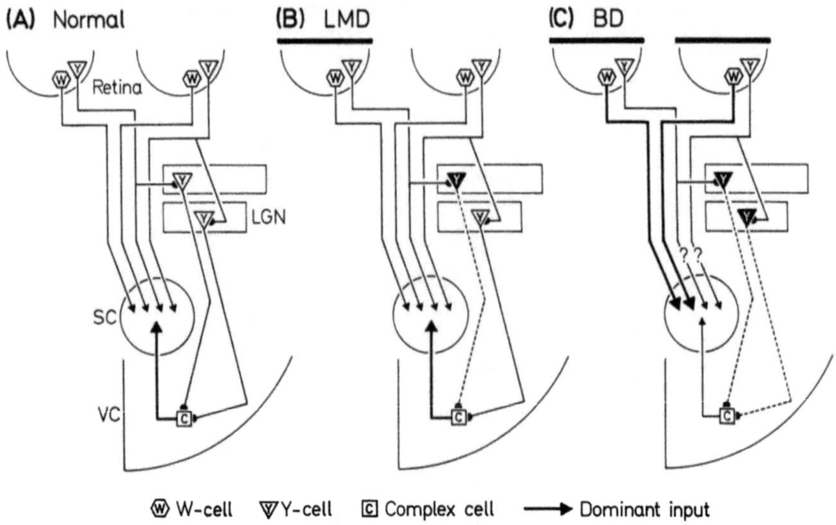

Fig. 14A–C. Summary schematic diagrams showing afferents to superior colliculus (SC) in normal (A), left-monocularly deprived (B), and binocularly deprived (C) cat. Diagrams are hypothetical explanations of much of data from visually deprived animals. W-cells, Y-cells, and complex cells shown as indicated. Unrecordable Y-cells in lateral geniculate nucleus (LGN) are shown as black. X-cells and cortical simple cells are omitted, since they play no major or obvious role in collicular afferentation. (A) pathways to SC in normal cat. Each eye provides W-direct and Y-direct retinotectal input plus Y-indirect corticotectal input. Last involves Y-cells from retina and LGN plus complex cells from visual cortex (VC). Y-indirect input seems to dominate many neuronal properties in SC. (B) pathways to SC in left-monocularly deprived cat. All three inputs (W- and Y-direct plus Y-indirect) are evident from nondeprived (right) eye, and retinotectal inputs (W- and Y-direct) are normal from deprived (left) eye. However, deprived eye provides no Y-indirect input, because Y-cells are no longer available in LGN. Although not illustrated, this deficit is limited to binocular segment of LGN. As in normal cat, Y-indirect input (only from nondeprived eye in binocular segment) is dominant among afferents to SC. (C) pathways to SC in binocularly deprived cat. From each eye, W-direct input is normal, Y-direct input is probably reduced, and Y-indirect input is completely missing. Y-indirect input loss is due to loss of recordable Y-cells throughout binocular and monocular segments of LGN. However, unlike normal and monocularly deprived cats, W-direct pathway dominates among afferents to SC. [Reproduced with permission from HOFFMANN and SHERMAN (1975), Fig. 6]

In summary, many of the effects of visual deprivation upon the development of the superior colliculus appear to reflect the dependence of Y-cells upon early visual experience for normal function (Fig. 14). The response properties of some collicular neurons, however, are determined innately, since the W-direct pathway appears normal in deprived animals. As with the development of Y-cells in the lateral geniculate nucleus, the precise effects of visual experience upon the development of Y-cell afferents to the superior colliculus are not known.

V. The Visual Cortex

1. Anatomical Studies

a) Binocular Deprivation

Morphological studies on the cortical effects of visual deprivation have focused upon the demonstration of changes in cell density (GYLLENSTEIN et al., 1965; FIFKOVA and HASSLER, 1969; BOAS et al., 1969; CRAGG, 1975a, b), synaptic density (CRAGG, 1967, 1969, 1972, 1975a, b; FIFKOVA, 1968; VRENSEN and DE GROOT, 1974), density of dendritic spines (GLOBUS and SCHEIBEL, 1967; VALVERDE, 1967, 1971; RUIZ-MARCOS and VALVERDE, 1969; PARNAVELAS et al., 1973), synaptic vesicle density (VRENSEN and DE GROOT, 1974; GAREY and PETTIGREW, 1974), and dendritic branching patterns (GLOBUS and SCHEIBEL, 1967; COLEMAN and RIESEN, 1968; WEST and GREENOUGH, 1972; GREENOUGH and VOLKMAR, 1973). The results of these studies are often contradictory, suggesting that a multitude of factors influence cortical development. The situation is also complicated by the fact that few studies have attempted to control the location within the visual cortex from which material to be evaluated is taken. Since a nonuniform cortical distribution of different physiological cell types has been reported (WILSON and SHERMAN, 1976; LEVENTHAL and HIRSCH, 1977), such controls are imperative. A number of common trends do, however, emerge from these studies and these are discussed below.

Although synaptogenesis begins well before birth, most of the synaptic connections characteristic of the adult visual cortex develop postnatally. In mouse, the number of dendritic spines present on the apical shafts of layer-V pyramidal cells is reported to increase slowly during the first week after birth. This is followed by a dramatic increase in the number of dendritic spines, which coincides with the time of normal eye opening (day 13–14). This first postnatal burst in the rate at which dendritic spines are formed is apparently related to eye opening, since it is not observed in animals reared in darkness. Then, from approximately day 20, the rate of increase adopts a steady value, which is maintained until the animal is mature (RUIZ-MARCOS and VALVERDE, 1969; VALVERDE and RUIZ-MARCOS, 1969; VALVERDE, 1971) (Fig. 15).

The development of spines in animals kept in total darkness parallels the development of spines in normal control animals only until day 15. After this time spine density in dark-reared animals remains constant for some 10 days and then begins to increase gradually but never reaches the levels characteristic of normal animals (VALVERDE, 1971). Returning dark-reared mice to a lighted environment,

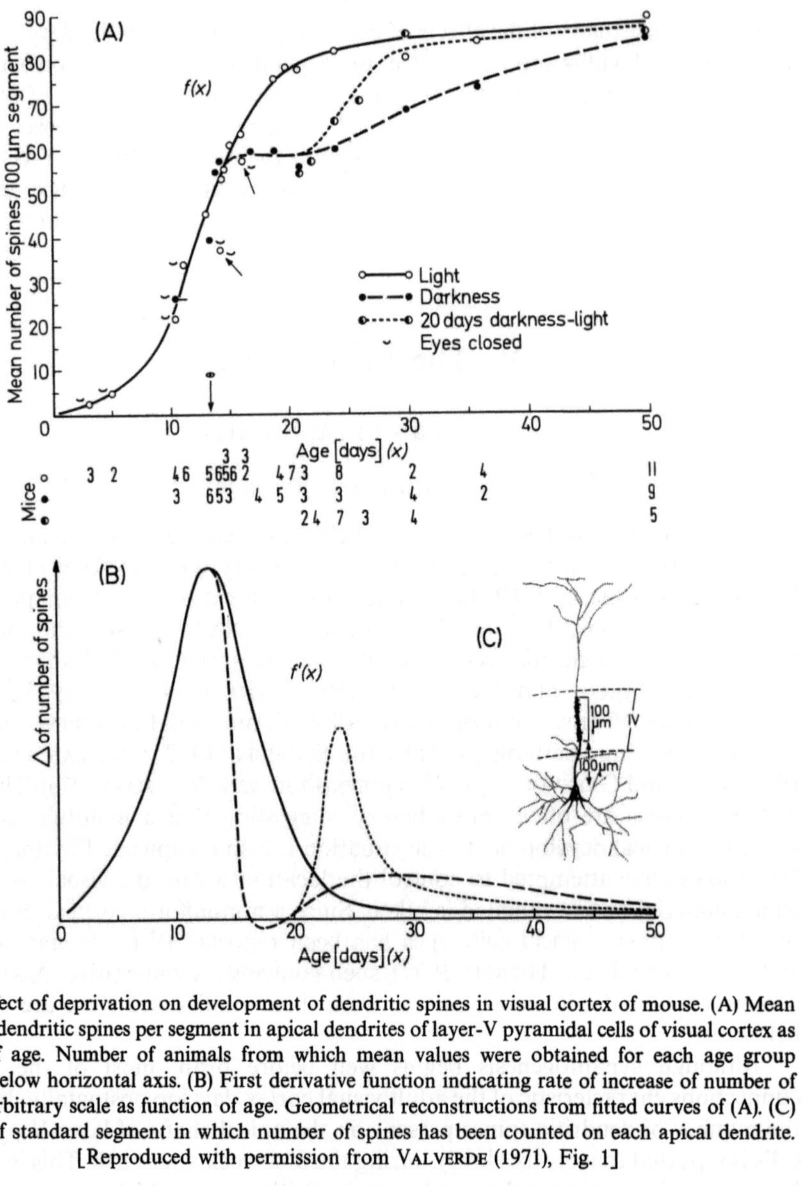

Fig. 15. Effect of deprivation on development of dendritic spines in visual cortex of mouse. (A) Mean number of dendritic spines per segment in apical dendrites of layer-V pyramidal cells of visual cortex as function of age. Number of animals from which mean values were obtained for each age group indicated below horizontal axis. (B) First derivative function indicating rate of increase of number of spines in arbitrary scale as function of age. Geometrical reconstructions from fitted curves of (A). (C) Location of standard segment in which number of spines has been counted on each apical dendrite. [Reproduced with permission from VALVERDE (1971), Fig. 1]

however, produces a sharp rise in the rate of formation of dendritic spines. This rise is sufficient to bring the number of spines up to nearly normal values within a week after the animals are moved into the light.

The results of these studies suggest that in the mouse there are at least two groups of dendritic spines. One group does not depend upon sensory stimulation for its development, since it is formed before the animals' eyes open. The second group of dendritic spines seems to require sensory stimulation for development or for maintenance of normal morphology; it is observed only after eye opening or when dark-reared animals are returned to a lighted environment[5]. Consistent with this, it has been reported that albino rats reared in constant but dim illumination (45 ft-Cd.) possess more dendritic spines on almost all parts of the dendritic field of layer-IV and layer-V pyramidal cells than do normal control rats (PARNAVELAS et al., 1973). Furthermore, WEST and GREENOUGH (1972) report that when rats are raised in an "enriched environment", increases can be seen in the length of the postsynaptic opaque layer of type-1 synapses (which have been reported to be characteristic of synapses on dendritic spines (GRAY, 1959)) in layers I and IV of the visual cortex.

The gradual increase in the number of dendritic spines observed after day 20 in animals reared in darkness suggests that visual deprivation retards the normal development of experience-sensitive dendritic spines but does not block it entirely. In support of this idea, BOAS et al. (1969) report finding no differences, using a variety of measures (cortical cell density, overall cortical thickness, thickness of different cortical layers, nuclear volume of cortical cells), between normal control rats and animals kept in darkness for extended periods (200 days) after birth. These findings extend the earlier work of GYLLENSTEIN et al. (1965), which showed that after four months of dark-rearing in mouse there is partial normalization of those deficits in cortical thickness and internuclear material that are produced by shorter periods (1–2 months) of deprivation.

Changes in the dendritic spines on the apical shafts of pyramidal cells have also been described in the visual cortex of rabbits deprived of visual stimulation for 30 days from birth (GLOBUS and SCHEIBEL, 1967) (Fig. 16). In this species, such a period of deprivation failed to affect either the total number of cells present or the relative proportions of the different cell types studied. Similarly, neither the overall length nor the branching patterns of dendrites of stellate cells was affected by this deprivation, although an increased variability in the length of these dendrites was reported. A similar increase in the variability of the dendritic apparatus of layer-IV stellate cells may also occur in the visual cortex of cats reared in darkness for six months from birth (GLOBUS and SCHEIBEL, 1967; COLEMAN and RIESEN, 1968). In addition, this period of deprivation was reported to result in a decreased tendency for the dendrites of stellate cells to branch. Finally, the average length of these dendritic branches was found to be greater in dark-reared animals than in normal control cats. Care should be taken in interpreting these results, however, since these authors report finding similar results upon examining the dendrites of layer-IV pyramidal cells in the cingulate gyrus of these same groups of deprived and control animals (COLEMAN and RIESEN, 1968).

Experience-dependent changes in synaptic density very similar to the changes in spine density observed in mice (VALVERDE, 1971) have been reported in studies of the developing visual cortex of cat (CRAGG, 1975a, b). In this species, synaptic development in the visual cortex begins three weeks before birth, but the number of synapses formed prenatally is quite small. Only 0.8% of the adult number of synapses is found on the average neuron when the kitten is born. This initial prenatal growth period is followed by a latent phase, during which few synapses

---

[5] In the primate visual cortex, the number of dendritic spines on the apical shafts of layer-III pyramidal cells is also reduced in a similar manner by visual deprivation (BOOTHE and LUND, 1976).

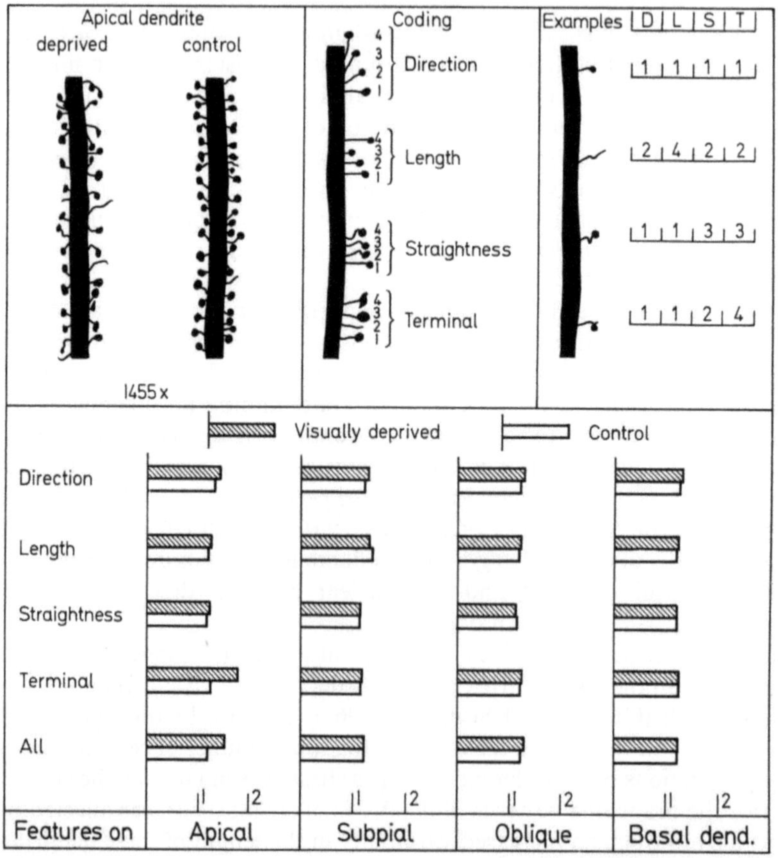

Fig. 16. Deformation of spines under visual deprivation. Upper left panel shows sections from two apical dendrites drawn at 1455X; left: dendrite from a visually deprived rabbit; right: from a control. Central panel illustrates method of coding spines for variation. Direction is coded from 1 to 4, 1 being perpendicular to dendrite, 4 at an acute angle to or lying against dendrite. Length is coded 1 to 4, 1 being mean length of spines, 2 and 3 progressively shorter, and 4 very much longer. Straightness is coded from 1 to 4 for increasing tortuosity of stem of spine. Terminal enlargement of spine is coded as follows: 1 represents terminal with diameter approx 2–3 times width of stem; 2 represents no visible enlargement of tip; 3 represents enlargement in excess of 3 times stem; 4 represents an irregular outline of terminal enlargement of spine. Far right upper panel shows examples. Spine at top, which is typical, is coded throughout as 1. Second is not quite normal to dendrite. Direction equals 2; its length is longer than average, equals 4; not quite straight, therefore coded as 2; has no terminal enlargement, therefore coded as 2. Third and fourth spines are coded similarly. Fourth spine is typical of spines along apical dendrite of experimental animal, showing deformed terminal enlargement. Lower half of diagram illustrates amount of spine pathology on various segments of dendritic tree of pyramids: apical, subpial, oblique, and basal. In direction, length, and straightness there is little or no difference between control and visually deprived cortex. Terminals along apical dendrites show considerable deformity, many of codes being 2 and 4. Lowest line of bar graphs is mean of all features coded, indicates that form of dendritic spines on apical dendrite of pyramidal cells in visual cortex varies more in deprived than in normal control animals. [Reproduced with permission from GLOBUS and SCHEIBEL (1967), Fig. 4]

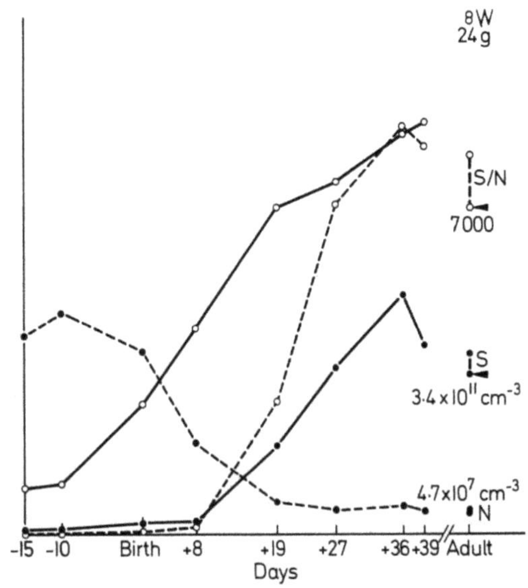

Fig. 17. Growth of brain weight (W), density of synapses (S), density of cell nuclei (N) and number of synapses per neuron (S/N) in 8 kittens and 2 adult cats. [Reproduced with permission from CRAGG (1972), Fig. 3]

are formed. Subsequently, the visual cortex undergoes a second major period of synaptogenesis, which occurs 8–37 days after birth. Significantly, this period of synaptic development coincides with the normal time of eye opening, as does the burst in the development of dendritic spines in the visual cortex of the mouse (Fig. 17). Unfortunately, there is no indication whether this growth phase is dependent upon sensory stimulation, as has been reported to be the case for mouse (VALVERDE, 1971).

The number of synapses per neuron in the cat's visual cortex is also reported to climb dramatically after the 8th day of life, until both synaptic density and the number of synapses per neuron reaches a maximum value at about seven weeks of age (CRAGG, 1975a). Subsequently, both decline somewhat from their peak values, especially in layers IV and V, until the somewhat lower values characteristic of adult cats are reached (Fig. 17). This decline in synaptic density may result from an expansion in cortical volume resulting from the development of glial cells as well as an actual regressive change in cortical synapses[6].

Binocular lid suture (or crushing of the optic nerves) from the time of normal eye opening to 45 days of age is reported to reduce the density of synapses in the cat's visual cortex, particularly in layers IV and V (CRAGG, 1975b). Similarly, the average number of synapses per standard counting screen, as well as the average number of synapses per neuron, was reduced in the deprived animals. In contrast,

[6] The development of synapses in the lateral geniculate nucleus is similar to the cortical development of synapses described, although development of the LGN precedes cortical development by some two days and the number of synapses in the LGN does not pass through a peak value at seven weeks of age (GAREY et al., 1973; CRAGG, 1975a). Furthermore, axon terminals in the LGN of deprived rats are larger than in control animals, but their density is lower (CRAGG, 1969) (Fig. 17).

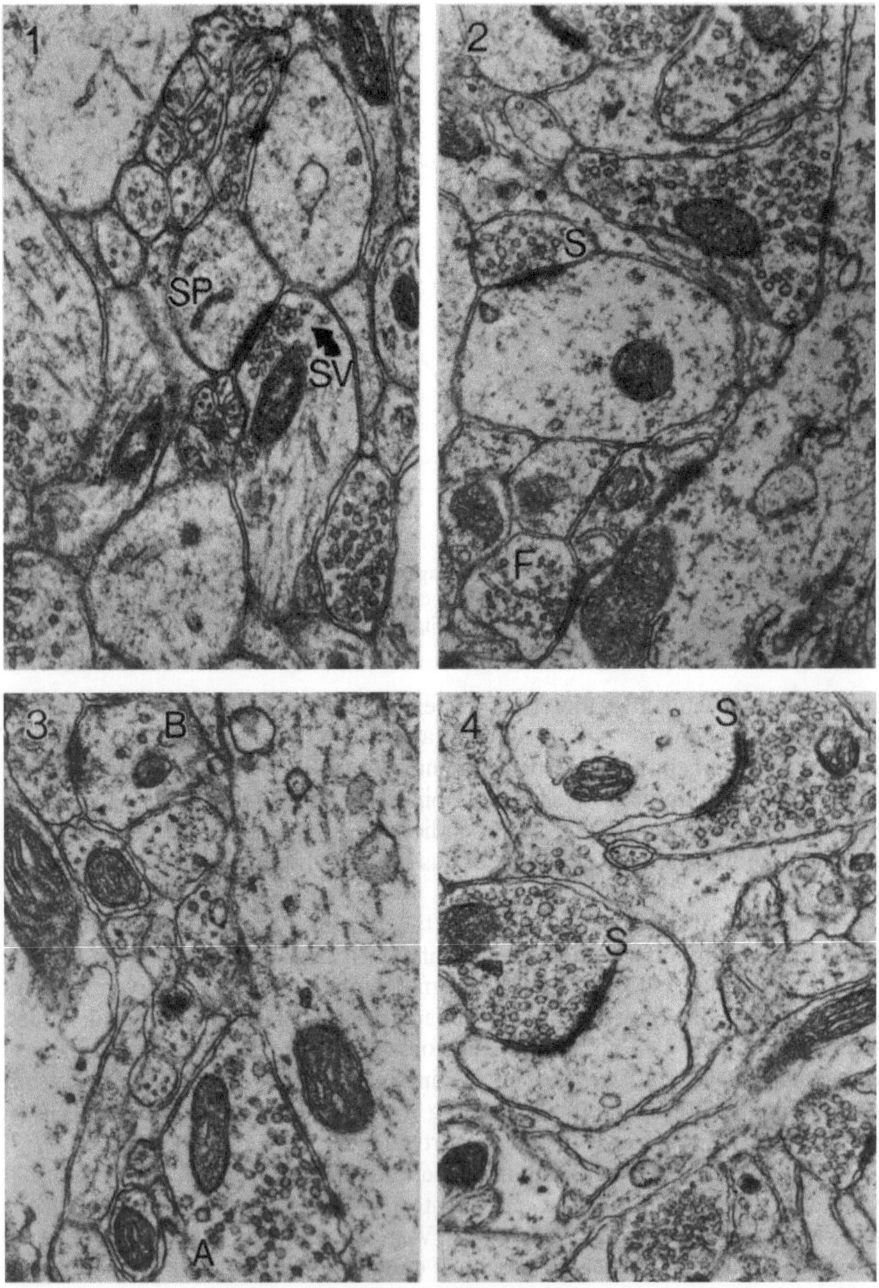

Fig. 18. Electron micrographs of area 17. ×28000. 1: dark-reared kitten (C6) with low density of synaptic vesicles (SV) in typical terminal with asymmetrical post-synaptic membrane onto dendritic spine (SP). 2: normal kitten (K43) with relatively high density of vesicles in terminals containing both spherical (S) and flattened (F) varieties. 3: unstimulated visual cortex in experiment FBP7. One terminal (A) ends at symmetrical thickening onto dendrite and contains moderate density of vesicles; another (B) has an asymmetrical thickening onto dendritic spine and contains only 3 clear vesicles. 4: stimulated cortex of experiment FBP7, showing uniformly high density of vesicles, both spherical (S) and flattened (F). [Reproduced with permission from GAREY and PETTIGREW (1974), Fig. 3]

the length of synaptic appositions appeared to be unaffected by the deprivation. In addition, in deprived animals the density of cells increased, while the diameter of neurons decreased throughout the different cortical layers. Finally, the thickness of the cortex was reduced in deprived cats. Thus, by 45 days of age it was reported that the connectivity in the visual cortex of kittens deprived of vision binocularly was some 30% less than the connectivity observed in normal animals of the same age (CRAGG, 1975b).

A question arising from this work is whether specific populations of cortical cells are affected by visual deprivation or, rather, whether deprivation affects the cortex in general. While the resolution of this question is for future work to determine, there is evidence suggesting that only certain interconnections may be dependent upon sensory stimulation for normal development. GAREY and PETTIGREW (1974) studied cats that were reared in darkness from birth to one month of age and subsequently received visual stimulation in only one hemifield. These cats exhibited a high overall concentration of synaptic vesicles in the axon terminals in the region of the cortex subserving the stimulated hemifield (the contralateral hemisphere). The terminals that contained the most vesicles were relatively small and were suggested to involve axodendritic or axospinous endings with asymmetrical thickenings (Fig. 18).

Similar findings have been reported by VRENSEN and DE GROOT (1974), who studied the effects of dark-rearing upon the synaptology of rabbit visual cortex. In this species, visual deprivation also resulted in a significant reduction in the number of synaptic vesicles in the visual cortex. This reduction was not reversed by returning the animals to a lighted environment for up to one year. Furthermore, in keeping with the observations of GLOBUS and SCHEIBEL (1967), VRENSEN and DE GROOT (1974) reported that rearing rabbits in darkness from birth to seven months of age did not affect the surface area, the length, or the number of synaptic contacts in either the visual or the motor cortex of this species.

b) Monocular Deprivation

Changes in neuronal density have been observed in the visual cortex of squirrels following monocular lid suture. In particular, neuronal density was observed to be significantly higher in the hemisphere contralateral to the deprived eye than in the hemisphere ipsilateral to the deprived eye. Furthermore, cell density was reported to be affected equally in the binocular and in the monocular segments of the visual cortex of this species (GUILLERY and KAAS, 1974b). In addition, the supragranular layer (bottom of layer I to top of layer IV) was affected at least as much as the granular layer (layer IV) (Figs. 19 and 20).

Cell densities were also examined in the cingulate region of the cortex, where no differences between the two hemispheres were demonstrated. Thus, in this species, in contrast to cat (COLEMAN and RIESEN, 1968), these effects of visual deprivation are specific to the visual cortex and are not simply a reflection of a generalized trophic effect upon the nervous system.

These findings demonstrate that changes in the cortex must, to some extent, be independent of changes in the geniculate, since the effects of monocular deprivation were observed to be uniform throughout the cortex, while the effects of

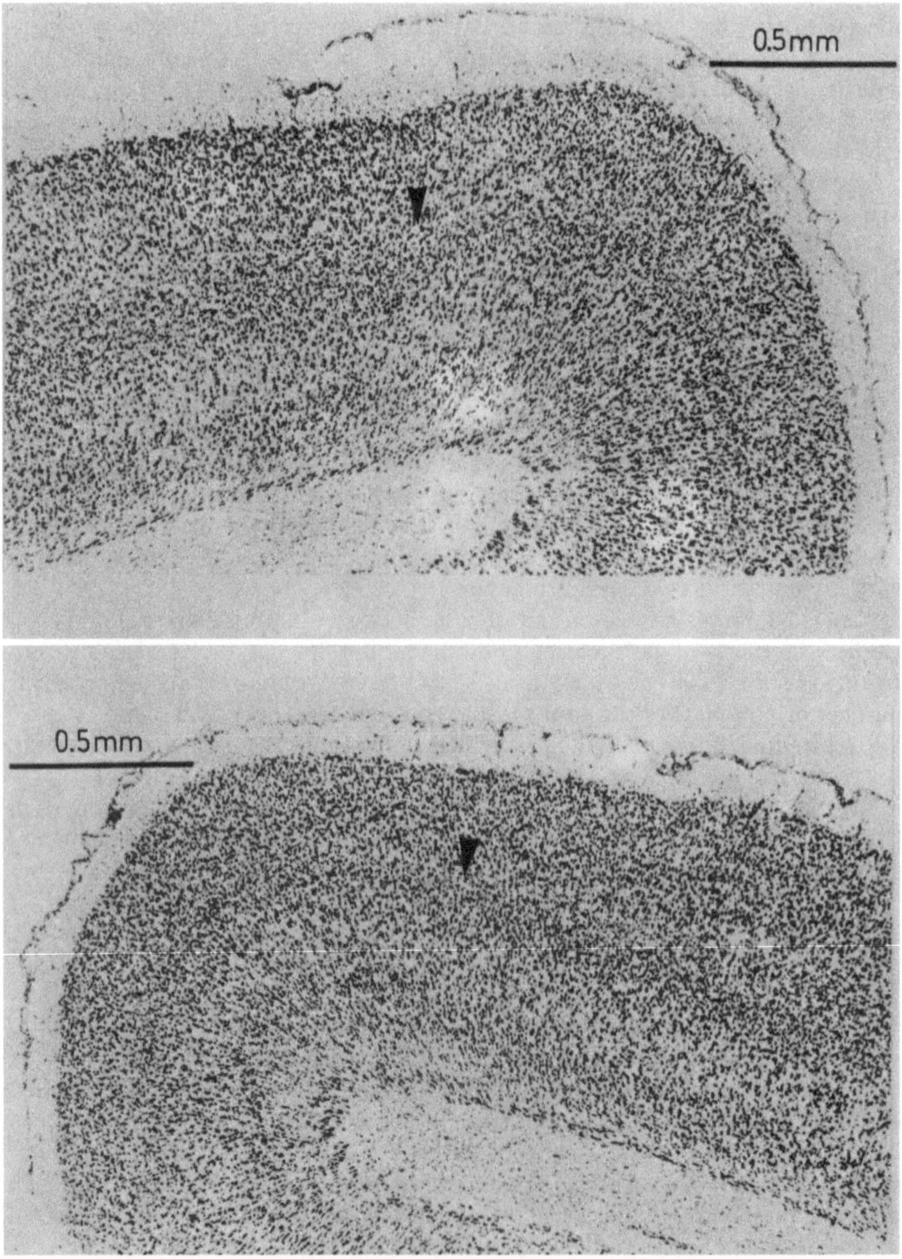

Fig. 19 (Top). Coronal section of visual cortex (area 17) ipsilateral to lid suture in squirrel 70–215. Small arrow shows part of layer IV that marks border between monocular and binocular segments of cortex. Note that layer IV is thinner in monocular than in binocular segment. Nissl method. (Bottom) Coronal section of visual cortex (area 17) contralateral to lid suture in squirrel 70–215. [Reproduced with permission from Guillery and Kaas (1974), Figs. 1 and 2]

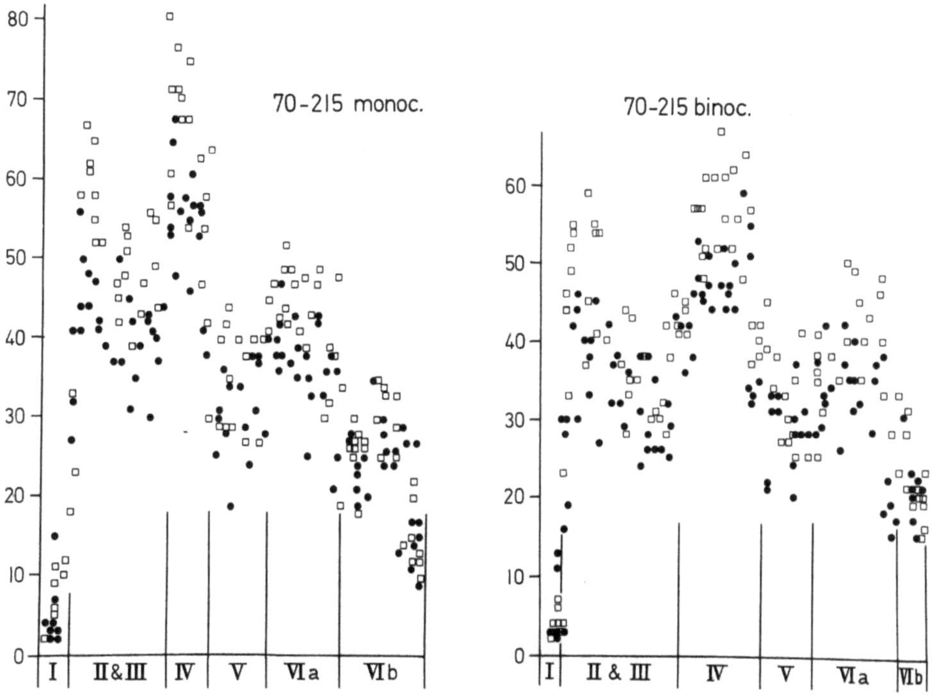

Fig. 20. Cell density in monocular (left) and binocular (right) region of visual cortex in monocularly deprived squirrel (70–215). Cell density measures (no. of cells per 10^4 μm^2) are plotted along ordinate, relative laminar depth along abscissa. Squares show measures obtained contralateral to suture and discs measures ipsilateral to suture. [Reproduced with permission from GUILLERY and KAAS (1974), Figs. 4 and 5]

such deprivation were most pronounced in the binocular segment of the LGN. Thus, while some of the changes observed in monocularly deprived squirrels may well depend upon competitive interaction between the two eyes, such a mechanism cannot account entirely for the cortical effects observed. There must be an additional effect of visual deprivation which explains why both the binocular and the monocular segments of the visual cortex are effected by monocular deprivation (GUILLERY and KAAS, 1973).

To complicate matters further, since many cells in the binocular segment of the mammalian cortex receive an input from both eyes at birth, closure of one eye results in only partial deprivation of these neurons. Specifically, even if cells lose their input from the deprived eye following monocular occlusion, they still receive an input from the normal eye. Despite this asymmetry, however, it is interesting that the observed cortical changes were nearly as large in the binocular segment of the cortex as in the monocular one. This suggests that the contralateral eye dominates cortical activity in the binocular segment of the cortex (GUILLERY and KAAS, 1974b). It is also possible that lateral effects (from the monocular segment) induce a deprivation effect in the binocular segment. Thus, a partial representation of a hemifield may develop less well than a complete one. Some support for this has been found in studies of Siamese cats (GUILLERY and KAAS, 1973).

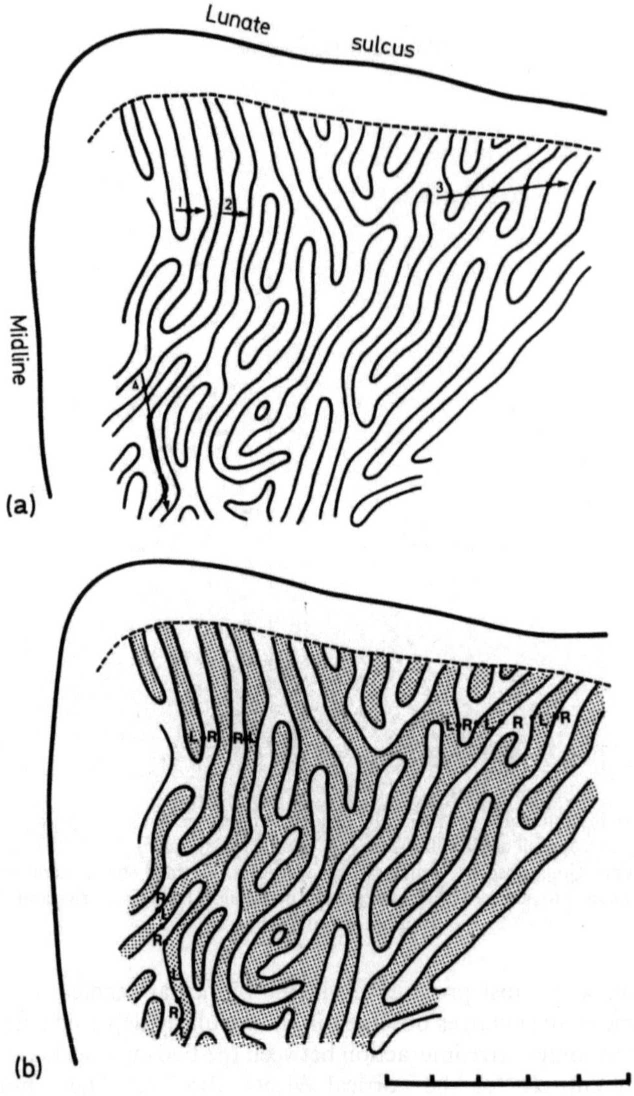

Fig. 21 (a). Reconstruction of banding pattern (solid lines = pale bands), electrode tracks (arrows), and lesions marking borders of ocular dominance columns in visual cortex of Macaque monkey (black dots). All 12 lesions, made in four penetrations, coincide with pale bands. Note also that for much of fourth penetration electrode was running along a dark band close to its edge. (b) Same reconstruction, indicating eye preference of columns crossed by electrode. The left-eye (contralateral) ocular dominance columns have been shaded. Dashed line: 17–18 border. Bar = 5 mm. Posterior is down in this figure. [Reproduced with permission from LeVay et al. (1975), Fig. 4a and b]

Fig. 22A–C. Ocular dominance columns in layer IVc in normal Macaque monkey and in two monkeys ▷ with monocular lid suture early in life. Each animal had one eye injected with a tritiated proline-fucose mixture (roughly 2 mCi) 2 weeks before perfusion. Plane of section tangential to convex, exposed surface of occipital lobe, passing at deepest point through layer IVc in (B) and through layer V in (C). Layer IVc is therefore seen as an oval in (B), as a ring in (A), and because of some dimpling, as a pair of rings in (C). (A) Normal monkey, right hemisphere. Anterior is up, medial to left. (B) Monkey whose right eye was closed at 2 weeks of age. Injection of left eye was done at 18 months. Left hemisphere (contralateral to closed eye, ipsilateral to injected eye). Surrounding oval IVc label is second thin ring of label; this is in "upper tier", layer IVa (see Hubel and Wiesel, 1970). Anterior is up, medial to right. (C) Monkey whose right eye was closed at 3 weeks of age; injection of right eye was done at 6 months of age. Right hemisphere (ipsilateral to closed, injected eye). Anterior is to left, medial is down. [Reproduced with permission from Hubel et al. (1976), Fig. 7]

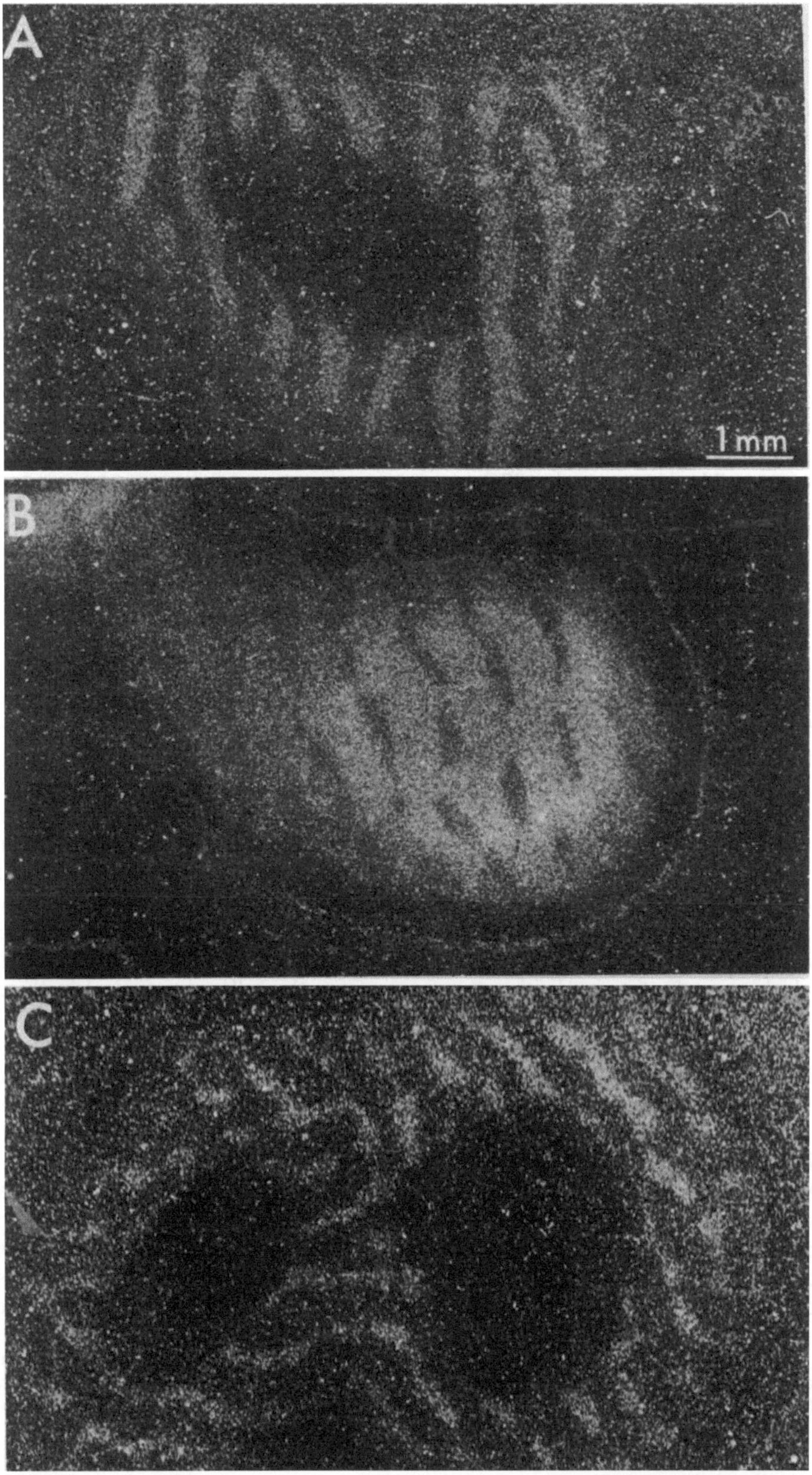
A
1 mm
B
C

In the visual cortex of normal cats and monkeys, neurons which receive their major source of afferent input from the same eye are grouped into 'columns' or 'sheets' which are oriented at right angles to the pial surface and which are bordered on both sides by similar columns which are innervated by the animal's other eye (HUBEL and WIESEL, 1962; HUBEL et al., 1976; WIESEL et al., 1974; SCHATZ et al., 1975; LEVAY et al., 1975) (Fig. 21). Recent autoradiographic studies of the visual cortex (layer IV) of monkeys that had been deprived of vision monocularly during rearing have provided insight into the effects of early asymmetric visual input upon the development of this aspect of cortical organization (HUBEL et al., 1976) (Fig. 22). In particular, the area of visual cortex innervated by the normal eye in monocularly deprived monkeys was observed to be greater than the corresponding area in normal animals[7]. While it is tempting to postulate that these effects are due to a sprouting of connections from the eye which was put at a competitive advantage during development, other explanations are also possible (HUBEL et al., 1976).

2. Physiological Studies

a) Binocular Deprivation

The earliest studies of the response properties of cells in the striate cortex of visually naive kittens suggested that much of normal cortical development is determined innately. Neurons in the cortex of young kittens were reported to display many of the response properties exhibited by cells in mature animals (HUBEL and WIESEL, 1963a). Specifically, cells in visually naive kittens were found to be activated binocularly, to respond selectively to lines of a particular orientation and to be grouped into columns in the cortex according to their ocular dominance and preferred orientation. Furthermore, apparently normal cells of various types (simple, complex, hypercomplex) were reported to be present. Only the sluggishness of many of the cells, the rapid rate at which their response to visual stimulation fatigued, and the rather broad range of stimulus orientations which were effective in eliciting a response from the cells was atypical in these animals.

While some recent studies have confirmed these findings (SHERK and STRYKER, 1976) and have extended them to the monkey (WIESEL and HUBEL, 1974), other authors report that few, if any, orientation-sensitive cells are present in the cortex of young, visually naive kittens (BARLOW and PETTIGREW, 1971; PETTIGREW, 1974b; IMBERT and BUISSERET, 1975; BUISSERET and IMBERT, 1976). In addition, BLAKEMORE and VAN SLUYTERS (1975) report that only a small proportion of a specific class of cortical neurons (simple cells) have well-defined orientation preferences in visually deprived kittens, while members of a second class (complex cells) possess only weak orientation biases or are nonselective for stimulus orientation in these animals (Fig. 23). Since it has been suggested that simple

---

[7] In the visual cortex of mouse and tree shrew, anatomical changes following monocular deprivation have not been observed, although physiological changes do result from this procedure (DRÄGER, 1976; CASAGRANDE et al., 1976).

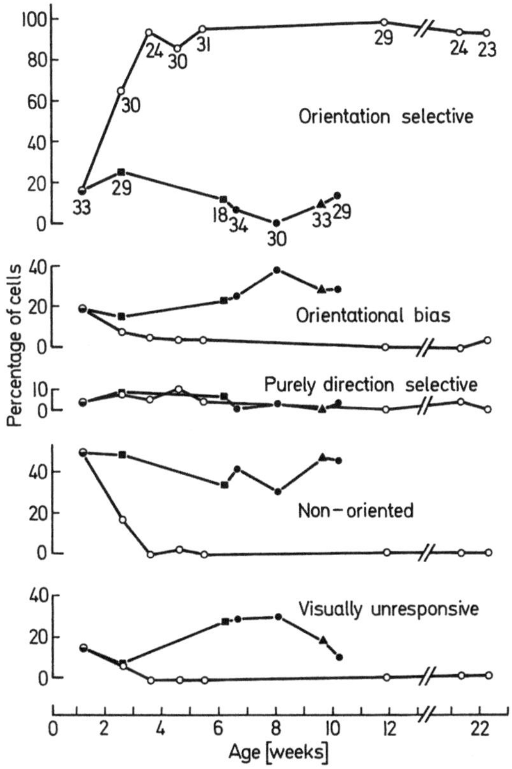

Fig. 23. Comparison of effects of binocular deprivation with those of normal visual experience. Five pairs of curves show percentage of each type of receptive field for 397 cells recorded from seven normal kittens (open circles) and six binocularly deprived kittens (filled symbols) of various ages. In each case, two curves originate from points representing data for 9-day-old kitten recorded at time of natural eye opening. For binocularly deprived kittens: lid suture = filled circles; dark-reared = filled squares; nictitating membrane suture = filled triangle. Numbers beneath data points for top pair of curves indicate total number of cells recorded in each animal. [Reproduced with permission from BLAKEMORE and VAN SLUYTERS (1975), Fig. 6]

cortical cells receive LGN X-cell afferents and complex cortical cells receive LGN Y-cell afferent inputs, these authors suggested that X-cell influenced cortical neurons possess orientation preferences which are determined intrinsically, while Y-cell influenced neurons do not[8].

Additional evidence that the orientation preferences of X-cell influenced cortical neurons are determined intrinsically is provided by a study of cortical neurons in a group of cats reared in darkness from birth to 10–12 months of age (HIRSCH and LEVENTHAL, 1976; LEVENTHAL and HIRSCH, 1977). In these animals, a majority of the orientation-sensitive cells were activated monocularly, had small receptive fields, responded only to slow-moving stimuli, and preferred either horizontal or vertical lines. Most cells displaying no obvious orientation selec-

[8] One study has failed to reveal a selective loss of cortical cells receiving fast afferents (LGN Y-cell afferent inputs) following binocular lid suture (SINGER and TRETTER, 1976). This discrepancy may reflect differences in the cortical effects of dark-rearing and lid suture.

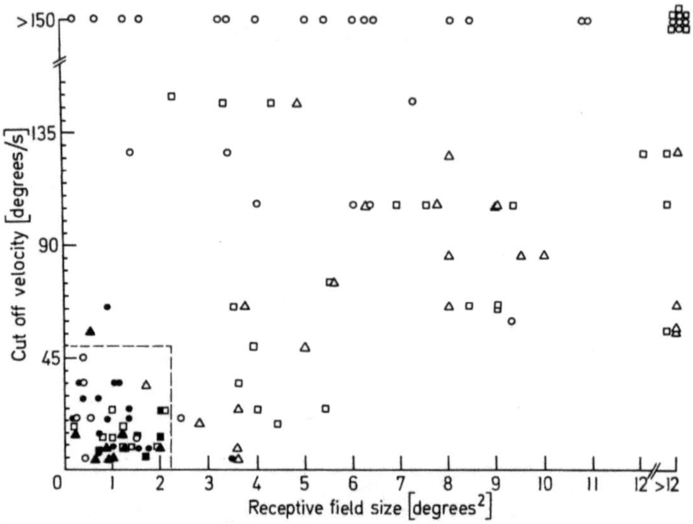

Fig. 24. Receptive field sizes and cutoff velocities of units sensitive to stimulus orientation (solid symbols) and of cells insensitive to stimulus orientation (open symbols) in six dark-reared cats (circles) and one diffuse-reared animal (triangles). Neurons that were studied quantitatively are indicated by squares. Preponderance of solid symbols in lower left of figure indicates that most of orientation-sensitive neurons in pattern-deprived cats have small receptive fields and low cutoff velocities. [Reproduced with permission from LEVENTHAL and HIRSCH (1977), Fig. 11]

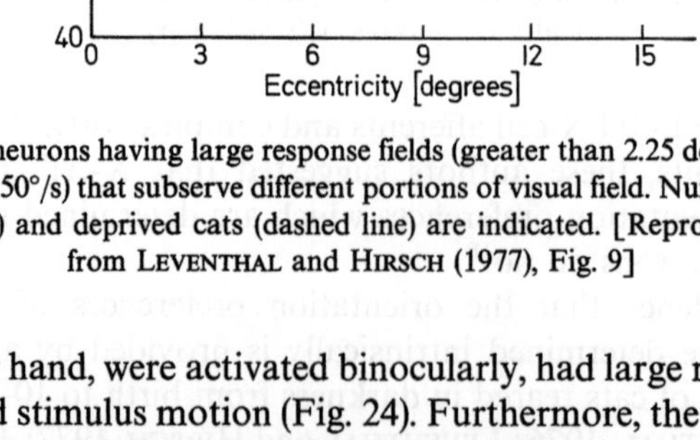

Fig. 25. Proportion of neurons having large response fields (greater than 2.25 degrees^2) and high cutoff velocities (greater than 50°/s) that subserve different portions of visual field. Number of units studied in normal cats (solid line) and deprived cats (dashed line) are indicated. [Reproduced with permission from LEVENTHAL and HIRSCH (1977), Fig. 9]

tivity, on the other hand, were activated binocularly, had large receptive fields, and responded to rapid stimulus motion (Fig. 24). Furthermore, the proportion of cells of this type was reduced following dark-rearing and this is consistent with the loss of LGN Y-cells that results from binocular deprivation (SHERMAN et al., 1972) (Fig. 25). These results suggest that certain cells receiving X-cell afferents have orientation preferences that are determined innately and can persist after visual deprivation, while Y-cell-influenced cortical neurons depend upon early visual experience for the maintenance or the development of normal orientation sensitivity.

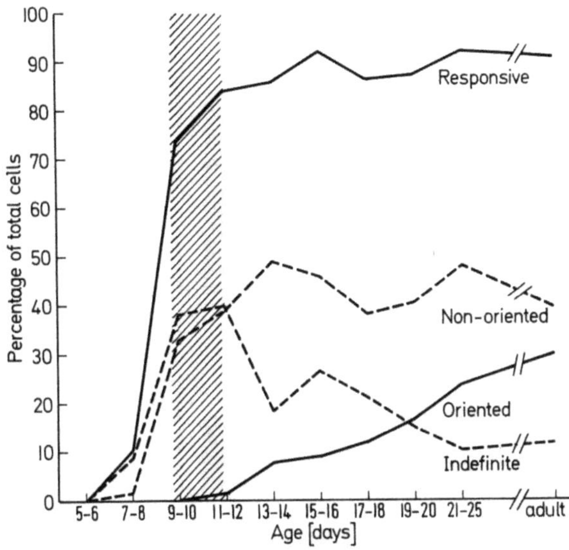

Fig. 26a. Normal development of receptive field organization in rabbit visual cortex. Four curves illustrate changes as function of age in the properties of samples of cortical neurons recorded in rabbits undergoing normal development. Abscissa represents age in days after birth. Ordinate represents percentage of total number of neurons recorded at any given age. Shading indicates age at which rabbits open eyes. Upper curve shows percentage of neurons encountered that were responsive to visual stimulation of any kind. Lower three curves indicate percentage of neurons encountered that were categorized as nonoriented, oriented, or indefinite. Direction-selective neurons are not shown. [Reproduced with permission from GROBSTEIN and CHOW, *Science*, Vol. 190, pp. 352—358, Fig. 1, 24 October 1975; Copyright 1975 by the American Association for the Advancement of Science]

The suggestion that Y-cells are particularly sensitive to visual deprivation receives further support from the observation that neurons in the striate cortex of visually deprived cats lack inhibitory sidebands (WATKINS and SHERMAN, 1975; SINGER and TRETTER, 1976). To be more specific, the excitatory regions of the receptive fields of simple cells may be mediated by X-cell afferents, while the inhibitory regions of the receptive fields of these cells may be mediated by Y-cell afferents (SINGER et al., 1975). A selective loss or dysfunction of Y-cells could thus result in the deprivation-induced loss of inhibitory sidebands of cortical neurons. Other explanations for the loss of inhibitory sidebands in visually deprived animals have also been proposed (SINGER and TRETTER, 1976).

A number of anatomical studies can also be interpreted as supporting the suggestion that visual deprivation affects neurons receiving Y-cell afferent input. In particular, visual deprivation has been shown to affect the afferent input onto dendritic spines of pyramidal cells in the visual cortex (GLOBUS and SCHEIBEL, 1967; VALVERDE and RUIZ-MARCOS, 1969; BOOTHE and LUND, 1976), while there is less clear-cut evidence for changes in stellate cells (GLOBUS and SCHEIBEL, 1967; COLEMAN and RIESEN, 1968). Since cortical pyramidal cells are reported to be complex (KELLY and VAN ESSEN, 1974) and therefore may receive Y-cell afferents (HOFFMANN and STONE, 1971; STONE and DREHER, 1973), while cortical stellate cells are reported to be simple (KELLY and VAN ESSEN, 1974) and thus may have X-

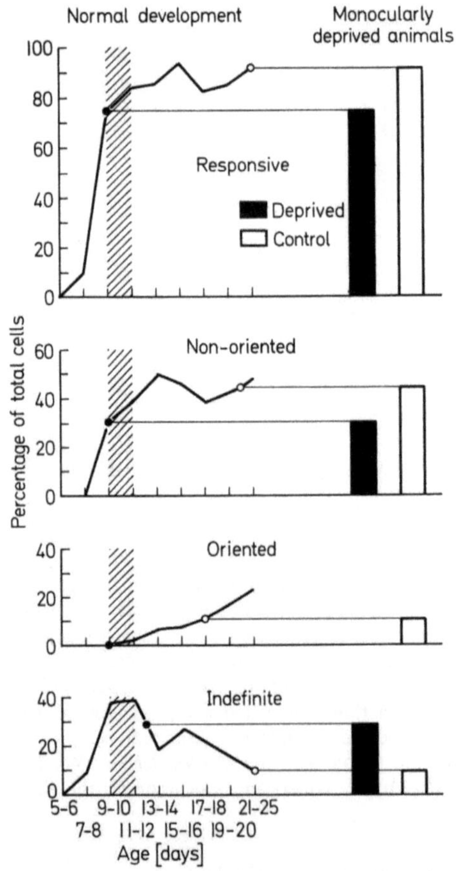

Fig. 26b. Effects of short-term delayed eye-opening on receptive field development in rabbit visual cortex. Histograms on right show percentages of responsive neurons and of neurons falling into nonoriented, oriented, and indefinite categories in deprived (closed bars) and control (open bars) cortices of 20- to 25-day-old rabbits in which one eye was prevented from opening by lid suture. Horizontal lines extending to left intersect curves redrawn from Fig. 26a, indicating ages during normal development when values indicated by histograms would be expected. Clustering of these intersections above 17 days (open circles) for control corticles and below 12 days for deprived cortices (closed circles) indicates retardation of development. [Reproduced with permission from GROBSTEIN and CHOW, *Science*, Vol. 190, pp. 352—358, Fig. 2, 24 October 1975; Copyright 1975 by the American Association for the Advancement of Science]

cell afferents (HOFFMANN and STONE, 1971; STONE and DREHER, 1973), these results suggest that visual deprivation affects the morphology of cortical cells receiving Y-cell afferents. Whether visual deprivation also affects the morphology of cells receiving X-cell afferents remains an open question.

The effects of early visual experience upon the development of the response properties of cells within the rabbit's visual cortex have also been investigated (GROBSTEIN and CHOW, 1975). These studies indicate that cortical cells in rabbits deprived of vision from birth for brief periods do not exhibit the selectivity of response characteristic of neurons in animals of the same age which are experienced visually. Many cells studied in inexperienced animals were unresponsive to visual stimulation or did not respond selectively to stimulus parameters such as orientation or direction (Fig. 26).

In contrast, the response properties of cortical cells in rabbits deprived of vision from birth for extended periods were reported to be much like those of cells found in the normal adult (GROBSTEIN et al., 1975). This suggests that in this species visual experience facilitates the development of normal cortical physiology, but that most aspects of cortical function will develop eventually in animals deprived of vision. These results differ from the results of similar studies of cortical development in cat, since the visual cortex of cats deprived for extended periods contains many unresponsive or nonselective neurons (WIESEL and HUBEL, 1965a; HIRSCH and LEVENTHAL, 1976; SINGER and TRETTER, 1976; KRATZ and SPEAR, 1976; LEVENTHAL and HIRSCH, 1977).

In summary, in rabbit, rat and mouse there is considerable evidence that much of cortical development is completed successfully in the absence of postnatal visual stimulation, albeit at a reduced rate. Furthermore, most studies dealing with cortical development in cat and monkey indicate that many cortical neurons do have orientation preferences that are determined innately. The various reports on the proportion of cortical cells that lack orientation specificity in young kittens, however, remain contradictory[9]. Finally, a number of studies in this area are consistent with the hypothesis that cortical cells receiving X-cell afferents from the lateral geniculate nucleus do not require sensory stimulation either for the development or for the maintenance of orientation-selectivity, while neurons receiving Y-cell afferent input require such stimulation for normal function.

The effects of binocular deprivation upon the development of the ocular dominance distribution of cortical neurons have been explored repeatedly. Studies

[9] We should call attention to several stumbling blocks inherent in any attempt to study the physiology of the visual cortex of young animals. For one, changes may occur during the course of the experiment itself; cells in the visual cortex may be susceptible to the uncontrolled stimulation they receive during these experiments (PETTIGREW et al., 1973a; PETTIGREW and GAREY, 1974; IMBERT and BUISSERET, 1975). Thus, factors such as the length of the recording session and the types of visual stimulation used may, at least partially, account for the different proportions of nonselective cells observed in these animals. Furthermore, the number of cells displaying responses reported characteristic of X-cell-influenced cortical neurons in the cat is highest in areas of the visual cortex subserving central vision and decreases rapidly in eccentric regions (WILSON and SHERMAN, 1976; LEVENTHAL and HIRSCH, 1977). This suggests that many orientation-sensitive cells may be found in the cortex of young kittens near the projection of the area centralis, while few such cells will be encountered in more peripheral regions. Thus, differences among the cortical regions from which experimenters have sampled neurons, not to mention differences in the cortical layers from which they were recorded, could account for some of the differences in the proportions of orientation-sensitive cells that have been reported in studies of the kitten visual cortex. In fact, such differences are to be expected, since the poor optics exhibited by the eyes of the young kitten make the task of locating the retinal landmarks used in assessing which region of the visual field is being studied very difficult (HUBEL and WIESEL, 1963a; PETTIGREW, 1974b). Because of these problems, studies utilizing conventional acute recording procedures have not been able to demonstrate convincingly whether the environment actually exerts a constructive effect upon developing cortical neurons. It may be that the resolution of this question will await the development of procedures for monitoring the activity of significant numbers of cortical neurons for prolonged periods during development. In fact, advances have been made to adapt an array of chronic recording microelectrodes (SALCMAN and BAK, 1973, 1974; SALCMAN et al., 1976) for use in young kittens (SALCMAN and HIRSCH, unpublished observations).

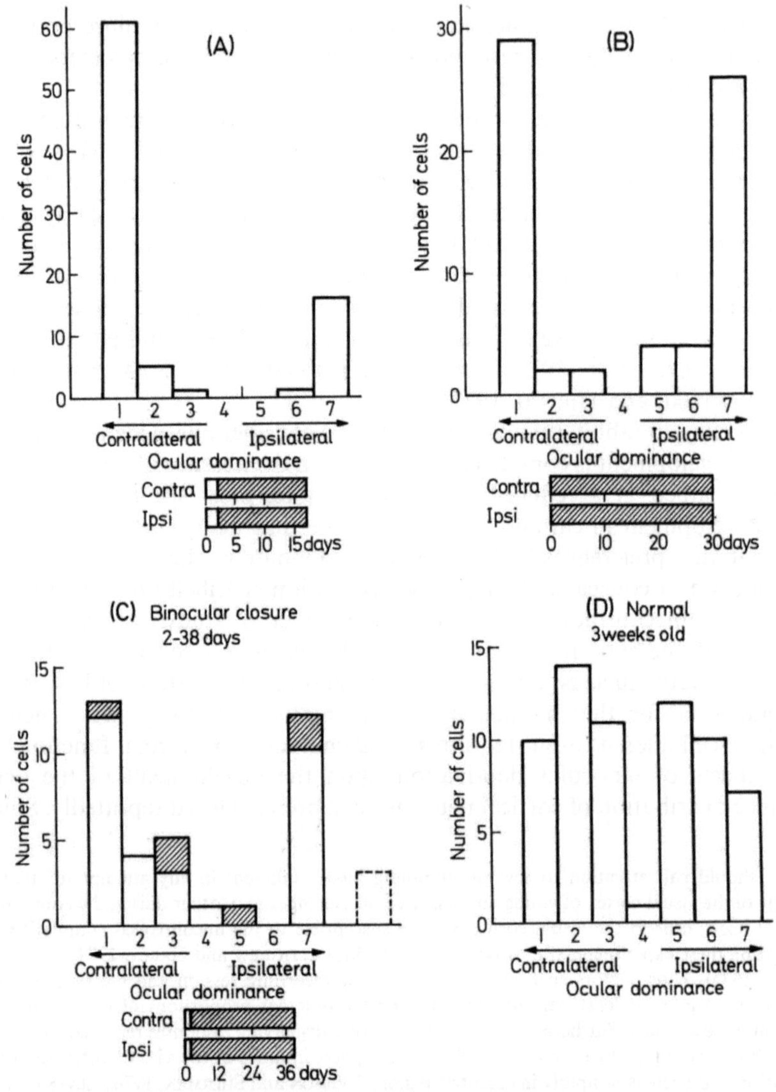

Fig. 27A–F. Ocular-dominance histograms for different visually deprived monkeys. (A) Monkey no. 1, binocular closure 2–17 days. (B) Monkey no. 2, binocular closure 0–30 days (Caesarian section). (C) Monkey no. 3, binocular closure 2–38 days. (D) Monkey no. 4, normal 21-day-old monkey. (E) Ocular-dominance histogram for 1116 cells in area 17 in 28 normal adult (and juvenile) monkeys. (F) Monkey no. 5, binocular closure 21–49 days. Definition of ocular-dominance groups: cells of group 1 driven only by contralateral eye; for cells of group 2, marked dominance of contralateral eye; for group 3, slight dominance. For cells in group 4, no obvious difference between two eyes. In group 5, ipsilateral eye dominated slightly; in group 6, markedly, and in group 7 cells driven only by ipsilateral eye. Shaded areas at bottom of each histogram indicate period during which one or other eye was closed. Shaded areas in histograms themselves represent cells that gave abnormal responses, e.g., lack of orientation specificity or unusual sluggishness. Dotted bins to right of histograms (C) and (F) represent cells that failed to respond to either eye. [Reproduced with permission from Wiesel and Hubel (1974), Fig. 3] ▷

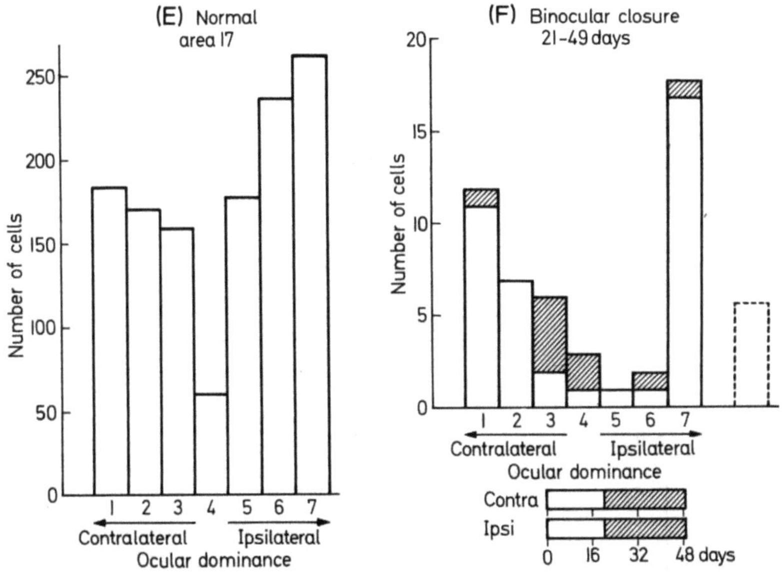

of the cortical physiology of cats deprived of vision in both eyes throughout the first three months of life indicated that many cortical cells in these animals could be activated by both eyes (WIESEL and HUBEL, 1965a). Recently, however, it has been shown that in both monkey and cat, binocular deprivation reduces the proportion of cortical cells that can be activated by both eyes (WIESEL and HUBEL, 1974; LEVENTHAL and HIRSCH, 1977; KRATZ and SPEAR, 1976) (Fig. 27). In deprived cats, the reduction in the proportion of binocularly-activated cells is greatest among that class of neurons that has small response fields and responds preferentially to slowly-moving stimuli (LEVENTHAL and HIRSCH, 1976, 1977). In normal adult cats only 12 % of the cells of this type are activated monocularly, while in the deprived animals 70 % are monocular, regardless of their location in the cortex (LEVENTHAL and HIRSCH, 1976, 1977) (Fig. 28). It is not clear whether these cells have lost binocularity or simply have failed to develop it, although three lines of evidence suggest that cells of this type are activated monocularly in young kittens: (1) only cells of this type are orientation selective in cats deprived of patterned visual stimulation for long periods of time (LEVENTHAL and HIRSCH, 1977); (2) most cells that are orientation selective in very young, deprived kittens are activated monocularly (BLAKEMORE and VAN SLYTERS, 1975; BUISSERET and IMBERT, 1976; KRATZ and SPEAR, 1976); (3) X-type cells in LGN are not affected by binocular competition, perhaps because they project onto cortical cells which are monocular in the visually-inexperienced kitten.

In contrast, deprivation causes little or no change in the proportion of binocularly-activated cells among the class of cells that has large response fields or responds preferentially to rapidly-moving stimuli. In normal adult cats 80 % of these cells are binocular, whereas 73 % are binocular in deprived animals. Cells of this type

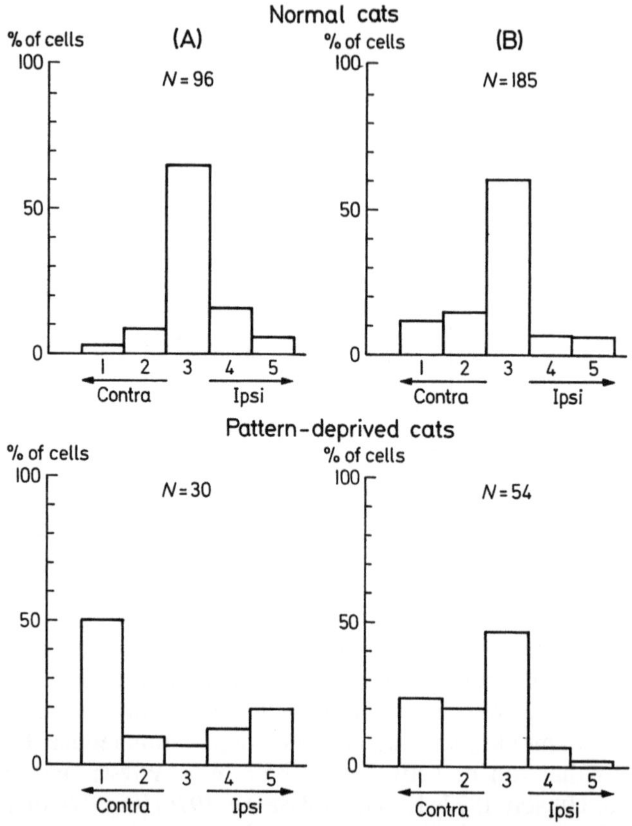

Fig. 28. Ocular dominance distributions of cells having receptive fields smaller than 2.25 degrees^2 and cutoff velocities less than 50°/s (left column-A), as well as ocular dominance distributions of cells with larger receptive fields and/or higher cutoff velocities (right column-B) in normal cats (top row) and in pattern-deprived cats (bottom row). Ocular dominance groups are defined as follows: group 1—cells activated only by contralateral eye; group 2—cells activated by both eyes, contralateral eye being strongly favored; group 3—cells activated by both eyes, responses evoked by the two eyes being comparable; group 4—cells activated by both eyes, ipsilateral eye being strongly favored; group 5—cells activated only by ipsilateral eye. [Reproduced with permission from LEVENTHAL and HIRSCH (1977), Fig. 5]

thus appear not to require sensory stimulation for the development of binocularity. Thus Y-cells in the LGN may be affected by binocular competition only because they project onto cortical cells which are binocular in the visually-inexperienced kitten.

b) Monocular Deprivation

The physiological consequences of monocular deprivation have been investigated intensively. When patterned visual stimulation is excluded from one eye during development, the ability of that eye to activate cortical cells is reduced severely (WIESEL and HUBEL, 1963b, 1965a; GANZ et al., 1968; HUBEL and WIESEL, 1970; HUBEL et al., 1976; SHAW et al., 1974; VAN SLUYTERS and STEWART, 1974;

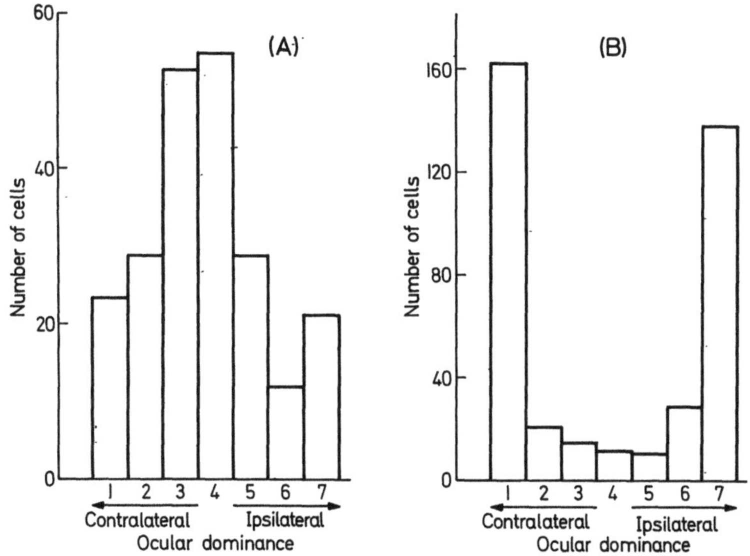

Fig. 29A and B. Effect of surgically induced strabismus on ocular dominance distribution of cells in the visual cortex of the cat. (A) ocular dominance of 223 cells recorded from a series of normal adult cats. (B) ocular dominance of 384 cells recorded from four cats with surgically induced strabismus. [Reproduced with permission from HUBEL and WIESEL (1965), Fig. 5]

DRÄGER, 1976; PETTIGREW and KONISHI, 1976) in cat, monkey, rabbit, rat, mouse, and owl. Similarly, in cat, when stimulation is excluded alternately from one eye and then the other (alternate monocular occlusion), the visual cortex contains normal numbers of orientation-sensitive neurons, but the number of cells that can be activated binocularly is reduced sharply (HUBEL and WIESEL, 1965; BLAKE and HIRSCH, 1975). If the correspondence between the two eyes is disrupted, in either cat or monkey, by severance of an extraocular muscle of one eye, there is also a reduction in the number of cortical cells that can be activated binocularly. Presumably, these effects are the result of a competitive interaction between the inputs from the two eyes (HUBEL and WIESEL, 1965; BAKER et al., 1974) (Fig. 29). This interpretation has been complicated by a recent report which indicates that similar changes occur in the cortex of dark-reared kittens when one eye is immobilized surgically (MAFFEI and BISTI, 1976).

Monocular stimulation effectively alters the ocular dominance distribution of cortical cells only during a limited portion of the cat's early postnatal development. Specifically, the period of susceptibility to monocular deprivation has been shown to begin at approximately four weeks of age and to continue for 8 to 12 weeks (HUBEL and WIESEL, 1970). Closing one eye prior to this critical period has no apparent effect upon this aspect of cortical organization. Similarly, evidence suggests that monocular lid-suture does not affect the binocularity of cortical cells if the animal is permitted normal binocular vision during the first three months of life (WIESEL and HUBEL, 1963).

The period of monocular stimulation sufficient to produce a loss of binocularity has been examined in several studies. In kittens a few days of monocular deprivation during the second month of life reduces the proportion of cortical cells

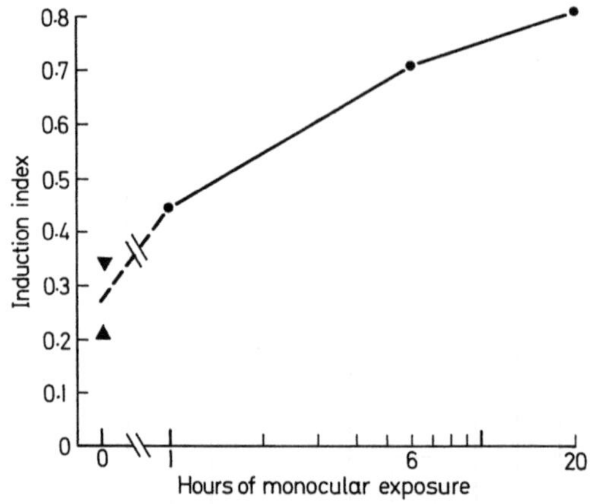

Fig. 30. Effects on ocular-dominance distribution of neurons in visual cortex of cat after brief monocular exposure. Graph illustrates proportion of cells activated predominantly or exclusively by the nondeprived eye as function of total number of cells studied. Induction index = numbers of cells in ocular dominance groups 5–7/total number of visually responsive cells. Three experimental animals (●) were exposed in ipsilateral eye for 1, 6 or 20 hrs on day 29. Control animals (▼: 6 hrs brief binocular exposure; ▲ normal 5.5-week-old) had no monocular deprivation. All these animals recorded in 7th week of life. In this ocular-dominance figure, data points are joined for convenience. Ocular-dominance groups used in this figure are as defined in Fig. 27. [Reproduced with permission from PECK and BLAKEMORE (1975), Fig. 2]

that can be activated by both eyes (HUBEL and WIESEL, 1970). In fact, very brief monocular exposure (6–20 hrs) during the same period is reported to be sufficient to reduce binocular interaction within the visual cortex (PECK and BLAKEMORE, 1975; OLSON and FREEMAN, 1975) (Fig. 30). Finally, evidence has been presented indicating that ocular dominance of cells can be modified repeatedly during the second month of life (BLAKEMORE and VAN SLUYTERS, 1974; MOVSHON, 1976).

Recent studies have suggested hypotheses to explain how monocular deprivation affects the developing visual cortex. In particular, KRATZ et al. (1976) have studied the visual cortex of cats that were deprived of vision monocularly from birth for extended periods and later had their normal eye removed. These authors report finding a significant number of cells in the binocular segment of the cortex in these cats which could be activated by the animal's remaining deprived eye. Many of these cells, however, were reported to display abnormal response properties. The authors suggest that monocular occlusion produces a tonic inhibitory imbalance, which ultimately results in suppression of the response to stimulation of the deprived eye.

Consistent with this hypothesis is the report that binocular response of cortical neurons can be restored in cats deprived of vision in one eye by the intravenous administration of bicuculine. This drug has been reported to block the effects of the inhibitory transmitter gamma-amino-butyric acid (GABA) and presumably interrupts the inhibitory supression of afferents from the deprived eye (SILLITO, 1975a, b; DUFFY et al., 1976). The increased size of the ocular dominance columns that subserve the normal eye in monocularly deprived monkeys, however, suggests that other mechanisms, such as the sprouting of conections, must also be involved in the cortical effects of monocular deprivation (HUBEL et al., 1976).

Insight into the mechanisms by which competitive interactions can affect the developing nervous system has been provided by studies of the cricket, *Acheta domesticus*. This animal has paired abdominal sensory appendages, the cerci, which provide a major source of afferent input to two medial giant interneurons (MGI). Each MGI receives excitatory input from the ipsilateral cercus and inhibitory input from the contralateral cercus. Recent evidence suggests that there is a competitive interaction between the excitatory and the inhibitory input to each MGI during development of the nervous system of this species (MATSUMOTO and MURPHEY, 1977). Specifically, permanent alterations result in the anatomy and the physiology of the medial giant interneurons if one cercus is removed surgically (and then allowed to regenerate) (PALKA and EDWARDS, 1974; MURPHEY et al., 1975) or if one cercus is covered by cream during development to reduce sensory input (MATSUMOTO and MURPHEY, 1977). The resulting diminished response of the deprived medial giant interneuron may reflect an imbalance between its excitatory and inhibitory input. Indeed, immobilization of the sensory hairs on the contralateral cercus results in an increase in the responsiveness of the deprived medial giant interneuron indicating that inhibition is enhanced. Thus this species may well serve as a model system for the study of functionally induced changes in the nervous system at the cellular level.

3. Cortical Effects of Early Selective Visual Exposure

To investigate how visual stimulation affects the development of the orientation sensitivity of cortical neurons, techniques have been developed for raising animals in a controlled visual environment (HIRSCH, 1970). Kittens were fitted with goggles that allowed them to view only a field of vertical stripes with one eye and only a field of horizontal stripes with the other eye during the first twelve weeks of life (HIRSCH and SPINELLI, 1970, 1971; PETTIGREW et al., 1973b; LEVENTHAL and HIRSCH, 1975; STRYKER et al., 1978). These animals developed cortical abnormalities reflecting their early visual experience[10].

[10] The initial results of experiments on the effects of early selective exposure to only lines of one orientation provoked a series of studies focusing upon several questions. The first of these was to determine the sensitive period for cortical effects of early exposure to controlled visual stimuli (BLAKEMORE and VAN SLUYTERS, 1974). Most of the evidence gathered to date indicates that the critical period for such changes is much the same (though perhaps slightly earlier for direction selectivity; DAW and WYATT, 1974) as that for stimulation-induced changes in occular dominance. The time course involved may also be comparable (BLAKEMORE and MITCHELL, 1973; BLAKEMORE, 1974). A second question raised is whether the response properties of individual neurons can be modified by sensory input. Efforts to answer this have involved rearing kittens in abnormal environments and correlating the stimuli which the animals viewed during rearing with the response characteristics of their cortical cells (SHLAER, 1971; SPINELLI and HIRSCH, 1971; SPINELLI et al., 1972; PETTIGREW and FREEMAN, 1973; VAN SLUYTERS and BLAKEMORE, 1973; ZERNICKI and MICHALSKI, 1974; MICHALSKI et al., 1975; YINON, 1975; BLAKEMORE et al., 1975). To study changes at the single cell level more directly, efforts have also been made to study stimulation-induced changes as they occur (PETTIGREW et al., 1973a; IMBERT and BUISSERET, 1975). To date this approach has perhaps succeeded in demonstrating changes in ocular dominance, but not in orientation selectivity. Finally, the effects of early visual exposure on disparity sensitivity (SHLAER, 1971), direction selectivity (TRETTER et al., 1975; CYNADER et al., 1975; DAW and WYATT, 1974), and motion sensitivity (CYNADER et al., 1973; OLSON and PETTIGREW, 1974) have been investigated. While the results of these studies do suggest that development of many cortical cell response properties can be affected by controlled early experience, the precise conditions and the mechanisms which give rise to these effects are unclear presently. In summary, although the number of studies investigating the consequences of early selective visual experience is growing at a rapid rate, it remains to be demonstrated that studies of this nature provide any insight into the mechanisms which govern the normal development of the visual system (GROBSTEIN and CHOW, 1975).

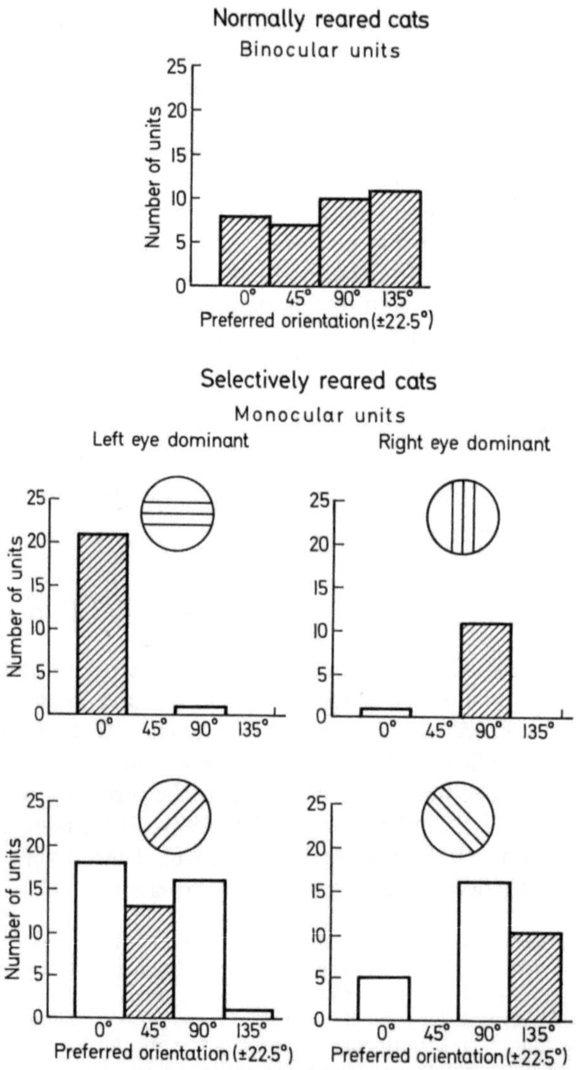

Fig. 31A–C. Number of neurons in visual cortex of normal and selectively reared cats with preferred orientations within each of four ranges. Shaded areas in each histogram indicate orientation of lines (±22.5°) to which animals were exposed during first 4 months of life. Lines inscribed in each circle illustrate patterns presented during rearing. (A) Full range of orientation preferences was found in animals raised in a normal visual environment. (B) In animals that viewed only horizontal and vertical lines, most orientation-sensitive neurons responded preferentially to horizontal or to vertical stimuli. (C) In animals that viewed only diagonal lines, many orientation-sensitive neurons responded preferentially to horizontal or to vertical lines. Rest were activated most strongly by lines with orientation matching that of patterns presented during rearing. [Reproduced with permission from LEVENTHAL and HIRSCH, *Science*, Vol. 190, pp. 902—904, Fig. 2b and c, 28 November 1975; Copyright 1975 by the American Association for the Advancement of Science]

Specifically, while many of the cortical neurons studied in these cats were either unresponsive to visual stimulation or not selective for stimulus orientation, most neurons which did respond selectively to elongated stimuli preferred either horizontal (±22.5°) or vertical (±22.5°) lines. In addition, those neurons that preferred vertical patterns were almost all activated by only the eye that had been exposed to vertical lines, while all but a few of the cells preferring horizontal patterns were activated exclusively by the eye that had viewed horizontal lines during rearing (Fig. 31). Finally, although the majority of orientation-sensitive cells observed in these experiments responded best to stimuli of the same orientation (horizontal and vertical) as the stripes the animals viewed during rearing, a number of cells in these cats responded most strongly to lines oriented at right angles to the patterns the cats had been exposed to (PETTIGREW and HIRSCH, unpublished observations; LEVENTHAL and HIRSCH, 1975).

A related series of reports (BLAKEMORE and COOPER, 1970; BLAKEMORE and MITCHELL, 1973; BLAKEMORE, 1974) describes the cortical physiology of kittens reared in cylinders with either vertical or horizontal stripes painted on their inside surfaces. Most of the cells recorded from these animals were reported to respond preferentially to lines oriented within 45° of the exposed orientation; few nonresponsive or nonselective cells were found. Results of one experiment have confirmed these findings (MIZE and MURPHY, 1973). Other experiments in which different electrophysiological procedures were used contradict these findings and suggest that striped cylinders provide an ineffective means of restricting early visual experience (STRYKER and SHERK, 1975; DAW and WYATT, 1976).

To provide further insight into the effects of early visual experience upon the development of the orientation sensitivity of cortical cells, LEVENTHAL and HIRSCH (1975) have investigated the cortical effects of early exposure to diagonal lines. These effects were in many respects similar to those produced by early selective exposure to horizontal and vertical lines (HIRSCH and SPINELLI, 1970, 1971; LEVENTHAL and HIRSCH, 1975). In cats exposed only to diagonal patterns, 71% of the orientation-sensitive cells responded preferentially to horizontal (±22.5°) or to vertical (±22.5°) lines, while only 29% responded most strongly to diagonal patterns. The visual environment to which the animals had been exposed was reflected only in the ocular dominance distribution of the orientation-sensitive cortical cells that responded preferentially to diagonal patterns. Of these neurons, those that responded to the eye that had viewed 45° stripes responded best to 45° lines, and those that were activated by the eye that was exposed to 135° lines tended to respond preferentially to that orientation (LEVENTHAL and HIRSCH, 1975).

In concluding, we will summarize some of the data on the role of sensory stimulation in the development of the visual cortex and identify a few common principles. (1) In the visual cortex of normal adult cats we have described one class of cells which has small receptive fields, responds only to stimuli moving slowly, and generally prefers either horizontal or vertical lines and a second class which has larger receptive fields and/or responds to stimuli moving rapidly (LEVENTHAL and HIRSCH, 1977). The first group shares many of the characteristics of cortical cells receiving X-cell (and possibly W-cell) afferents from the LGN, while the second group has response properties characteristic of cortical cells receiving Y-cell afferents (STONE and DREHER, 1973; HOFFMANN and STONE, 1971). (2) When kittens are raised in total darkness and thus deprived of patterned visual stimulation, some of their cortical cells remain orientation sensitive. These neurons are monocular, respond preferentially to horizontal or to vertical lines, have small receptive fields and respond only to stimuli moving slowly (LEVENTHAL and HIRSCH, 1977). (3) The orientation-sensitive cells in the visual cortex of kittens that were reared with one eye viewing vertical stripes and the other eye viewing horizontal stripes are for the most part monocular and respond preferentially to lines of the same orientation as those presented during the rearing (HIRSCH and SPINELLI, 1970, 1971; LEVENTHAL and HIRSCH, 1975; STRYKER et al., 1978). (4) In contrast, after kittens are exposed to diagonal lines, most of the orientation-sensitive cortical cells respond preferentially not to diagonal lines but to horizontal or to vertical patterns, and some of these are binocular. Approximately one third of the orientation-sensitive cells respond preferentially to diagonal stripes, and these respond only to stimulation of the eye that viewed stimuli matching the cell's preferred orientation during early development (LEVENTHAL and HIRSCH, 1975).

We interpret these findings to mean that cortical neurons receiving primarily X-cell afferents (and possibly those receiving W-cell afferents) are insensitive to visual experience for development or maintenance of orientation sensitivity but appear to be sensitive to such experience for the development or maintenance of binocularity. On the other hand, cortical cells which receive primarily Y-cell afferents appear to be sensitive to early visual experience for the development or maintenance of orientation selectivity but appear not to require it for the development of binocularity. Thus, there is now evidence at different levels along the visual pathways that the three cell types (X, Y, and W) which have been described differ in their dependence upon early visual experience.

D. Conclusion: Variant and Invariant Components of the Visual System

Many experimenters have investigated the role played by early experience in the development of the mammalian primary visual areas. While there is considerable agreement concerning the dependence upon sensory stimulation of certain response characteristics of visual neurons, such as binocularity, conflicting results have stirred controversy concerning the development of other properties, such as the orientation specificity of cortical cells. The factors mediating the development of the visual system as a whole have thus been clouded by contradictions concerning the specific conditions necessary for the development of its normal function (GROBSTEIN and CHOW, 1975). In spite of this, however, it is clear that the visual system depends heavily upon early experience for its normal structure as well as function.

To clarify the mechanisms underlying visual system development, experimenters have recently suggested that the visual system contains distinct pathways (X-cells, Y-cells, W-cells), which can be affected in different ways by early experience. The results of studies in this area have indicated that these pathways give rise to a visual system which contains *both* experience-sensitive and experience-insensitive components. On the basis of the evidence accumulated to date, it appears that, in many cases, the invariant component of the mammalian visual system consists of retinal W-cells and X-cells and their central projections and terminations, while the variable component, within which the environment exerts its (still poorly understood) effect, consists of Y-cells and their central projections and terminations. Such an invariant, intrinsically determined framework, coupled with a variable, experience-sensitive component, may provide the constancy in the organization of the visual system that is necessary for development and function, without sacrificing the flexibility of response necessary for survival in an unpredictable environment.

E. Acknowledgments

We thank J. STONE, D. G. TIEMAN, and S. B. TIEMAN for critical comments on the manuscript. J. LEVENTHAL, R. LOOS, C. MANDIGO, R. SPECK, and L. STERN assisted in the preparation of the manuscript and the figures. Support was provided by U.S. Public Health Service Research Grant R01 EY-01268 and Alfred P. Sloan Foundation Fellowship BR 1677.

References

Arduini,J., Pinneo,L.R.: The tonic activity of the lateral geniculate nucleus in dark and light adaptation. Arch ital. biol. **101**, 493—507 (1963).

Baker,F.H., Grigg,P., von Noorden,G.K.: Effects of visual deprivation and strabismus on the response of neurons in the visual cortex of the monkey, including studies on the striate and prestriate cortex in the normal animal. Brain Res. **66**, 185—208 (1974).

Barlow,H.B., Pettigrew,J.D.: Lack of specificity of neurones in the visual cortex of young kittens. J. Physiol. (Lond.) **218**, 98—100 (1971).

Bennett,M.H., Dyer,R.F., Dunn,J.D.: Light induced retinal degeneration: effect upon light-dark discrimination. Exp. Neurol. **34**, 434—445 (1972).

Bennett,M.H., Dyer,R.F., Dunn,J.D.: Visual deficit following long-term continuous light exposure. Exp. Neurol. **38**, 80—89 (1973a).

Bennett,M.H., Dyer,R.F., Dunn,J.D.: Visual dysfunction after long-term continuous light exposure. Exp. Neurol. **40**, 652—660 (1973b).

Bishop,P.O., Coombs,J.S., Henry,G.H.: Interaction effects of visual contours on the discharge frequency of simple striate neurones. J. Physiol. (Lond.) **219**, 659—687 (1971a).

Bishop,P.O., Coombs,J.S., Henry,G.H.: Responses to visual contours: spatio-temporal aspects of excitation in the receptive fields of simple striate neurones. J. Physiol. (Lond.) **219**, 625—657 (1971b).

Bishop,P.O., Henry,G.H., Smith,C.J.: Binocular interaction fields of single units in the cat striate cortex. J. Physiol. (Lond.) **216**, 39—68 (1971c).

Bishop,P.O., Coombs,J.S., Henry,G.H.: Receptive fields of simple cells in the cat striate cortex. J. Physiol. (Lond.) **231**, 31—60 (1973).

Blake,R., Hirsch,H.V.B.: Deficits in binocular depth perception in cats after alternating monocular deprivation. Science **190**, 1114—1116 (1975).

Blakemore,C.: Developmental factors in the formation of feature-extracting neurons. In: The Neurosciences, 3rd Study Program (F.O.Schmidt, Ed.).Cambridge, Mass.: MIT Press 1974.

Blakemore,C., Cooper,G.F.: Development of the brain depends on the visual environment. Nature **228**, 477—478 (1970).

Blakemore,C., Mitchell,D.E.: Environmental modification of the visual cortex and the neural basis of learning and memory. Nature **241**, 467—468 (1973).

Blakemore,C., Van Sluyters,R.C.: Reversal of the physiological effects of monocular deprivation in kittens: further evidence for a sensitive period. J. Physiol. (Lond.) **237**, 195—216 (1974).

Blakemore,C., Van Sluyters,R.C.: Innate and environmental factors in the development of the kitten's visual cortex. J. Physiol. (Lond.) **248**, 663—716 (1975).

Blakemore,C., Van Sluyters,R.C., Peck,C., Hein,A.: Development of the cat visual cortex following rotation of one eye. Nature **257**, 584—586 (1975).

Boas,J.A.R., Ramsey,R.L., Riesen,A.J., Walker,J.P.: Absence of change in some measures of cortical morphology in dark-reared adult rats. Psychonomic Sci. **15**, 251—252 (1969).

Boothe,R.G., Lund,J.S.: A quantitative study of pyramidal cell spine density in developing visual cortex of normal and dark-reared macaque monkeys. Paper presented to the Society for Neuroscience, New York, 1976.

Boycott,B.B., Waessle,H.: The morphological types of ganglion cells of the domestic cat's retina. J. Physiol. (Lond.) **240**, 397—419 (1974).

Buisseret,P., Imbert,M.: Visual cortical cells: their developmental properties in normal and dark-reared kittens. J. Physiol. (Lond.) **255**, 511—525 (1976).

Burke,W., Hayhow,W.R.: Disuse in the lateral geniculate nucleus of the cat. J. Physiol. (Lond.) **194**, 485—519 (1968).

Caldwell,J., Daw,N.W.: X, Y, and W cells in the rabbit retina. Paper presented at Arvo, Sarasota, Fla, (1976).

Casagrande,V.A., Norton,T.T., Guillery,R.W., Harting,J.K.: Studies of binocular competition in the development of the visual pathways of the tree shrew. Paper presented to the Society for Neuroscience, Toronto, 1976.

Chernenko,G.A., West,R.W.: A re-examination of anatomical plasticity in the rat retina. J. comp. Neurol. **167**, 49—62 (1976).

Chow,K.L., Spear,P.D.: Morphological and functional effects of visual deprivation on the rabbit visual system. Exp. Neurol. **42**, 429—447 (1974).

Chow,K.L., Stewart,D.L.: Reversal of structural and functional effects of long-term visual deprivation in cats. Exp. Neurol. **34**, 409—433 (1972).

Cleland,B.G., Dubin,M.W., Levick,W.R.: Sustained and transient neurones in the cat's retina and lateral geniculate nucleus. J. Physiol. (Lond.) **217**, 473—496 (1971).

Cleland,B.G., Levick,W.R., Sanderson,K.J.: Properties of sustained and transient ganglion cells in the cat retina. J. Physiol. (Lond.) **228**, 649—680 (1973).

Coleman,P.D., Riesen,A.H.: Environmental effects on cortical dendritic fields. J. Anat. **102**, 363—374 (1968).

Cragg,B.G.: Changes in the visual cortex on first exposure of rats to light. Nature **215**, 251—253 (1967).

Cragg,B.G.: The effects of vision and dark-rearing on the size and density of synapses in the lateral geniculate nucleus measured by electron microscopy. Brain Res. **13**, 53—67 (1969).

Cragg,B.G.: Plasticity of synapses. In: The Structure and Function of Nervous Tissue, Vol. 4 (G.H.Bourne, Ed.). New York: Academic 1972a.

Cragg,B.G.: The development of synapses in cat visual cortex. Invest. Ophthalmol. **11**, 377—383 (1972b).

Cragg,B.G.: The development of synapses in the visual system of the cat. J. comp. Neurol. **160**, 147—166 (1975a).

Cragg,B.G.: The development of synapses in kitten visual cortex during visual deprivation. Exp. Neurol. **46**, 445—451 (1975b).

Cynader, M., Berman, N., Hein, A.: Cats reared in stroboscopic illumination: effects on receptive fields in visual cortex. Proc. nat. Acad. Sci. (Wash.) **70**, 1353—1354 (1973).
Cynader, M., Berman, N., Hein, A.: Cats raised in a one-directional world: effects on receptive fields in visual cortex and superior colliculus. Exp. Brain Res. **22**, 267—280 (1975).
Daniels, J.D., Pettigrew, J.D.: Development of neuronal responses in the visual system of cats. In: Studies on the Development of Behavior and the Nervous System, Vol. 3 (G. Gottlieb, Ed.). New York: Academic 1976.
Daniels, J.D., Norman, N.L., Pettigrew, J.D.: Modulation of single unit LGN responses by visual cortex feedback. Paper presented to the Society for Neuroscience. New York, 1975.
Daw, N.W., Wyatt, H.J.: Raising rabbits in a moving visual environment: an attempt to modify directional sensitivity in the retina. J. Physiol. (Lond.) **240**, 309—330 (1974).
Daw, N.W., Wyatt, H.J.: Kittens reared in a unidirectional environment: evidence for a critical period. J. Physiol. (Lond.) **257**, 155—170 (1976).
de Monasterio, F.M., Gouras, P.: Functional properties of ganglion cells of the Rhesus monkey retina. J. Physiol. (Lond.) **251**, 167—195 (1975).
Dräger, U.C.: The effects of monocular deprivation on mouse striate cortex. Paper presented to the Society for Neuroscience, Toronto, 1976.
Dreher, B., Sandersen, K.J.: Receptive field analysis: response to moving visual contours by single lateral geniculate neurones in the cat. J. Physiol. (Lond.) **234**, 95—118 (1973).
Dreher, B., Fukuda, Y., Rodieck, R.M.: Identification, classification and anatomical segregation of cells with X-like and Y-like properties in the lateral geniculate nucleus of old world primates. J. Physiol. (Lond.) **258**, 433—452 (1976).
Duffy, F.H., Snodgrass, S.R., Burchfiel, J.L., Conway, J.L.: Bicuculline reversal of deprivation amblyopia in the cat. Nature **260**, 256—257 (1976).
Enroth-Cugell, C., Robson, J.G.: The contrast sensitivity of retinal ganglion cells of the cat. J. Physiol. (Lond.) **187**, 517—552 (1966).
Eysel, U.T., Gaedt, C.: Maintained activity in the lateral geniculate body of the cat and the effects of visual deprivation. Pflügers Arch. **327**, 68—81 (1971).
Fifkova, E.: Changes in the visual cortex of rats after unilateral deprivation. Nature **220**, 379—381 (1968).
Fifkova, E.: Effect of visual deprivation and light on synapses of the inner plexiform layer. Exp. Neurol. **35**, 458—469 (1972a).
Fifkova, E.: Effect of light and visual deprivation on the retina. Exp. Neurol. **35**, 450—457 (1972b).
Fifkova, E.: Effect of light on the synaptic organization of the inner plexiform layer of the retina in albino rats. Experientia **29**, 851—854 (1973).
Fifkova, E., Hassler, R.: Quantitative morphological changes in visual centers in rats after unilateral deprivation. J. comp. Neurol. **135**, 167—178 (1969).
Fukada, Y.: Receptive field organization of cat optic nerve fibers with special reference to conduction velocity. Vision Res. **11**, 209—226 (1971).
Fukuda, Y., Stone, J.: Retinal distribution and central projections of Y-, X-, and W-cells of the cat's retina. J. Neurophysiol. **37**, 749—772 (1974).
Fukuda, Y., Stone, J.: Direct identification of the cell bodies of Y-, X-, and W-cells in the cat's retina. Vision Res. **15**, 1034—1036 (1975).
Fukuda, Y., Sugitani, M.: Cortical projections of two types of principal cells of the rat lateral geniculate body. Brain Res. **67**, 157—161 (1974).
Ganz, L., Fitch, M., Satterberg, J.A.: The selective effect of visual deprivation on receptive field shape determined neurophysiologically. Exp. Neurol. **22**, 614—637 (1968).
Garey, L.J., Pettigrew, J.D.: Ultrastructural changes in kitten visual cortex after environmental modification. Brain Res. **66**, 165—172 (1974).
Globus, A., Scheibel, A.B.: The effect of visual deprivation on cortical neurons: a Golgi study. Exp. Neurol. **19**, 331—345 (1967).
Gorn, R.A., Kuwabara, T.: Retinal damage by visible light: a physiologic study. Arch. ophthal. **77**, 115—118 (1967).
Gouras, P.: Identification of cone mechanisms in monkey ganglion cells. J. Physiol. (Lond.) **199**, 533—547 (1968).
Gouras, P.: Antidromic responses of orthodromically identified ganglion cells in monkey retina. J. Physiol. (Lond.) **204**, 407—419 (1969).

Gray, E.G.: Axosomatic and axodendritic synapses of the cerebral cortex: an electron microscopic study. J. Anat. **93**, 420—433 (1959).
Greenough, W.T., Volkmar, F.R.: Pattern of dendritic branching in occipital cortex of rats reared in complex environments. Exp. Neurol. **40**, 491—504 (1973).
Grignolo, A.N., Castellazzo-Orzalesi, R., Vittone, P.: Retinal damage by visible light in albino rats. Ophthalmologica **157**, 43—59 (1969).
Grobstein, P., Chow, K.L.: Receptive field development and individual experience. Science **190**, 352—358 (1975).
Grobstein, P., Chow, K.L.: Receptive field organization in the mammalian cortex: the role of individual experience in development. In: Studies on the Development of Behavior and the Nervous System (G.Gottlieb, Ed.). New York: Academic 1976.
Grobstein, P., Chow, K.L., Fox, P.C.: Development of receptive fields in rabbit visual cortex: changes in time course due to delayed eye opening. Proc. nat. Acad. Sci. (Wash.) **72**, 1543—1545 (1975).
Guillery, R.W.: Binocular competition in the control of geniculate cell growth. J. comp. Neuol. **144**, 117—127 (1972).
Guillery, R.W.: The effect of lid suture upon the growth of cells in the dorsal lateral geniculate nucleus of kittens. J. comp. Neurol. **148**, 417—422 (1973a).
Guillery, R.W.: Quantitative studies of transneuronal atrophy in the dorsal lateral geniculate nucleus of cats and kittens. J. comp. Neurol. **149**, 423—438 (1973b).
Guillery, R.W.: On structural changes that can be produced experimentally in the mammalian visual pathways. In: Essays on the Nervous System (R.Bellairs and E.G.Gray, Eds.). Oxford: Clarendon 1974.
Guillery, R.W., Kaas, J.H.: The transfer of abnormal visual field representations from the dorsal lateral geniculate nucleus to the visual cortex in Siamese cats. Brain Res. **59**, 51—95 (1973).
Guillery, R.W., Kaas, J.H.: The effects of monocular lid suture upon the development of the lateral geniculate nucleus in squirrels *(Sciureus corolinensis)*. J. comp. Neurol. **154**, 433—442 (1974a).
Guillery, R.W., Kaas, J.H.: The effects of monocular lid suture upon the development of the visual cortex in squirrels *(Sciureus carolinensis)*. J. comp. Neurol. **154**, 443—452 (1974b).
Guillery, R.W., Stelzner, D.J.: The differential effects of unilateral lid closure upon the monocular and binocular segments of the dorsal lateral geniculate nucleus in the cat. J. comp. Neurol. **139**, 413—422 (1970).
Gyllenstein, L., Malmfors, T., Norrlin, M.: Effects of visual deprivation on the optic centers of growing and adult mice. J. comp. Neurol. **124**, 149—160 (1965).
Hansson, H.A.: Ultrastructural studies on rat retina damaged by visible light. Virchows Arch. B **6**, 247—262 (1970).
Haug, H.: Remarks on the determination and significance of the Gray cell coefficient. J. comp. Neurol. **104**, 473—492 (1956).
Hendrickson, A., Boothe, R.: Morphology of the retina and dorsal lateral geniculate nucleus in dark-reared monkeys *(Macaca nemestrina)*. Vision Res. **16**, 517—521 (1976).
Henry, G.H., Dreher, B., Bishop, P.O.: Orientation specificity of cells in cat striate cortex. J. Neurophysiol. **37**, 1394—1409 (1974).
Hickey, T.L., Spear, P.D., Kratz, K.E.: Quantitative studies of cell size in the cat's dorsal lateral geniculate nucleus following visual deprivation. J. comp. Neurol. **172**, 265—282 (1977).
Hirsch, H.V.B.: Controlled visual stimulation and deprivation. In: Genesis of neuronal Patterns (M.V.Edds, Jr, Ed.). Brooklin, Mass.: NRP Work Session 1970.
Hirsch, H.V.B., Jacobson, M.: The perfectible brain: principles of neuronal development. In: Handbook of Psychobiology (M.Gazzaniga and C.Blakemore, Eds.). New York: Academic 1975.
Hirsch, H.V.B., Leventhal, A.G.: X-cell and Y-cell influenced neurons in the cat's visual cortex following long-term pattern deprivation. Paper presented at Arvo, Sarasota, Fla, 1976.
Hirsch, H.V.B., Spinelli, D.N.: Visual experiences. Modified distribution of horizontally and vertically oriented receptive fields in cats. Science **168**, 869—871 (1970).
Hirsch, H.V.B., Spinelli, D.N.: Modification of the distribution of receptive field orientation in cats by selective visual exposure during development. Exp. Brain Res. **12**, 509—527 (1971).
Hoffmann, K.P.: Conduction velocity in pathways from retina to superior colliculus in the cat: a correlation with receptive-field properties. J. Neurophysiol. **36**, 409—424 (1973).
Hoffmann, K.P., Cynader, M.: Recovery in the LGN of the cat after early visual deprivation. Brain Res. **85**, 179 (1975).

Hoffmann,K.P., Sherman,S.M.: Effects of early monocular deprivation on visual input to cat superior colliculus. J. Neurophysiol. **37**, 1276—1286 (1974).
Hoffmann,K.P., Sherman,S.M.: Effects of early binocular deprivation on visual input to cat superior colliculus. J. Neurophysiol. **38**, 1049—1059 (1975).
Hoffmann,K.P., Stone,J.: Conduction velocity of afferents to cat visual cortex: a correlation with cortical receptive field properties. Brain Res. **32**, 460—466 (1971).
Hoffmann,K.P., Stone,J., Sherman,S.M.: Relay of receptive-field properties in dorsal lateral geniculate nucleus of the cat. J. Neurophysiol. **35**, 518—531 (1972).
Hubel,D.H., Wiesel,T.N.: Receptive fields of single neurons in the cat's striate cortex. J. Physiol. (Lond.) **148**, 574—591 (1959).
Hubel,D.H., Wiesel,T.N.: Receptive fields, binocular interaction, and functional architecture in the cat's visual cortex. J. Physiol. (Lond.) **160**, 106—154 (1962).
Hubel,D.H., Wiesel,T.N.: Receptive fields of cells in striate cortex of very young, visually inexperienced kittens. J. Neurophysiol. **26**, 994—1002 (1963a).
Hubel,D.H., Wiesel,T.N.: Shape and arrangement of columns in cat's striate cortex. J. Physiol. (Lond.) **165**, 559—568 (1963b).
Hubel,D.H., Wiesel,T.N.: Binocular interaction in striate cortex of kittens reared with artificial squint. J. Neurophysiol. **28**, 1041—1059 (1965).
Hubel,D.H., Wiesel,T.N.: Receptive fields and functional architecture of monkey striate cortex. J. Physiol. (Lond.) **195**, 215—243 (1968).
Hubel,D.H., Wiesel,T.N.: The period of susceptibility to the physiological effects of unilateral eye closure in kittens. J. Physiol. (Lond.) **206**, 419—436 (1970).
Hubel,D.H., Wiesel,T.N., Levay,S.: Functional architecture of area 17 in normal and monocularly deprived Macaque monkeys. Cold Spr. Harb. Symp. quant. Biol. **XL**, 581—589 (1976).
Ikeda,H., Wright,M.J.: Receptive field organization of "sustained" and "transient" retinal ganglion cells which subserve different functional roles. J. Physiol. (Lond.) **227**, 769—800 (1972).
Imbert,M., Buisseret,P.: Receptive field characteristics and plastic properties of visual cortex cells in kittens reared with or without visual experience. Exp. Brain Res. **22**, 25—36 (1975).
Jacobson,M.: Developmental Neurobiology. New York: Holt, Rhinehart and Winston 1970.
Jacobson,M.: A plenitude of neurons. In: Studies on the Development of Behavior and the Nervous System, Vol. 2 (G.Gottlieb, Ed.). New York: Academic 1974.
Jacobson,M.: Developmental Neurobiology. New York: Plenum 1978.
Kelly,J.P., Van Essen,D.C.: Cell structure and function in the visual cortex of the cat. J. Physiol. (Lond.) **238**, 515—547 (1974).
Kratz,K.E., Spear,P.D.: Effects of visual deprivation and alterations in binocular competition on responses of striate neurons in the cat. J. comp. Neurol. **170**, 141—151 (1976).
Kratz,K.E., Webb,S.V., Sherman,S.M.: The effects of monocular deprivation on the medial interlaminar nucleus of cats. Paper presented to the Society for Neuroscience, Toronto, 1976a.
Kratz,K.E., Spear,P.D., Smith,D.C.: Postcritical-period reversal of effects of monocular deprivation on striate cortex cells in the cat. J. Neurophysiol. **39**, 501—511 (1976b).
Kuffler,S.W.: Discharge patterns and functional organization of mammalian retina. J. Neurophysiol. **16**, 37—68 (1953).
Kuwubara,T., Gorn,R.A.: Retinal damage by visible light. Arch. ophthal. **79**, 69—78 (1968).
Kuwubara,T., Weidman,T.A.: Development of the prenatal rat retina. Invest. Ophthal. **13**, 725—739 (1974).
Lawwill,T.: Effects of prolonged exposure of rabbit retina to low-intensity light. Invest. Ophthal. **12**, 45—51 (1973).
Levay,S., Hubel,D.H., Wiesel,T.N.: The pattern of ocular dominance columns in Macaque visual cortex revealed by a reduced silver stain. J. comp. Neurol. **159**, 559—576 (1975).
Leventhal,A.G., Hirsch,H.V.B.: Cortical effect of early selective exposure to diagonal lines. Science **190**, 902—904 (1975).
Leventhal,A.G., Hirsch,H.V.B.: Response characteristics of monocular and of binocular neurons in the visual cortex of pattern-deprived cats. Paper presented to the Society for Neuroscience, Toronto, 1976.
Leventhal,A.G., Hirsch,H.V.B.: Effects of early experience upon the orientation sensitivity and the binocularity of neurons in the cat's visual cortex. Proc. nat. Acad. Sci. (Wash.) **74**, 1272—1276 (1977).

Lund,R.D., Lund,J.S.: Development of synaptic patterns in the superior colliculus of the rat. Brain Res. **42**, 1—20 (1972a).

Lund,R.D., Lund,J.S.: The effects of varying periods of visual deprivation on synaptogenesis in the superior colliculus of the rat. Brain Res. **42**, 21—32 (1972b).

Maffei,L., Bisti,S.: Binocular interaction in strabismic kittens deprived of vision. Science **191**, 579—580 (1976).

Maffei,L., Fiorentini,A.: Geniculate neural plasticity in kittens after exposure to periodic gratings. Science **186**, 447—449 (1974).

Mason,R.: Cell properties in the medial interlaminar nucleus of the cat's lateral geniculate complex in relation to the transient/sustained classification. Exp. Brain Res. **22**, 327—329 (1975).

Matsumoto,S.G., Murphey,R.K.: Experience sensitivity of cricket neurons. J. Physiol. (Lond.) **268**, 533—548 (1977).

Michalski,A., Kossut,M., Zernicki,B.: Single-unit responses to natural objects in area 19 of cats with different early visual experiences. Acta neurobiol. exp. **35**, 77—83 (1975).

Mize,R.R., Murphy,E.H.: Selective visual experience fails to modify receptive field properties of rabbit striate cortex neurons. Science **180**, 320—323 (1973).

Movshon,J.A.: The velocity tuning of single units in cat striate cortex. J. Physiol. (Lond.) **249**, 445—468 (1975).

Movshon,J.A.: Reversal of the physiological effects of monocular deprivation in the kitten's visual cortex. J. Physiol. (Lond.) **261**, 125—174 (1976).

Murphey,R.K., Mendenhall,B., Palka,J., Edwards,J.S.: Deafferentation slows the growth of specific dendrites of identified giant interneurons. J. comp. Neurol. **159**, 407—418 (1975).

Noell,W.K., Walker,D.S., King,B.S., Berman,S.: Retinal damage by light in rats. Invest. ophthal. **5**, 450—483 (1966).

Norman,J.L., Daniels,J.D., Pettigrew,J.D.: Absence of surround antagonism in unit responses of the LGN in young kittens. Paper presented to the Society for Neuroscience, New York, 1975.

Norton,T.T.: Receptive-field properties of superior colliculus cells and development of visual behavior in kittens. J. Neurophysiol. **37**, 674—690 (1974).

Norton,T.T., Casagrande,V.A., Sherman,S.M.: Absence of Y-cells in deprived laminae of the tree shrew LGN after monocular lid suture. Paper presented at Arvo, Sarasota, Fla, 1976.

Olson,C.R., Freeman,R.D.: Progressive changes in kitten striate cortex during monocular vision. J. Neurophysiol. **38**, 26—32 (1975).

Olson,C.R., Pettigrew,J.D.: Single units in visual cortex of kittens reared in stroboscopic illumination. Brain Res. **70**, 189—204 (1974).

O'Steen,W.K.: Retinal and optic nerve serotonin and retinal degeneration as influenced by photoperiod. Exp. Neurol. **27**, 195—205 (1970).

O'Steen,W.K., Lytle,R.B.: Early cellular disruption and phagocytosis in photically-induced retinal degeneration. Amer. J. Anat. **130**, 227—234 (1971).

Palka,J., Edwards,J.S.: The cerci and abdominal giant fibres of the house cricket, *Acheta domesticus*. II. Regeneration and effects of chronic deprivation. Proc. roy. Soc. Lond. B **185**, 105—121 (1974).

Palmer,L.A., Rosenquist,A.C.: Visual receptive fields of single striate cortical units projecting to the superior colliculus in the cat. Brain Res. **67**, 27—42 (1974).

Parnavelas,J.G., Globus,A., Kaups,P.: Continuous illumination from birth affects spine density of neurons in the visual cortex of the rat. Exp. Neurol. **40**, 742—747 (1973).

Peck,C.K., Blakemore,C.: Modification of single neurons in the kitten's visual cortex after brief periods of monocular visual experience. Exp. Brain Res. **22**, 57—68 (1975).

Pettigrew,J.D.: The effects of selective visual experience on stimulus trigger features of kitten cortical neurons. Ann. N.Y. Acad. Sci. **228**, 393—405 (1974a).

Pettigrew,J.D.: The effect of visual experience on the development of stimulus by kitten cortical neurones. J. Physiol. (Lond.) **237**, 49—74 (1974b).

Pettigrew,J.D., Freeman,R.D.: Visual experience without lines: effect on developing cortical neurons. Science **182**, 599—601 (1973).

Pettigrew,J.D., Garey,L.J.: Selective modification of single neuron properties in the visual cortex of kittens. Brain Res. **66**, 160—164 (1974).

Pettigrew,J.D., Konishi,M.: The owl and the pussycat. Paper presented to the Society for Neuroscience, Toronto, 1976.

Pettigrew,J.D., Nikara,T., Bishop,P.O.: Responses to moving slits by single units in cat striate cortex. Exp. Brain Res. **6**, 373—390 (1968).
Pettigrew,J.D., Olson,C., Barlow,H.B.: Kitten visual cortex: short-term, stimulus-induced changes in connectivity. Science **180**, 1202—1203 (1973a).
Pettigrew,J.D., Olson,C., Hirsch,H.V.B.: Cortical effect of selective visual experience: degeneration or reorganization? Brain Res. **51**, 345—351 (1973b).
Ptacek,J.M., Fagan-Dubin,L.: Developmental changes in neuron size and density in the visual cortex and superior colliculus of the postnatal golden hamster. J. comp. Neurol. **158**, 237—242 (1974).
Rosenquist,A.C., Palmer,L.A.: Visual receptive field properties of cells of the superior colliculus after cortical lesions in the cat. Exp. Neurol. **33**, 629—652 (1971).
Rowe,M.H., Stone,J.: Conduction velocity groupings among axons of cat retinal ganglion cells, and their relationship to retinal topography. Exp. Brain Res. **25**, 339—357 (1976).
Ruiz-Marcos,A., Valverde,F.: The temporal evolution of the distribution of dendritic spines in the visual cortex of normal and dark-reared mice. Exp. Brain Res. **8**, 284—294 (1969).
Salcman,M., Bak,M.J.: Design, fabrication and *in-vivo* behavior of chronic recording intracortical microelectrodes. IEEE Trans. biomed. Eng. BME **20**, 253—260 (1973).
Salcman,M., Bak,M.J.: Evaluation of a new chronic recording intracortical microelectrode. Fed. Proc. **33**, 331 (1974).
Salcman,M., Whetsell,W., Bak,M.J.: Histopathologic evaluation of parylene-coated chronically implanted microelectrodes. Paper presented to the Society for Neuroscience, Toronto, 1976.
Schiller,P.H., Finlay,B.L., Volman,S.F.: Quantitative studies of single-cell properties in monkey striate cortex. I. Spatiotemporal organization of receptive fields. J. Neurophysiol. **39**, 1288—1319 (1976a).
Schiller,P.H., Finlay,B.L., Volman,S.F.: Quantitative studies of single-cell properties in monkey striate cortex. II. Orientation specificity and ocular dominance. J. Neurophysiol. **39**, 1320—1333 (1976b).
Schiller,P.H., Finlay,B.L., Volman,S.F.: Quantitative studies of single-cell properties in monkey striate cortex. III. Spatial frequency. J. Neurophysiol. **39**, 1334—1351 (1976c).
Schiller,P.H., Finlay,B.L., Volman,S.F.: Quantitative studies of single-cell properties in monkey striate cortex. IV. Multivariate statistical analysis and models. J. Neurophysiol. **39**, 1362—1374 (1976d).
Shariff,G.A.: Cell counts in the primate cerebral cortex. J. comp. Neurol. **98**, 381—400 (1953).
Shatz,C., Lindstrom,S., Wiesel,T.N.: Ocular dominance columns in the cat's visual cortex. Paper presented to the Society for Neuroscience, New York, 1975.
Shatz,C., Lindstrom,S., Wiesel,T.N.: The distribution of afferents representing the right and left eyes in the cat's visual cortex. Brain Res. **131**, 103—116 (1977).
Shaw,C., Yinon,U., Auerbach,E.: Diminution of evoked neuronal activity in the visual cortex of pattern-deprived rats. Exp. Neurol. **45**, 42—49 (1974).
Sherk,H., Stryker,M.P.: Quantitative study of cortical orientation selectivity in visually inexperienced kitten. J. Neurophysiol. **39**, 63—70 (1976).
Sherman,S.M., Sanderson,K.J.: Binocular interaction on cells of the dorsal lateral geniculate nucleus of visually deprived cats. Brain Res. **37**, 126—131 (1972).
Sherman,S.M., Stone,J.: Physiological normality of the retina in visually deprived cats. Brain Res. **60**, 224—230 (1973).
Sherman,S.M., Hoffmann,K.P., Stone,J.: Loss of a specific cell type from dorsal lateral geniculate nucleus in visually deprived cats. J. Neurophysiol. **35**, 532—541 (1972).
Sherman,S.M., Guillery,R.W., Kaas,J.H., Sanderson,K.J.: Behavioral, electrophysiological and morphological studies of binocular competition in the development of the geniculo-cortical pathways of cats. J. comp. Neurol. **158**, 1—18 (1974).
Sherman,S.M., Wilson,J.R., Guillery,R.W.: Evidence that binocular competition affects the postnatal development of Y-cells in the cat's lateral geniculate nucleus. Brain Res. **100**, 441—444 (1975).
Sherman,S.M., Wilson,J.R., Kaas,J.H., Webb,S.V.: X- and Y-cells in the dorsal lateral geniculate nucleus of the owl monkey *(Aotus trivirgatus)*. Science **192**, 475—476 (1976).
Shlaer,R.: Shift in binocular disparity causes compensating change in the cortical structure of kittens. Science **173**, 638—641 (1971).
Sillito,A.M.: The effectiveness of bicuculline as an antagonist of GABA and visually evoked inhibition in the cat's striate cortex. J. Physiol. (Lond.) **250**, 287—304 (1975a).
Sillito,A.M.: The contribution of inhibitory mechanisms to the receptive field properties of neurones in the striate cortex of the cat. J. Physiol. (Lond.) **250**, 305—329 (1975b).

Singer, W., Bedworth, N.: Inhibitory interaction between X and Y units in the cat lateral geniculate nucleus. Brain Res. **49**, 291—307 (1973).
Singer, W., Tretter, F.: Receptive-field properties and neuronal connectivity in striate and parastriate cortex of contour-deprived cats. J. Neurophysiol. **39**, 613—630 (1976).
Singer, W., Tretter, F., Cynader, M.: Organization of cat striate cortex: a correlation of receptive-field properties with afferent and efferent connections. J. Neurophysiol. **38**, 1080—1098 (1975).
Sosula, L., Glow, P.H.: Increase in number of synapses in the inner plexiform layer of light-deprived rat retinae: quantitative electron microscopy. J. comp. Neurol. **141**, 427—452 (1971).
Spinelli, D.N., Hirsch, H.V.B.: Genesis of receptive field shapes in single units of cat's visual cortex. Fed. Proc. **30**, 615 (1971).
Spinelli, D.N., Hirsch, H.V.B., Phelps, R.W., Metzler, J.: Visual experience as a determinant of the response characteristics of cortical receptive fields in cats. Exp. Brain Res. **15**, 289—304 (1972).
Stein, B.E., Magalhaes-Castro, B.: Effects of neonatal cortical lesions upon the cat superior colliculus. Brain Res. **83**, 480—485 (1975).
Stein, B.E., Labos, E., Kruger, L.: Sequence of changes in properties of neurons of superior colliculus of the kitten during maturation. J. Neurophysiol. **36**, 667—679 (1973a).
Stein, B.E., Labos, E., Kruger, L.: Determinants of response latency in neurons of superior colliculus in kittens. J. Neurophysiol. **36**, 680—689 (1973b).
Stone, J., Dreher, B.: Projection of X- and Y-cells of the cat's lateral geniculate nucleus to areas 17 and 18 of visual cortex. J. Neurophysiol. **36**, 551—567 (1973).
Stone, J., Fukuda, Y.: Properties of cat retinal ganglion cells: a comparison of W-cells with X- and Y-cells. J. Neurophysiol. **37**, 749—772 (1974).
Stone, J., Hoffmann, K.P.: Very slow-conducting ganglion cells in the cat's retina: a major, new functional type? Brain Res. **43**, 610—616 (1972).
Stryker, M.P., Sherk, H.: Modification of cortical orientation selectivity in the cat by restricted visual experience: a reexamination. Science **190**, 904—905 (1975).
Stryker, M.P., Sherk, H., Leventhal, A.G., Hirsch, H.V.B.: Physiological consequences for the cat's visual cortex of effectively restricting early visual experience with orientation. J. Neurophysiol. in press (1978).
Tretter, F., Cynader, M., Singer, W.: Modification of direction selectivity of neurons in the visual cortex of kittens. Brain Res. **84**, 143—149 (1975).
Valverde, F.: Apical dendritic spines of the visual cortex and light deprivation in the mouse. Exp. Brain Res. **3**, 337—352 (1967).
Valverde, F.: Rate and extent of recovery from dark rearing in the visual cortex of the mouse. Brain Res. **33**, 1—11 (1971).
Valverde, F., Ruiz-Marcos, A.: Dendritic spines in the visual cortex of the mouse. Introduction to a mathematical model. Exp. Brain Res. **8**, 269—283 (1969).
Van Sluyters, R.C., Blakemore, C.: Experimental creation of unusual neuronal properties in visual cortex of kitten. Nature **246**, 506—508 (1973).
Van Sluyters, R.C., Stewart, D.L.: Binocular neurons of the rabbit's visual cortex: effects of monocular sensory deprivation. Exp. Brain Res. **19**, 196—204 (1974).
Vrensen, G., de Groot, D.: The effects of dark rearing and its recovery on synaptic terminals in the visual cortex of rabbits. A quantitative electron microscopic study. Brain Res. **78**, 263—278 (1974).
Watkins, D.W., Berkley, M.A.: The orientation selectivity of single neurons in cat striate cortex. Exp. Brain Res. **19**, 433—446 (1974).
Watkins, D.W., Sherman, S.M.: Effects of binocular deprivation on cat striate cortex. Paper presented to the Society for Neuroscience, New York, 1975.
Weidman, T.A., Kuwabara, T.: Postnatal development of the rat retina. Arch. Ophthalmol. **79**, 470—484 (1968).
Werblin, F.S., Dowling, J.E.: Organization of the retina of the mudpuppy, *Necturus maculosus*. II. Intracellular recording. J. Neurophysiol. **32**, 339—355 (1969).
West, R.W., Greenough, W.T.: Effect of environmental complexity on cortical synapses of rats: preliminary results. Behav. Biol. **7**, 279—284 (1972).
Wickelgren, B.G., Sterling, P.: Influence of visual cortex on receptive fields in the superior colliculus of the cat. J. Neurophysiol. **32**, 16—23 (1969a).
Wickelgren, B.G., Sterling, P.: Effect on the superior colliculus of cortical removal in visually deprived cats. Nature **224**, 1032—1033 (1969b).

Wiesel, T. N., Hubel, D. H.: Effects of visual deprivation on morphology and physiology of cells in the cat's lateral geniculate body. J. Neurophysiol. **26**, 978—993 (1963a).

Wiesel, T. N., Hubel, D. H.: Single-cell responses in striate cortex of kittens deprived of vision in one eye. J. Neurophysiol. **26**, 1003—1017 (1963b).

Wiesel, T. N., Hubel, D. H.: Comparison of the effects of unilateral and bilateral eye closure on cortical unit responses in kittens. J. Neurophysiol. **28**, 1029—1040 (1965a).

Wiesel, T. N., Hubel, D. H.: Extent of recovery from effects of visual deprivation in kittens. J. Neurophysiol. **28**, 1050—1072 (1965b).

Wiesel, T. N., Hubel, D. H.: Ordered arrangement of orientation columns in monkeys lacking visual experience. J. comp. Neurol. **158**, 307—318 (1974).

Wiesel, T. N., Hubel, D. H., Lam, D. M. K.: Autoradiographic demonstration of ocular-dominance columns in the monkey striate cortex by means of transneuronal transport. Brain Res. **79**, 273—279 (1974).

Wilson, J. R., Sherman, S. M.: Receptive-field characteristics of neurons in cat striate cortex: changes with visual field eccentricity. J. Neurophysiol. **39**, 512—533 (1976).

Wilson, P. D., Stone, J.: Evidence of W-cell input to the cat's visual cortex via the C laminae of the lateral geniculate nucleus. Brain Res. **92**, 472—478 (1975).

Wilson, P. D., Rowe, M. H., Stone, J.: Properties of relay cells in the cat's lateral geniculate nucleus: a comparison of W-cells with X-cells and Y-cells. J. Neurophysiol. **39**, 1193—1209 (1976).

Yinon, U.: Eye rotation in developing kittens: the effect on ocular dominance and receptive field organization of cortical cells. Exp. Brain Res. **24**, 215—218 (1975).

Zernicki, B., Michalski, A.: Single-unit responses to natural objects in visual areas 17 and 18 of cats reared under different visual experiences. Acta neurobiol. exp. **34**, 697—712 (1974).

Chapter 8

Development of Cutaneous Sensory Receptors in Birds

RAYMOND SAXOD, Grenoble, Isère, France

With 117 Figures and 4 Tables

Contents

A. Introduction

In recent studies, refined physiological and cytological techniques have permitted a number of precise observations to be made on the structure of peripheral cutaneous sensory receptors in vertebrates and their role in somatic sensation (for reviews, see CAUNA, 1962; QUILLIAM and ARMSTRONG, 1963, 1966; POLACEK, 1966; MUNGER, 1971; ANDRES and DÜRING, 1973; GOGLIA, 1973; SCHWARTZKOPFF, 1973; IGGO, 1974; ANDRES, 1974).

However, there is still one fundamental problem that has rarely been considered concerning the relationship between the sensory nerves and their end-organs, namely the problem of the factors involved in the development and maintenance of the peripheral sensory receptors and particularly of the cutaneous sensory corpuscles.

It is well known that in each kind of sensory receptor two types of structures can be distinguished. There is the nervous structure, formed by the peripheral process of the spinal sensory ganglionic neuron, and also a cellular structure, more or less specialized, surrounding this nerve ending. The specific function of the receptors depends largely on the mode of association of these two types of structures.

There are two phases of development of these terminal organs, namely an initial "phase of determination", during which the kind of association between the nerve terminal and the surrounding cells is determined, followed by a "phase of histogenesis", during which the structure is built up. Thus the development of diverse types of cutaneous sensory receptors results from a morphogenetic interaction between the nerve terminals and the innervated cutaneous tissue.

Two methods can be used to analyze these interactions. The first consists of experimenting with the adult animal, by sectioning or poisoning the peripheral nerve, or making grafts of cutaneous tissue, and then studying the degeneration and regeneration of the receptors. The second method, which is more suitable in this author's opinion, consists of analyzing the morphogenesis of these terminal organs during the course of their embryonic or fetal development, using the techniques of experimental embryology.

For the latter purpose, the tetrapods are the most interesting to study because they possess many types of different sensory cutaneous receptors with a regional and specific distribution. In particular, birds offer excellent material for this type of experiment, because of easy access to the embryo, the short period of embryonic development, and the availability of the material. Historically, birds were the first objects of this kind of histogenetic (SZYMONOWICS, 1897; HERINGA, 1918) and experimental (BOTEZAT, 1910; BOEKE, 1926; KLEIN, 1932; DIJKSTRA, 1933) research.

The principal questions posed about the development of cutaneous sensory receptors are: (1) How do the different types of receptors differ? (2) What is their histogenesis? (3) What is the developmental origin of the different components? (4) What are the factors involved in their morphogenesis, and particularly, what are the roles of the nerve ending and of the innervated cutaneous territory?

B. Morphological Data

I. Types of Cutaneous Sensory Receptors

There are four main types of cutaneous sensory receptors in birds: Herbst corpuscles, Grandry corpuscles, Merkel corpuscles, and free nerve endings. Of these, only the first two, described respectively by HERBST (1848) and WILL (1850) and by GRANDRY (1869), are generally considered to be characteristic of birds and are found in abundance in the skin of the beak of aquatic species, especially in the duck. The other two types of receptors, free nerve endings and the corpuscles of Merkel, are not peculiar to birds and are found in all kinds of tetrapods. For several reasons, it is difficult to make comparisons of these receptors between the different classes of vertebrates. One reason is that transitional forms exist between the principal types of corpuscles. There are also differences between species in the same class as well as between different classes. Moreover, morphological homology may not mean that neurophysiological functions are the same. In what follows, these receptors in birds are designated by their eponyms without prejudging the real homologies that might eventually be established. Finally, for simplicity, the morphological data that follow concern principally the three species of birds in which these receptors have been most thoroughly studied structurally and experimentally, namely, duck, chicken, and quail.

1. Free Nerve Endings

The dermis of the skin contains an extremely complex network of sensory, motor, somatic and visceral nerve fibers. Because of the extreme difficulty of establishing the existence of free nerve endings in the dermis, there has been considerable discussion about their existence there (MUNGER, 1971a, for review). By contrast, it is far easier, at least at the level of the light microscope, to find evidence of free nerve endings in the epidermis. Such endings have been described in numerous vertebrates (for reviews see QUILLIAM, 1966; IGGO, 1974a, b), and with electron microscopy particularly in mammals (KADANOFF, 1971; CHOUCHKOV, 1971, 1972, 1974; TSUJI, 1971). By contrast, they have rarely been mentioned in birds, except in the beak of the chicken (MALINOVSKY and ZEMANEK, 1971) and in the beak and tongue of the duck (SAXOD, unpubl., cited in this paper). In duck the free nerve endings are particularly abundant in the upper beak at the level of the lateral ridge, at the tip of the beak, and around the nostrils. Innervation is provided by myelinated nerve fibers of the dermal network, which penetrate the epidermis principally in the region between the ridges of the epithelium where it is thin (Figs. 1–4). The myelin sheath and the Schwann cells are lost when the nerves penetrate the basement membrane. The nerve fibers have a generally oblique and winding trajectory through the epidermis; their diameter is fine (1 μm or less); they are often very long and sometimes run as far as the superficial cornified desquamating layer of cells (Fig. 2). Electron microscopy often reveals a plexus of fine myelinated nerve fibers in the region of the epidermal ridges, but it has not been possible to find evidence of these fibers inside the epidermis.

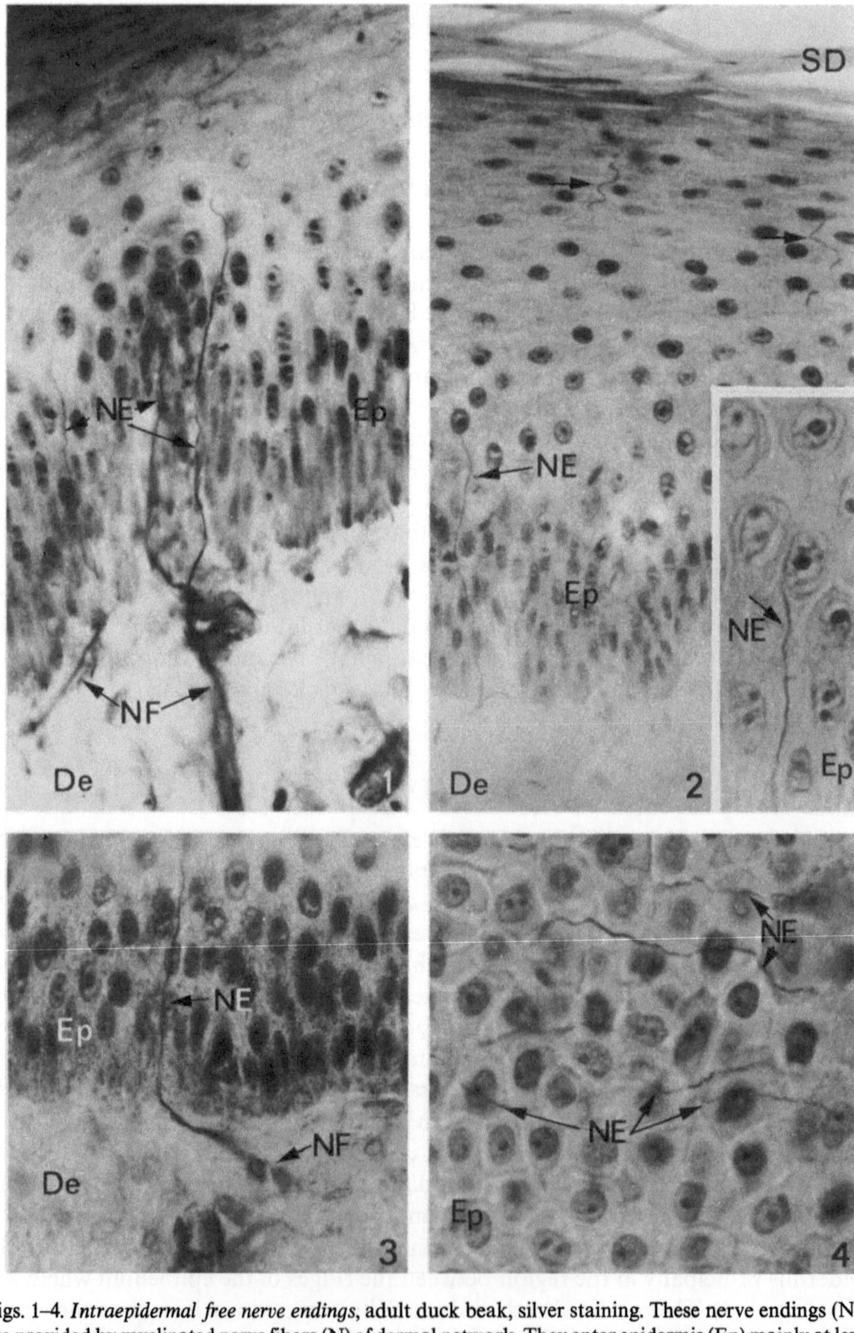

Figs. 1–4. *Intraepidermal free nerve endings*, adult duck beak, silver staining. These nerve endings (NE) are provided by myelinated nerve fibers (N) of dermal network. They enter epidermis (Ep) mainly at level of *rete pegs*. Myelin sheath and Schwann cells are lost at level of basal membrane (Fig. 1). These thin free endings, sometimes numerous (Fig. 4), follow winding path in epidermis (Fig. 2, inset, and Fig. 4). They sometimes terminate (Fig. 2, arrows) in immediate vicinity of the *stratum disjunctum* (SD). Figs. 1–3, transverse sections; Fig. 4, oblique section. Figures 1 and 2, × 500 (inset × 1000); Fig. 3, × 900; Fig. 4, × 1200. De, dermis; Ep, epidermis; NE, nerve ending; NF, nerve fiber; SD, stratum disjunctum

2. Herbst Corpuscles

Herbst corpuscles are the most common cutaneous receptors in birds. They have been observed in all species that have been studied, are present in various regions of the skin, and are also found in the muscles, between the muscles, and in the region of the tendons and the articular capsules (WINKELMANN and MYERS, 1961; POLACEK et al., 1966; SAXOD, 1967; MALINOVSKY and ZEMANEK, 1969, 1970; BURNS and WIGHT, 1971). The Herbst corpuscles have a certain variability of structure, size and density (MALINOVSKY, 1967a; ZISWILER and TRNKA, 1972). These are mainly species-specific differences, but are also dependent upon the precise location in the skin. Consequently many kinds of Herbst corpuscles have been described.

The general structure of Herbst corpuscles is as follows. Each corpuscle is an ellipsoidal body, which is quite large (up to 300 × 110 µm) and is formed of three parts (Figs. 5–9): a *central inner bulb* formed to two rows of specialized, so-called sensory cells that are symmetrically arranged around the central sensory nerve terminal; an *inner space* or peribulbar space, occupied by a loose lamellar system formed by flat cells in association with collagen fibers; and an *outer* lamellar *capsule.*

Examination of the different regions of the skin of duck, chicken, and quail show that it is possible to distinguish two varieties of Herbst corpuscles (Figs. 5–9).

The duck beak type is exclusively confined to the upper and lower beak of duck, where it is found in abundance, and in the tongue of duck. It is ovoid, with fairly constant dimensions (on average 160 × 100 µm; extreme values 130 × 80 µm and 190 × 120 µm). The internal bulb is composed of about 20 cells with nuclei arranged symmetrically on each side of the nerve terminal, and the internal space if obviously lamellar (Fig. 5–7).

The chicken beak type is present in the beak of chicken, and in the dermis of diverse cutaneous regions of chicken, quail, and duck, but is not found in the beak or tongue of the duck. This type is thus found in feathered skin (in particular beneath the erector muscles of the feathers (Fig. 8) and around the base of the follicles), nonfeathered skin, scaly skin, and plantar skin, including the foot webs of duck (SAXOD, 1967). Chicken beak corpuscles are elongated and may have quite different dimensions within the same zone of the skin (ranging from 25 × 12 µm to 300 × 110 µm); their internal bulb is composed of about 50 cells, which have a tendency to be arranged in several rows, and the internal space is less lamellar than it is in the duck beak type (Figs. 8 and 9). Often these bulbs branch internally and form a biaxial corpuscle.

The ultrastructural description given at this point applies to the Herbst corpuscle of the type found in the beak of the duck (QUILLIAM and ARMSTRONG, 1963; SAXOD, 1968, 1970a; GOGLIA, 1969a; HALATA, 1971a). This differs ultrastructurally somewhat from the corpuscles of Herbst in chicken (ANDERSON and NAFSTAD, 1968; NAFSTAD and ANDERSON, 1970).

The inner bulb cells (Figs. 10–12) are very asymmetric. Each cell characteristically has 20–50 processes, which form lamellae that interdigitate with the lamellae of neighboring cells (Figs. 10 and 11). Between these lamellae there are some junctions of the *zonula occludens* and *macula adherens* type (Figs. 16 and 17). In the

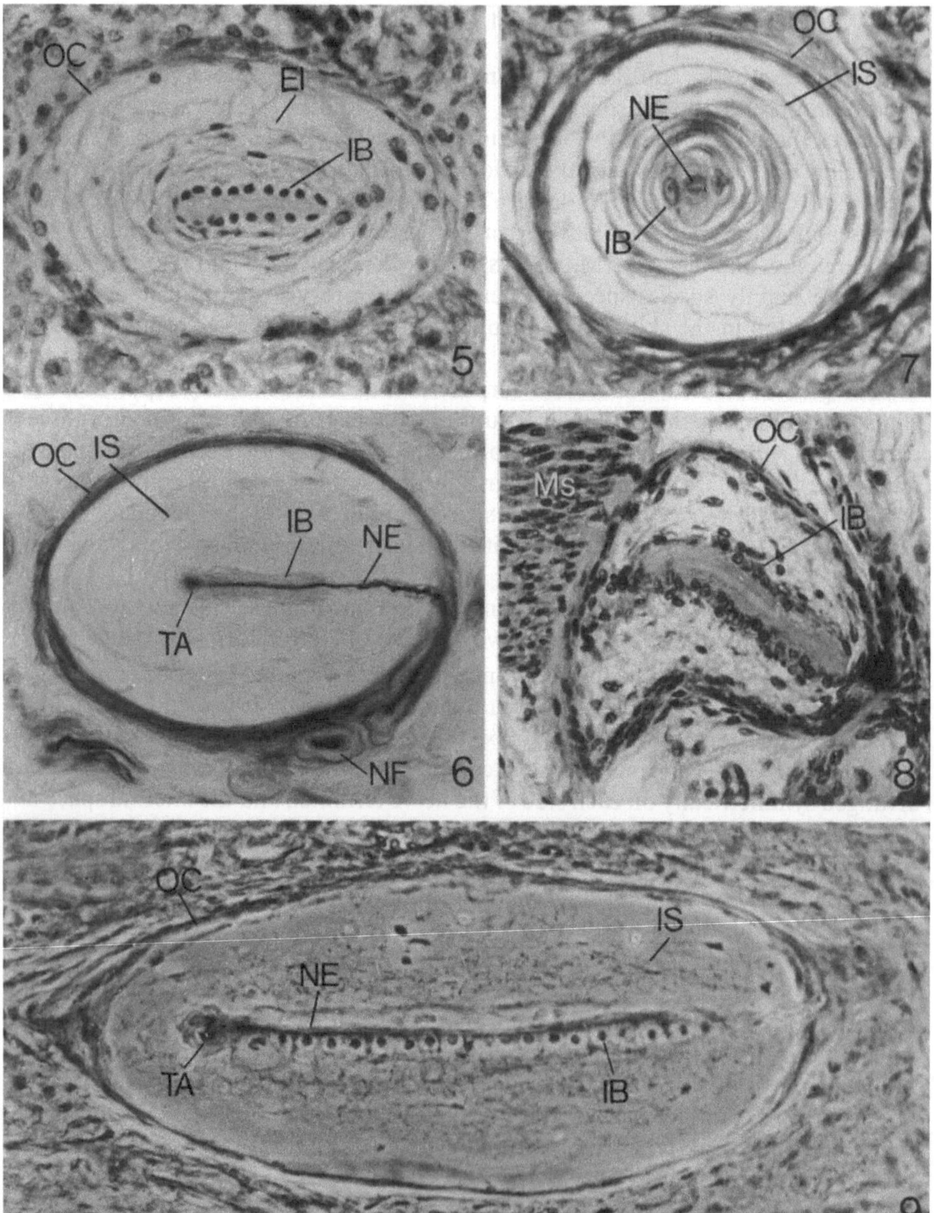

Figs. 5–7. *Herbst corpuscles, adult duck beak*, × 600. Longitudinal sections: Fig. 5 (Feulgen staining) and Fig. 6 (silver staining). Transverse section: Fig. 7 (Masson staining). Inner bulb is formed by two rows of cells bordering axial nerve ending (NE), distally dilated into terminal ampulla (TA). Inner space (IS) is lamellar. An outer capsule (OC) surrounds whole structure

Figs. 8 and 9. *Herbst corpuscles, adult chicken.* Longitudinal sections. Fig. 8. (Masson staining, × 400): this corpuscle of wing is placed close to an erector muscle (Ms) of a flight feather. Fig. 9. Corpuscle in corium of superior beak (silver staining, phase contrast × 500); inner bulb (IB) includes numerous cells, inner space (IS) is not well structured. IB, inner bulb; IS, inner space; Ms, muscle; NE, nerve ending; NF, nerve fiber; OC, outer capsule; TA, terminal ampulla

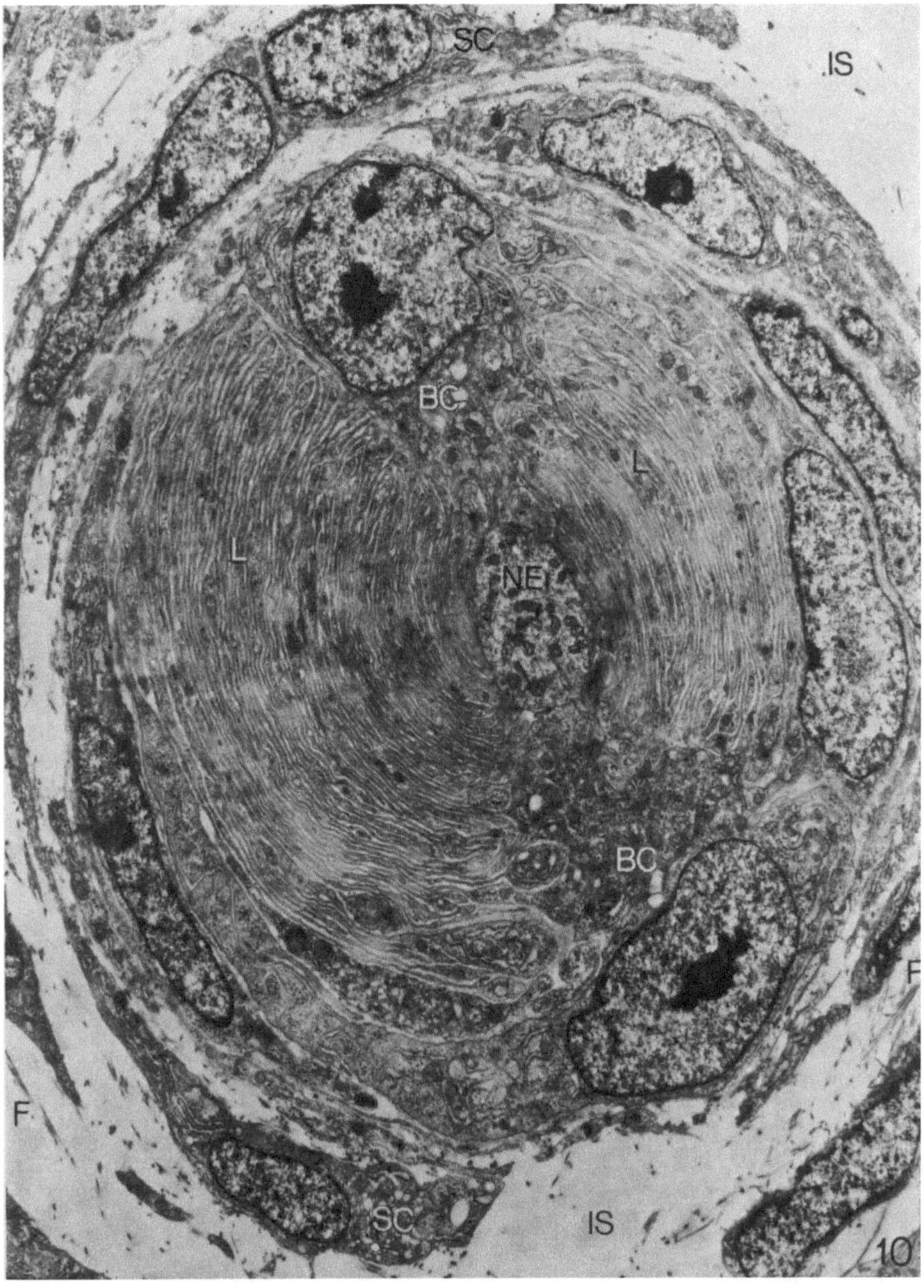

Fig. 10. Herbst corpuscle, adult duck beak. Transverse section of inner bulb. Inner bulb cells (BC) arranged symmetrically with respect to axial sensory nerve ending (NE), lamellar systems (L) closely imbricated. Inner space (IS) contains fibroblast-like cells (SC) and numerous bundles of collagen fibers (F) × 7000. BC, inner bulb cell; F, collagen fibers; IS, inner space; L, lamellae; NE nerve ending; SC, inner space cells

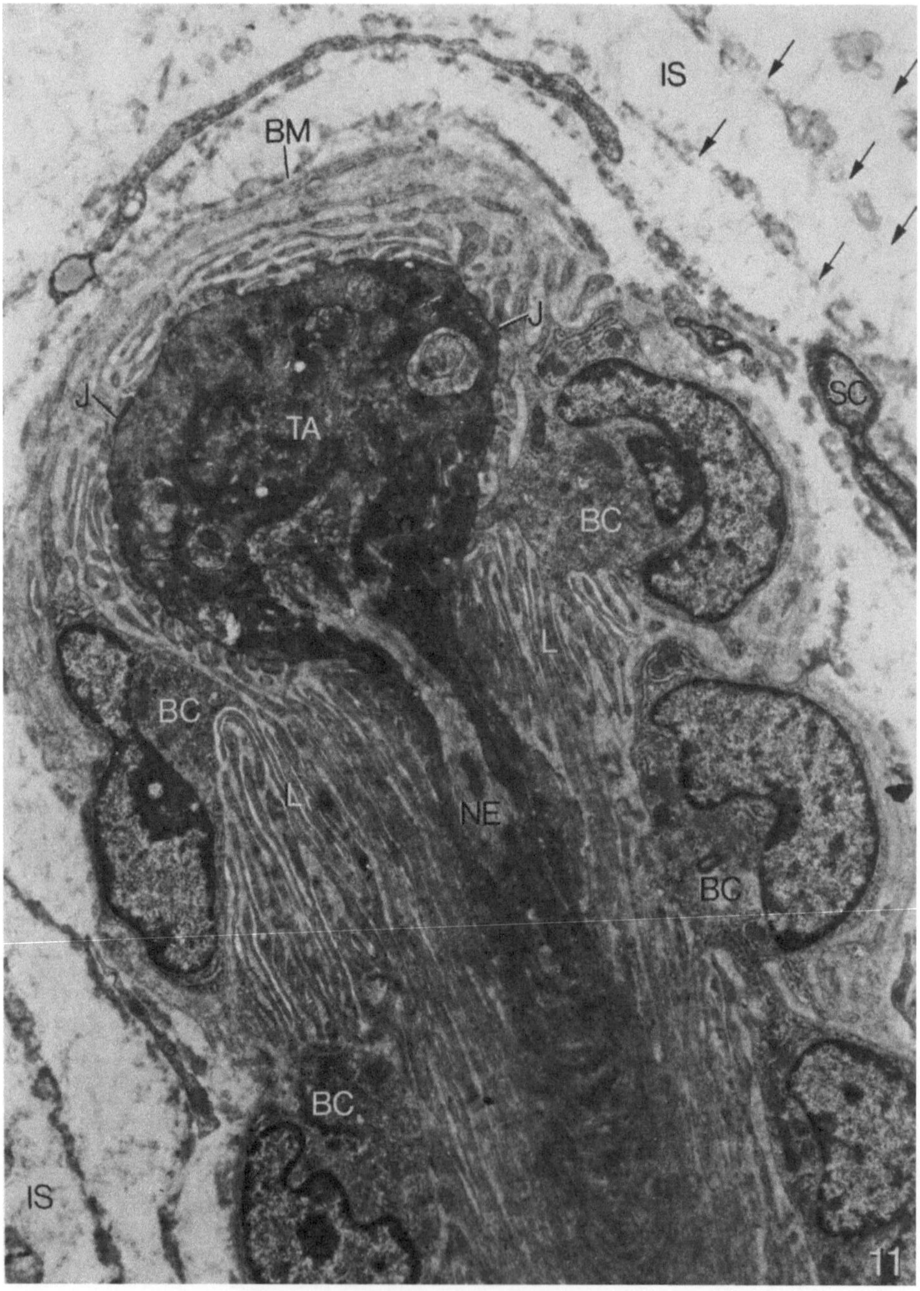

Fig. 11. *Herbst corpuscle*, adult duck beak. Longitudinal section of inner bulb at level of terminal ampulla (TA). Basal lamina (BM) is present on outer surface of inner bulb. Inner bulb formed of two rows of cells (BC) with lamellar processes (L). Some junctions (J) are present in certain zones of contact between axolemma and lamellae. Concentric lamellae of inner space (IS) are perforated (arrows). × 8800. BC, inner bulb cell; BM, basal lamina; IS, inner space; J, junction; L, lamellae; NE, nerve ending; SC, inner space cell; TA, terminal ampulla

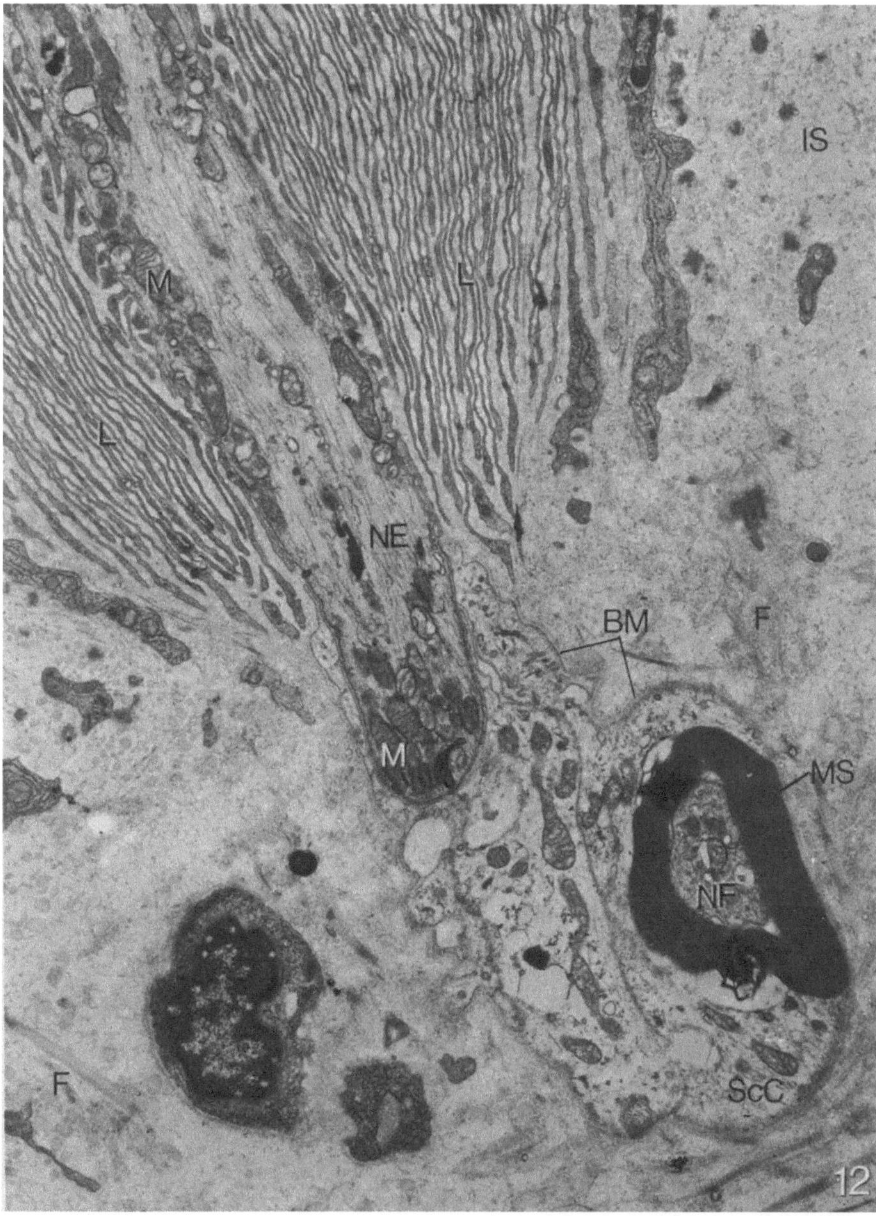

Fig. 12. *Herbst corpuscle*, adult beak. Longitudinal section of proximal part of inner bulb (peduncular zone. Nerve fiber (NF) goes through outer capsule and loses its myelin sheath (MS) at level of inner bulb. Nerve ending (NE) contains numberous mitochondria at periphery, and axial neurotubules and neurofilaments; it is enclosed by inner bulb cells (BC). × 16000. BM, basal lamina; F, collagen fibers; IS, inner space; L, lamellae; M, mitochondria; MS, myelin sheath; NE, nerve ending; NF, nerve fiber; ScC, Schwann cell

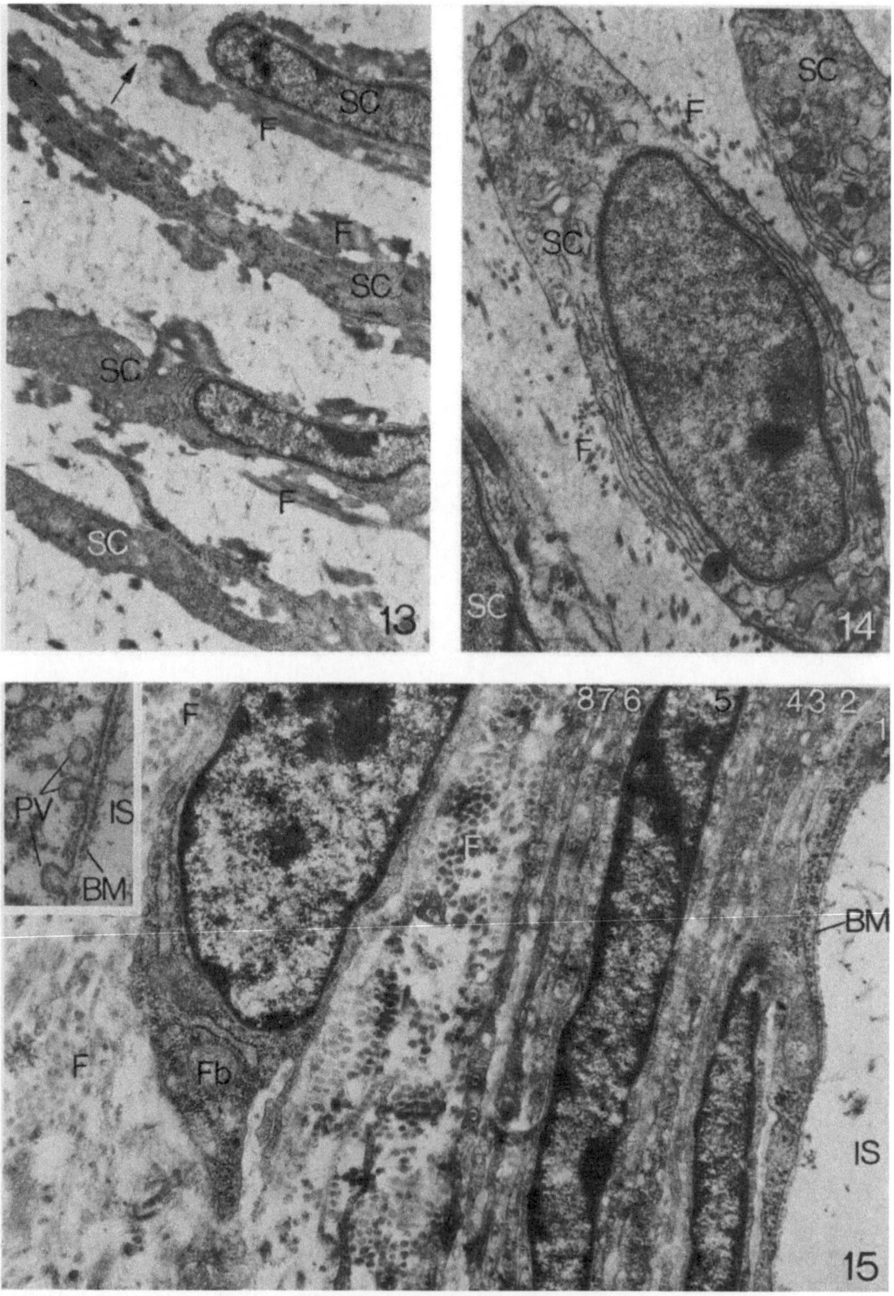

Figs. 13–15. *Herbst corpuscle*, adult duck beak. Inner space and outer capsule. Fig. 13. Perforated lamellae (arrows) of inner space are formed by flat cells (SC) and bundles of collagen fibers (F). Transverse section. × 12000. Fig. 14. Inner space cells (SC) lack basal lamina and look like fibroblasts. Longitudinal section. × 12000. Fig. 15. Outer capsule is formed of 8–10 layers (in this figure 8 layers, numbered 1–8) of very flat cells with fibrillar material in between. Basal lamina (BM) is present on side facing inner space (IS) × 25000. *Inset*: detail of inner capsular layer showing basal lamina and some pinocytotic vesicles (PV). × 55000. BM, basal lamina; F, collagen fibers; Fb, fibroblast; IS, inner space; SC, inner space cells; PV, pinocytotic vesicles

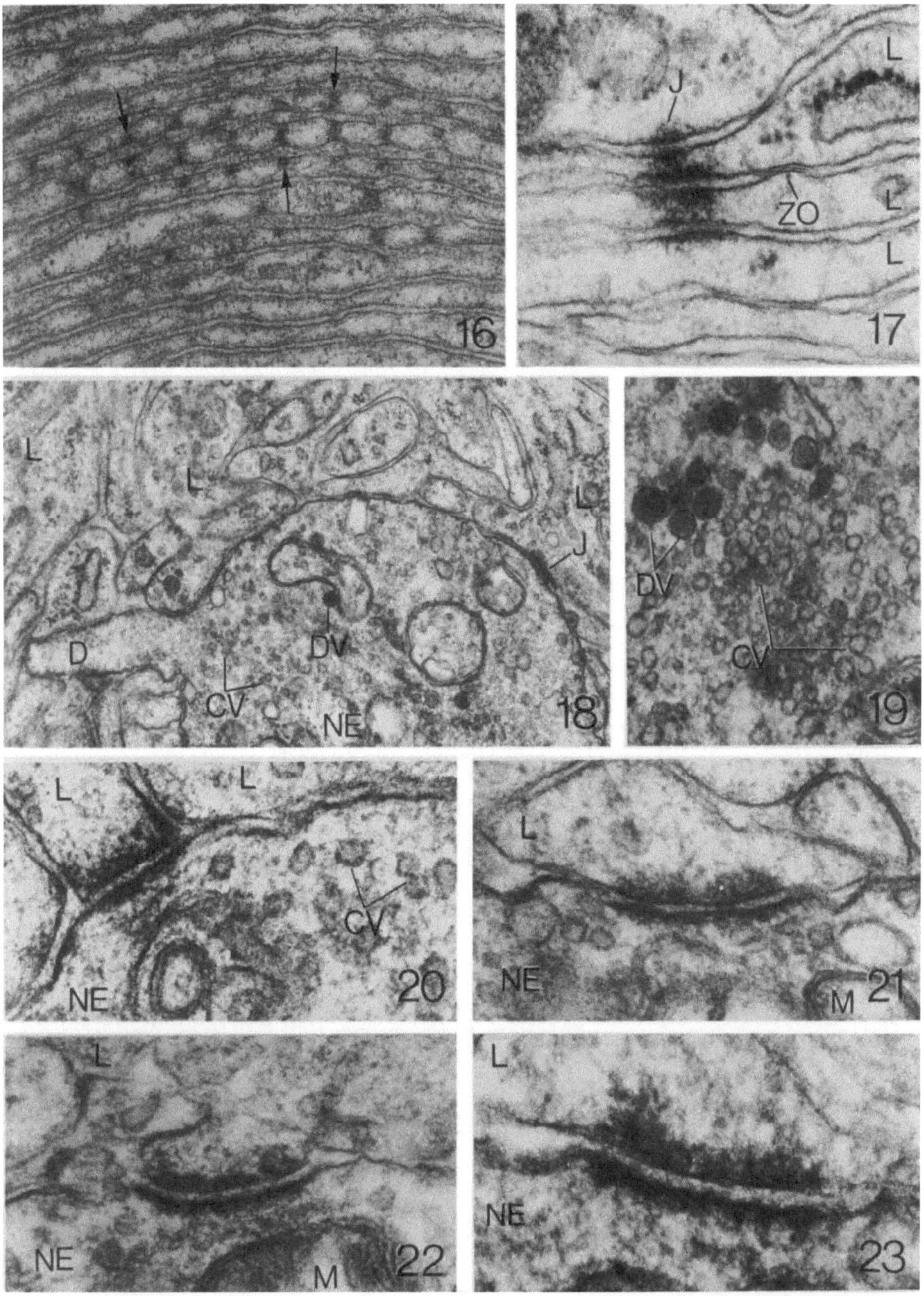

Figs. 16–23. ***Herbst corpuscle***, duck beak. ***Junction and vesicles.*** Fig. 16. Lamellae of inner bulb cells in transverse section. Numerous junctional complexes are present (arrows). ×40000. Fig. 17. *Zonula occludens* (ZO) and symmetrical junctional complexes (J) between lamellae (L) of inner bulb cells. ×85000. Fig. 18. Transverse section of axial part of inner bulb. Junctional complex (J) present between inner lamella (L) and nerve ending (NE). Nerve ending forms a digitation (D) and contains clear (CV) and dense-core vesicles (DV). ×30000. Fig. 19. Clear vesicles (CV) and dense-core vesicles (DV) of nerve ending are particularly numerous at level of the terminal ampulla. ×60000. Figs. 20–23. Asymmetric junctions between nerve ending (NE) and inner bulb lamellae (L), with thickened membranes and accumulation of granular material in vicinity. Figs. 20–22 ×85000; Fig. 23 ×125000. CV, clear vesicles; D, digitation; DV, dense core vesicles; J, junctional complex; L, lamella; M, mitochondria; NE, nerve ending; ZO, *zonula occludens*

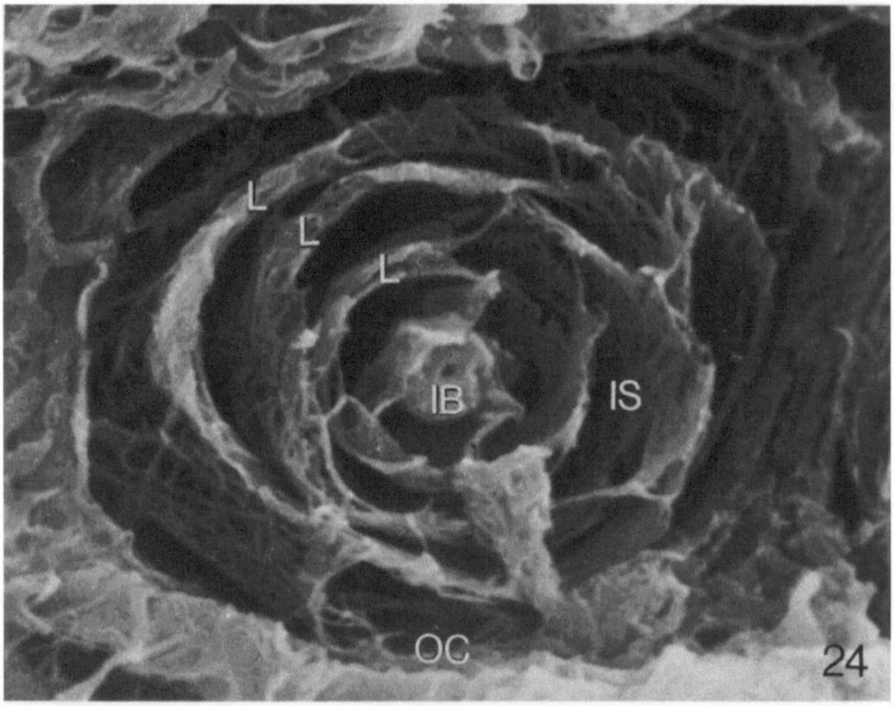

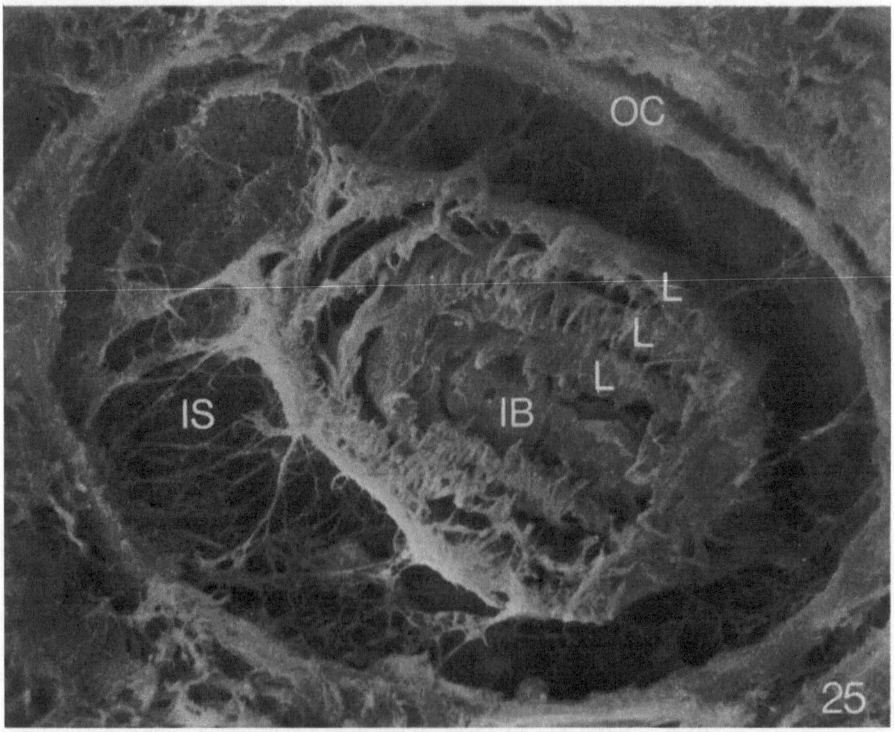

Figs. 24 and 25. *Herbst corpuscle*, adult duck beak. Scanning electron microscope, gold coating. × 1500. Fig. 24. Transverse section, CO_2 critical point drying. Fig. 25. Longitudinal section, air drying. Perforated lamellae (L) of inner space (IS) have concentric arrangement around inner bulb (IB). These lamellae are formed of flat cells interconnected by cytoplasmic extensions and are strengthened by large bundles of collagen fibers. IB, inner bulb; IS, inner space; L, lamella; OC, outer capsule

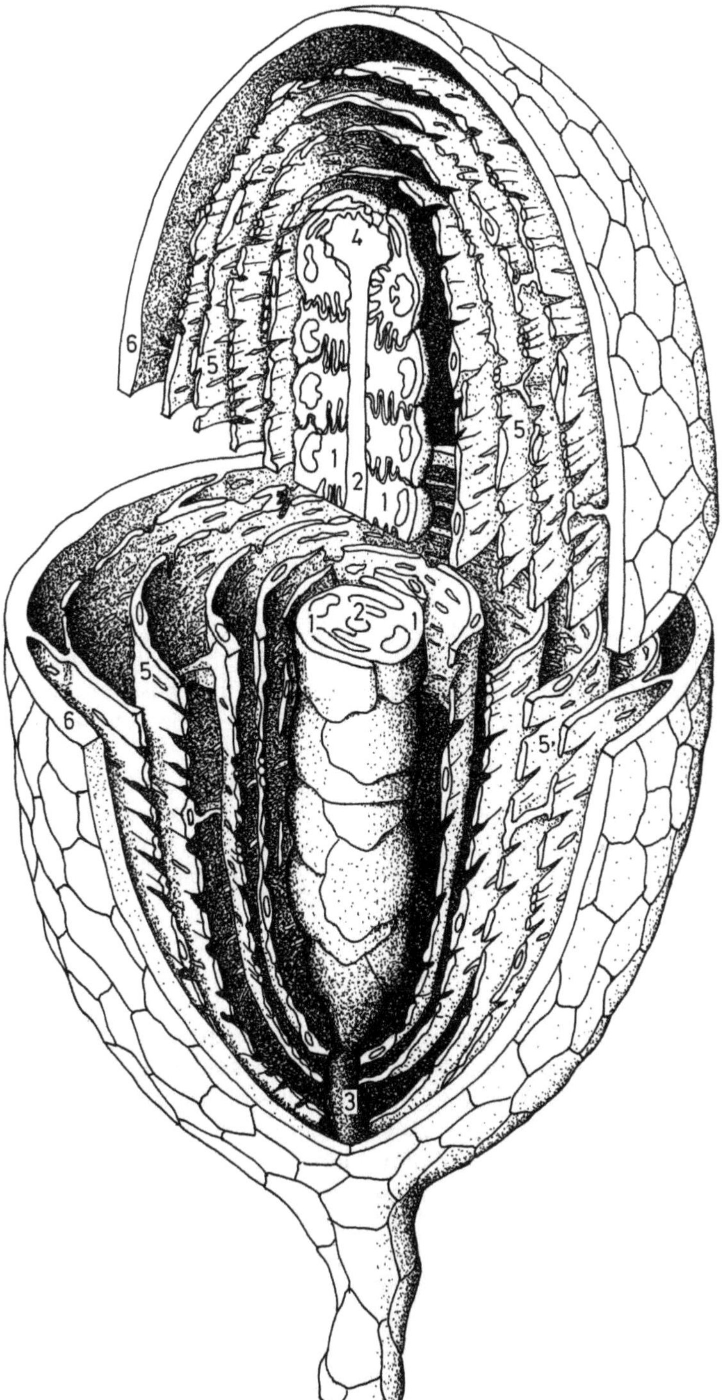

Fig. 26A–C. *Spatial diagram of the Herbst corpuscle* (simplified). Duck beak. Three parts can be distinguished: (A) A central *inner bulb* formed of two rows of imbricated lamellar cells (inner bulb cells, 1) along axial nerve ending (2). Myelinated proximal part of nerve ending forms peduncle (3) of inner bulb. Distally, nerve is enlarged into terminal ampulla (4). (B) An *inner space* (or peribulbar space) (very simplified in this drawing) containing 15–20 perforated lamellae (5) formed of flat cells and collagen fibers. (C) An *outer capsule* (6) with lamellar structure, formed by joined flat cells

Herbst corpuscle of the beak of the chicken, nonmyelinated nerve fibers have been observed between the lamellae (NAFSTAD and ANDERSON, 1970). A basal lamina is present on the external surface of these inner bulb cells (Fig. 11). It is continuous with the basal lamina of the Schwann cells, which form the intracorpuscular myelinated part of the nerve terminal (Fig. 12). The myelin sheath is lost shortly after the nerve penetrates through the capsule and just before it reaches the inner bulb cells. The unmyelinated nerve ending is oval in cross-section, measures 3×5 μm on average, and ends distally in a terminal ampulla with a diameter of 8–10 μm (Fig. 10). Spines up to 3 μm in length are often observed on the terminal ampulla (Fig. 11). In the nerve endings there are two types of vesicles: clear vesicles with a diameter of 400–800 Å, looking like synaptic vesicles which are particularly abundant in the region of the terminal ampulla, and dense core vesicles with a diameter of 600–1000 Å (Figs. 18 and 19). Some specialized contacts can be seen between the membranes of the lamellae and the axolemma. These consist of a thickening of the membrane and an accumulation of granular material in their immediate neighbourhood (Figs. 20–23). The structure of this junction is generally asymmetric.

The inner space is composed of 15 to 20 concentrically arranged lamellae, irregularly perforated, and separated from one another by a fluid-filled space. The cells of the lamellae are fibroblastic in type and do not possess a basal lamina (Fig. 14). These flattened cells are reinforced by collagen fibers in the interlamellar space (Figs. 11 and 13).

The outer capsule (Fig. 15) is 2–4 μm thick and is composed of 6–10 concentrically arranged lamellae reinforced by collagen fibers. These cells have numerous structures resembling pinocytic vesicles (Fig. 15). A basal lamina is very obvious on the inner face of the capsule.

3. Grandry Corpuscles

These receptors are found characteristically in aquatic birds and are exclusively confined to the superficial dermis of the beak and the tongue, where they are found in great numbers intermingled with the corpuscles of Herbst. They are absent in chicken, pigeon, and quail (for review, MALINOVSKY, 1967a). Typical Grandry corpuscles (Figs. 27–30) are formed of two large specialized cells called the *cells of Grandry* (also called sensory cells). These cells are hemispherical, with dimensions of approximately 50×15 μm; their flat surfaces are parallel to the epidermis and they enclose a discoid *nerve ending*. This arrangement is enclosed by *satellite cells*. Around each corpuscle there is an incompletely formed capsule of collagen fibers and dermal fibroblasts. Fine unmyelinated nerve fibers are frequently seen around the capsule. Some Grandry corpuscles contain up to seven sensory cells. A single nerve fiber may branch to supply more than one Grandry corpuscle (Fig. 31).

There have been several studies of the ultrastructure of Grandry corpuscle in duck (QUILLIAM, 1966a, b; SAXOD, 1968, 1970b, 1973d and unpubl.; GOGLIA, 1969b; MUNGER, 1971a, b) and in goose and swan (HALATA, 1971b). The Grandry cells (Fig. 32) are characterized by numerous membrane-bound electron-dense granules (dense-core vesicles) with diameter between 1200 and 2800 Å which appear

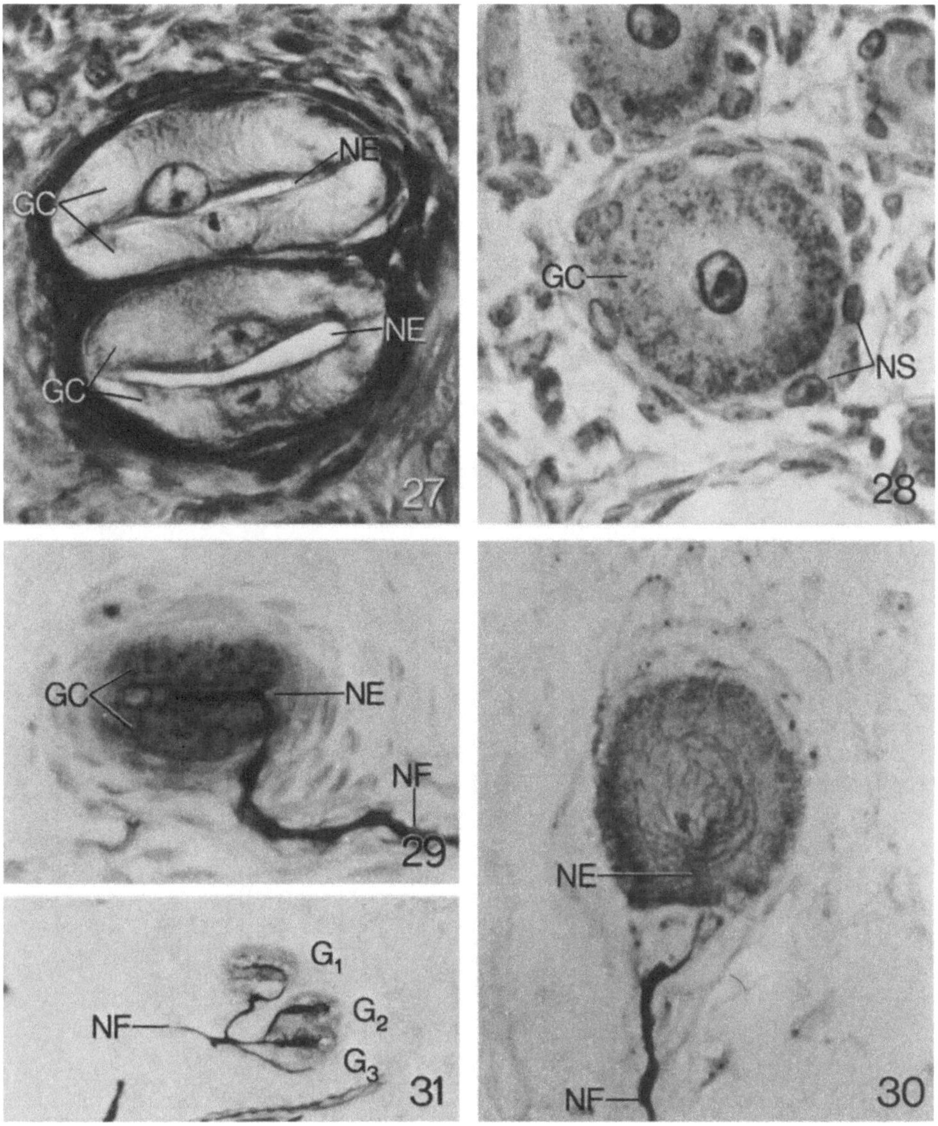

Figs. 27–31. *Grandry corpuscle*, adult duck beak. Fig. 27. (Masson staining), longitudinal sections of two Grandry sensory corpuscles. Each corpuscle is formed of two Grandry sensory cells (GC) embracing nerve ending (NE). Note "sheaf-like" perinuclear striation. Fig. 28. (Frontal section, silver staining): Nuclei of satellite cells (NS) are conspicuous. Note numerous argyrophil granules at periphery of Grandry cell. Fig. 29. (Longitudinal section, silver staining): The two Grandry cells embrace nerve ending (NE). Fig. 30. (Silver staining): Frontal section of terminal nervous disk. Neurofibrillar network is conspicuous. Fig. 31. (Silver staining): several Grandry corpuscles (in this figure three: G_1, G_2, G_3) may be innervated by terminal branches of a single nerve fiber (NF). Figs. 27–30 × 1200; Fig. 31 × 500. GC, Grandry cell; NE, nerve ending; NF, nerve fiber; NS, nucleus of a satellite cell

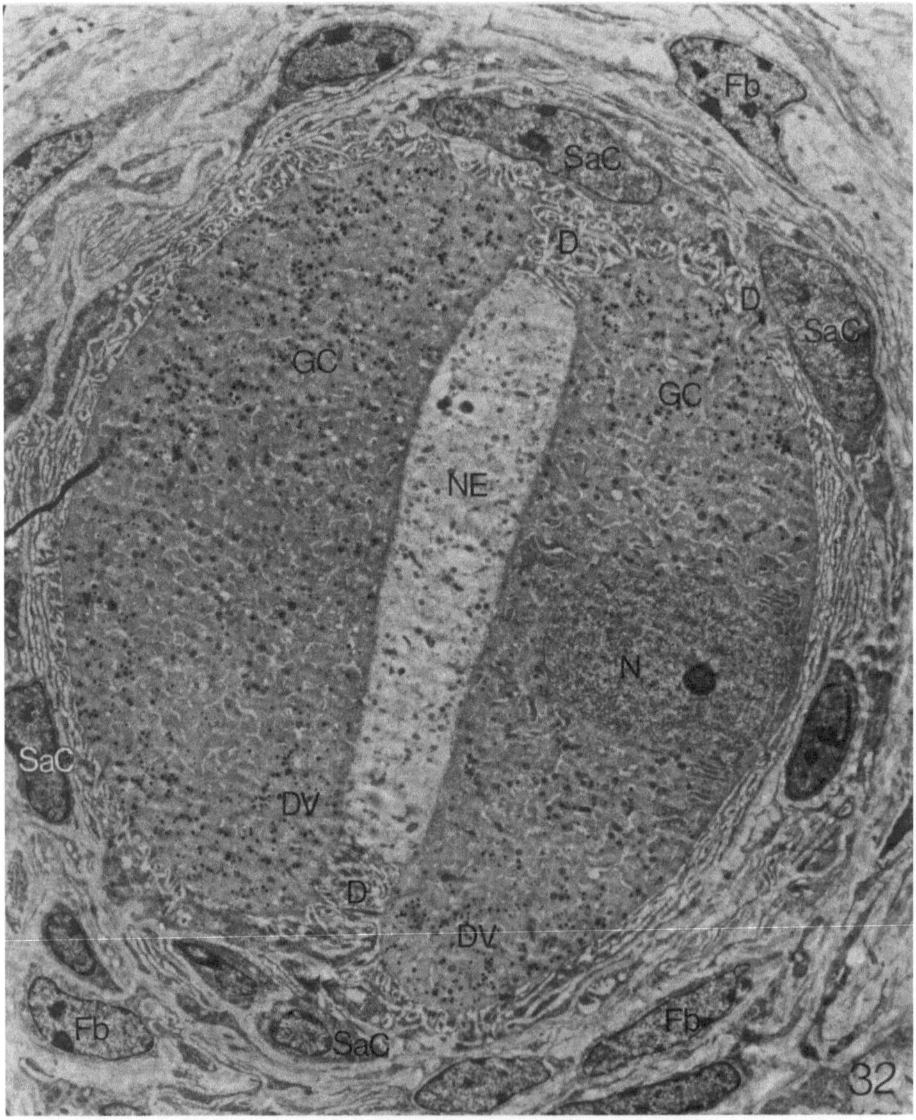

Fig. 32. *Grandry corpuscle*, longitudinal section. The two Grandry cells (GC) are hemispherical. Flat side is associated with nerve ending (NE). Cytoplasm of Grandry cell is characterized by numerous dense-core vesicles (DV), mainly localized at periphery of cell. Satellite cells (SaC) display numerous digitations (D) imbricated with finger-like processes of Grandry cells. Some dermal fibroblasts and bundles of collagen fibers form an incomplete capsule around the corpuscle. × 3200. D, digitations; DV, dense-core vesicles; Fb, fibroblast; GC, Grandry cell; N, nucleus; NE, nerve ending; SaC, satellite cell

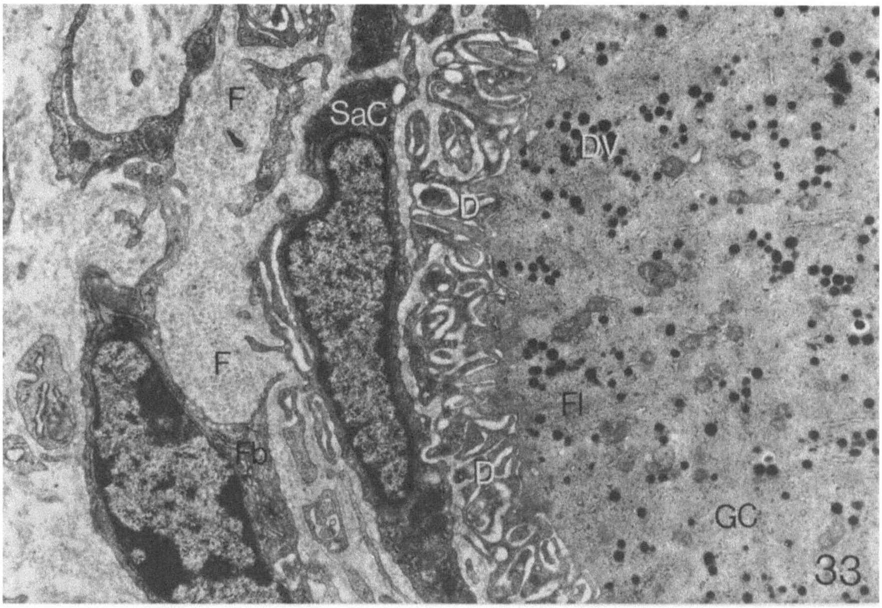

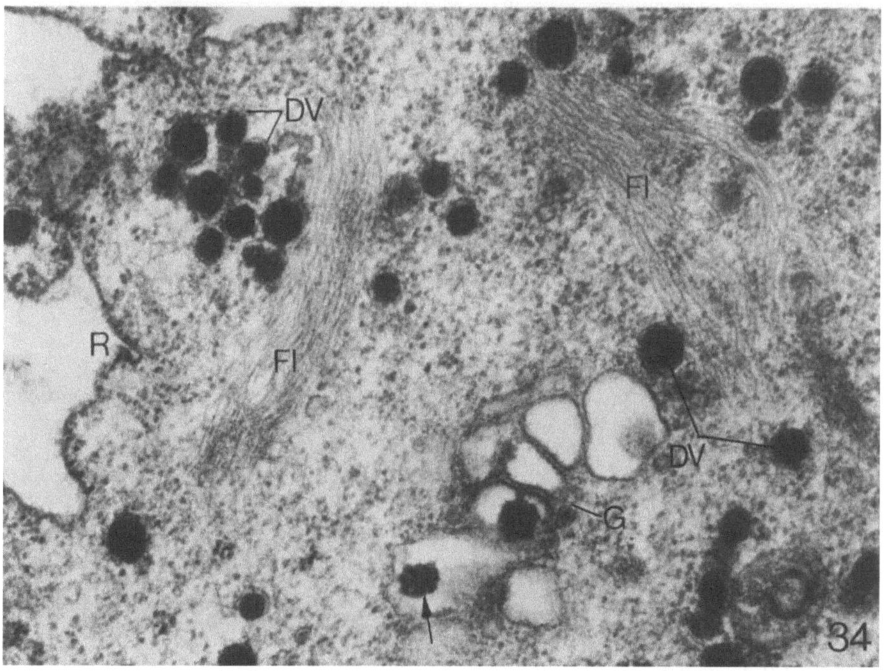

Figs. 33 and 34. *Grandry corpuscle*, adult duck beak. Fig. 33. Imbrication of digitations (D) of satellite cells (SaC) and Grandry cells (GC). Presence of numerous dense-core vesicles (DV) and microfilaments (Fl) in Grandry cell. Dermal fibroblasts (Fb) and collagen fibers (F) form incomplete capsule around corpuscle. × 14000. Fig. 34. Grandry cell. Dense-core vesicles (DV) are elaborated by Golgi apparatus (G). Dense core is formed by binding of elemental granules (arrow). Presence of characteristic bundles of microfilaments (Fl) in cytoplasm of Grandry cell. × 60000. D, digitations; DV, dense core vesicles; F, collagen fibers; Fb, fibroblast; Fl, microfilaments; G, Golgi apparatus; GC, Grandry cell; R, rough reticulum; SaC, satellite cell

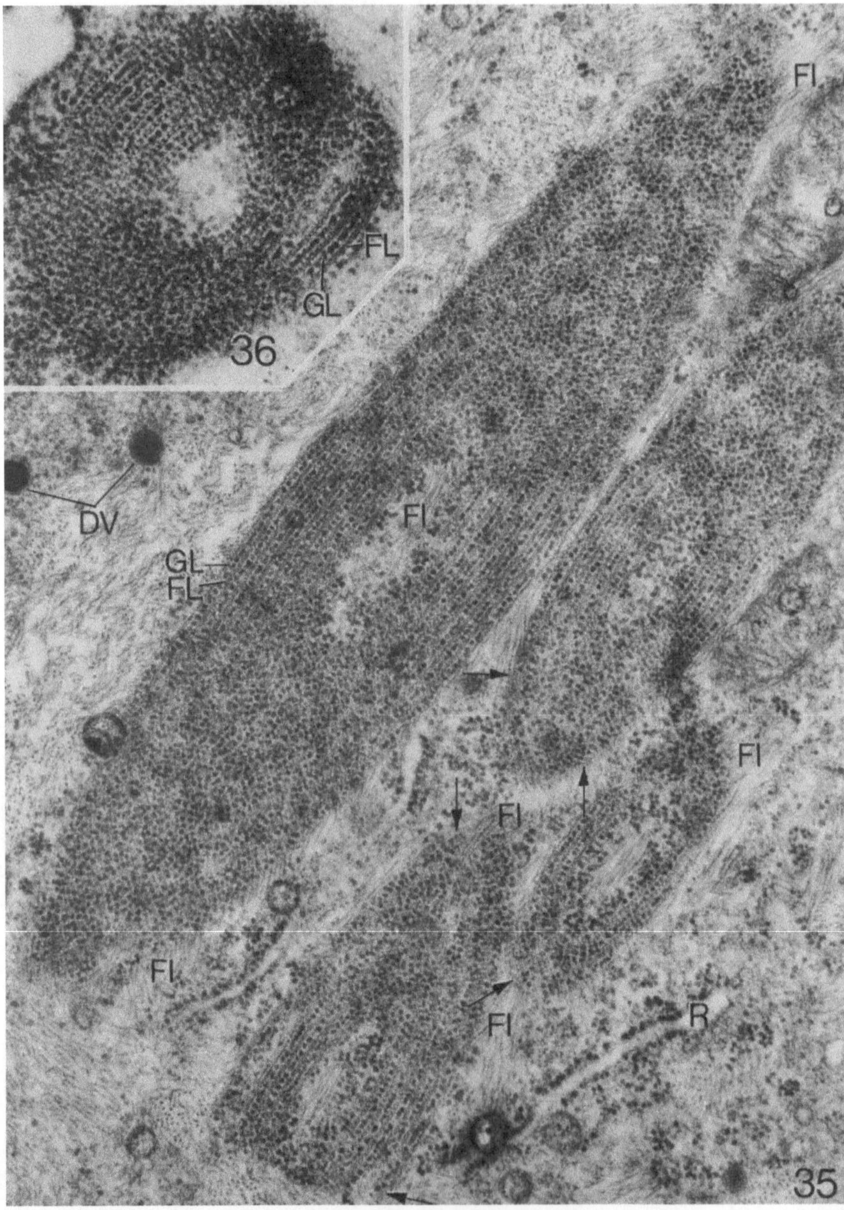

Figs. 35 and 36. *Perinuclear organelles* of the Grandry cell. Fig. 35. Longitudinal section of fusiform and conical organelles at different stages of development. × 50000. Fig. 36. Conical organelles. × 70000. These perinuclear organelles are formed by regular alternation of granular layers (GL) and fibrous layers (FL). The microfilaments of the fibrous layers are in continuity with the intracytoplasmic microfilaments (arrows). DV, dense core vesicles; FL, fibrous layer; Fl, microfilaments; GL, granular layer; R, rough reticulum.

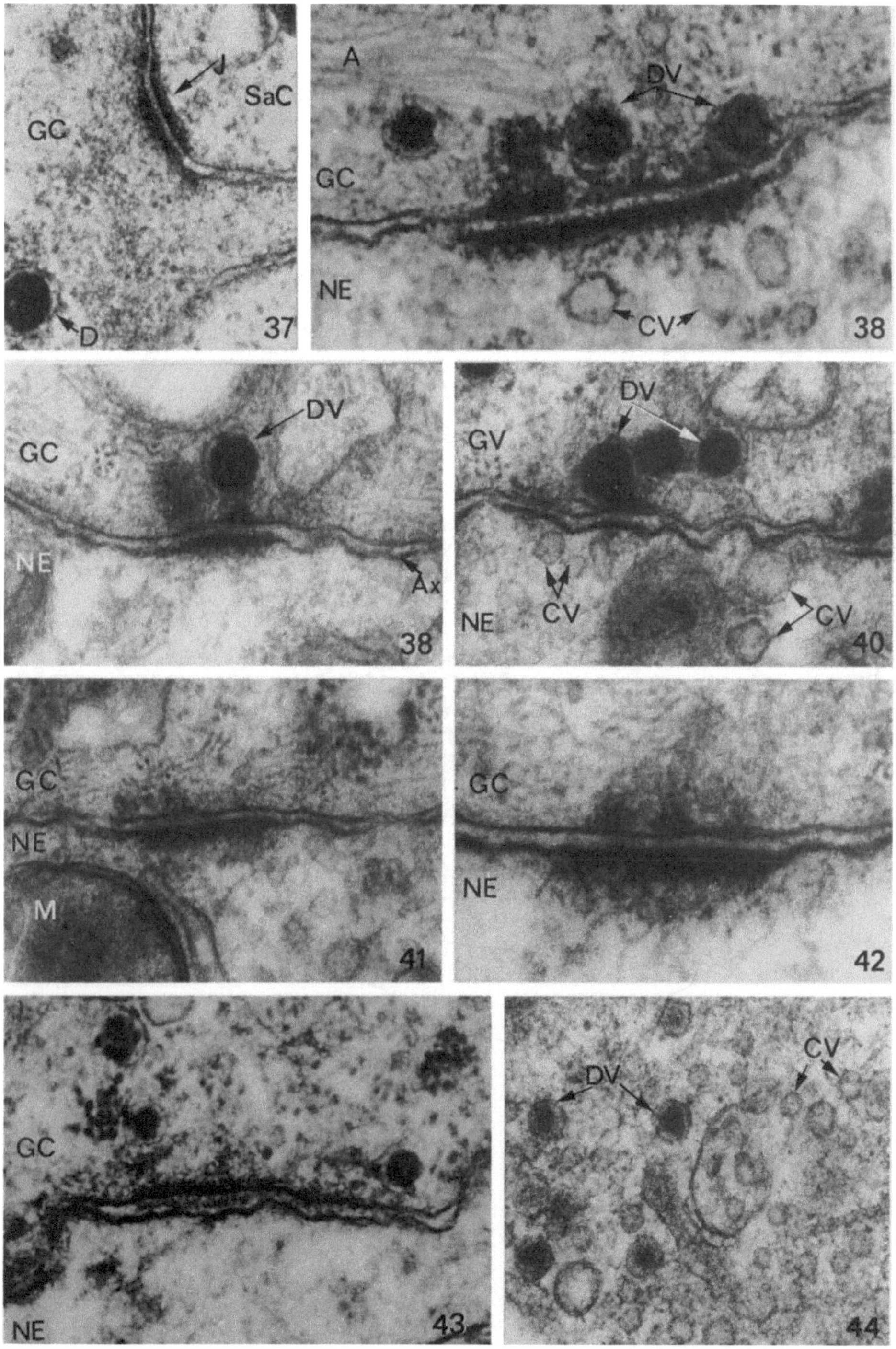

Legends to this Figures see p. 357

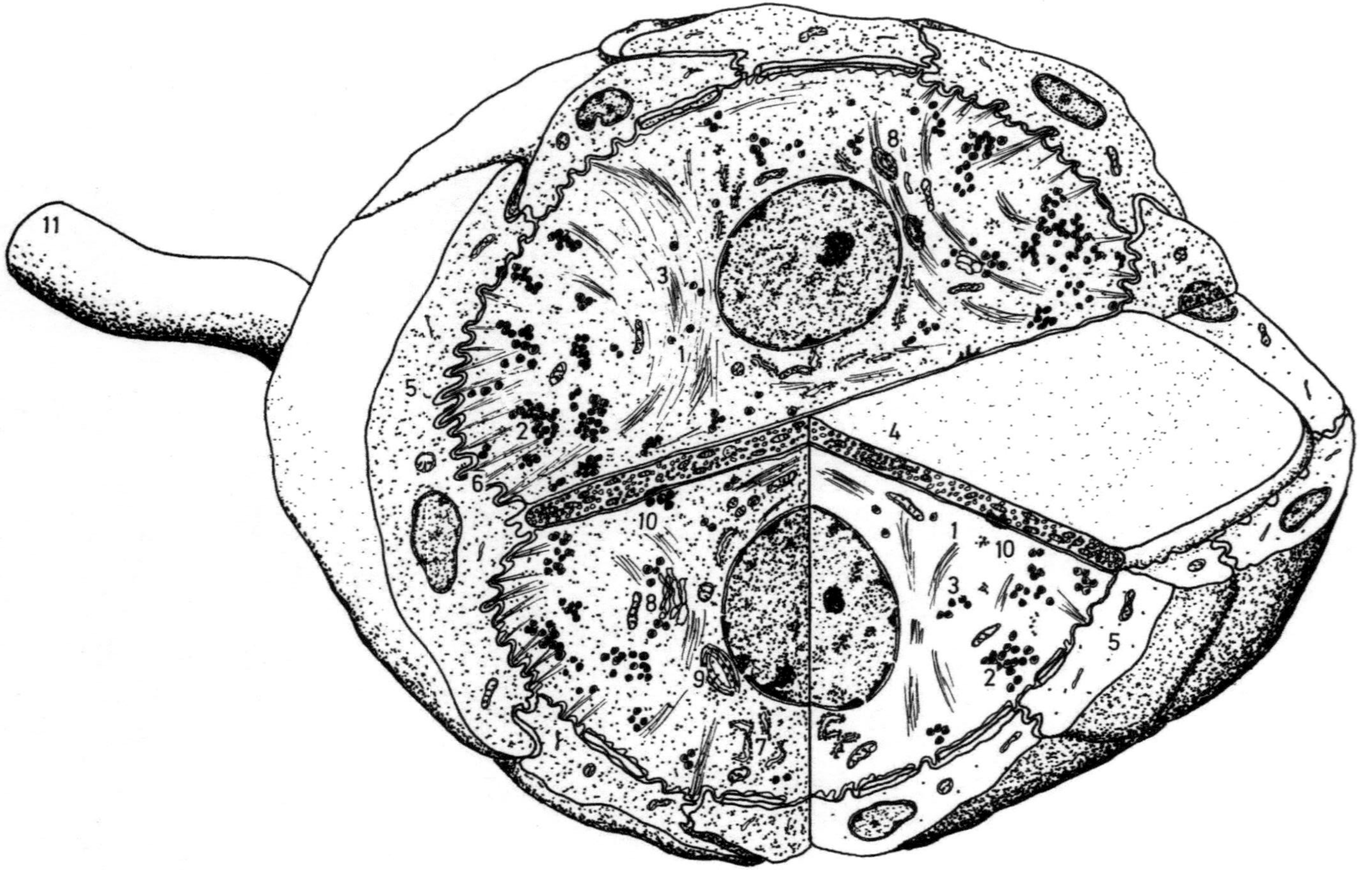

Fig. 45. *Spatial diagram of the Grandry corpuscle* (duck beak). A typical Grandry corpuscle is formed by: (A) Two large hemispherical cells, called *Grandry cells* (1), with cytoplasm characterized by numerous dense-core vesicles (2) and bundles of microfilaments (3); (B) A disk-like *nerve ending* (4), associated with flat side of Grandry cells; (C) *Satellite cells* (5) displaying numerous digitations (6) imbricated with finger-like processes of Grandry cells. Thin incomplete capsule surrounding corpuscle is not shown in this drawing. (7) granular endoplasmic reticulum; (8) Golgi apparatus; (9) perinuclear organelles; (10) specialized junctions; (11) afferent nerve fiber

to originate from the Golgi apparatus (Fig. 34), and by large bundles of 60–80 Å diameter microfilaments (Fig. 34). The large bundles of microfilaments may also be seen with the light microscope as streaks near the nucleus (Fig. 27). Unique to these cells are cytoplasmic organelles termed "perinuclear organelles" by the present author (SAXOD, 1973c). These are avoid fusiform structures 5 × 1 μm in size (Figs. 35 and 36). They are formed by a regular alternation of layers of granules rich in RNA, which are thought to be ribosomes, and layers of cytoplasmic microfilaments. The functions of these perinuclear organelles are not known.

Each Grandry corpuscle is innervated by a single myelinated nerve fiber (Figs. 29 and 30). The nerve loses its myelin sheath shortly after penetrating the capsule and ends as a discoid enlargement 40 μm in diameter and 2 μm thick. The nerve terminal is rich in mitochondria, contains neurofilaments, neurotubules, clear vesicles with a diameter of 400—800 Å, which look like synaptic vesicles, and a few dense-core vesicles that are generally 600–1000 Å in diameter (Fig. 44).

Various types of junctions can be observed between the sensory cells of Grandry and the nerve terminal (Figs. 38–43). The most common type is asymmetric and in some cases the dense-core vesicles of the Grandry cells may be seen in close apposition to a membrane thickening (Figs. 38–42).

The satellite cells (Figs. 32, 33, and 37) have a basal lamina on the side of the cell facing the capsule. The inner surface has numerous folds, which interdigitate with finger-like processes of the Grandry cells (Figs. 32, 33, and 45).

4. Merkel Corpuscles

In birds, the so-called "Merkel corpuscles" are found exclusively in the dermis. They are observed only in nonaquatic birds and are confined to the beak, in the region of the palate, where they are found in groups between the palatal epithelial digitations (ANDERSEN and NAFSTAD, 1968; NAFSTAD, 1971a; SAXOD, 1973d, 1978b) (Figs. 46–49). Merkel corpuscles consist of three parts (Fig. 43): one or several specialized cells called *Merkel cells*, discoid in form (10 × 3 μm) and arranged parallel to the surface of the epidermis; discoid *nerve terminals*; and *lamellar cells* enclosing the entire structure. Merkel cells are characterized by the presence of bundles of microfilaments and by electron-dense core vesicles (Figs. 50 and 51).

The nerve terminal, in addition to containing numerous mitochondria, neurofilaments and neurotubules, also contains clear vesicles of the synaptic type as

Figs. 37–44. *Grandry corpuscle. Junctions and vesicles.* Fig. 37. Symmetric junctional complex (J) between a Grandry cell (GC) and a satellite cell (SaC). × 80000. Figs. 38–43. Different types of asymmetric junctions between Grandry cell (GC) and nerve ending (NE), with thickened membranes and accumulation of granular and fibrillar material in vicinity. In most cases, dense zone is thicker on nervous side than on Grandry cell side of junction. On Grandry cell side, thickened membrane often forms triangular expansions (Fig. 42). Sometimes some dense-core vesicles are in close contact with these expansions and seem to fuse with them (Figs. 38–40). In a few cases (Fig. 43), dense zone of the junction is thicker on the Grandry cell side, and there is granular material found in the intermembrane space. Figs. 38–41 and 43 × 100000; Fig. 42 × 140000. Fig. 44. Nerve ending contains clear vesicles (CV) and dense-core vesicles (DV). × 100000. CV, clear vesicles; DV, dense core vesicles; Fl, microfilaments; GC, Grandry cell; J, junctional complex; NE, nerve ending; SaC, satellite cell

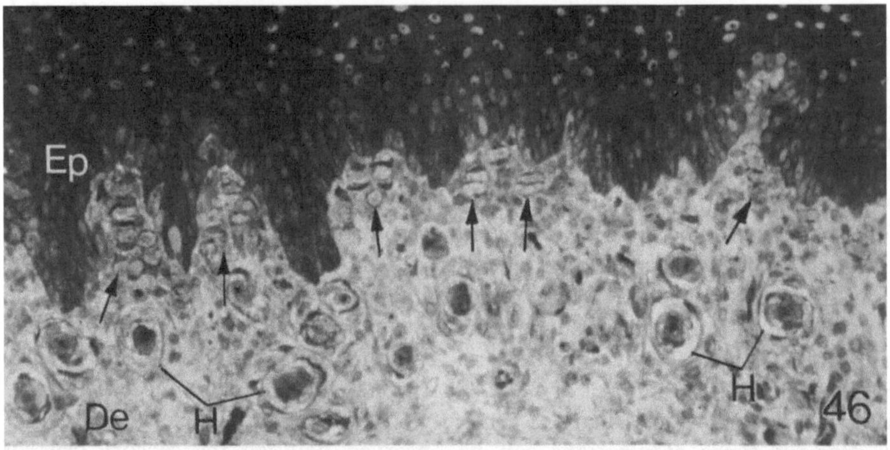
Ep
De
H
H
46

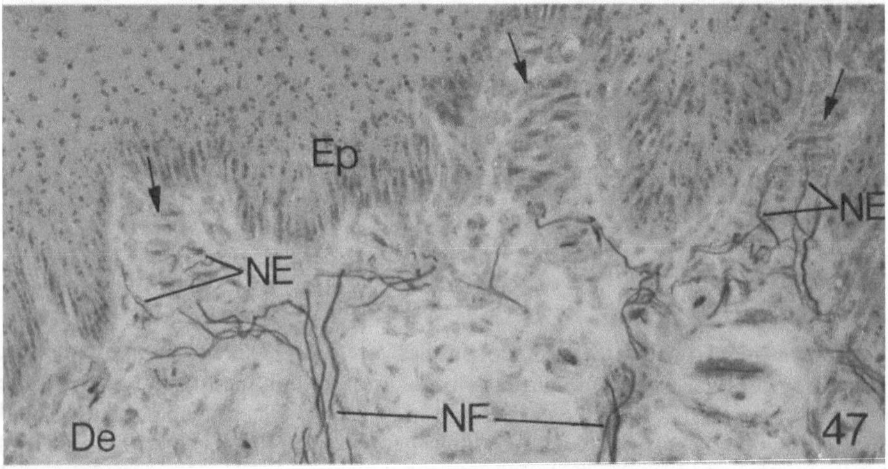
Ep
NE
NE
NF
De
47

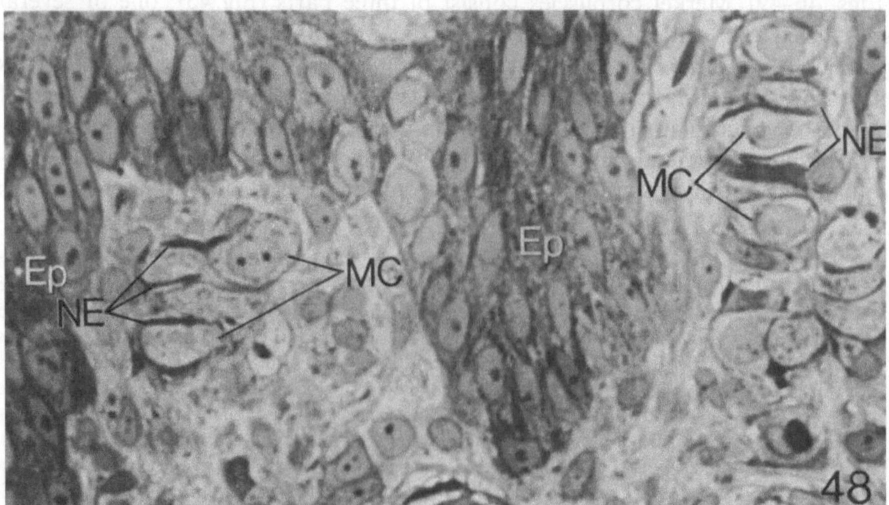
NE
MC
Ep
Ep
NE
MC
48

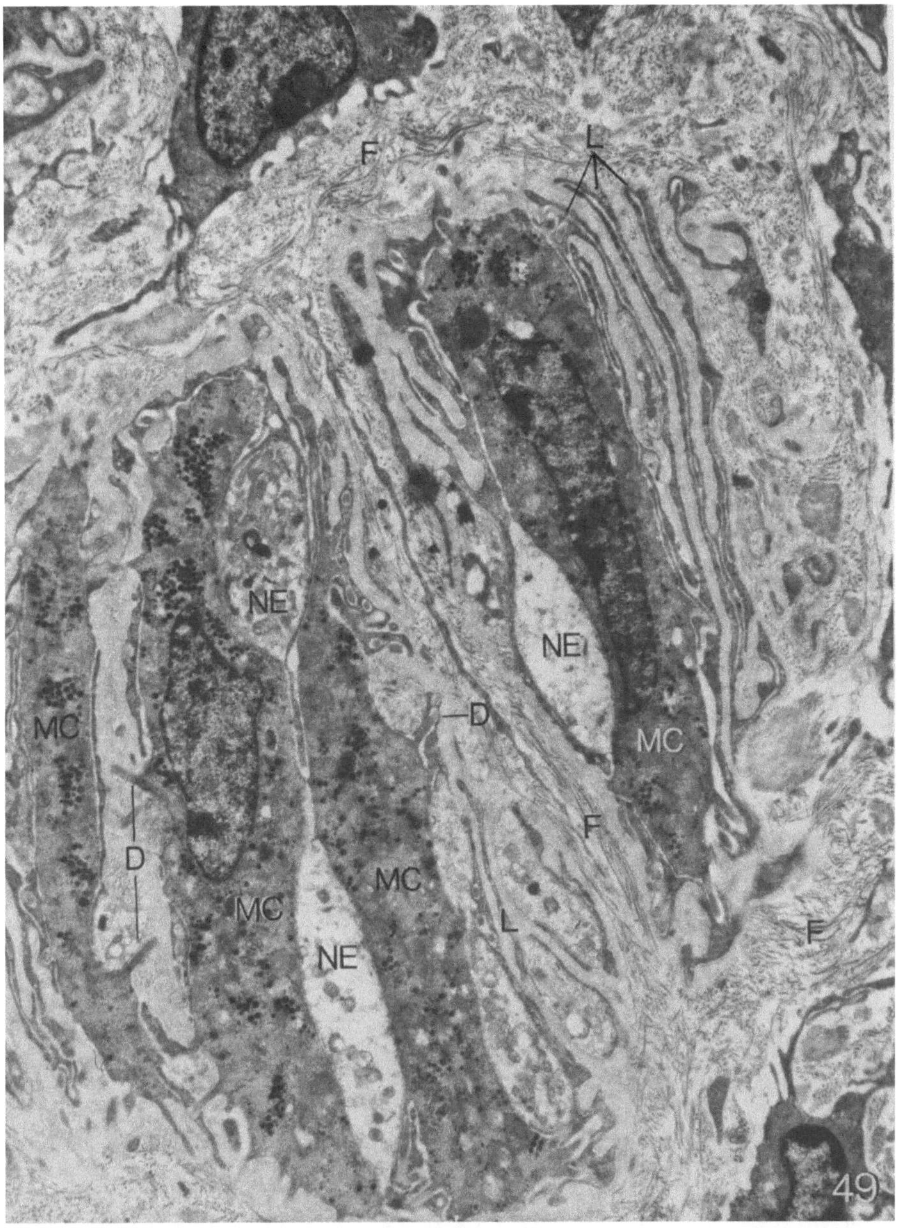

Fig. 49. *Merkel corpuscles*, adult quail beak. The Merkel cells (MC) are characterized by numerous dense-core vesicles, bundles of microfilaments and digitations (D). They are in contact with nerve endings (NE). Every group of Merkel cells and nerve endings is surrounded by lamellae (L) of lamellar cells, and whole structure constitutes Merkel corpuscle. Note fibrillar and collagen fibers (F) between Merkel corpuscles. D, digitations; F, collagen fibers; L, lamellae; MC, Merkel cells, NE; nerve ending

◁ Figs. 46–48. *Merkel corpuscles*, adult chicken beak. Fig. 46. Merkel corpuscles (arrows) located in dermis (De) and form irregular groups between "rete pegs" of palatine epithelium (Ep). Note small Herbst corpuscles (H). Semithin transverse section. × 240. Fig. 47. Numerous nerve fibers (NF) innervate groups of Merkel corpuscles (arrows). Transverse section, silver staining. × 400. Fig. 48. Merkel cells (MC) have spherical nucleus and granular cytoplasm. They are in contact with nerve endings (NE, dark zones). Semithin transverse section. × 1000. Ep, palatine epithelium; De, dermis; H, Herbst corpuscles; MC, Merkel cells; NE, nerve ending; NF, nerve fibers

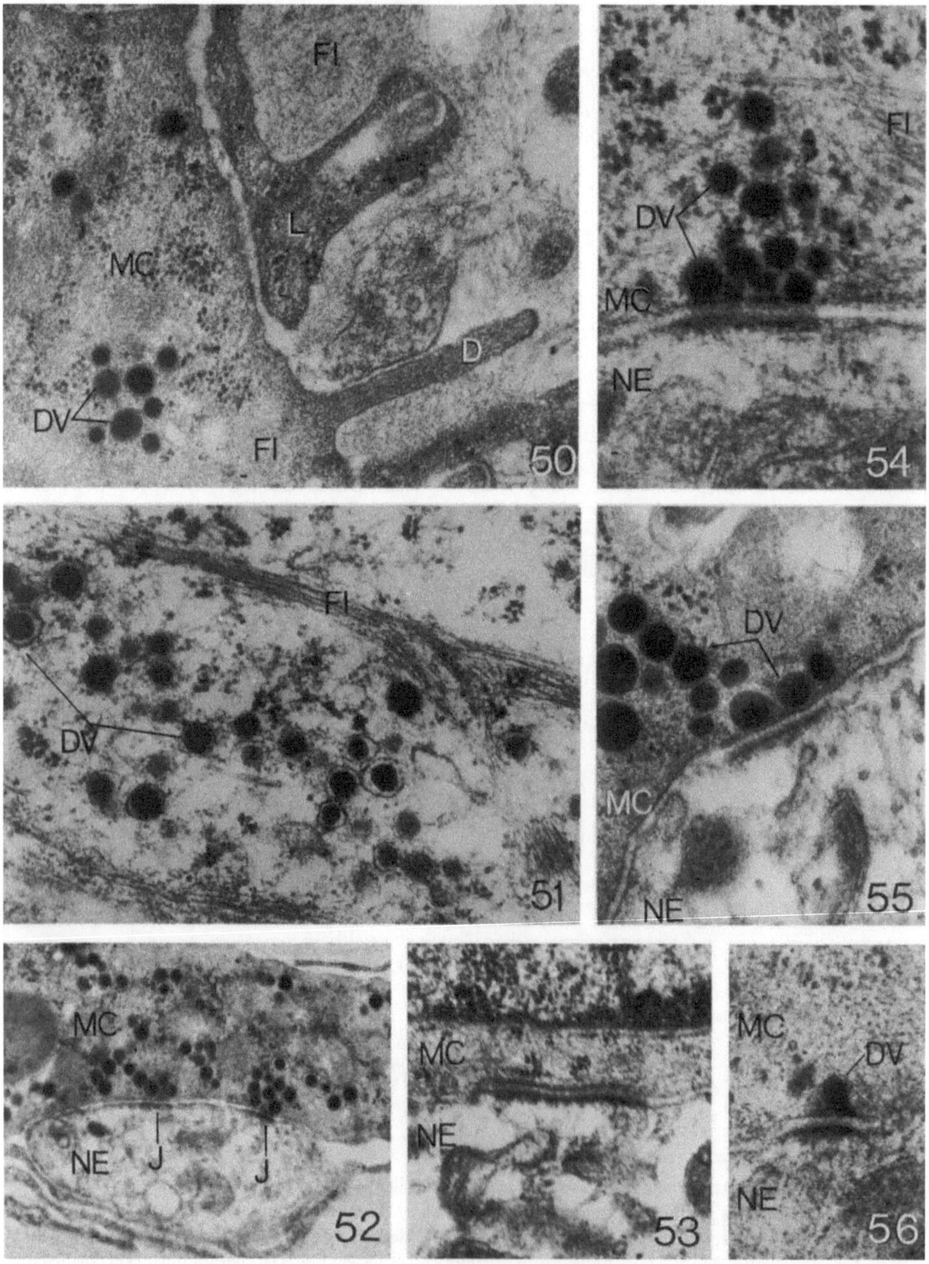

Figs. 50–56. *Merkel corpuscles*, adult chicken beak. Fig. 50. Zone of contact between lamellae (L) of lamellar cells and Merkel cells (MC). Merkel cells display digitations (D) with a fibrillar axis (Fl) and cytoplasmic dense-core vesicles (DV). Note network of fibrils (Fl) around lamellar cells. × 50000. Fig. 51. Detail of bundles of microfilaments (Fl) and dense-core vesicles (DV) of a Merkel cell. × 50000. Figs. 52–56. Different types of asymmetric junctions between Merkel cell (MC) and nerve ending (NE). In most cases, dense zone is thicker on nervous side of junction than on Merkel cell side. Sometimes Merkel cell displays dense-core vesicles (DV) in close contact with thickened junctional membrane (Figs. 54–56). Fig. 52 × 18000; Fig. 53 × 35000; Figs. 54–56 × 70000. D, digitation; DV, dense core vesicles; Fl, fibrils; J, junction; L, lamella of the lamellar cells; MC, Merkel cell; NE, nerve ending

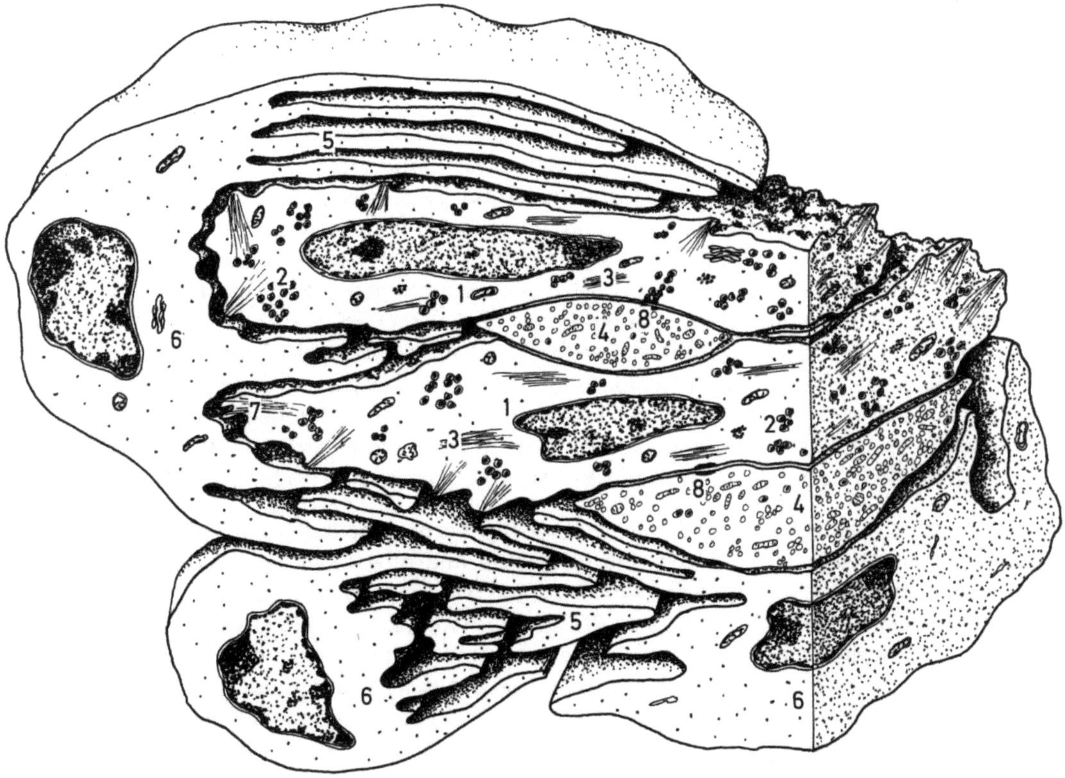

Fig. 57A–C. *Spatial diagram of the Merkel corpuscle* (chicken beak). A Merkel corpuscle is formed of: (A) *Merkel cells* (1), characterized by numerous dense-core vesicles (2) and bundles of microfilaments (3); (B) Afferent *nerve endings* (4), usually rich in mitochondria and clear vesicles; (C) *Lamellar cells* (6) displaying numerous folds and digitations (5). (7) Finger-like processes of the Merkel cell (8) Specialized junctions

well as a few dense-core vesicles. Asymmetric membrane specializations can be found between the Merkel cells and the nerve terminal, and sometimes the dense-core granules are seen in close proximity to the membrane thickening (Figs. 52–57).

II. Relationship between Morphology and Physiology

1. General Consideration on the Structure, Physiology, and Phylogeny of Corpuscles

There is a considerable body of knowledge about the organization of cutaneous sensory receptors in birds. In addition to morphological studies, including those with the electron microscope, physiological investigations have determined the functions of some of the receptors in the skin of birds. These studies have made it possible to compare the functions and structures of receptors in birds with those in other vertebrates, particularly tetrapods, from a structural, functional, and phylogenetic point of view.

a) Structure and Function

Free Nerve Endings are present mainly in the beak and the tongue of birds. From a functional point of view, their intraepidermal nerve endings in birds probably correspond with the thermosensitive units described by GREGORY (1973) and, according to the definition of HENSEL et al. (1960), they are cold receptors (LEITNER et al., 1973; LEITNER and ROUMY, 1974a). These receptors in birds are less sensitive to cold than the corresponding receptors in mammals, and in birds are not at all sensitive to vibrations. It seems that these kinds of nerve terminals are very widely found in vertebrates. They have been studied mainly in mammals, where they are particularly numerous in the skin of the face and the tongue (see review by IGGO, 1974). In reptiles, two types of intraepidermal afferent endings have been described (DÜRING, 1973a, 1975). These are simple endings distributed in the touch papillae and enlarged endings distributed throughout the scales. In amphibians and fishes the free nerve endings do not seem to show any particular localized distribution (NAFSTAD and BAKER, 1973; WHITEAR, 1971a, b, 1974a, b).

Herbst corpuscles of birds have classically been compared to the Pacinian corpuscles of mammals. Recent morphological and physiological investigations have again shown the analogies between these two kinds of receptors (see review by QUILLIAM, 1966a, b; MUNGER, 1971a; SCHWARTZKOPFF, 1973; IGGO, 1974). From a morphological point of view, the inner core of the Pacinian corpuscle may be compared with the inner bulb of the Herbst corpuscle. The capsular space of the former is similar to that in the Pacinian corpuscle, although it is more lamellar and divided into channels in the Herbst corpuscle (ANDRES and DÜRING, 1973). Functionally, they are both rapidly adapting mechanoreceptors sensitive to vibration, responding to frequencies from 60 to 600 Hz, with a maximal sensitivity at about 400 Hz (LOEWENSTEIN, 1971) for the Pacinian corpuscle, and with a sensitivity range of 40 to 1500 Hz (maximal sensitivity at about 300 to 600 Hz) for the Herbst corpuscle (QUILLIAM and ARMSTRONG, 1963; DORWARD, 1970; GREGORY, 1973; LEITNER and ROUMY, 1974b; GOTTSCHALDT, 1974). The functional differences between the two kinds of receptors probably depend upon the differences in their lamellar organization. These structures are probably acting like band pass filters, particularly for low-frequency mechanical displacements. Lamellar mechanoreceptors are not only found in mammals and birds; they are also found in other tetrapods. In reptiles three kinds of corpuscles with a lamellar inner bulb have been found (PAC, 1972; DÜRING, 1973b, 1975). The two first kinds are simple, and one of these can be compared to the Golgi-Mazzoni corpuscles in mammals. The third type, found particularly in the caiman, is more complex and has evolved an external capsule and an inner bulb, so that it has an appearance similar to the Herbst corpuscle of birds. Functionally, it is likely to be a rapidly adapting mechanoreceptor (SIMINOFF and KRUGER, 1968).

In amphibians, encapsulated lamellar receptors of various sizes have been found in the skin of anurans (DÜRING and SEILER, 1974). The inner bulb has a simple structure with few lamellae. However, the biggest corpuscles have a well developed inner space (or capsular space) with a lamellar structure, and this space is rich in collagen fibers. Thus these encapsulated lamellar corpuscles can be compared to the Herbst and Pacinian corpuscles found in the higher order of vertebrates. They probably correspond to the rapidly adapting mechanoreceptors found physiologically in anurans (GRAY and MALCOM, 1951; LOEWENSTEIN, 1956; CATTON, 1958). The encapsulated lamellar receptors do not appear to be present in fishes or in the urodele amphibians. By contrast, in the anuran amphibians and in reptiles, in addition to the simple lamellar corpuscles there are some corpuscles with a complex structure, which have the basic three main components of such receptors. The components are: an inner bulb or core; an inner or capsular space; and an outer capsule. These corpuscles in anurans and in reptiles are similar both structurally and functionally to the corpuscles in birds and mammals.

Grandry and Merkel Corpuscles. These two types of receptors are both located in the dermis, and the distinction between them depends upon their morphology and their distribution. However, from a cytological point of view, it is obvious that there are numerous similarities between these two kinds of receptors. Grandry's cells and Merkel's cells both have dense-core vesicles and microfilaments, and the same relationship with the nerve ending, and the satellite and lamellar cells are similar in many respects, including the presence of a basal lamina. These two types of receptors are never found together in the same species of birds. They have the same physiology, but their structure and distribution apparently depend on dermal factors. It is likely that the Grandry and Merkel corpuscles of birds are two varieties of the same receptor. Classically, avian Grandry and Merkel corpuscles have been compared to the Merkel corpuscles of other vertebrates, particularly of mammals (for reviews, see QUILLIAM, 1966a, b; MUNGER, 1971a, b, 1976; ANDRES and DÜRING, 1973; GOGLIA, 1973; WINKELMANN and BREATHNACH, 1973; IGGO, 1974). Little research has been devoted to these corpuscles in lower vertebrates. In fishes and urodeles Merkel corpuscles appear to be absent. In anurans, intraepithelial Merkel cells have recently been described that are cytologically very similar to the Merkel cells of birds and mammals (NAFSTAD and BAKER, 1973). In reptiles they have been found in the dermis, where they are usually arranged in groups called "Merkel cell-neurite complexes" under the touch papillae (DÜRING, 1974b).

Almost all our knowledge of the structure and function of Merkel cells has been gained from research performed in mammals. Most of these results show that these receptors have an intraepidermal location and that they have desmosomal junctions with other epidermal cells (IGGO, 1963; MUNGER, 1965; ANDRES, 1966; IGGO and MUIR, 1969; HALATA, 1970, 1972; MUNGER et al., 1971; CHOUCHKOV, 1974). It is generally agreed that these cells have an epidermal origin (MUNGER, 1965; SMITH, 1967; LYNE and HOLLIS, 1971; NAFSTAD, 1971, 1972; ENGLISH, 1974). However, there is considerable controversy about this conclusion, particularly as it can be shown that during embryonic stages Merkel cells are located in the dermis (BREATHNACH and ROBINS, 1970; HASHIMOTO, 1972; WINKELMANN and BREATHNACH, 1973). Therefore, a new hypothesis has been proposed, namely that the Merkel cell migrates from the dermis into the epidermis with an extension of the cutaneous nerves from the dermis. Merkel cells have also been found in both the dermis and the epidermis of adult humans (MAHRLE and ORFANOS, 1974). The dermal Merkel cells are enclosed in neural corpuscles, while the epidermal Merkel cells are found in the follicular epithelium. However, no final and definitive conclusions can be drawn from these morphological data, so the origin of Merkel cells still remains unknown.

Cytological similarities between Merkel cells in many tetrapods have led many authors to conclude that they might have the same functional properties. They are thus thought to be slowly adapting mechanoreceptors (IGGO and MUIR, 1969; HORCH et al., 1974), as they are known to be in the case of mammals. In anurans (GRAY and MALCOLM, 1951; LOEWENSTEIN, 1956; CATTON, 1958) and in reptiles (MILLER and KASHARA, 1967; SIMONOFF and KRUGER, 1968; BAILEY, 1969; KENTON et al., 1971), both rapidly adapting and slowly adapting mechanoreceptors have been found, and it is generally thought that the second type of receptor corresponds to the Merkel corpuscle (DÜRING and SEILER, 1974; DÜRING, 1974b).

Contrary to expectation, recent neurophysiological studies show that Grandry corpuscles in the duck and goose, like Merkel corpuscles in the chicken, give rise to a rapidly adapting discharge (LEITNER et al., 1973; LEITNER and ROUMY, 1974; GOTTSCHALDT, 1974b). It thus appears that the functional properties of these receptors are not the same in birds as in mammals. It would also be most interesting to know what the physiology of the Merkel corpuscle is in reptiles and amphibians. Thus, in spite of similarities in their structure, it appears that we may have two physiologically and functionally different types of receptors (GOTTSCHALDT and LAUSMANN, 1974), and thus it would be very interesting to know their developmental origin.

b) Phylogeny

To summarize, two kinds of cutaneous receptors exist in birds: thermoreceptors (free intraepidermal nerve endings) and mechanoreceptors (Herbst, Grandry, and Merkel corpuscles). While endings of the first type are very widely distributed in vertebrates, the situation is different for the second type. In this regard, from a phylogenetic point of view, it is interesting to consider the lower vertebrates (fishes, urodeles, anuran larvae), all of which are aquatic species. In these, cutaneous mechanoreception is largely subserved with the phylogenetically ancient lateral line system. In this system, transduction occurs when displacement of cilia attached to specialized receptor cells produces a change in the ionic permeability of the membrane, giving rise to a local potential.

In the pattern common to higher vertebrates, this system has evolved to form a new organ (the labyrinth) mediating a specialized type of mechanoreception (hearing). In the course of evolution, another system was developed for cutaneous mechanosensitivity in land tetrapods, which led to the formation of new types of receptors. They can be found in a simple form in adult anurans and are also found in all the other, more highly evolved, vertebrates. They include two kinds of basic mechanoreceptors, which are very different in their morphology and their physiology, namely, the laminated corpuscles and the Merkel-type corpuscles.

In reptiles, these receptors have evolved a higher complexity in the organization of the lamellar corpuscles (formation of well structured inner space) and a tendency to a preferential localization of the nerve endings (particularly at the level of the touch papillae).

In higher vertebrates, and particularly in mammals, the high diversification and, at the same time, the high specialization of cutaneous receptors are apparent (see review by IGGO, 1974). The present author feels that this evolution has a strong adaptive character. The development of receptors that are sensitive to mechanical stimulation surely occurred along with the development of active movement, which established new types of contacts with the environment. The kind of receptors formed depends on the kinds of movements and the life-style of the species. Thus, adaptation of the limbs to make quick displacements seems to be correlated with the development of a rich mechanical sensitivity, mainly at the level of articular capsules, allowing a better control of movement of the limbs. This adaptation provokes a higher diversification of the receptors (see review by POLACEK, 1966) and/or a higher density of mechanosensitive endings, as in the wing of birds (POLACEK et al., 1966; SCHWARTZKOPFF, 1973).

One can also see that in higher vertebrates there is a close relation between the various modalities of feeding and a high development of the faculty of prehension, that is, the capacity to manipulate objects. This aptitude requires the acquisition of good tactile discrimination, and this has been obtained by the development of various kinds of detectors with different functional modalities and by increasing the number of receptors per unit area of skin. In mammals, this evolutionary process is most obvious in the hand of primates (KADANOFF, 1970). In birds, things are a little different. Because the anterior limbs of birds are specialized for locomotion, it is the beak that is used for prehension and can be considered to perform the functions of the primate hand. Consequently, it is in the beak that we can see the higher variety and the higher density of corpuscles (up to some 10000 in the skin of one beak), allowing the good tactile discrimination needed for searching for and manipulating food and other objects.

Note also that in many mammals we can observe a big development of facial sensitivity, particularly in the trigeminal system, in parallel with the development of diverse types of receptors inside and around the oral cavity. In the case of certain species with a particular life-style, this sometimes leads to the formation of highly specialized organs, such as Eimer's organs in the snout of the mole. It is also important to note that mammals present a high development of mechanosensitivity in cutaneous regions related to the genital apparatus (erogenous zones). This development has very significant evolutionary adaptive value, as has been shown by research conducted in primates (WINKELMANN, 1965; MALINOVSKY, 1968; POLACEK and MALINOVSKY, 1971; MALINOVSKY and SOMMEROVA, 1972b, 1973).

Moreover, the evolution of cutaneous mechanoreceptors in higher vertebrates has been strongly influenced by the formation of erect cutaneous appendages (feathers, hairs) protruding from the epidermis and playing the role of amplifier levers. This evolution is attested to in birds by a high density of Herbst corpuscles around the follicles of contour feathers and around the feather erector muscles. In mammals, mechanoreceptors are often localized around the hairs (STRAILE, 1969; HASHIMOTO, 1973), and diversified types of hair follicle receptors have been developed for the various categories of hairs (tylotrichs, guard and down hairs) (see review by IGGO, 1974). An extreme example of this evolution is shown by the development in nonprimate mammals of a very complex organ of a new type, the sinus hair, grouping several kinds of mechanoreceptors.

Thus, numerous factors are involved in the evolution of cutaneous mechanoreception. This evolution leads to the formation of more specialized receptors, the localization of these receptors in functionally advantageous regions, giving increasingly more sensitive detection and accurate spatial discrimination of mechanical stimuli, and, finally, allowing better adaptation of the species to their environment.

However, this evolution is surely not restricted to a simple specialization and diversification of the receptors. Evolution has also acted at a higher lever, the level where the information about cutaneous stimuli given by the peripheral nerves is processed, i.e. the central nervous system. Integration and perception of these stimuli occurs in the central nervous system, and it is at this level that the response is elaborated. But here we are entering the field of behavioral psychology, which is outside the scope of this paper.

2. The Problem of the Junctions

From all these data it is now possible to attempt correlations between the cytology of the receptors and the phenomena of transduction. One question is whether and by what mechanisms the structures close to the nerve endings participate directly in the excitatory processes, that is, in the generation of a local potential by mechanical deformation. To answer this question it is necessary to know the cytochemistry of the receptors, particularly regarding the chemicals involved in nervous activity (neurotransmitters and associated enzymes), and the nature of the junctions and vesicular structures associated with the nerve endings.

a) *Cholinesterase Activity in Herbst and Grandry Corpuscles*

Cholinesterase activity in Herbst and Grandry corpuscles was studied by the method of KARNOVSKY (1964) in the beak of duck from the 20th day of incubation to adult hood. Acetylthiocholine iodide and butyrylthiocholine iodide were used as substrates, and BW 284C51, iso-OMPA, and eserine as inhibitors (SAXOD, 1973a).

Figs. 58–62. *Cholinesterase activity in the Herbst corpuscle* (duck beak). Figs. 58 and 59. Light microscope, × 140. Fig. 58, longitudinal section, BuThCh substrate. Figure 59, transverse section, AcThCh substrate. End-product of reaction localized in external lamellae of inner bulb cells and around nuclei (NB). No reaction in inner space (IS) and outer capsule (OC). Figs. 60–62. Electron microscope, substrates: AcThCh (Figs. 60 and 61) and BuThCh (Fig. 62). Fig. 60. Hemi-inner-bulb in transverse section. Cholinesterase activity is mainly localized in rough reticulum (R), in perinuclear cisternae (arrows), in vesicles associated with Golgi apparatus and along plasma membranes of lamellar cells of inner bulb. Cytoplasm and plasma membrane of inner space cells (SC) are devoid of reaction product. × 8000. Fig. 61. Cholinesterase activity localized in rough reticulum (R) of external lamella (L_1) and along plasma membrane of lamellae (L) of inner bulb cells. × 32000. Fig. 62. Zone of contact between terminal ampulla of nerve ending (NE) and lamella (L) of inner bulb cell. End-product of reaction is localized in intermembrane space (arrows) and in outer part of membranes. No reaction in clear vesicles (CV) of nerve ending. CV, clear vesicles; G, Golgi apparatus; IB, inner bulb; IS, inner space; L, lamellae; M, mitochondria; NB, nucleus of the inner bulb cell; NE, nerve ending; OC, outer capsule; R, rough reticulum; SC, inner space cell ▷

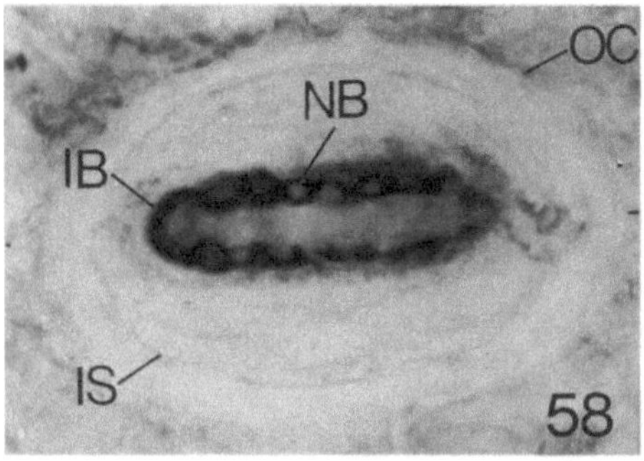
OC
NB
IB
IS
58

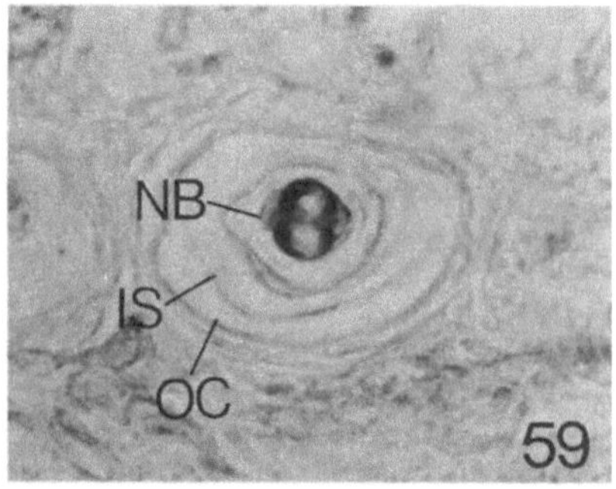
NB
IS
OC
59

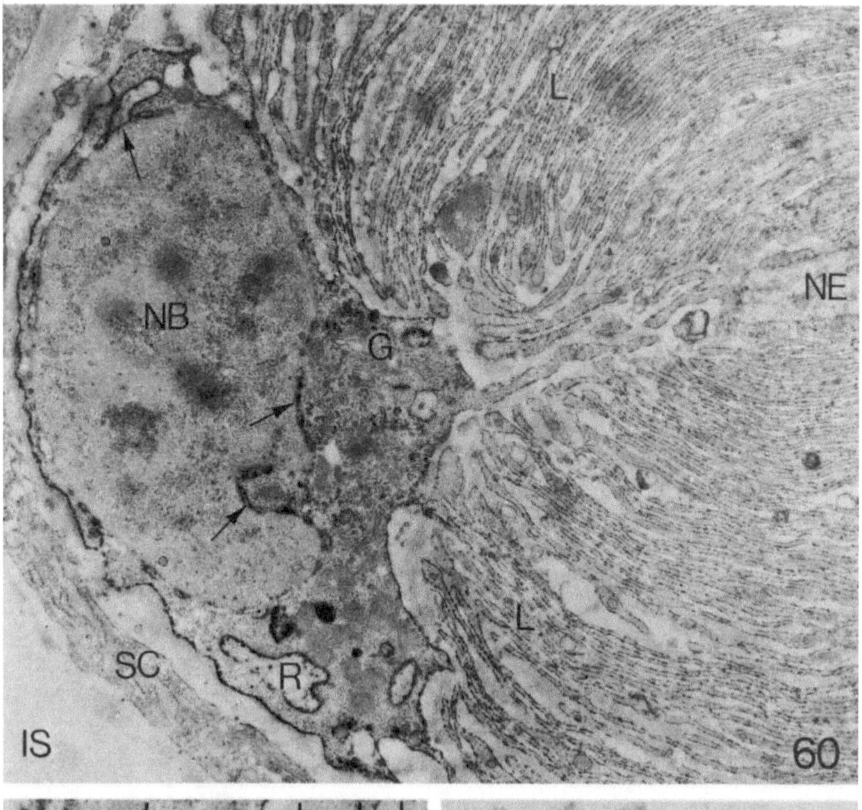
L
NE
NB
G
L
SC
R
IS
60

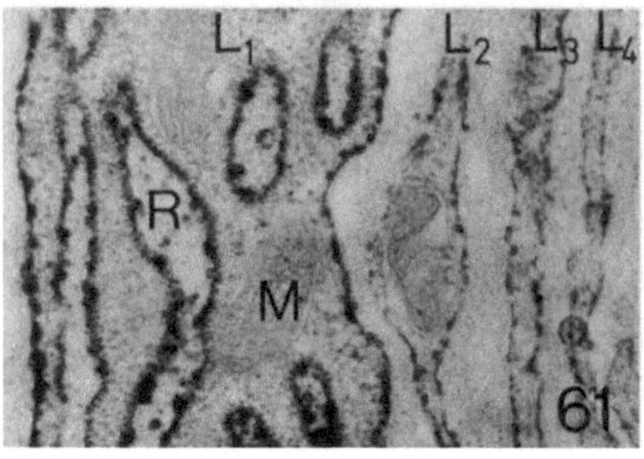
L1
L2
L3
L4
R
M
61

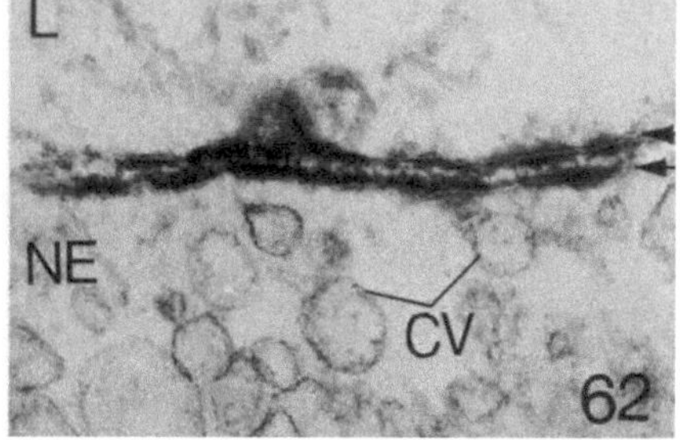
L
NE
CV
62

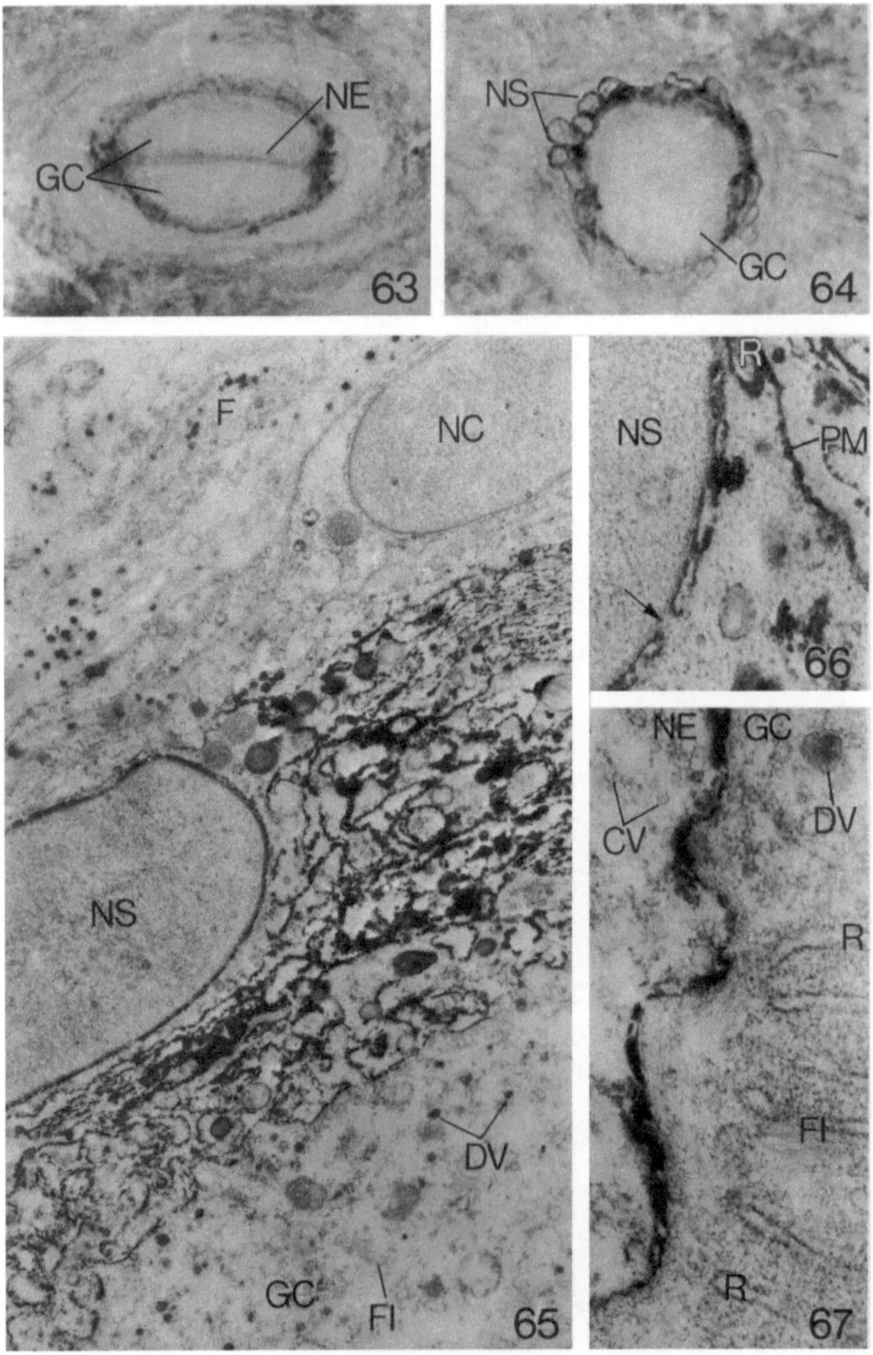
NE
GC
63
NS
GC
64
F
NC
NS
DV
GC
FI
65
R
NS
PM
66
NE
GC
CV
DV
R
FI
R
67

Identification of Cholinesterases. The results of inbibitory experiments show that corpuscles of Herbst and Grandry are characterized by nonspecific cholinesterase activity. In effect, the reaction is positive when acetylthiocholine and butyrylthiocholine are used as substrates with or without BW 284C51 (inhibitor of acetylcholinesterase). It is negative when iso-OMPA (inhibitor of nonspecific cholinesterases) or eserine (inhibitor of all cholinesterases) are added to the incubation medium. For the same incubation time, the reaction is stronger with butyrylthiocholine than with acetylthiocholine. The reaction consequently seems to be due mainly to butyrylcholinesterases. However, there may be some specific cholinesterase activity, particularly at the level of nerve endings.

Localization of Cholinesterases.

Corpuscles of Herbst: Cholinesterase activity is restricted to the inner bulb. No reaction can be seen in the cells of the outer capsule or in the inner space (Figs. 58 and 59). A strong cholinesterase activity was found in the cells of the inner bulb. It is localized in the perinuclear system, and the inner surface of the rough endoplasmic reticulum (mainly in the very large outer lamellae) and in vesicles associated with Golgi complexes (Figs. 60 and 61). Granules of reaction product are also present on the surface of the plasma membrane of the inner bulb cells and nerve endings. In the contact zones between bulbar lamellae and the nerve ending, a reaction is seen on the surface of the plasma membranes. There is little reaction in the proximal and medial zone of the inner bulb. But, at the level of the terminal ampulla, a strong nonspecific cholinesterase activity is evident in the zones of contacts. The reaction product is localized in the space between the two plasma membranes (Fig. 62). No reaction can be seen in the clear vesicles of the synaptic type in the nerve ending. Some mitochondria show a positive reaction.

Grandry Corpuscles: Cholinesterase activity is restricted to the satellite cells, the nerve ending, and the contact zone between satellite cells, Grandry cells, and nerve ending. There is no activity in the capsular cells, the perinuclear system, the rough endoplasmic reticulum or the Golgi apparatus of the Grandry sensory cells (Figs. 63 and 64). The reaction product can be seen only on the surface of the plasma membrane in zones of contact between the satellite cells and the nerve

◁ Figs. 63–67. *Cholinesterase activity in the Grandry corpuscle* (duck beak). Figs. 63 and 64. Light microscope, × 200, substrate AcThCh. Figure 63, longitudinal section; Fig. 64, frontal section. Product of reaction is localized at level of contacts between Grandy cells (GC) and satellite cells and around nuclei of satellite cells (NS). Weak reaction in nerve ending (NE). Figs. 65–67. Electron microscope, substrate AcThCh. Fig. 65. Satellite and Grandry cells (GC). Reaction is localized in perinuclear cisternae, rough reticulum and plasma membrane of satellite cells, and also along plasma membrane of Grandry cell, which forms digitations. No reaction in capsular cells. × 13000. Fig. 66. Detail of a satellite cell. Cholinesterase activity is localized in perinuclear cisternae (arrow: nuclear pore), rough reticulum (R) and at surface of plasma membrane (PM). × 45000. Fig. 67. Contact zone between a Grandy cell (GC) and a nerve ending (NE). Enzymatic activity is visible at surfaces of axolemma and of plasma membrane of Grandry cell and in intermembrane space. No reaction in clear vesicles (CV) of nerve ending or in rough reticulum (R) of Grandry cell. × 35000. CV, clear vesicles; DV, dense-core vesicles of Grandry cell; F, collagen fibers; Fl, fibrils; GC, Grandry cell; NC, nucleus of capsular cell; NE, nerve ending; NS, nucleus of satellite cell; PM, plasma membrane; R, rough reticulum ending (Figs. 65–67). The cholinesterase activity of satellite cells is localized in the perinuclear system and rough endoplasmic reticulum and on the surface of the plasma membrane. Some reaction can also be seen between the membranes of some mitochondria. In the nerve ending, the cholinesterase activity is restricted to the surface of the plasma membrane and the space between the nerve ending and the Grandry cells, and also between the nerve ending and the satellite cells. It is sometimes strong in certain zones of contact between the nerve ending and the Grandry cells. The reaction product was also seen in some of the vesicular structures in the nerve ending, particularly in the smooth endoplasmic reticulum system and at the level of the membrane of some mitochondria. No reaction can be seen in clear vesicles of the synaptic type in the nerve ending.

In this method there is a reliable, restricted and precise localization of the reaction product. The lack of diffusion of enzymes during preparation and incubation of the section is shown in Fig. 60, where two very close membranes are seen, one with a strong positive reaction (the membrane of the inner bulb cell) and the other without any reaction (the membrane of the inner space cell). Thus, it seems that this method really shows the physiological site of cholinesterases. The appearance of cholinesterase activity in Herbst and Grandry corpuscles is related to the cytological differentiation of corpuscular cells, and can be seen from the beginning of their development (21 days of incubation). This activity begins in the perinuclear zone of the inner bulb cells and satellite cells, and it is localized mainly in rough endoplasmic reticulum and on the membranes. The location of cholinesterase activity is the same during all stages of development of the corpuscles.

In conclusion, we see that the cholinesterase activity varies according to the type of corpuscle and that cholinesterases are not specifically associated with axoplasm or with structures adjacent to the nerve endings. Thus, in the Herbst corpuscles, it is the cells around the nerve ending (inner bulb cells) that show cholinesterase activity; but in the Grandry corpuscles the cells close to the nerve ending (Grandry cells) have no cholinesterase activity; rather these enzymes are localized in the satellite cells, which are not in direct contact with the nerve ending.

These results, which show that nonspecific cholinesterase activity is mainly contained in the nonnervous structures of Herbst and Grandry corpuscles, complement the results which have been obtained in mammalian corpuscles, which are also characterized by a nonspecific cholinesterase activity with the same kind of localization (for review see CAUNA, 1960–1962; WINKELMANN, 1965; BOURLOND, 1968). This is the case particularly for Pacinian corpuscles (LOEWENSTEIN and MOLINS, 1958; MONTAGNA, 1960; CHOUCHKOV, 1968; GUARINO and MILAZZO, 1974), for the Meissner corpuscles (MACHITA et al., 1967; VALENZUELA and CHACON, 1968), and for other kinds of capsular lamellated corpuscles (HURLEY, 1958; CAUNA, 1960; WALTER and HEBEL, 1966).

b) Specialized Junctions and Associated Vesicles

Several types of junctions can be seen in the Herbst and Grandry corpuscles. Some are present when development of the corpuscles begins, while some appear later. They can be classified into two main groups according to their structure.

The first group of junctions is formed by symmetric contact either of the *zonula occludens* or of the *zonula* or *macula adherens* type (according to the definition of

FARQUHAR and PALADE, 1963; KELLY, 1967). They are seen in the outer capsule and between the lamellae of the inner bulbar cells in the Herbst corpuscles (Figs. 16 and 17), between satellite cells and sensory cells in the Grandry corpuscles (Fig. 37), and between lamellar cells and Merkel cells. Junctions of the desmosome type are also seen in mammals between epidermal cells and Merkel cells (for review, see WINKELMANN and BREATHNACH, 1973).

The second group of junctions is formed by contacts seen between nerve endings and associated cells (Grandry, Merkel, and inner bulb cells). Usually these junctions are of the asymmetric type (Figs. 20–23, 38–43, 52–56). Some junctions with almost similar structures have also been seen in the Merkel cells of mammals (MUNGER, 1965; ANDRES, 1966; HALATA, 1970; WINKELMANN and FORTMAN, 1972). The main problem is to know whether they are attachment plaques or chemical synapses. From a structural point of view the main criterion for recognition of synapses has been the type of vesicles associated with these junctions (PALAY, 1950; GRAY, 1959; COUTEAUX, 1961; JONES, 1969). In the three kinds of corpuscles in birds, clear vesicles of the synaptic type are present in the nerve endings. Such vesicles are also seen in almost all cutaneous receptors in vertebrates (CAUNA, 1962; MUNGER, 1971; SPENCER and SCHAUMBURG, 1973; CHOUCHKOV, 1974; WHITEAR, 1974a, b). Contrary to what is usually seen in chemical synapses, in the corpuscles there is no accumulation of vesicles at the level of the junction, although they are sometimes very numerous in certain zones of the nerve ending (e.g., in the terminal ampulla of the Herbst corpuscles, Fig. 19). Moreover, from a functional point of view, these vesicles seem to be in a postsynaptic position. Thus it is very difficult to know what the functional role of these vesicles is in the generation of ion fluxes. However, neither are there good arguments to support the other two common hypotheses: trophic function (SINGER, 1964) and "relation to other dynamic processes" (WHITEAR, 1974b).

However, one must remember that our knowledge of the mechanisms of transduction is derived mainly from work on Pacinian corpuscles (see LOEWENSTEIN, 1960, 1965; GOLDMAN, 1965). It must be pointed out that LOEWENSTEIN, in his experiments, was not able to remove all the lamellae of the inner core surrounding the nerve endings, and the question of direct participation of these structures (with their specialized junctions) in excitatory processes remains open. Likewise, the role of nonspecific cholinesterases in mechanoreceptive function has not yet been established precisely. According to some authors, the inhibition of cholinesterase in the human skin has no effect on sensation (HURLEY and KOELLE, 1958), and the amount of enzyme activity in the corpuscles remains the same some weeks after section of afferent nerves (WINKELMANN, 1962; CHOUCHKOV, 1970). Thus, one could conclude that this class of enzymatic activity is not associated with nervous function. However, according to other authors, it is difficult to attain good inhibition of cholinesterases in the different kinds of corpuscles (CAUNA, 1960) and the system acetylcholine-cholinesterase could, in fact, play an important role in the transformation of mechanical stimuli into nerve impulses (see reviews by DIAMOND et al., 1956; CAUNA, 1960, 1962; DUNCAN, 1963; PERL, 1964; GOLDMAN, 1965; LOEWENSTEIN, 1965; BOURLOND, 1968; HORRIDGE, 1968).

Another problem is that of the nature and role of membrane-bound dense-core granules, which are very numerous in Grandry and Merkel cells. We have seen that dense core vesicles are sometimes localized close to the junctions and in some cases they seem to be coalescent with the thickening of the plasma membrane of the so-called "sensory cells" (Figs. 38–40 and 54–56). In the Merkel cells of mammals, vesicles can sometimes be seen in close contact with junctions (ENGLISH, 1974), and recently some fusion between granules and a synapse-like structure has been reported (CHEN et al., 1973). It is tempting to think that these secretory-type, dense-core granules contain a neurotransmitter, perhaps a monoamine. This hypothesis was originally proposed just after the discovery of these dense-core vesicles at an ultrastructural level (for review see WINKELMANN and BREATHNACH, 1973). In addition, there is indirect physiological evidence for synaptic transmission occurring in type-I cutaneous mechanoreceptors in mammals (HORCH et al., 1974). Such a synapse must be between the Merkel cell and the nerve ending.

To establish the precise nature of the different kinds of dense core vesicles seen in the corpuscles and particularly in Grandry corpuscles, the present author used several kinds of histochemical methods:

(1) The method of Hillarp and Hokfeld (see PEARSE, 1961) to demonstrate adrenaline and noradrenaline.

(2) The histo-autoradiographic method, with light microscopy after injection of tritiated Dopa[1].

(3) An enzymatic digestion method (pronase, trypsin) on ultra-thin sections, according to the technique of MONNERON and BERNHARD (1966) (SAXOD, 1973d).

The first and second methods gave negative results in the corpuscles studied. The third method showed that there are proteins in dense-core granules of the sensory cell of Grandry. These granules are digested by pronase and trypsin. This digestion is apparent within two hours of treatment and it is complete after five hours. These experiments show that the dense-core vesicles of the Grandry sensory cells contain proteins, but they do not seem to contain catecholamines.

These results should be compared with those obtained in Merkel cells in mammals, which are very confused. According to certain authors (IGGO and MUIR, 1969), reserpine did not change the function of corpuscles and produced no change in the receptor at the ultrastructural level. In contrast to this, KUROSUMI and DUZUKI (1971) found that reserpine caused a significant diminution in number of the granules. Contradictory results have also been obtained concerning the action of histamine, serotonin and acetylcholine: according to FJÄLLRANT and IGGO (1961), these agents had an initial stimulating and later a depressant effect on type-I, slowly adapting receptors, while according to SMITH and CREECH (1967), these chemicals had no effect on the physiology of tactile hair disks.

However, it appears that biogenic amines can play a role in the physiology of corpuscles, at least by way of the sympathetic endings which are frequently associated with corpuscles. Thus NILSSON (1972) showed a sympathetic modulation of the dynamic responses of the slowly adapting mechanoreceptors of the vibrissae by the direct effect of locally released noradrenaline. Among the rapidly adapting mechanoreceptors, the Pacinian corpuscles are known to be sensitive to adrenergic

[1] The experiments were performed on young 5-day-old ducks. The monoaminoxidase activity (M.A.O.) was first inhibited by intraperitoneal injection of 15 mg of niamide. Three hours later, each animal received 250 µCi of tritiated Dopa (specific activity 28 Ci/mM). The beaks were fixed 1 and 3 hrs after the injection of marker.

drugs (LOEWENSTEIN and ALTAMIRANO-ORREGO, 1956; TUTTLE and MCCLEARY, 1974). Sympathetic fibers with dense-core vesicles have been shown at the level of the inner core and it has been demonstrated that they contain catecholamines (CHOUCHKOV, 1968, 1970; SANTINI, 1968; SANTINI et al., 1971). Recent experiments (SCHIFF, 1974) have shown that the presence of norepinephrine, maintained by sympathetic tonus, is required for the afferent nerve terminal of the Pacinian corpuscle to be mechanosensitive.

In conclusion, it has not been possible to determine the functional significance of junctions that are present between the nerve endings and the specialized associated cells in mechanoreceptors. Similar uncertainty exists about the nature and the role of the different kinds of vesicles (clear vesicles, dense-core vesicles) present not far away from the junctions. Moreover, the physiological role of different kinds of neurotransmitters and associated enzymes (particularly cholinesterase) in the excitatory processes is also not yet known.

C. Morphogenesis

Before proceeding with an experimental study of the factors involved in the morphogenesis of the sensory corpuscles of birds, it is necessary to be familiar with the steps in morphogenesis during their normal development. The formation of these terminal organs has been known until now only by the old work of SZYMONOWICZ (1897), HERINGA (1918), and TELLO (1922), and more recently by the present author's electron microscopic observations of the development of Herbst and Grandry corpuscles in the beak of the duck (SAXOD, 1967, 1970a, b).

I. Description of the Normal Development of Herbst and Grandry Corpuscles

In birds, the sensory nerve endings arrive in the skin long before histogenesis of the corpuscles. Using a cholinesterase staining technique in whole mounts of the back skin of the chick embryo (SAXOD, 1978a), the present author has found that the first sensory nerves reach the skin as early as the fourth day of incubation, whereas the corpuscles begin to develop about the 17th day. From the 4th to the 10th day of incubation, the nerve fibers grow fast and at 10 days they form a hexagonal type of gross pattern. During this time the dermal condensations of the young feathers are forming in the dermis. Later the particular micro-pattern of innervation of each feather develops, and at 15 days it characteristically consists of two interconnected rings of nerves surrounding the feather follicle. The rings supply nerves to the dermal papilla and shaft of the follicle. Then, at 17 days the development of the cutaneous corpuscles begins synchronously in the skin of the different regions of the body.

The above description applies to the chick. In duck, the thickness and high fat content of the skin make it difficult to study the pattern of innervation in whole mounts, but the same steps in the development of the sensory endings in the skin must exist; and the formation of the cutaneous corpuscles also takes place late in the embryonic development. The following description concerns the development of the Herbst and Grandry corpuscles in the beak of the duck. For each type of corpuscle, five stages can be defined:

1. Grandry Corpuscles

At 20–21 days of incubation (stage 1), large multipolar polygonal cells form beneath the epidermis of the beak. These cells measure 10–15 μm, and they are already associated with nerve fibers 1.5–2 μm in diameter, which are non-myelinated at the terminals (Figs. 68, 69, and 75). At this stage, these large cells already possess some microfilaments and dense-core vesicles, which are characteristic of the sensory cells of Grandry. Specialized contacts are also already established between the young sensory cells and the nerve terminals (Fig. 75). I have never seen this kind of cell with these characteristic inclusions not in contact with nerve terminals.

At 22 to 23 days of incubation (stage 2), these cells have grown considerably, measuring 20 μm by 8 μm; they are hemispherical and resemble the adult sensory cells of Grandry. The nerve ending has a discoidal form at this stage and measures 15 μm by 2 μm (Figs. 70 and 76). The satellite cells can first be recognized at this stage.

At 24 to 25 days of incubation (stage 3), the satellite cells have completely enveloped the structure formed by the cells of Grandry and the nerve terminal (Figs. 71 and 72). They form numerous folds and digitations, which are associated with the expansions of sensory cells.

During the final three days of embryonic development (26, 27, and 28 days of incubation, stage 4), the full development of the Grandry corpuscles is achieved (Fig. 77). At stage 5, the final size of the adult structure (50 μm × 30 μm) is achieved by the growth of the sensory cells and the terminal nerve disk. The perinuclear organelles and the thin external capsule are formed after stage 4 (Figs. 73, 74, and 78).

2. Herbst Corpuscles

The earliest signs of formation of Herbst corpuscles in the beak of the duck can be seen at 20–21 days of incubation (stage 1). At this stage, characteristic large polygonal multipolar cells with a massive basophilic nucleus 4–6 μm in diameter appear around the nerve terminal (Figs. 79, 80, and 87). These cells rapidly develop a number of lamellar processes (Fig. 87).

At 22–23 days of incubation (stage 2) the inner bulb and the external capsule are formed (Figs. 81 and 88). At this stage the inner bulb cells become symmetrically

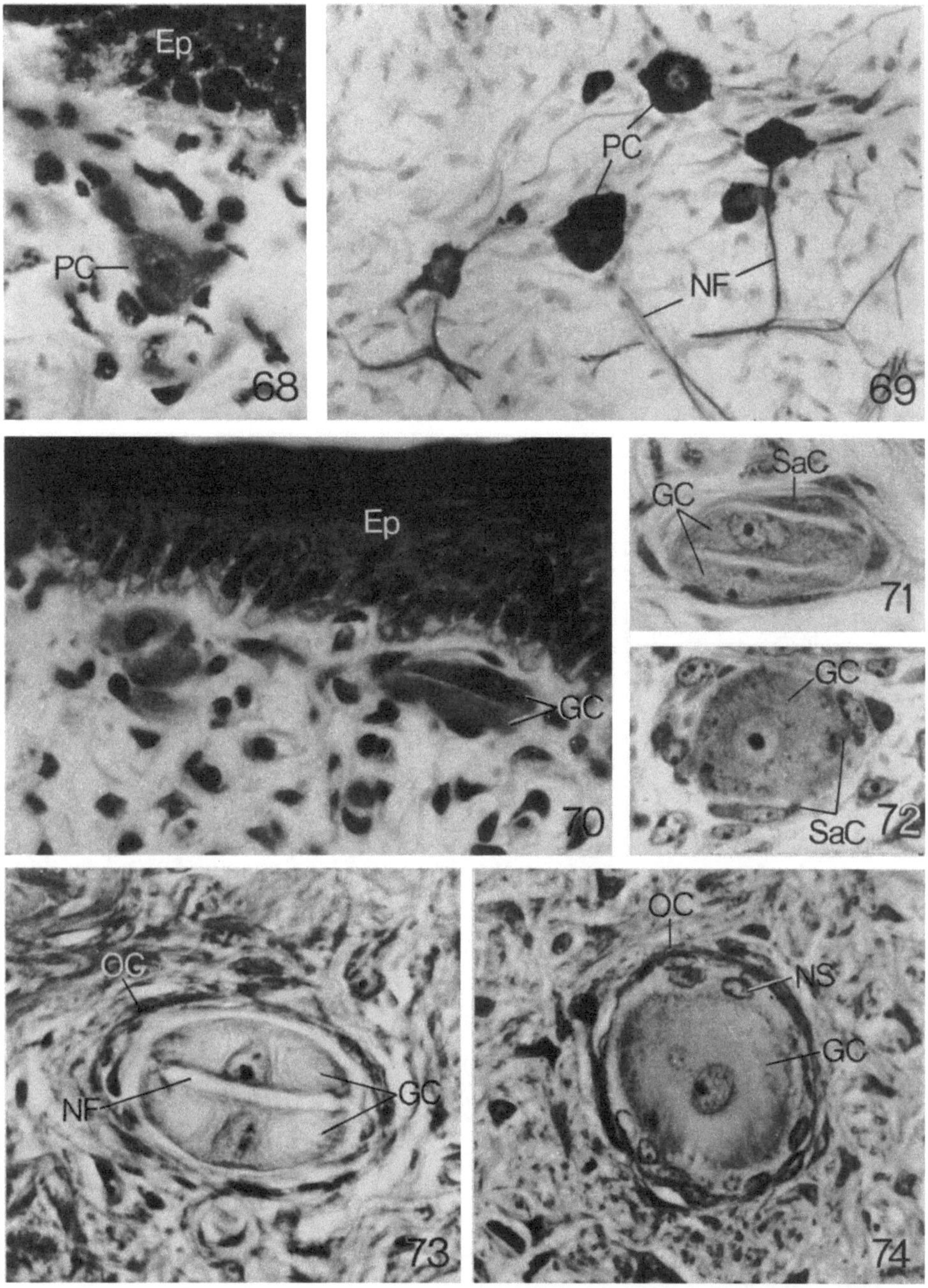

Figs. 68–74. *Stages of development of the Grandry corpuscle*. Duck beak, light microscope. Figures 68, 70, 71, 73, and 74, Masson staining; Fig. 69, silver staining; Fig. 72, semithin section. Fig. 68. *Stage 1* (embryo 20 days): large polygonal cells become visible in dermis, beneath epidermis (Ep). ×1200. Fig. 69. *Late Stage 1* (embryo 21 days): large polygonal cells are connected with nerve fibers (NF). Cells strongly argyrophil. ×800. Fig. 70. *Stage 2* (embryo 23 days): polygonal cells have grown into hemispherical Grandry cells (GC) and are surrounded by fusiform cells. ×1200. Figs. 71 and 72. *Stage 3* (embryo 25 days): longitudinal and frontal sections: fusiform cells become satellite cells (SaC). ×1200. Figs. 73 and 74. *Stage 5* (adult duck), longitudinal and frontal sections: Grandry cells (GC) display perinuclear striation looking like a "sheaf". They are connected with a nerve ending (NF) and surrounded by satellite cells with conspicuous nuclei (NS). An incomplete outer capsule (OC), formed of fibroblasts and collagen fibers, is present. ×1000. Ep, epidermis; GC, Grandry cells; NE, nerve ending; NF, nerve fiber; NS nucleus of a satellite cell; OC, outer capsule; PC, polygonal cell; SaC, satellite cell

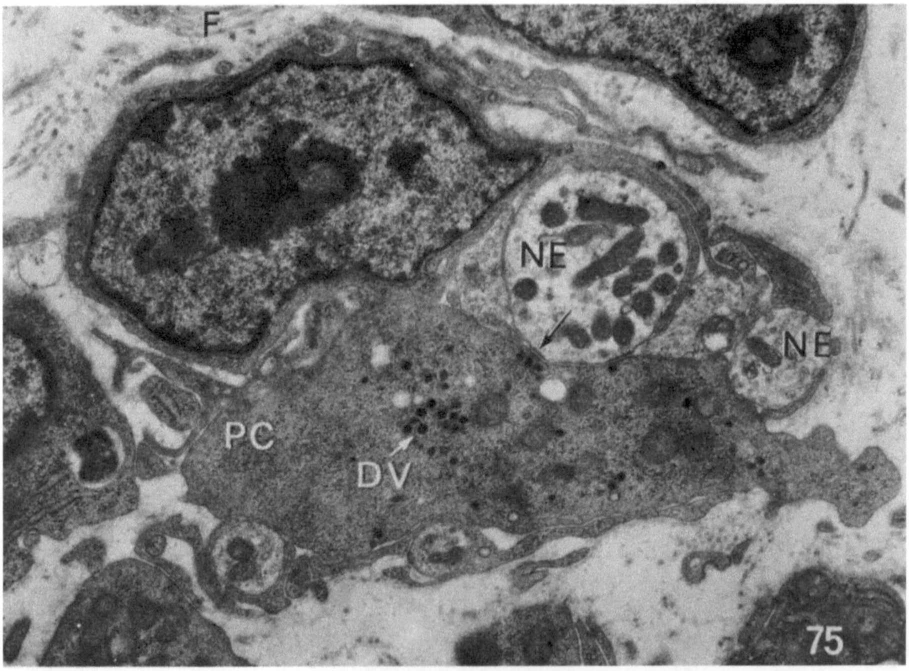

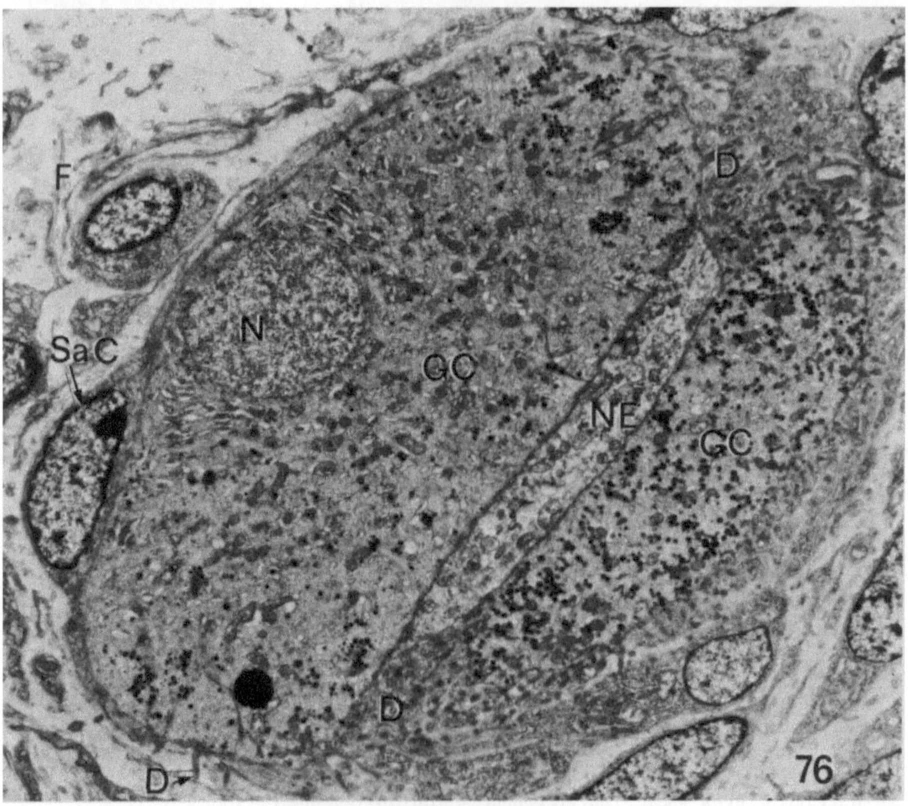

Legends to this Figures see p. 378

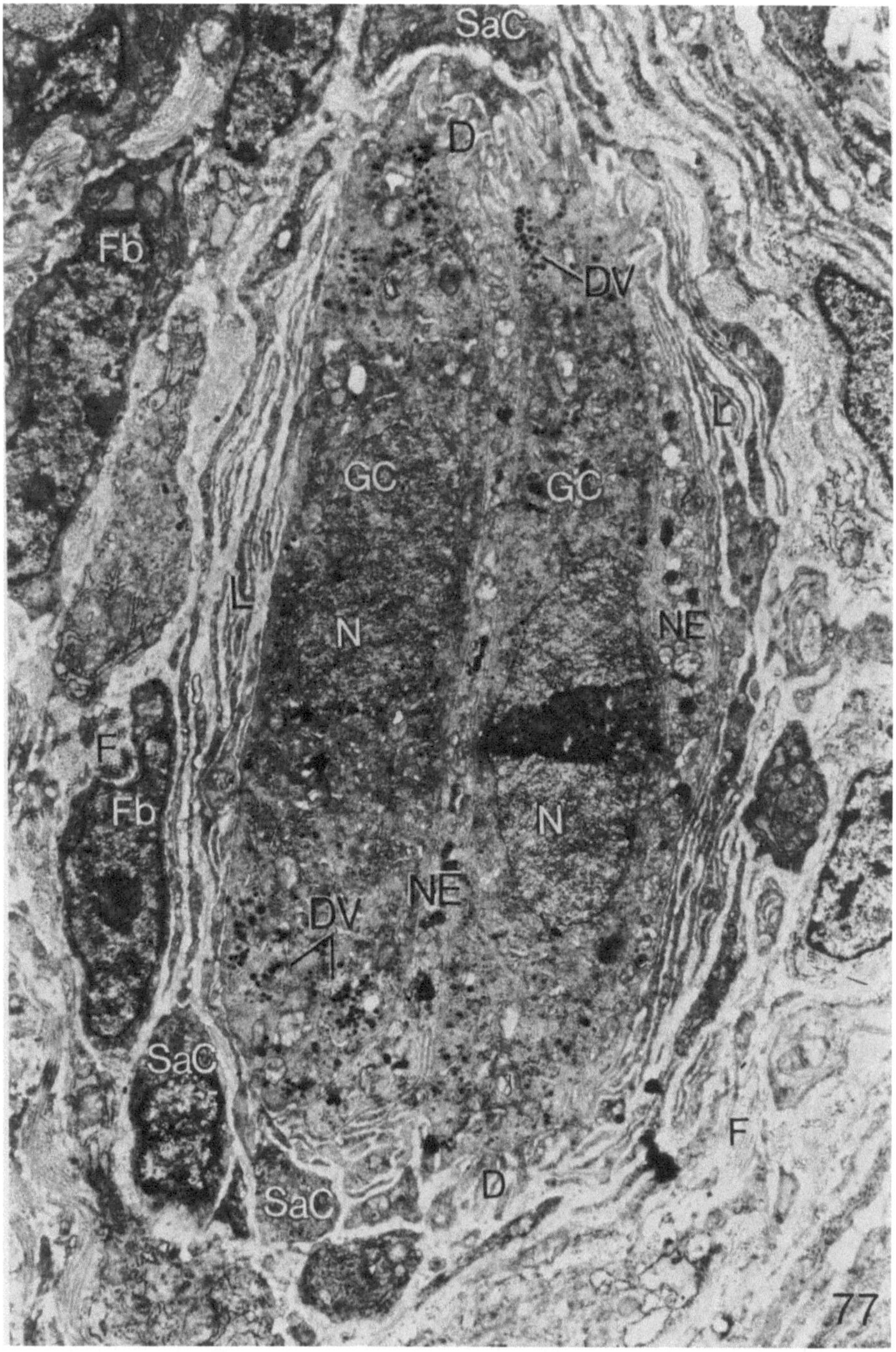

Fig. 77. *Development of the Grandry corpuscle* (duck beak, longitudinal section). *Stage 4.* (hatching duck). The Grandry corpuscle is well formed. Two Grandry cells (GC), characterized by numerous dense-core vesicles (DV) and microfilaments, are connected with two nerve endings (NE). Satellite cells display numerous digitations (D) and lamellar processes (L). D, digitations; DV, dense core vesicles; F, collagen fibers; Fb, fibroblast; GC, Grandry cell; L. lamellar processes; NE, nerve ending; SaC, satellite cell

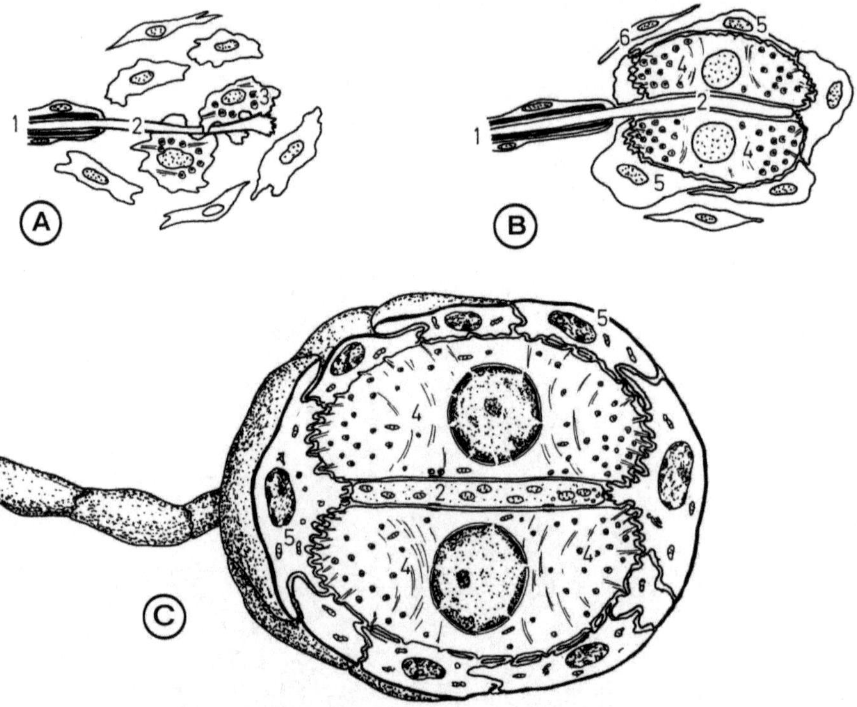

Fig. 78A–C. *Schematic representation of the three main stages in the development of the Grandry corpuscle* (duck beak, longitudinal sections). (A) Large polygonal cells (3) containing dense-core vesicles and bundles of microfilaments are present around sensory nerve endings (2) (20th day of incubation). (B) These cells grow into sensory Grandry cells (4), around which satellite cells (5) become arranged and an incomplete capsule of dermal fibroblasts (6) is forming (23rd to 25th day of incubation). (C) At hatching (28th day of incubation), development of Grandry corpuscle is almost complete (the incomplete capsule is not shown in this drawing). (1) nerve fiber; (2) nerve ending; (3) polygonal cell; (4) Grandry cell; (5) satellite cell; (6) dermal fibroblast

arranged around the nerve terminal, and the external capsule is formed from two or three layers of flattened cells with fusiform nuclei.

At the 24th to 25th days of incubation (stage 3), the capsule grows in size to reach 70 µm × 40 µm and the internal space (or peribulbar space) is forming (Figs. 82, 83, and 89). Flattened fibroblastic type cells can be seen in this space. At this stage the terminal ampulla is formed and the nerve terminal contains the usual organelles as well as clear vesicles and dense-core vesicles. The specialized membrane contacts between the axolemma and the lamellae are also present at this stage.

Figs. 75 and 76. Development of the Grandry corpuscle (duck beak). Fig. 75. *Stage 1* (embryo 21 days): A polygonal cell (PC) with dense-core vesicles (DV) and microfilaments is connected with a nerve ending (NE). Note junction (arrows). × 14000. Fig. 76. *Stage 2* (embryo 23 days): Hemispherical Grandry cells (GC) embrace nerve ending (NE). Satellite cells (SaC), not numerous at this stage, stretch around young corpuscle. D, digitations; DV, dense-core vesicles; F, collagen fibers; GC, Grandry cell; N, nucleus; NE, nerve ending; PC, polygonal cell; SaC, satellite cell

During the last three days of embryonic development (26, 27, and 28 days of incubation, stage 4), the organization of the Herbst corpuscle is completed. The internal space is well developed as a result of the growth of the external capsule (Fig. 84). By 28 days of incubation, the Herbst corpuscle has attained an adult form and measures 80 μm × 100 μm.

The final adult structure (stage 5) is achieved mainly by growth of the corpuscle to a final size of 160 μm × 180 μm, largely as a result of the growth of the capsule. This results in the inclusion within the corpuscle of the terminal myelinated segment of the nerve (Figs. 85, 86, and 91).

In conclusion, formation of the Grandry and Herbst corpuscles occurs very late during the last quarter of embryonic development in the duck. The constituent elements are assembled progressively from the center to the periphery around the nerve fiber, which is present from the beginning (Figs. 78 and 91). The cells that have the most characteristic highly differentiated form (the internal bulb cells, the cells of Grandry) are the first cells to take their place in relation to the nerve fiber. The less specialized cells (satellite and capsular cells, cells of the inner space) take their place and differentiate later around the more specialized cells and the nerve terminal. Thus there is an inside-out order of assembly as well as differentiation of the corpuscles. Although few studies have been devoted to the formation of sensory corpuscles, it appears that other types follow a similar sequence of development. Thus, for the Meissner corpuscle, MUNGER (1976) has described the change from an initial stage of "large polygonal cells associated with the nerve terminal" to a stage "of large lamellar cells". Likewise, for the Pacinian corpuscle, MALINOVSKY (1970) and MALINOVSKY and SOMMEROVA (1972a), with the light microscope, and MUNGER and NOVA (pers. comm., publication in preparation), using the electron microscope, observed stages of formation that are closely comparable to those described for the corpuscle of Herbst. In addition, in adult animals they have found Pacinian corpuscles in early stages of their development, which implies a postnatal formation of the corpuscles and probably a certain degree of turnover of the terminal organs.

Figs. 79–86. *Stages of development of the Herbst corpuscle.* Duck beak, light microscope. Figs. 79–82 and 85, silver staining; Fig. 84, Masson staining; Figs. 83 and 86, semithin sections. Fig. 79. *Stage 1* (embryo 20 days): nerve ending (NE) in dermis surrounded by dark cells with a massive nucleus. Longitudinal section. ×800. Fig. 80. *Late Stage 1* (embryo 21 days): elongated cells (EC) surrounding dark cells. Longitudinal section. ×800. Fig. 81. *Stage 2* (embryo 22–23 days): formation of inner bulb (IB). It is encircled by several layers of elongated cells (EC). Longitudinal section. ×80. Fig. 82. *Early Stage 3* (embryo 24 days): peripheral layers of elongated cells form outer capsule (OC) and are separated from inner bulb (IB) by inner space (IS) (or peribulbar space). Longitudinal section. ×600. Fig. 83. *Stage 3* (embryo 25 days): inner space (IS) is well formed. Transverse section. ×720. Fig. 84. *Late Stage 4* (young duck 4 days): middle-sized corpuscle is entirely formed. Inner space (IS) displays concentric lamellae. Note short peduncle (between arrows) of inner bulb. Longitudinal section. ×600. Figs. 85 and 86. *Stage 5* (adult duck): corpuscle is large, inner space well structured, peduncle of inner bulb is long (between two arrows). Nerve ending (NE) is surrounded by lamellae of inner bulb cells (BC) and dilates into terminal ampulla (TA). Figure 85, longitudinal section, ×450; Fig. 86, transverse section, ×550. BC, inner bulb cells; EC, elongated cells; Ep, epidermis; IB, inner bulb; IS, inner space; NE, nerve ending; NF, nerve fibers; OC, outer capsule; TA, terminal ampulla ▷

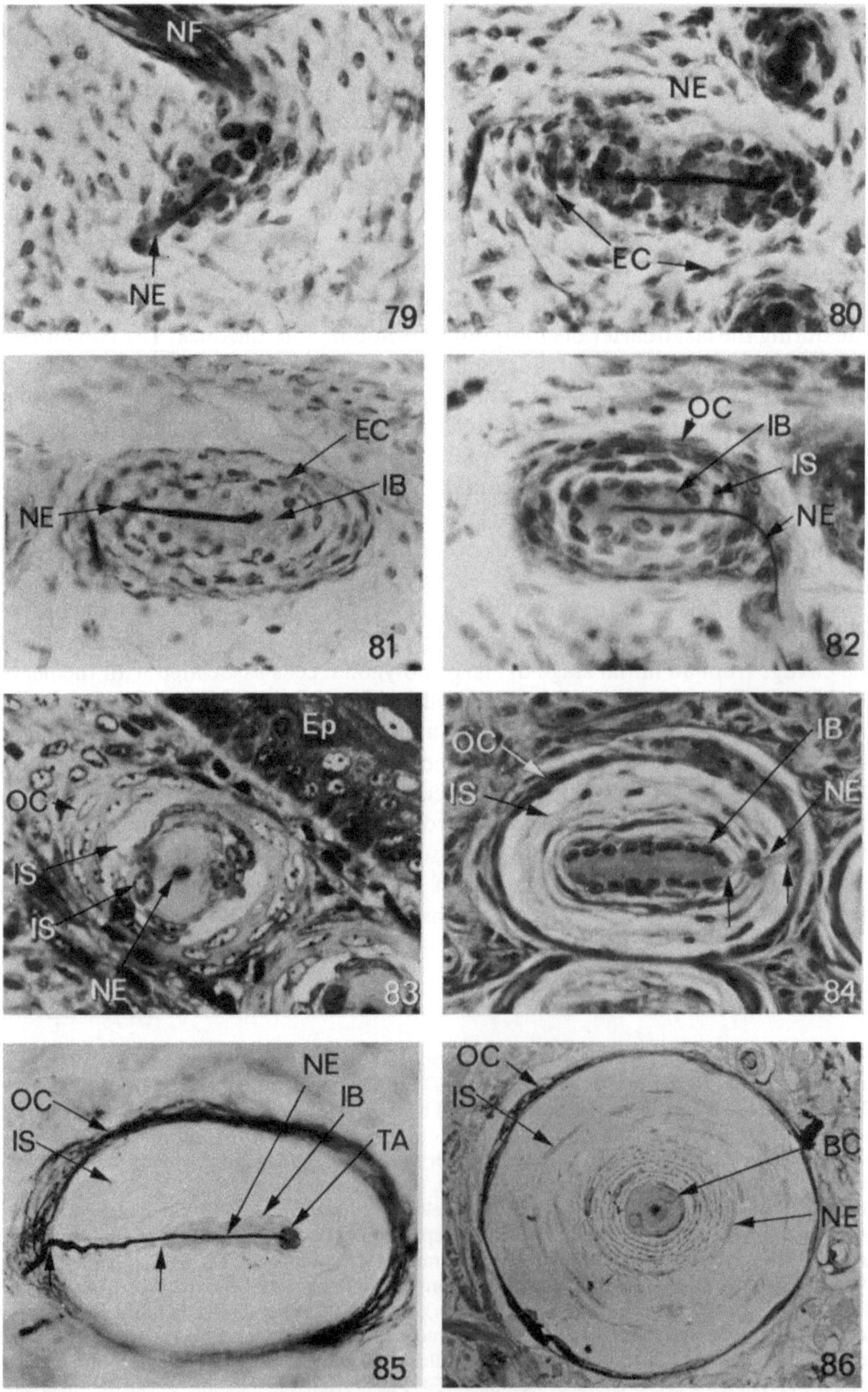
NF
NE
79
NE
EC
80
EC
NE
IB
81
OC
IB
IS
NE
82
Ep
OC
IS
IS
NE
83
OC
IB
IS
NE
84
NE
IB
OC
IS
TA
85
OC
IS
BC
NE
86

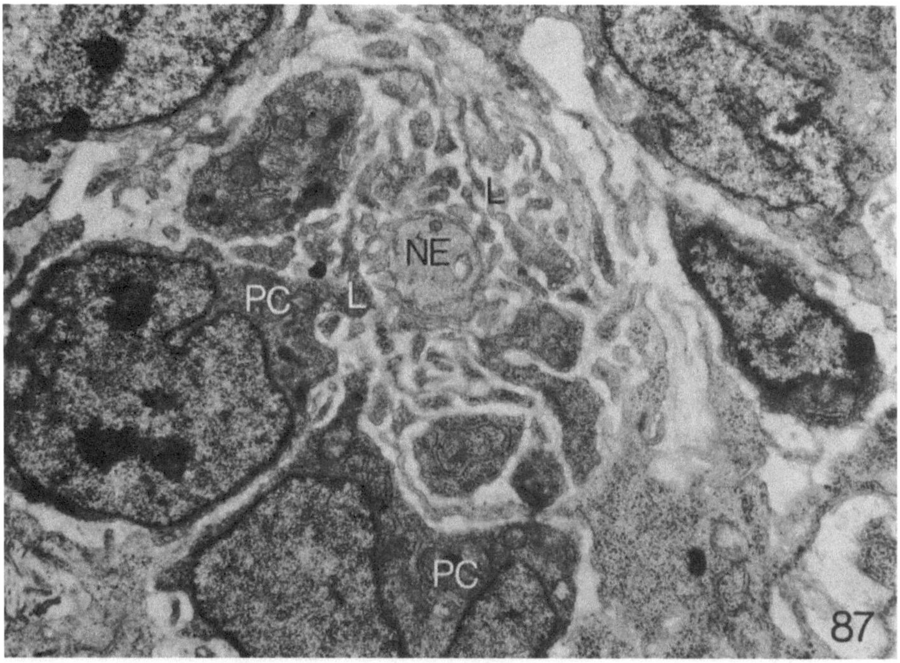

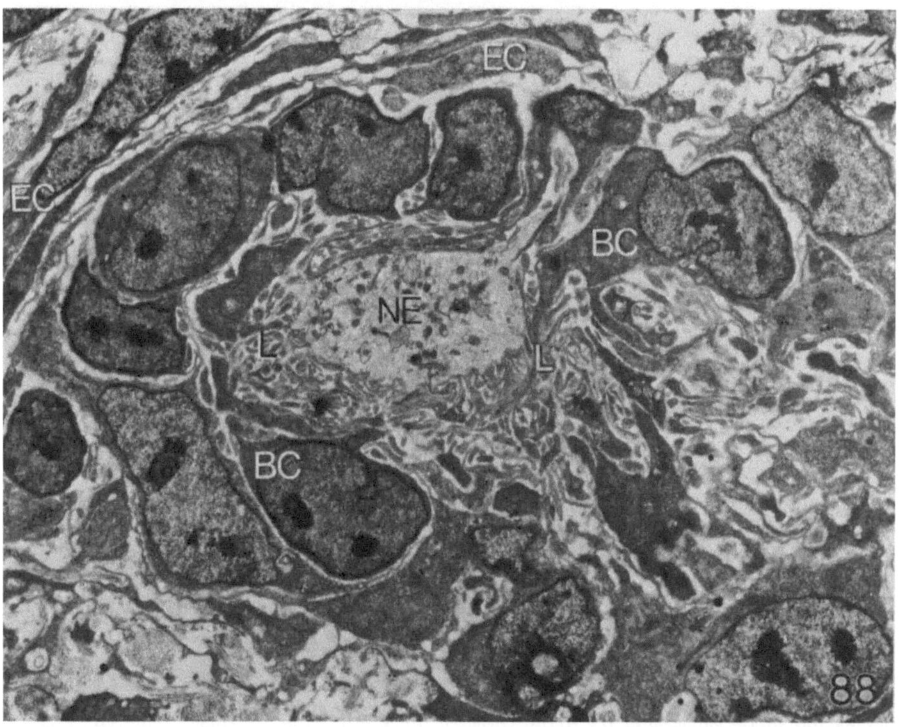

Figs. 87 and 88. *Development of the Herbst corpuscle* (duck beak). Fig. 87. *Stage 1*. (embryo 20 days): polygonal cells (PC) with lamellar processes (L) surrounding the nerve ending (NE). ×11000. Fig. 88. *Stage 2*. (embryo 22–23 days): lamellar (L) cells become inner bulb cells (BC) along nerve ending. Elongated cells (EC) become arranged around young inner bulb. × 5000. BC, inner bulb; EC, elongated cell; L, lamellae; NE, nerve ending; PC, polygonal cell

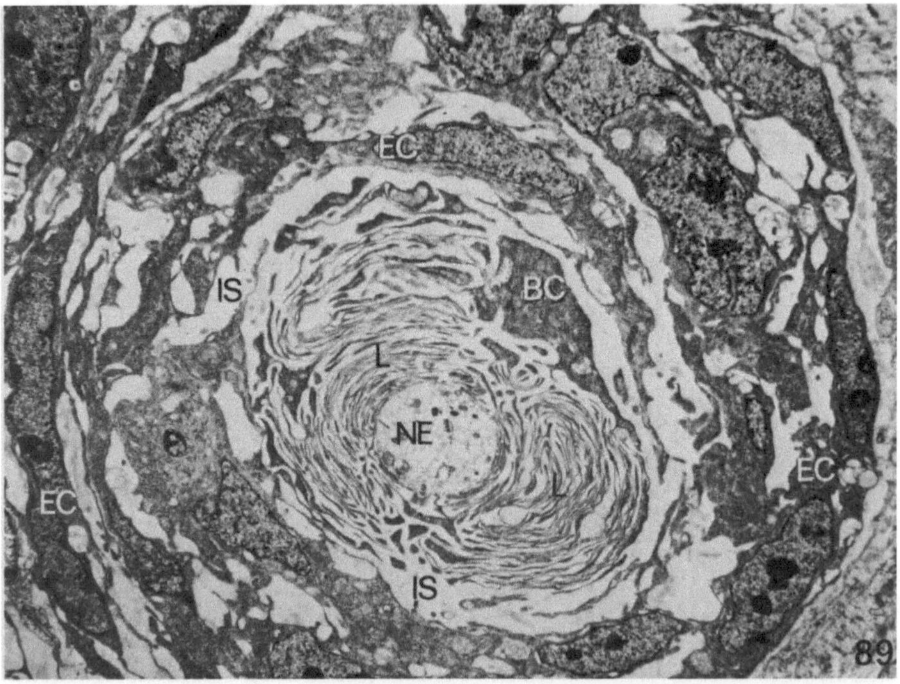

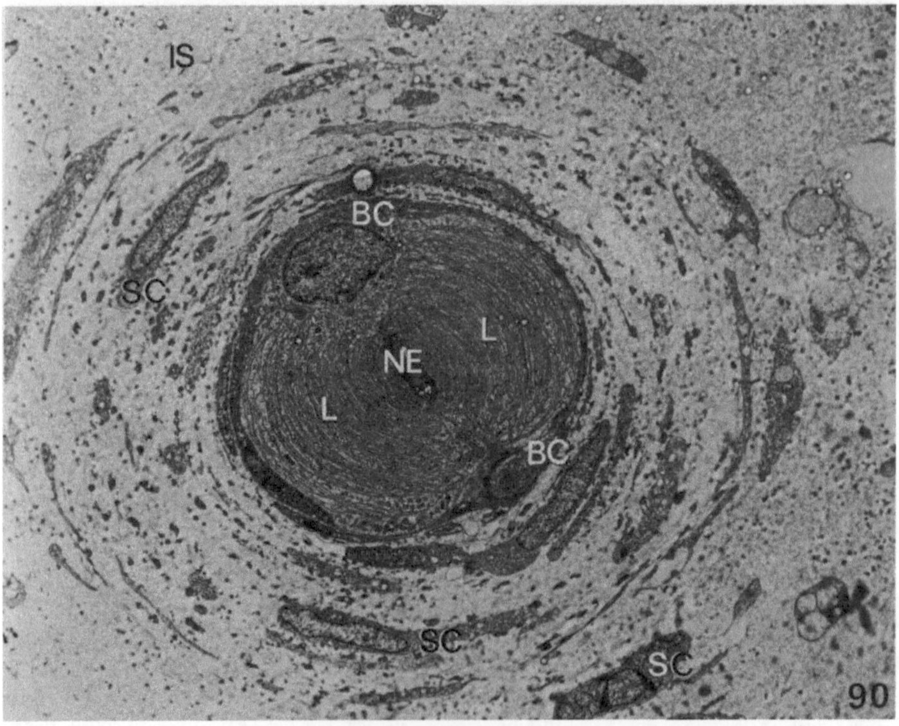

Figs. 89 and 90. *Development of the Herbst corpuscle* (duck beak). Fig. 89. *Stage 3* (embryo 24 days): transverse section. Lamellar (L) inner bulb cells (BC) form two symmetrical rows along axial nerve ending (NE). An inner space (EC) is forming between inner bulb and outer elongated cells (EC). × 5000. Fig. 90. *Late Stage 4* (young duck 4 days): transverse section of central part of a corpuscle. Herbst corpuscle is well formed. Lamellae (L) of inner bulb cells (BC) are numerous and closely interconnected around nerve ending (NE), now oval in transverse section. Inner space (IS) is well developed, and inner space cells (SC) form perforated lamellae. × 3600. BC, inner bulb cell; EC, elongated cell; IS, inner space; L, lamellae; NE, nerve ending; SC, inner space cell

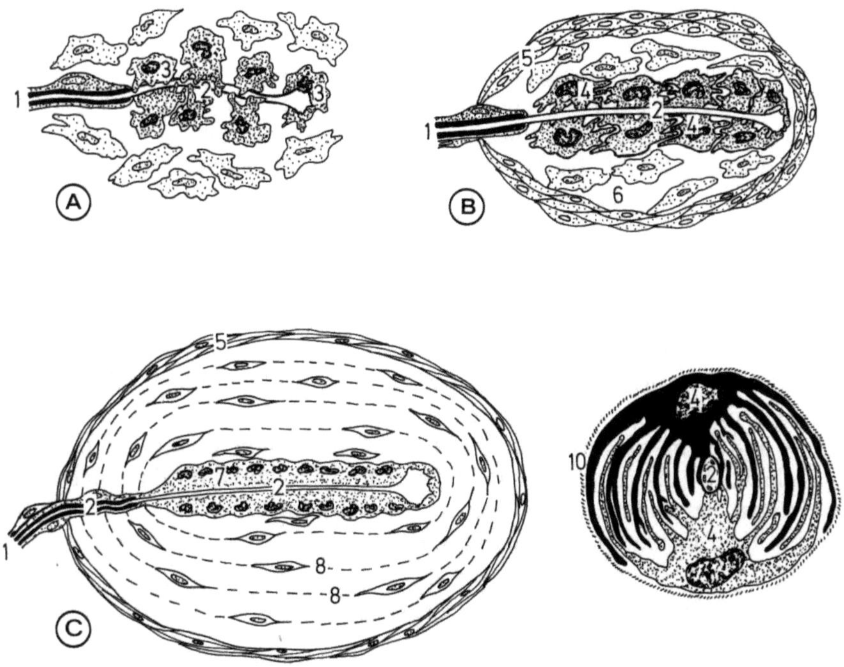

Fig. 91A–C. *Schematic representation of the three main stages in the development of the Herbst corpuscle* (duck beak, longitudinal sections). (A) Large multipolar cells with lamellar processes (3) are present around sensory nerve endings (2) (20th day of incubation). (B) These cells form lamellar cells (4) of inner bulb and are arranged in two symmetric rows along nerve ending. Flat cells form an outer capsule (5). Inner space (6) is narrow at this stage (23–26 days of incubation). (C) At hatching (28th day), development of Herbst corpuscle is almost complete; inner bulb (7) is made up of about 20 lamellar cells (on right, in transverse section). Inner space contains concentric perforated lamellae (8). Outer capsule (5) has included basal myelinated part (9) of nerve ending into corpuscle. (1) nerve fiber; (2) nerve ending; (3) multipolar cells; (4) lamellar cells of inner bulb; (5) outer capsule; (6) inner space; (7) inner bulb; (8) perforated lamella of inner bulb; (9) myelinated peduncle of inner bulb; (10) basal lamina

II. Experimental Studies

There are very few experimental studies dealing with the development of sensory cutaneous corpuscles in birds. The only ones that have been done are studies of degeneration and regeneration of the corpuscles in adult animals after sectioning or poisoning of the peripheral nerve (BOTEZAT, 1910; BOEKE, 1926; KLEIN, 1932; QUILLIAM and ARMSTRONG, 1961) or after heterotopic grafts in the adult (DIJKSTRA, 1933). The results obtained were very fragmentary and are difficult to apply to the embryo. The factors involved in the morphogenesis of these corpuscles are also not known, because there is considerable uncertainty about corpuscular structure during development in the embryo. Now that the precise ultrastructural organization and morphogenesis of the Herbst and

Grandry corpuscles are known, it is possible to make an experimental study of their differentiation during development (SAXOD, 1972a, b, 1973b, d).

From all these morphological data, two points emerge that are relevant to the questions about the morphogenesis of corpuscles that were posed at the beginning of the paper. The first point is that, in birds, there is a regional as well as a species-specific distribution of the corpuscles. The second is that the nerve ending associated with the specialized cells is present from the beginning of the formation of the corpuscles. The following questions might now be asked: Is the nerve ending indispensable for the development and maintenance of the corpuscles? If so, what categories of nerve fibers are able to ensure their development? Is the type of corpuscle that is formed determined by the nerve or by the skin that is innervated? To answer these questions, several types of experiments have been done. First the principal methods used will be explained. Next the proliferation of the cells during the course of development of the corpuscle will be described. Then the results showing the embryological origin of the different kinds of cells of the Herbst corpuscle will be discussed. Finally, the conditions required for the differentiation of the corpuscles and the morphogenetic tissue interactions that occur during the course of their formation will be analyzed.

1. Methods

The first experiment consisted of studying the formation of the corpuscles after modifying the normal relationship between the cutaneous territory and the peripheral nerve fiber during the course of embryonic development (suppression of innervation, innervation by heterotopic nerve fibers, etc.). The technique used was embryonic transplantation of the skin of the beak. This tissue was chosen because the skin of the beak contains different types of corpuscles in the three species that were studied. It should be pointed out that in the duck beak there are Herbst corpuscles of the "duck beak type" and Grandry corpuscles, while in the beak of chicken and quail there are only "chicken beak-type" corpuscles of Herbst and Merkel corpuscles.

The development of grafts without any innervation was achieved by making chorioallantoic grafts (Fig. 92). The grafts were explanted either before the beginning of formation of corpuscles (the stage of the frontal bud and young embryonic beak) or after the beginning of their formation (embryonic beak at other ages).

Two techniques were used to obtain innervation of grafts by nerve fibers from various sources. The first (Figs. 92, 94, and 95) consisted of implanting a graft on a very young host embryo, in which the nerve fibers were at the beginning of their development. These implantations were performed on embryos of a few days of age, after excision of a fragment of ectoderm, either in an orthotopic position (graft of the frontal bud: innervation by the trigeminal nerve, Fig. 94) or in a heterotopic position (graft on the limb bud or in the flank: innervation by a spinal nerve, Fig. 95). The second technique (Figs. 93 and 96) consisted of association, in a coelomic graft, of a cutaneous region with an isolated source of innervation: the spinal ganglion, the sympathetic ganglion, or a portion of the neural tube. These coelomic grafts were put into the embryo *via* the extra-embryonic coelom. The

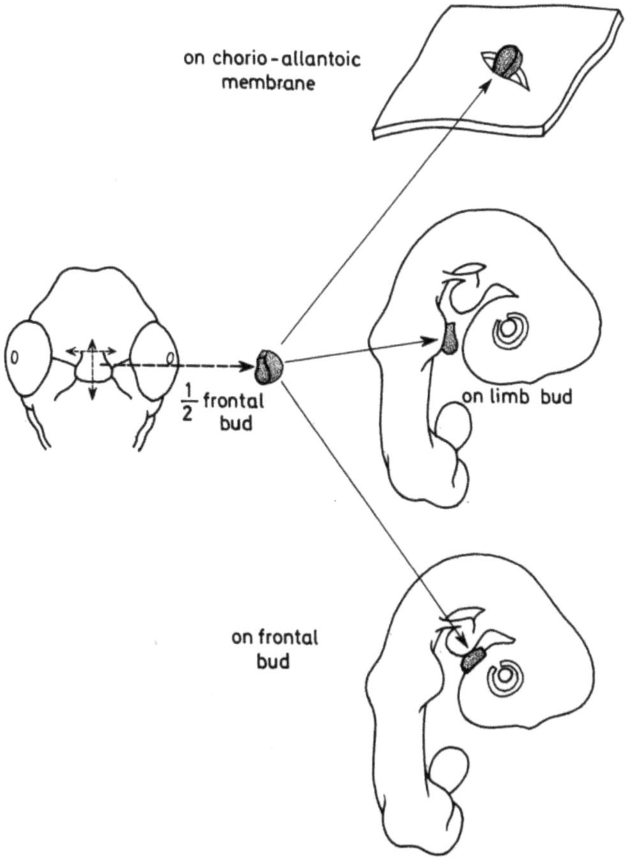

Fig. 92. *Three types of grafting*: on chorio-allantoic membrane, on wing bud, and on frontal bud. Drawings of face of donor embryo (duck 8 days) and lateral view of host embryos (duck 4, 5 days). Hemifrontal buds are either explanted on the chorio-allantoic membrane (development without innervation), or grafted on frontal bud (orthotopic graft) or on wing bud (heterotopic graft), so that they become respectively innervated by trigeminal nerve or brachial plexus

grafts became attached to the intestinal mesentery of the host embryo (Fig. 96) and they had no somatic innervation. Because the corpuscles developed very late, these grafts were removed at the end of development of the host, or sometimes after hatching. This is the explanation for the low percentage of successful grafts in the experiments.

Four types of experiments were performed:

(1) Chorioallantoic grafts: chicken/chicken, duck/chicken and duck/duck (73 grafts studied).

(2) Orthotopic grafts (to the frontal bud): duck/duck, duck/chicken and chicken/duck (24 grafts studied).

(3) Heterotopic grafts (to the flank or a wing-bud): chicken or quail/chicken, chicken or quail/duck, duck/chicken and duck/duck (272 grafts studied).

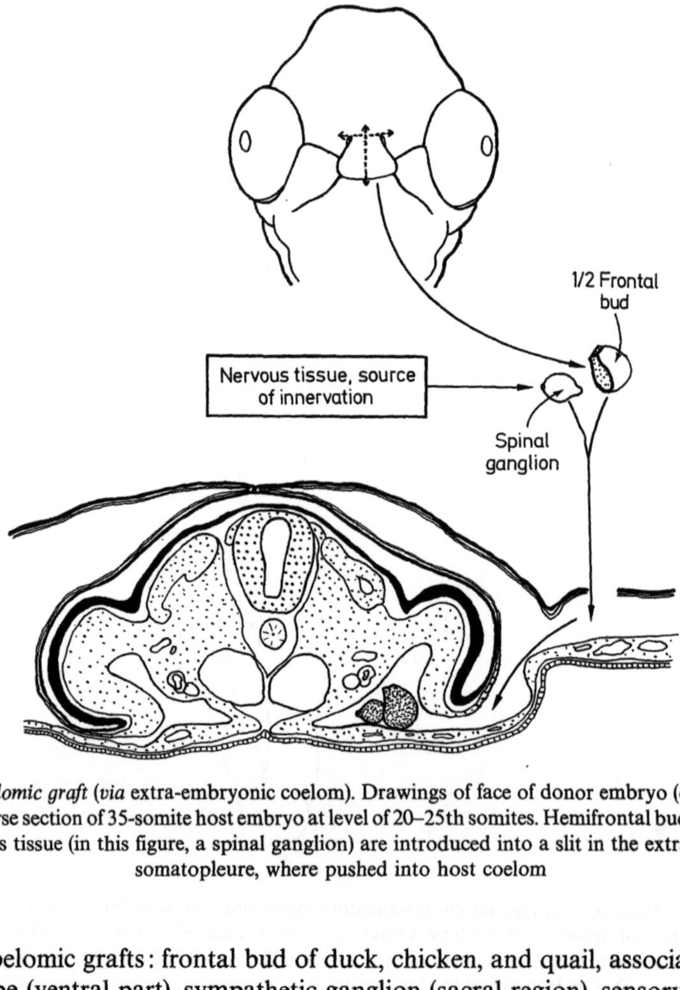

Fig. 93. *Coelomic graft* (*via* extra-embryonic coelom). Drawings of face of donor embryo (duck 8 days) and transverse section of 35-somite host embryo at level of 20–25th somites. Hemifrontal buds associated with nervous tissue (in this figure, a spinal ganglion) are introduced into a slit in the extra-embryonic somatopleure, where pushed into host coelom

(4) Coelomic grafts: frontal bud of duck, chicken, and quail, associated with: neural tube (ventral part), sympathetic ganglion (sacral region), sensory ganglion (cephalic or spinal) of duck, chicken, quail, mouse, lizard (268 grafts studied).

In all the grafts studied, the morphological and histological differentiation was in conformity with the origin of the graft (Figs. 94–96 and 112). When half a frontal bud was grafted, there was always regulation resulting in the formation of a reduced but fairly complete beak with, for example, in duck, the presence of bilateral filtration lamellae (Figs. 95 and 96).

2. Cellular Proliferation during Histogenesis

Development of corpuscles begins late during embryonic life, and a large number of corpuscles are made in a short time. Thus, in the duck beak, formation of several thousands of Herbst and Grandry corpuscles takes place in a few days. So in a very short time a very large number of cells (future corpuscular cells), which at this time can not be distinguished at the ultrastructural level from the other cells

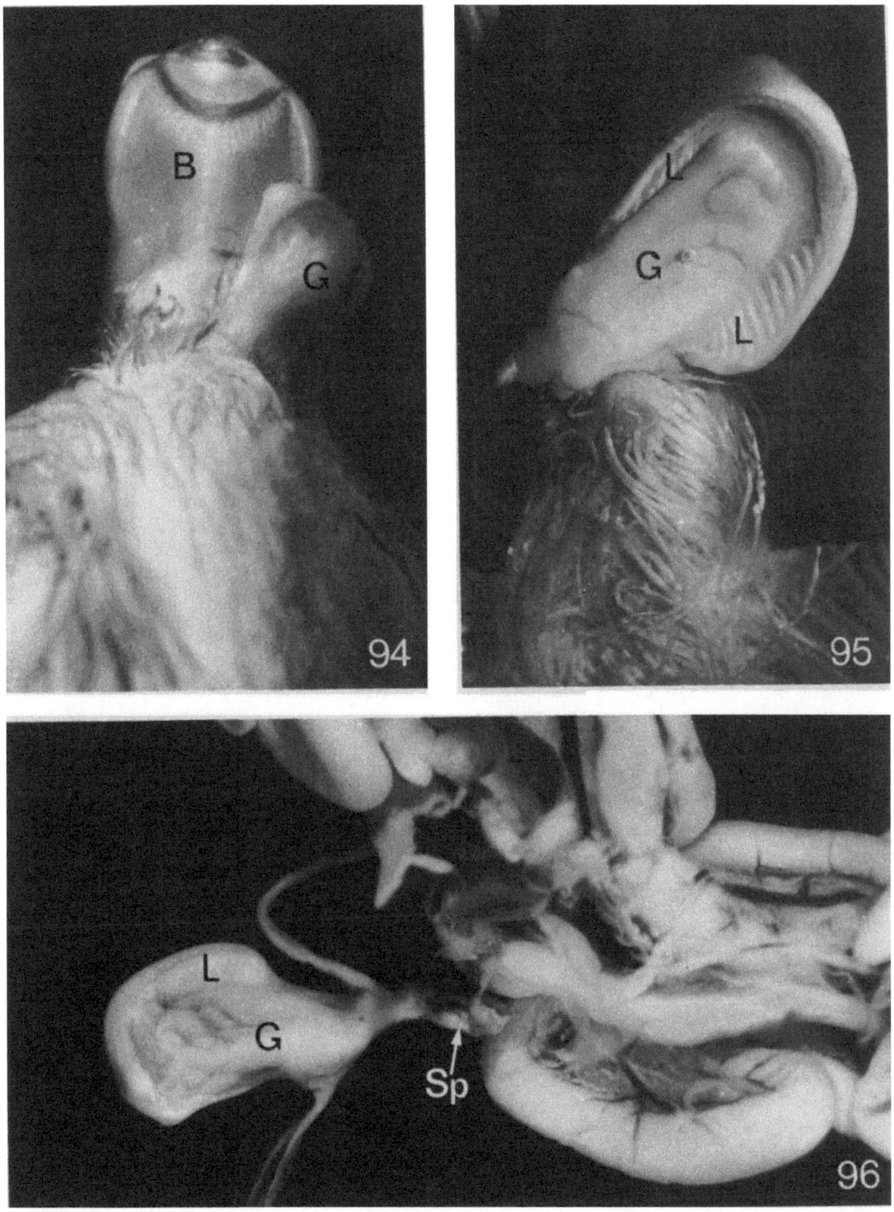

Figs. 94–96. *Macroscopic view of the grafts* that develop in the three types of grafting. Homografts of hemi-frontal buds of 6-day duck embryos, fixation 21 days after grafting. Fig. 94. Orthotopic graft on frontal bud (host embryo 4.5 days). Fig. 95. Heterotopic graft on wing bud (host embryo 4.5 days). Note that filtering lamellae are present on both sides (regulation). × 7. Fig. 96. Coelomic graft: hemifrontal bud associated with a spinal ganglion (host embryo 4 days). × 5. B, beak of host embryo; G, graft; L, lamellae; Sp, spinal ganglion; W, basal part of the wing

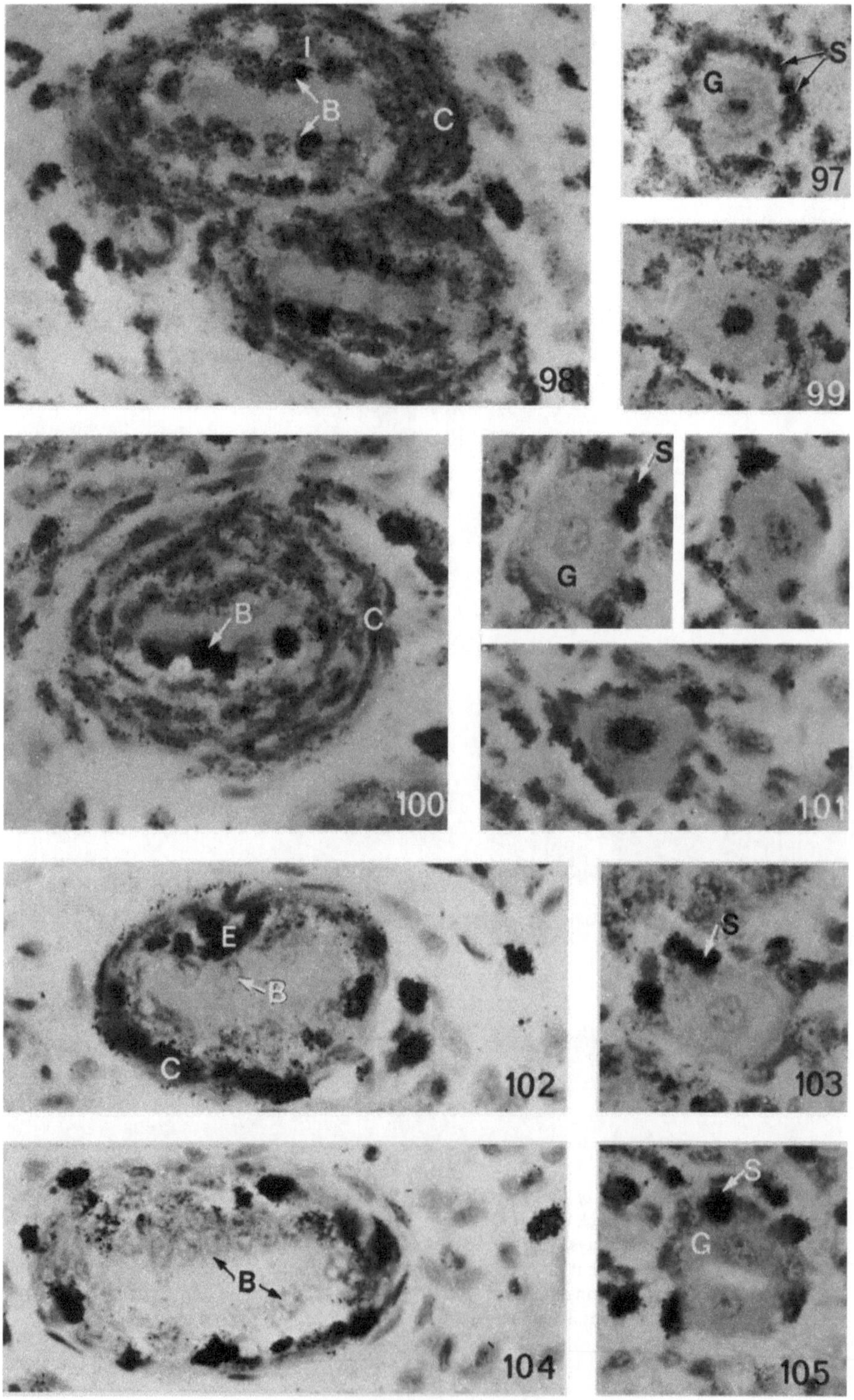
I
B
B
C
98
G
S
97
99
S
G
B
C
100
101
E
B
C
102
S
103
B
104
S
G
105

of the dermal mesenchyme, are forming and differentiating around the sensory nerve endings. The question is how these cells are mobilized. Is there any localized groups of cells waiting, a kind of "stock" of corpuscular cells placed in the skin from the beginning of embryonic development, which are able to proliferate very actively just before the beginning of formation of corpuscles? Or do the future corpuscular cells divide continuously during embryonic development so that they are not distinguishable before the formation of the corpuscles from the other mesenchymal dermal cells?

To answer these questions, the proliferation of different cellular categories participating in the formation of Herbst and Grandry corpuscles was studied by autoradiography (SAXOD, 1973d). The method consisted of administration of tritiated thymidine (50–100 μCi, specific activity 5 Ci/mM) at single embryonic stages in order to obtain good labeling of all the dividing cells for each embryonic stage. The location of the different types of labeled cells in the corpuscles was finally studied at the end of the development after the formation of the corpuscles. The injection of the label was made either before the beginning of development of corpuscles (embryos of 4.5–18 days) or during their development (embryo 20 and 23 days old). With this method, the labeling of all the proliferating cells can be obtained only if the duration of availability of label is at least equal to the length of cellular cycles of all the cellular categories studied. The average length of the generation time being close to 16 hrs for embryos between 6 and 12 days (MALAMUD, 1971), the availability time of tritiated thymidine was restricted to 24 hrs by giving a chase of pure thymidine (× 100 the original amount) the day after injection of the label. By this method, all the cells dividing during this 24-hrs period were strongly labeled. It has not yet been possible to perform a quantitative study of proliferation and, consequently, these results are obtained from a statistical estimation of the relative intensity of labeling for each type of cell in the corpuscle: strong labeling, medium labeling, light labeling. The attenuation of the labeling between the time of injection and the time of fixation reflects the mitotic activity of the cells during the interval. The weaker the label, the more divisions will have occurred during this period.

◁ Figs. 97–105. *Autoradiographs of Herbst and Grandry corpuscles* of duck beak after tritiated thymidine labeling of embryos between 6 and 23 days of incubation. Chase by cold thymidine 24 hours later. Beaks fixed at 25–27 days of incubation. × 720. Fig. 97. Labeling at 6 days—Grandry corpuscles: all satellite cells (S) and Grandry cells (G) are labeled. Labeling of corpuscle cells and of dermal cells is about same. Figs. 98–99. Labeling at 14 days. Herbst corpuscle (Fig. 98): all capsular (C), inner space (I) and inner bulb (B) cells are labeled. Grandry corpuscle (Fig. 99): satellite (S) and Grandry cells (G) are labeled. Figs. 100–101. Labeling at 18 days. Herbst corpuscle (Fig. 100): all corpuscle cells are labeled; some inner bulb cells (B) display heavier labeling than capsular cells (C). Grandry corpuscle (Fig. 101): satellite cells are heavily labeled; some Grandry cells (G) are heavily labeled (bottom), some lightly (top right), others not at all (top left). Figs. 102–103. Labeling at 20 days—Herbst corpuscle (Fig. 102): most capsular cells (C) and inner space cells (I) are heavily labeled, but inner bulb cells (B) are not (arrows). Grandry corpuscle (Fig. 103): most satellite cells (S) labeled, but Grandry cells (G) are not. Figs. 104–105. Labeling at 23 days. Herbst corpuscle (Fig. 104): some dermal cells and some capsular (C) and inner space cells are labeled, but inner bulb cells not (arrows). Grandry corpuscle (Fig. 105): some dermal and satellite cells (S) are labeled, but Grandry cells (G) are not. B, inner bulb cell; C, outer capsule; G, Grandry cell; I, inner space cell; N, nerve ending; S, satellite cell

The results show that three main periods can be distinguished in the embryonic development of duck during histogenesis of the corpuscles. The first period extends from the beginning of embryonic development to the 18th day of incubation (Figs. 97–99). When the marker is injected at stages occurring from the 4th to the 16th day of incubation, there is almost uniform labeling of all the cells in Herbst and Grandry corpuscles. The younger the embryo at the time of injection, the lighter the labeling at the time of fixation, and in all cases the labeling of the corpuscular cells is of the same order of magnitude as the labeling in the other dermal cells. Thus all presumptive corpuscular cells have a high mitotic activity between the time of injection and the time of fixation, and we cannot see, with this method, any differences between these cells and the other cells of the dermal mesenchyme. If quantitative differences exist in the proliferation of these cells, they can be shown only by a more refined method.

The second period extends from 18 to 20 days of incubation (Figs. 100 and 101). When the embryos are injected at 18 days of incubation, the capsular cells (Herbst corpuscles) and the satellite cells (Grandry corpuscles) are lightly labeled, reflecting a high mitotic activity. On the other hand, the inner bulbar cells and Grandry cells show great heterogeneity of labeling. Thus these cells are not uniform in their mitotic activity between the time of administration of the marker and fixation. Some are dividing very actively (light labeling), other are undergoing few divisions (heavy labeling), and others are ceasing to divide (no labeling). This indicates a decrease in the proliferation of these cells, which will be associated with the nerve ending (the future so-called "sensory cells"), and these cells will be the first to differentiate.

The third period begins at about the 20th day of incubation (Figs. 102–105). The experiments performed with embryos more than 20 days old showed that the inner bulb cells and Grandry cells are the first to stop dividing (no labeling). At the same time, the mitotic activity of capsular cells and inner space cells (Herbst corpuscle) and of the satellite cells (Grandry corpuscles), which is very high at the beginning of this period (light labeling), decreases progressively (heavy labeling). It is precisely at this stage, 20 days of incubation, that the constituent elements of the corpuscles are forming from the inner part to the periphery around the nerve endings.

In conclusion, it seems that there are only slight differences in the rate of proliferation of the different kinds of cells in the dermal mesenchyme of the duck beak during the first part of embryonic development (up to 18 days of incubation). During this period, the future corpuscular cells seem to divide in a regular and continuous way; the hypothesis of "stocks" of corpuscular cells must be rejected. The first period of proliferation of the cells forming the corpuscles corresponds to a phase of high proliferation for all these cells. The second period (18–20 days of incubation), just before the beginning of development of corpuscles, is a phase of decreasing proliferation of the future sensory cells preceding their differentiation. The third period (starting at 20 days of incubation) is a phase of differential and progressive cessation of proliferation for all the types of cells of the corpuscles, advancing from the center of the organs to the periphery parallel with their differentiation.

3. Developmental Origin of the Herbst Corpuscle

We have seen the difficulties involved in the interpretation of the few experiments devoted to the development of cutaneous sensory corpuscles. These difficulties arose because the embryonic origin of the cells constituting these corpuscles was not known. Numerous hypotheses have been proposed, particularly for the Herbst and Pacinian corpuscles, which have been studied for more than a century. It was generally admitted that the capsule of these two kinds of receptors are formed by endothelial cells (Polacek, 1966) or fibroblastic cells (Pease and Quilliam, 1957; Cauna and Mannan, 1958; Quilliam, 1966a; Shanta and Bourne, 1968) which are mesenchymal origin, but there was great discordance concerning the origin of the inner space and the inner bulb or inner core cells. For some authors (Szymonowicz, 1897; Pease and Quilliam, 1957), these cells were modified fibrocytes with a mesodermal origin; for other authors (particularly Heringa, 1918; Rhodin, 1963; Polacek, 1966; Chouchkov, 1971), the inner bulb cells were of Schwann type and had an ectodermal origin (neural crest). For Shanta and Bourne (1962, 1963, 1968), the sensory corpuscles were no more than terminal enlargements of sensory nerves and all these cells had the same origin as the sheath cells of peripheral nerves.

The present author's own experimental studies show that it is possible to obtain the formation of corpuscles by homoplastic and xenoplastic association of frontal buds and sensory ganglia in bird embryos (these experiments will be detailed below). The origin of the different categories of cells in the Herbst

Table 1. Types of grafting and recombination experiments of frontal bud with specific sources of sensory innervation

Grafts			No. of cases: Histologically examined	No. of cases: With corpuscles
A. Grafts on the wing-bud (innervation by brachial plexus)				
Frontal bud		On wing-bud		
Quail 6–7 days		Chick 3.5 days	17	13
		Duck 4.5 days	8	5
B. Coelomic grafts				
Associations: Frontal bud + source of sensory innervation		Into coelom		
Quail 5–7 days +	duck Gasserian ganglion 5–7 days	Duck 4 days	13	5
	chick Gasserian ganglion 5–6 days	Duck 4 days	19	12
	chick Gasserian ganglion 5–6 days	Chick 3 days	12	10
Chick 6–7 days +	quail Gasserian ganglion 5–7 days	Chick 3 days	25	15
Duck 5–7 days +	quail Gasserian ganglion 5–7 days	Duck 4 days	24	14
	quail spinal ganglion 5–7 days	Duck 4 days	8	7
Total			126	81

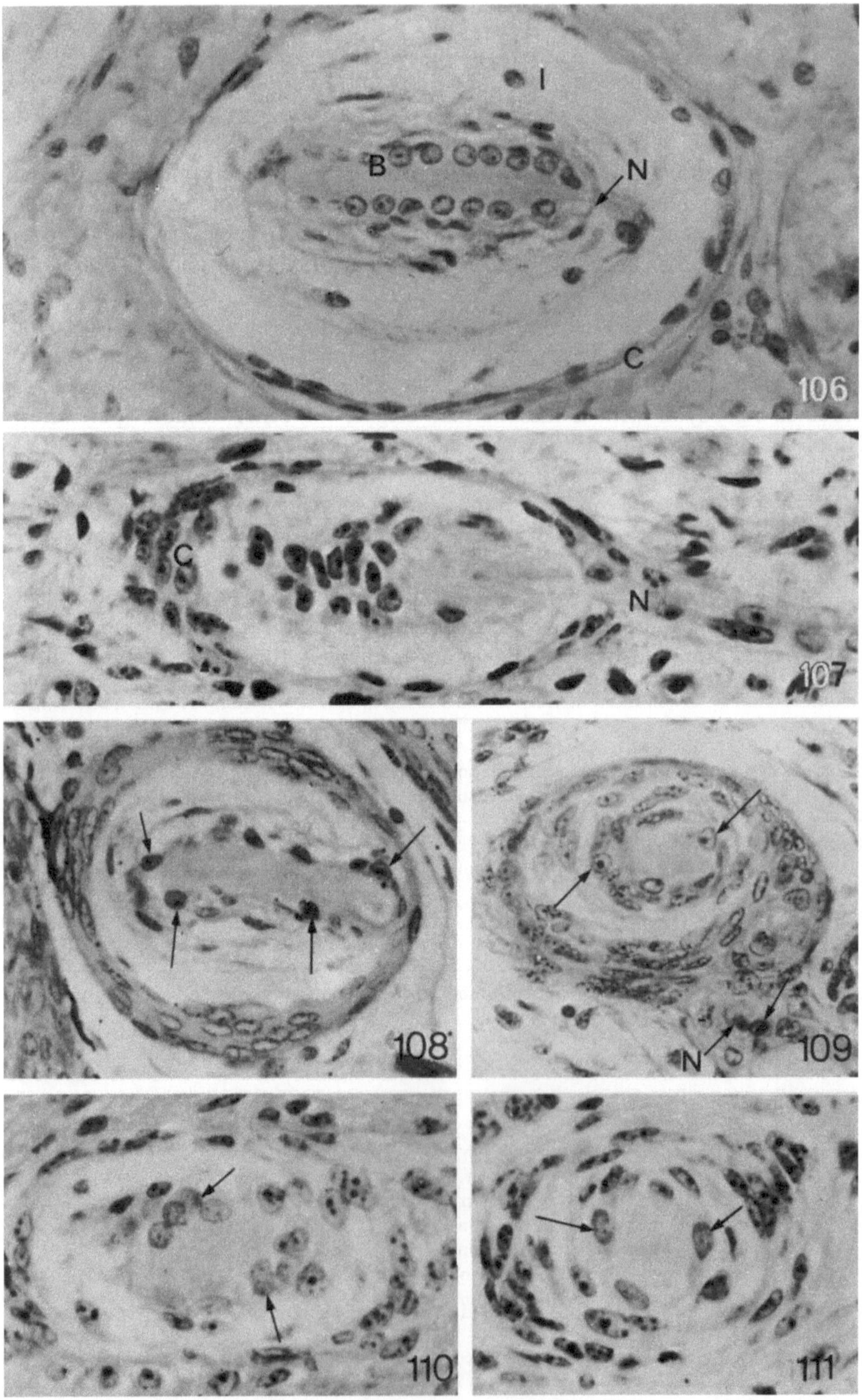
I
B
N
C
106
C
N
107
108
N
109
110
111

corpuscles has proved accessible to study with this association technique and labeling by means of cellular markers.

The classic technique of labeling with tritiated thymidine gives results that are difficult to interprete because of the dilution of the label at the time of the formation of corpuscles in the graft (SAXOD, 1973b). The present author has used the biological markers introduced by LE DOUARIN and BARQ (1969) and LE DOUARIN (1971, 1973). These authors showed that in quail interphasic nucleus, the chromatin is condensed in characteristic large masses (nucleolus-associated chromatin), whereas in chicken and duck the chromatin is scattered throughout the nucleus. Thus it is possible in quail-chicken and quail-duck associations to recognize the quail cells by Feulgen staining of the sections. The following xenoplastic associations were performed (Table 1).

1. Grafts of quail frontal bud on the wing-bud of chicken and duck.
2. Coelomic association of frontal buds of quail with duck of chicken sensory ganglion, and vice versa.

As a control, sections of the beak of quail, chicken, and duck were stained with Feulgen reagent (Figs. 106 and 107). Only the quail cells characteristically have one and sometimes two large nuclear granules in the Herbst corpuscles (Fig. 107). One must point out that some quail cells, particularly the outer capsular cells, can contain several small nuclear granules. The absence of a unique granule in the nucleus of the cells does not permit us to claim that this cell is not a quail cell. Therefore, for each kind of association, it was necessary to also make the reverse association.

Since the different grafting techniques used do not affect the results, we shall only analyze the experiments on the xenoplastic associations between quail and duck and quail and chicken performed in coelomic grafts.

a) Association between Quail Sensory Ganglion and Duck or Chicken Frontal Bud (Figs. 108 and 109)

In all the cases where the grafts were innervated, chimeric Herbst corpuscles developed. The inner bulb was formed largely, if not totally, of quail cells. Some

◁ Figs. 106 and 107. Control Herbst corpuscles stained with Feulgen reagent. Fig. 106. Duck beak, 15 days posthatching, longitudinal section. × 700. Fig. 107. Quail beak, 3 days posthatching, oblique section. Note the large nuclear characteristic Feulgen-positive granule. Certain nuclei of the outer capsule contain 2 or 3 granules. × 850

Figs. 108 and 109. Associtation of quail embryo sensory ganglion and duck (Fig. 108) or chick (Fig. 109) frontal bud in coelomic graft. In the chimeric Herbst corpuscles obtained, the inner bulb is composed almost entirely, if not totally, of quail cells (arrows). The inner space and capsular cells are of the same type as the dermal cells, and lacked the quail type nuclear granule. In the Fig. 109, the nerve (N), seen in transverse section, shows a quail type Schwannian cell. Fig. 108, longitudinal section. Figure 109, transverse section. × 700

Figs. 110 and 111. Associations of quail frontal bud and chick (Fig. 110) or duck (Fig. 111) sensory ganglion in coelomic graft. In the chimeric Herbst corpuscles obtained, the cells of the outer capsule and of the inner space contain the characteristic quail type nuclear granule. Most inner bulb cells are devoid of it (arrows). Figure 110, transverse section. Figure 111, oblique section. × 800. B, inner bulb; C, outer capsule; I, inner space; N, nerve ending quail cells were sometimes seen in the inner space, but in most cases none of the cells of the inner space and none of the cells of the outer capsule contained the characteristic large granules of the quail cells.

b) Association between Quail Frontal Bud and Duck or Chicken Sensory Ganglion (Figs. 110 and 111)

This reverse association also leads to the formation of chimeric Herbst corpuscles. The outer capsular cells and almost all the inner space cells were of quail type, whereas the nuclei of the inner bulb cells did not contain large Feulgen-positive granules.

These results show that the Herbst corpuscle is formed by cells with two distinct origins:

1. The large majority, if not all of the inner bulb cells and probably some inner space cells originate from cells that accompany the nerve during its outgrowth. These cells are present in the spinal ganglion and in the Gasserian ganglion of 5-, 6-, and 7-day-old embryos.
2. All the other cells of the corpuscles, namely the capsular cells and the majority, if not all, of inner space cells are provided by the predermal mesenchyme.

These results contradict the views of authors who have proposed that the cells of the Herbst corpuscles were only of mesodermal origin. They confirm the hypotheses of many authors concerning a mesodermal or mesenchymal origin of capsular cells. Probably these cells are in continuity with epineural and perineural connective tissue. The inner bulb cells, therefore, seem to have the same origin and the same kind of migration as the Schwann cells. The ultrastructural and histochemical research on Herbst corpuscles has shown that the inner bulb cells can be related to Schwann cells by the following properties: unique association with the nerve ending, presence of a basal lamina, nonspecific cholinesterase activity. These results suggest that the inner bulb cells are no more than specialized and modified Schwann cells.

The problem of the origin of the two other types of corpuscles in birds, namely the Grandry and Merkel corpuscles, can not yet be resolved by these cellular marker techniques. For the Grandry corpuscle this is because this type of receptor can be obtained only in one kind of association (see below), so the results are not complete and cannot be confirmed by the reverse association. The Merkel corpuscles are present in the chicken and in the quail, but their small size makes it difficult to study them by light microscopy. Studies with the electron microscope are needed to solve this problem.

4. Role of Nerve and Cutaneous Tissues in the Development of the Corpuscles

a) Neuronal Influence on Corpuscle Formation

The first problem is whether the nerve ending is needed for the development of nerve corpuscles. The answer to this question is given by comparing grafts that develop without any innervation with those that develop in the presence of nerve fibers (Tables 1–4).

Table 2. Explantations of frontal buds and beaks on the chorio-allantoic membrane: development in absence of innervation

Donor embryos of frontal buds and beaks[a]	No. of cases	
	Histologically examined	With corpuscles
Explantation before the formation of corpuscles		
Chick 8–14 days	7	0
Duck 8–18 days	30	0
Explantation after the formation of corpuscles		
Duck 20–24 days	36	0

[a] Age in number of days of incubation.

Development of Grafts without any Innervation: Two kinds of experiments were performed and the grafts were removed from the embryo either before or after the beginning of the development of the corpuscles (Tables 2–4). It must be remembered that their formation begins at 16 days in the chicken and at 20 days in the duck. All the noninnervated grafts studied were devoid of corpuscles. Thus, corpuscles of Herbst and Grandry do not form without innervation. Moreover, removal of the innervation of sensory corpuscles during their development not only arrests their development, but also causes their disorganization and disappearance. The next question is whether development of the corpuscle is possible if the frontal bud is grafted onto a host embryo and innervated by the host.

Innervation of Frontal Buds by Trigeminal Nerve and by Spinal Nerve (Table 3): To determine whether corpuscles in frontal buds develop when innervated by the normal source of innervation, i.e., the trigeminal nerve, orthotopic grafts of the frontal bud were performed (Figs. 92 and 94 and Table 3). Corpuscles developed in the vast majority of these grafts. These results and those from the grafts without innervation show that the presence of nerve endings is a prerequisite for the development of cutaneous corpuscles in birds. Since the development of corpuscles in a frontal bud innervated by trigeminal nerves occurs experimentally, we may ask what occurs with other sources of innervation.

By performing heterotopic grafts (Table 3), it was possible to innervate frontal buds by spinal nerves: lumbar nerve for grafts in the flank and nerves from the brachial plexus for grafts of the wing-bud (Figs. 92 and 95). Corpuscles developed in both kinds of grafts (Table 3). Thus a heterotopic source of innervation, and even a xenoplastic one (between chicken and duck), allows the development of sensory corpuscles. This innervation being provided by mixed nerves, the question is whether the different kinds of fibers in these nerves (sensory, motor, somatic, sympathetic fibers) have the same capacities to support development of corpuscles.

Innervation of Frontal Buds by Restricted Sources of Nerve Fibers: To test the role of different categories of nerve fibers in the development of corpuscles, coelomic grafts were performed to produce an association between frontal buds and isolated sources of innervation (Table 4 and Fig. 92). In a control experimental series, coelomic grafts of frontal bud alone were performed. The grafts, which were fixed on the intestinal mesentery (Fig. 96), contained light sympathetic innervation

Table 3. Orthotopic and heterotopic grafting of frontal buds

	Donor embryo[a]	Host embryo[a]	No. of cases examined	Types of corpuscles formed in the graft			
				Herbst, duck-beak type	Grandry	Herbst, chicken-beak type	Merkel
Grafts onto the frontal bud Innervation by trigeminal nerve	Chick 5 days	Duck 4.5 days	4	—	−	+	+
	Duck 5–14 days	Duck 4.5 days	13	+	+	−	−
	Duck 6.5 days	Chick 3.5 days	7	+	+	−	−
Grafts into the flank Innervation by a lumbar nerve	Chick 8–13 days	Chick 2 days	11	−	−	+	+
	Duck 10–12 days	Chick 2 days		+	+	−	−
Grafts on the wing-bud Innervation by brachial plexus	Chick 5–9 days	Chick 3.5 days	4	−	−	+	+
	Chick 5–9 days	Duck 4.5 days	4	−	−	+	+
	Duck 5–9 days	Chick 3.5 days	18	+	+	−	−
	Duck 3–12 days	Duck 4.5 days	70	+	+	−	−
	Duck frontal mesenchyme 5–8 days	Duck 4.5 days	12	+	+	−	−

[a] Age in number of days of incubation.

Table 4. Coelomic grafting of frontal bud associated with nervous tissue

Association[a] Nervous tissue	+ Frontal bud or beak	No. of cases examined	Types of corpuscles formed in the graft: Herbst, duck-beak type	Grandry	Herbst, chicken-beak type	Merkel
Spinal or Gasserian ganglion						
Duck 6 days	0	8	0	0	0	0
0	Duck 6–14 days	24	0	0	0	0
Duck 6 days	Duck 5–6 days	19	+	+		
Chick 6 days	Duck 5–6 days	11	+	+		
Duck 6 days	Chick 5–6 days	9			+	+
Neural tube						
Duck 6 days Chick 7 days	Duck 5–14 days	18	0	0	0	0
Sympathetic ganglion						
Duck 7 days Chick 7 days	Duck 5 days	9	0	0	0	0
Mouse spinal of Gasserian ganglion						
Mouse embryo 12–15 days Mouse 4 days	Duck 5 days	41	0	0	0	0
Lizard spinal or Gasserian ganglion						
Lizard embryo 22 days[b]	Duck 6 days	25	0	0	0	0

[a] The host is always a 4-day duck embryo, age in number of days of incubation or gestation.
[b] Embryos of *Lacerta muralis* Laur. (eggs incubated at 26° C; Dhouailly and Saxod, 1974).

of mesenteric origin but were devoid of sensory corpuscles. This control series was completed by experiments on the association of frontal buds and sympathetic ganglia (Table 4). No corpuscles developed in this kind of combination, so one can conclude that sympathetic nervous fibers are unable to ensure the differentiation of cutaneous sensory corpuscles. Moreover, these results show that the technique of coelomic grafts can be used to test other sources of innervation such as neural tube and sensory ganglia.

Association of frontal buds and ventral parts of the neural tube gave rise to good somatomotor innervation of the grafts (Saxod, 1972a). All the grafts studied were devoid of corpuscles (Table 4), so one can conclude that the somatomotor nerve endings are unable to ensure the development of cutaneous sensory corpuscles. On the other hand, when frontal buds and bird sensory ganglion (Gasserian ganglion or spinal ganglion) are associated, corpuscles are found in all the innervated grafts (Table 4; Figs. 112–114). Although these ganglia contain both somatosensory neurons and sympathetic neurons, the results of control experiments and of combination between frontal bud and sympathetic ganglion showed that the somatosensory nerve endings alone are able to promote the development of cutaneous sensory corpuscles in birds, even in the absence of central connections.

All these results were obtained from experiments on homoplastic and xenoplastic association involving chicken, duck, and quail (Tables 1–4), and thus are reasonably valid for all birds. One can ask whether the somatosensory fibers of the other classes of vertebrates, and particularly of amniotes, are able to promote the development of corpuscles in cutaneous tissue of birds. To answer this question, associations in coelomic grafts of frontal buds of duck with Gasserian or spinal ganglia from mouse or lizard embryos were performed (Table 4). This technique allowed a good development of the innervation in both kinds of combinations, although there was a slight immune necrosis of the grafts. However, all the results were negative: this kind of interclass combination does not allow the formation of corpuscles.

In conclusion, the onset of development of sensory corpuscles in birds can be obtained only in the presence of a nerve ending. This nerve ending is also necessary for maintenance of the structural integrity of corpuscles already formed and for their further development. Only the somatosensory nerve endings are able to ensure the development of corpuscles, but they can do so even in the absence of any central connection. The sympathetic and somatic motor fibers do not have these properties, nor do the somatosensory fibers of the other classes of vertebrates.

These results, which demonstrate that the presence of nerve endings is necessary for the development and maintenance of bird sensory corpuscles, extend those from the classic experiments performed in adult ducks (BOEKE, 1926; KLEIN, 1932; DIJKSTRA, 1933), which showed that Herbst and Grandry corpuscles degenerate after denervation and reappear after regeneration of the nerves. The same phenomena have been seen in several kinds of sensory corpuscles in mammals. Thus, for Pacinian corpuscles, recent ultrastructural studies (CHOUCHKOV, 1971) show that after sectioning of the afferent nerve there is a quick degeneration of the entire inner core and replacement of this by collagen fibers. This shows that the nerve fiber is essential for support of the nonneuronal receptor structures. Likewise, the Merkel tactile disks (also called "domes") degenerate after sectioning of the nerve and reappear when the nerve regenerates (BROWN and IGGO, 1963; PALMER, 1965; BURGESS et al., 1974), which indicates that the innervation is necessary for maintenance or differentiation of the Merkel cells (ENGLISH, 1974). Similar results, after sectioning of afferent nerves in the adult, were obtained for taste buds and sensory organs of the lateral-line system in fish and amphibians and for the gustatory cells of mammals (for reviews see WRIGHT, 1951; ZELENA, 1964; HUGHES, 1968; GUTH, 1969; JACOBSON, 1970; ZALEWSKI, 1974). In some vertebrates, however, the formation of receptor organs is independent of innervation, at least at a certain stage of embryonic or larval development. This is true particularly for the primordia of the lateral line in amphibians, which if removed sufficiently early, can develop without any innervation; the dependence on innervation appears at about at the time of metamorphosis (for reviews see HUGHES, 1968; JACOBSON, 1970).

We have seen that in birds only the somatosensory nerve endings were able to ensure development of cutaneous corpuscles and that these fibers can originate from a heterotopic and/or xenoplastic source of innervation (with the condition that the species are phylogenetically closely related). Moreover, this development can be obtained in the absence of any central nervous connection. This is a new example of the relative independence of peripheral receptors and central nervous system during development (VERLEY et al., 1969), and extends to cutaneous corpuscles the conclusion concerning the taste buds in fishes (KAMRIN and SINGER, 1953) and mammals (ZALEWSKI, 1974). In mammals, it is also possible to obtain the development of cutaneous corpuscles in the presence of heterotopic nervous fibers. Thus, for Pacinian corpuscles, recent experiments (ILYINSKY et al., 1973) performed in cat showed that additional innervation of the mesentery by a somatic sensory nerve (N. *saphenus*) foreign to this tissue led to the formation of histologically and physiologically normal Pacinian corpuscles.

It seems, however, that for other kinds of receptor organs differences can exist between classes and even species regarding the capacity of heterotopic fibers to promote receptor end-organ formation, although few experiments have been performed in this field. Thus, for taste buds in *Triturus*, their development and maintenance can be ensured by a heterotopic source of innervation (PORITSKY and SINGER, 1963), while it seems that in rat their maintenance can be only ensured by gustatory nerves (GUTH, 1958, 1963; ZALEWSKI, 1969). However, recent neuronal transplantation experiments in rat (ZALEWSKI, 1972) have demonstrated that the negative results previously obtained were due to the technique used (nerve anastomoses) and that it was possible to get the development of taste buds in a graft of the tongue's vallate papilla innervated by cranial or spinal sensory ganglia, but that sympathetic neurons were ineffective. In all the experiments performed on terminal organs, it appears that these organs can develop and persist only in the presence of nerve fibers of the same category as the normal ones: that is, sensory nerves for sensory organs, motor nerves for motor endplates.

Classically, the role of nerves in the maintenance of the structural integrity of peripheral sense organs (PARKER, 1937; SPEIDEL, 1948; SINGER, 1960, 1964; HUGHES, 1968; GUTH, 1969; JACOBSON, 1970) is believed to be neurotrophic, a word with a loose meaning, which is not committed to a single mechanism. So, for the results of the heterotopic innervation experiments, two kinds of explanation can be given. Firstly, all the sensory ganglia may contain all the different kinds of specialized neurons able to promote the formation of the different kinds of terminal organs (sensory corpuscles, taste buds, etc.), only some of which are expressed during normal development (influence of tissue environment). Secondly, the neurons may have little specificity, so that consequently the peripheral territory plays the determinant role. So we have to ask the questions regarding the determination of the specificity of the corpuscles, the respective roles of nerve endings and cutaneous tissue, and finally, the sequence and mechanisms of morphogenetic interactions.

b) Tissue Influence on Corpuscle Formation

The question of specificity of corpuscles and their distribution was analyzed by determining what kind of corpuscles are formed in grafts between chicken and duck. Then the grafts were analyzed for the presence or absence of chicken-beak type or duck-beak type Herbst corpuscles, Grandry corpuscles, and Merkel corpuscles: the results are very clear and consistent (Tables 3 and 4). The types of corpuscles formed in the explants are always in keeping with the origin of the frontal bud or beak which had been grafted: Herbst corpuscles of the duck-beak

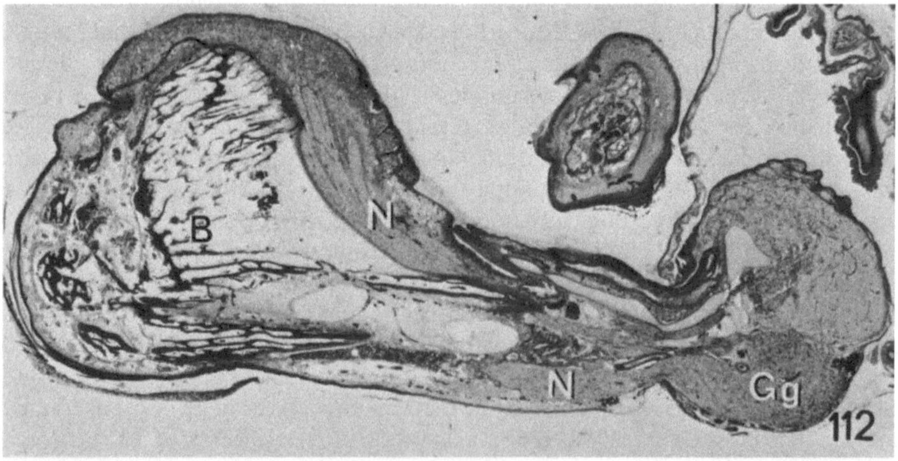
B
N
N
Cg
112

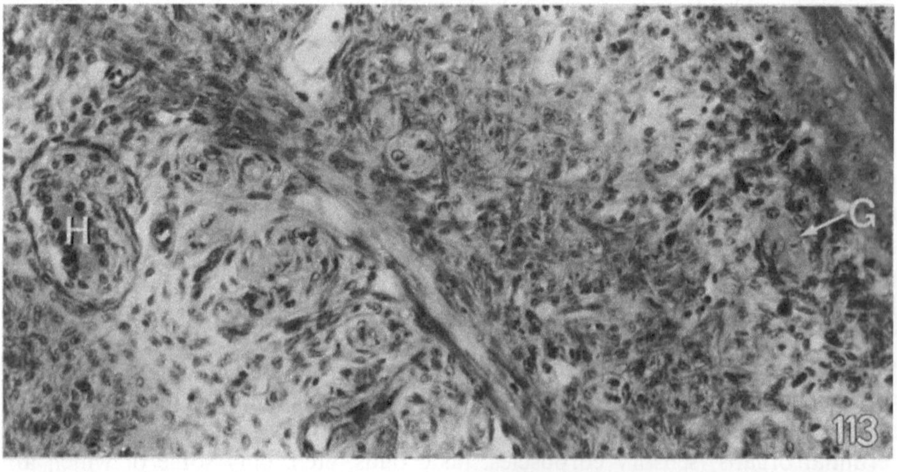
H
G
113

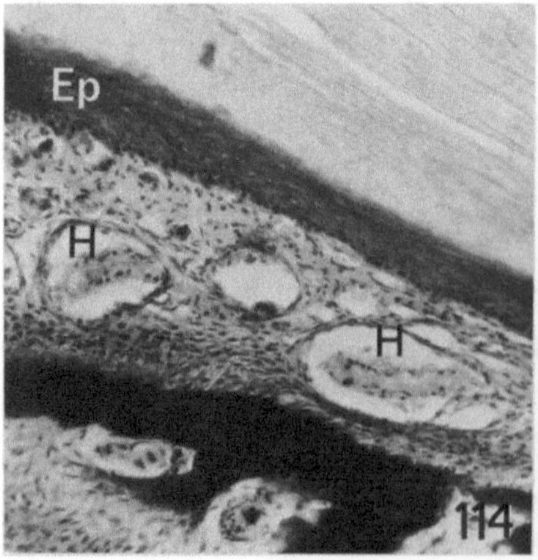
Ep
H
H
114

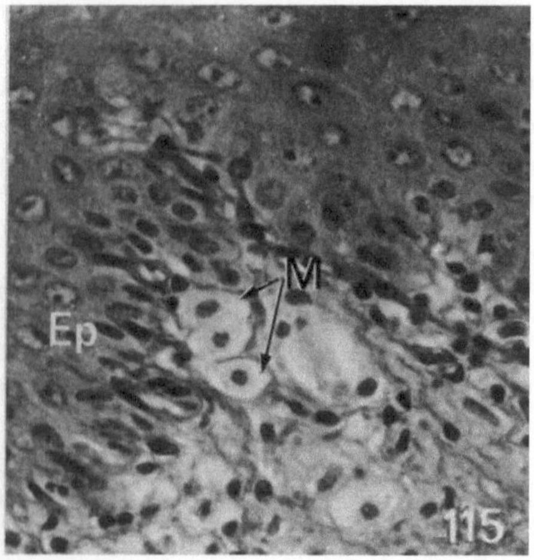
M
Ep
115

type and Grandry corpuscles in a duck graft (Figs. 112 and 113), and Herbst corpuscles of the chicken-beak type and Merkel corpuscles in a chicken graft (Fig. 114). The innervation can be provided by the trigeminal nerve, by brachial plexus or by a lumbar nerve, and can even be of heterospecific origin (between chicken and duck) without any change in the differentiation of corpuscles, which always correspond in type to the origin of the skin. Moreover, the specific origin of the graft determines not only the type of corpuscles formed but also their number and distribution.

In the case of chicken grafts, Merkel corpuscles are located between the epithelial digitations and Herbst corpuscles, are not numerous and vary in size. In the case of the duck grafts, the Herbst corpuscles are very numerous and are distributed in two layers, the superficial one being in the subepidermal zone mixed with Grandry corpuscles and the second one being in the deep dermis. In the case of grafts of frontal mesenchyme alone (from duck, Table 3), the results are not changed by the absence of their own ectodermis in the graft. Thus the ectodermis is not needed, at the stages studied, for the differentiation of Herbst and Grandry corpuscles. Consequently, it is the regional and specific quality of the innervated dermal mesenchyme that determines the corpuscles' type and distribution. This determination occurs very early in development and is already established in the frontal buds of three-day-old duck embryos (see results from grafts of frontal bud presumptive territory (Table 3). Moreover, this determination can develop in a piece of beak removed from an old embryo, grafted on a 4.5-day-old host, and innervated again (see below).

These results extend the research of KLEIN (1932) and DIJKSTRA (1933) dealing with regeneration of corpuscles in adult duck after the exchange of cutaneous territory between the beak and the foot (DIJKSTRA) or after sectioning of the ophthalmic and maxillary branches of the trigeminal nerve, reinnervation by these branches of the beak skin, or innervation by them of the skin of the lateral part of the head (KLEIN). The experiments showed that Herbst corpuscles and Grandry corpuscles can be formed only in the skin of the beak. However, in such regeneration experiments, two different kinds of mechanisms can be envisaged to explain why the type of corpuscles formed corresponds to the origin of the skin. The first mechanism is that the regenerating fibers are reinnervating the previously existing corpuscles. The second one is that these fibers "induce" the formation of new corpuscles, whose structure is determined by the skin.

Our experiments performed on embryos long before the beginning of development of corpuscles show that the second type of mechanism operates. Comparable

---

◁ Figs. 112–115. Homoplastic and xenoplastic associations of frontal buds and sensory ganglia in coelomic graft in 4-day duck embryo. Fixation 21 days later. Masson staining. Fig. 112. Hemifrontal bud and Gasserian ganglion of 6-day duck embryo associated in coelomic graft. Nerves (N) provided by ganglion (Gg) observed in the beak (B). Herbst and Grandry corpuscles present in dermis of beak. Longitudinal section, × 16. Fig. 113. Hemifrontal bud of 6-day duck embryo associated with Gasserian ganglion of 6-day chick embryo. Grandry (G) and Herbst (H) corpuscle seen in dermis of graft. × 330. Fig. 114. Hemifrontal bud of 5-day chick embryo associated with Gasserian ganglion of 5-day duck embryo. Graft displays Herbst (H) corpuscles of "chicken" type (on left), and Merkel (M) corpuscles (on right) between the epidermal (Ep) digitations. × 175 and × 100. B, beak; Ep; epidermis; G, Grandry corpuscle; Gg, ganglion; H, Herbst corpuscle; M, Merkel corpuscle; N, nerve results were obtained for corpuscles in mammals. Thus, the experiments of ILYINSKY et al. (1973) on the Pacinian corpuscles of the cat show that development of Pacinian corpuscles occurs in the zone of mesocolon supplied by nerves foreign to this tissue (these nerves, according to the authors, do not normally form corpuscles of this type). These authors conclude also that it is the tissue that determines the process of development of the receptor endings. [RIDLEY (1970), performing grafts of human skin between the forearm and the fingertip, arrived at the same conclusion regarding the Meissner corpuscles.] Comparable conclusions regarding epithelial specificity have also been reached for the development of taste buds in mammals (ZALEWSKI, 1972, 1974).

Before further progress can be made in the analysis of the mechanisms of corpuscle formation, it is necessary to know the precise developmental origin of the various cell categories of the corpuscles. At present this information is known only for the Herbst corpuscle (SAXOD, 1973), and I shall deal mainly with this kind of receptor. By means of cellular markers it has been shown that the inner bulb cells of the Herbst corpuscle originate from the sensory ganglion and accompany the nerve during its outgrowth, whereas the capsular cells and the majority of the inner space cells, if not all the inner space cells, are provided by the dermal mesenchyme.

The specificity of the dermal mesenchyme is thus determined by cells that do not participate in the formation of the more morphologically and physiologically specialized part of the corpuscles, namely the inner bulb. Consequently, at a certain stage of its development, the dermal mesenchyme acts necessarily and specifically (including a species-specific influence) on the nerve endings and the cells that accompany the nerves, and makes them able to form an inner bulb that is in keeping with the specific origin of the innervated skin. This specificity is determined very early in development, before three days of incubation in the case of the frontal bud of the duck. This determination seems to occur at the time of formation of the subectodermal mesenchyme, and is perhaps associated with a very early segregation of cells with a particular origin (neural crest?)[2].

In embryology there are many examples of the fundamental role played by the mesenchyme during morphogenesis (see review by HOLTFRETER, 1968). Restricting the discussion to the integument of birds, it has been demonstrated that the morphogenesis of the feathers requires dermal-epidermal interactions. It is the dermis that imposes the regional quality of the feathers by determining the time of onset of development, the distribution, the architectural organization, and the growth of the feathers (SENGEL, 1958, 1971; DHOUAILLY, 1970). In our experiments, however, the results of associations between different classes of animals were always negative, whereas the different kinds of xenoplastic combinations between dermis and epidermis in amniotes (DHOUAILLY, 1975) have led to the formation of buds of cutaneous appendages corresponding to the origin of the epidermis. However, these are unable to develop further in the absence of dermis originating

---

[2] This hypothesis was tested experimentally by performing coelomic grafts, associating spinal ganglia (excised from old embryos to avoid the presence of nondifferentiated cells in the ganglia) with presumptive territory of limbs removed before and after the migration of neural crest. Although a good development of the limbs was obtained, all these experiments were negative in that the nerves grew poorly in this very young graft.

from the same class of vertebrates. In the case of cutaneous sensory corpuscles in birds, it seems that the nerve endings of reptiles and mammals are not able to respond to the morphogenetic influence of the bird dermal mesenchyme or that the nerve is incapable of stimulating the mesenchyme. Such competence is restricted to the somatosensory nervous fibers from the same class of animals as the dermal mesenchyme.

c) Morphogenetic Sequence

The respective roles of cutaneous tissue and nerve endings being established, two points of chronology remain to be dealt with before we are able to establish the sequence of morphogenetic interáctions necessary for the development of corpuscles, and more particularly for the Herbst corpuscles. It is necessary to know from what stage and during what period the mesenchymal cells are acting on the nerve, and the time limits of the nerve competence.

To answer these questions the normal chronological relationship between the nerve ending and the integument has been changed by performing heterochronic associations (SAXOD, 1972b). These heterochronic experiments were performed by transplanting frontal buds and pieces of duck beak from embryos of various ages to hosts at a fixed age (4.5 days). Such transplants allow beaks of increasing age to be innervated by nerve endings of a fixed age. The development of the corpuscles of the grafts is then studied and compared with the normal development. One must remember that in duck the development of the corpuscles is synchronous for the various cutaneous regions, and begins at 20 days of incubation. The frontal buds and beak pieces were grafted either onto the frontal bud (orthotopic grafts) or onto the wing-bud (heterotopic grafts). At the end of the experiment, i.e., at the time of fixation, the age of the graft (its absolute age expressed as the number of incubation days) is the sum of the age of the donor at the time of removing the transplant plus the duration of growth on the host. Although both Herbst and Grandry corpuscles were obtained in these grafts, only the Herbst corpuscles were studied in these experiments because the five stages of its development are easier to determine (Figs. 79–86) and because its embryological origin is known.

The control experiments showed that when the donors and host embryos (these were always 4.5 days old) were almost of the same age (synchronous association) there was no delay in the development of the corpuscles of the graft, so at the time of fixation they were at the same stage as the corpuscles of the host (Fig. 115, beginning of curve). In the experiments with heterochronic association (transplant 6–27 days old) there were two kinds of results according to the age of the transplant. When the transplants were taken from 6- to 14-day-old embryos, the corpuscles of the grafts were more advanced in development than the corpuscles of the hosts (Fig. 115, beginning first part of curves). This advanced development, which increased with the age of the donor, attained a maximum of 6–7 days, advancement for explants provided by 13- to 14-day-old embryos. The difference in absolute age between the graft and the host at this time was about 8.5–9.5 days. When the transplant was provided by embryos older than 14 days, the advance in development of the corpuscles of the graft with respect to the host is very low (less than two days for explants of 15 and 16.5 days); and this advance was nonexistent when the donor was 19 days or more old (Fig. 115, second part of curves).

This study of the development of corpuscles in these grafts shows that they differentiate in the graft at the same speed as during normal development. As the amount by which the development of the corpuscles is advanced at the time of fixation is known, one can calculate the age of the graft and the age of the host (that is the age of nerve) at the time when the corpuscles began to develop in the graft. For example, when the advance is of 6–7 days (maximal value obtained), one can calculate that it was at about 14 days of incubation of the host that the development of corpuscles began in the graft and at this time the graft had an absolute age of about 23 days. A study of all the values so calculated shows that from the age of 14 days the nerves are able to participate in the formation of Herbst corpuscles but their development can only be obtained when the integument is older than 20 days.

Thus it seems that the onset of formation of corpuscles depends principally on the stage of development of cutaneous tissue. However, one may ask whether the duration of innervation is important in the process of maturation of the dermal mesenchyme. To answer this question, it is necessary to know the total duration of innervation of the graft and to study the development of corpuscles in comparison with the preceding experiments. For this purpose experiments were performed by putting frontal buds (provided by 5- to 6-day-old embryos) in coelomic grafts for 7–20 days, where they developed without any somatic innervation (aneurogenic development). These frontal buds were then grafted onto the wing-bud of a 4.5-day-old embryo. The results of these experiments were exactly similar to those obtained in grafts made directly onto the wing-bud (Fig. 116, dashes). While the absolute age of the transplant at the time of grafting onto the wing-bud was 12 days (5-day-old frontal bud and 7 days in coelomic graft), the corpuscles of the graft had an advance in development of 5–6 days with respect to the corpuscles of the host. When the absolute age of the transplant was more than 15 days, the corpuscles of the graft and of the host developed in a synchronous manner and the development of the corpuscles began at the 20th postembryonic day. One can conclude from this experiment that the formation of corpuscles in a graft does not depend on the total duration of innervation of these grafts.

To summarize, if the transplant is less than 15 days old at the time of grafting, the stage of development of corpuscles in the graft is consistent, sometimes with a slight delay, with the absolute age of the graft and is not related to the age of the host (dyschronism). Thus grafting does not change the sequence of morphogenetic interactions leading to the formation of Herbst corpuscles in the graft. When the transplant is older than 15 days, the stage of development of corpuscles in the graft is in conformity with the age of the host (synchronism). In this case the morphogenetic sequence has been perturbed and corpuscles can develop in the graft only if the graft repeats the whole morphogenetic sequence synchronously with the integuments of the host. It is conceivable that the young nerve fibers of the host growing into the graft may start this new morphogenetic sequence. A similar stimulation of the dermal mesenchyme by young nerves during their outgrowth can be envisaged during normal development and during regeneration of corpuscles in adults.

The embryonic stage at which development of corpuscles begins depends on the absolute age of the cutaneous tissue and is not related to the total length of innervation. Consequently, the specific action of the dermal mesenchyme on the nerve endings and the cells that accompany the nerve occurs at a precise stage of its development and during a reasonably short period. The study of proliferation of

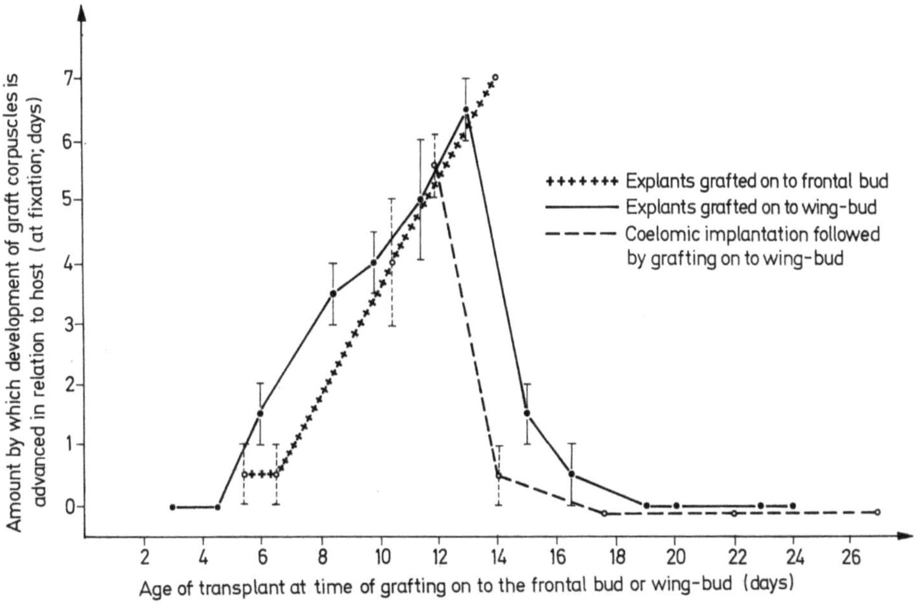

Fig. 116. *Heterochronic transplantations:* relationship between stage of corpuscles developed in graft (by comparison with that of host corpuscles) and age of explant at time of grafting. Homografts of frontal bud or beak on frontal bud (innervation by trigeminal nerve) or wing bud (innervation by brachial plexus) of 4- to 4.5-day duck embryos. In one experimental series (dotted line) frontal buds were at first explanted in coelom of a 4-day host (where development proceeded in absence of somatic innervation). Seven to twenty days later, they were removed (absolute age of these transplants was then 12 to 26.5 days) and grafted onto wing bud of 4.5-day embryo. There they became innervated by brachial plexus of host

cells participating in Herbst corpuscle formation shows that the phase of decrease of mitotic activity of future inner bulb cells occurs just before the beginning of development of the corpuscles. During this short preparatory phase, the specific number of cells that will form the inner bulb is established and the machinery for synthesis needed for cytodifferentiation is prepared. It seems that the dermal mesenchyme exerts a continuous action, both during the preparatory phase and during the phase of development of corpuscles, i.e. from the age of 19–20 days until about 25 days. However, the differentiation of corpuscles is obtained only when the nerves are more than 14 days old, so it seems that the cells accompanying the nerve are able to respond to the action of the mesenchyme (acquisition of "competence" by the nerve) and are able to differentiate only if they are older than 14 days. It is interesting to note that it is exactly at this stage of embryonic development that the skin develops complete antigenicity (for review, see May, 1966) and that the formation of motor end-plates and the concentration of cholinesterases at this level occurs (see review by Hughes, 1968).

In conclusion, the sequence of morphogenetic interactions between the dermal mesenchyme, which determines the types of corpuscles formed, and the somatosensory nerve, which is indispensable to their development, can be described for the Herbst corpuscle according to the following scheme. From the time of its formation, the subectodermal mesenchyme seems to contain cells, perhaps of a particular origin, which, after stimulation by the nerve during their outgrowth, act in a specific way at a certain stage of their development on the somatosensory nerve ending and

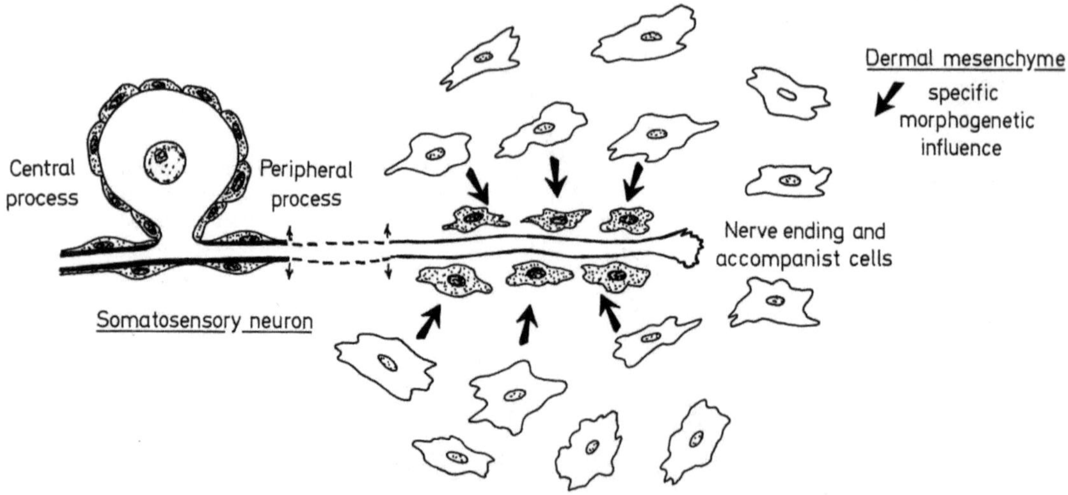

Fig. 117. *Morphogenesis of Herbst corpuscle.* Type of Herbst corpuscle found in skin is determined by dermal mesenchyme. Latter, at a certain stage of its development, acts specifically on somatosensory nerve endings and their accompanying cells, and renders them able to form an inner bulb in conformity with specific and regional origin of innervated integument

on the cells that accompany the nerve (Fig. 117). According to the origin of the dermal mesenchyme, there are formations of lamellar inner bulbs of the duck-beak or chicken-beak type that differ in the number of their cells and in their spatial organization. The dermal mesenchyme exerts a triple influence: it controls the proliferation of the future inner bulb cells, causes their cytodifferentiation (formation of lamellae, etc.), and determines their manner of spatial organization. Then the inner space is formed by interaction between inner bulb and mesenchymal cells, which divide and form the outer capsule. At this time, this inner space contains mesenchymal cells, which are the more numerous, and some cells provided by the nerve ending. After completion of the development of the corpuscles, however, the influence of the nerve endings remains necessary to ensure the structural integrity of the inner bulb cells, which are no more, in the present author's opinion, than modified and specialized Schwann cells. The inner bulb cells also have the same kind of dependence on the nerve ending. Such interactions may occur during the morphogenesis of Grandry and Merkel corpuscles and other types of cutaneous sensory corpuscles, but this remains to be demonstrated.

D. Conclusion

All these results show that we now have a good knowledge of the ultrastructural organization and the physiology of the avian cutaneous sensory receptors: free nerve endings, Herbst, Grandry, and Merkel corpuscles. However, some problems, which are common to the different kinds of peripheral sensory receptors in vertebrates, have not yet been solved. They concern particularly the nature of the different kinds of junctions and vesicles seen at the level of the nerve endings, the nature of the associated specialized cells, and their role in the mechanisms of transduction. From the data available, it is tempting to try to establish some homology between corpuscles of different classes of vertebrates and to sketch a phylogeny of cutaneous sensory corpuscles. However, one must be careful not to jump to conclusions. Moreover, the strong adaptive character of this evolution can give rise to a convergence in which receptor structures of different origins can be similar in appearance. Likewise, corpuscles with almost similar structures may correspond to physiologically different types of receptors, as is the case for the Merkel cells in birds and mammals.

Having reviewed the state of our knowledge about the structure and physiology of different types of corpuscles in birds, we can discuss the problem of the modes of their development. Two remarks can be made concerning normal development. The first is that the corpuscles form very late during embryogenesis and are among the latest organs to undergo their morphogenesis. The second is that the character of their development is spectacular in spite of the small size of these organs, and includes both the remarkable differentiation of the cells associated with the nerve ending and the number of corpuscles formed (several thousand in the skin of the duck beak).

From the experimental data reported here, it is obvious that the development of corpuscles requires interaction and collaboration between two tissues of different developmental origin. One, the nervous tissue (the somatosensory nerve endings) has a stimulatory role on the tissue it innervates, and is able, in return (at least for the Herbst corpuscles) to respond to the morphogenetic action of these cutaneous tissues. The other, the mesenchymal tissue, contains the specific information that determines the quality of the differentiation and elaborates the morphogenetic factors. However, numerous problems are still unsolved, the most important of which include: when and how is the early determination of the dermal mesenchyme established? By what mechanisms are the nerve fibers able to stimulate the dermal mesenchyme? What is the biochemical nature of morphogenetic messages and how are they produced, exchanged, interpreted? The last is clearly one of the fundamental problems of developmental biology, which will be only solved by the joint efforts of many disciplines.

E. Summary

1. Four principal types of cutaneous sensory receptors are present in birds: free nerve endings, Herbst corpuscles, Grandry corpuscles, and Merkel corpuscles.

2. The free nerve endings are intra-epidermal and are cold receptors, whereas the three kinds of corpuscles are located in the dermis and are rapidly adapting mechanoreceptors.

3. The Herbst corpuscle is formed of three parts: (1) a central inner bulb formed by two rows of lamellar specialized cells surrounding the axial nerve endings; (2) an inner space (or peribulbar space) with a lamellar structure; (3) an outer capsule.

4. Grandry corpuscles are usually formed of two large specialized cells called Grandry cells (with characteristic microfilaments and dense-core vesicles) enclosing a discoid nerve ending, and by satellite cells around the Grandry cells. The corpuscles are incompletely enveloped by a capsule.

5. The Merkel corpuscle in birds is formed from one or several specialized cells called Merkel cells (also with characteristic microfilaments and dense-core vesicles), closely connected with flat nerve endings and by enveloping lamellar cells.

6. Herbst and Grandry corpuscles are characterized by nonspecific cholinesterase activity, contained mainly in the nonnervous structures.

7. The formation of Herbst and Grandry corpuscles takes place during the last quarter of embryonic development and the different elements are progressively organized from the center toward the periphery around the nerve ending, which is present from the beginning of corpuscle differentiation. Their development is almost complete at hatching.

8. Experiments involving explantation of embryonic cutaneous tissue (frontal bud of pieces of young beak) in the absence or in the presence of nervous fibers show that the nerve endings are required for the onset of the histogenesis of the corpuscles, their subsequent development, and the maintenance of their structural integrity.

9. Experiments involving heterotopic and xenoplastic grafts and the association of frontal buds with isolated sources of innervation show that only the somatosensory nerve endings are able to ensure the development of corpuscles, even in the absence of any central connection. The sympathetic and somatomotor nerve endings are unable to promote this development, and this lack of competence also applies to somatosensory fibers from other classes of vertebrates (reptiles, mammals).

10. The types and distribution of corpuscles are consistent with the origin of the integument in which they develop and are determined by the specificity and regional quality of the innervated dermal mesenchyme. The determination of the cutaneous tissue occurs very early in embryogenesis.

11. The experiments with cellular markers demonstrate that the Herbst corpuscle is formed of cells with two distinct developmental origins: the inner bulb cells are derived from cells that accompany the nerve during its outgrowth, and the other categories of cells of the corpuscles are provided by the dermal mesenchyme.

12. Moreover, the heterochronic association of tissues required for the development of corpuscles shows that the stage at which development of corpuscles begins depends on the absolute age of cutaneous tissue, but is not related to the total length of its innervation.

13. The development of cutaneous sensory corpuscles results from morphogenetic interactions between the dermal mesenchyme, which determines the types of corpuscles formed, and the somatosensory nerve, which is indispensable to their histogenesis and differentiation. The sequences of these interactions has been determined for the Herbst corpuscles. Similar morphogenetic sequences can be envisaged for the morphogenesis of the other types of sensory corpuscles.

F. Appendix Added in Proof

The review of the literature, included in this chapter, extends through July 1975. Since submission of the manuscript, several papers have been published dealing with various aspects of structure, function and development of cutaneous sensory receptors. Some of the more significant summarized here and bibliographic references cited.

New data on morphology and physiology of cutaneous sensory receptors concern mainly lower vertebrates. Thus, it has been demonstrated that fishes possess mechanoreceptors. In selachians, two distinct classes of specific mechanoreceptors have been differentiated physiologically in perioral skin (NIER, 1976). In teleost fishes, LANE and WHITEAR (1977) have described small intraepidermal cells (about 4 μm in diameter) resembling Merkel cells of tetrapods. These cells, which were always convected with nerve fibers, were not restricted to the basal layer of the epidermis but were found also in the outer layers. The presence of intraepidermal cells of the Merkel type have been also confirmed in anurans. Thus, DÜRING and ANDRES (1976) have described complexes of nerve endings and Merkel cells at the base of taste buds of hard palate of frogs; and these new findings justify a fundamental revision of the physiological behavior of the taste disc of frogs. In mammals also, authors have pointed out that in taste buds type III cells are very similar in their morphology to the Merkel cells (CIGES et al., 1976), and their role is questionable.

Concerning morphogenesis of avian cutaneous receptors, the developmental sequence of Merkel corpuscles of chicken has been traced in the toe skin (IDÉ and MUNGER, in press)[3] and in the hard palate of the beak (SAXOD, in preparation). Both studies have shown that developing Merkel cells can first be identified at about 10–12 days of incubation in the deep dermis, and that they are always associated with nerve fibers. During next stage (14–15 days of incubation) groups of Merkel cells are present in the sub-epithelial dermis and lamellar cells begin to differentiate. At hatching (21 days) Merkel corpuscles have most morphological characteristics of adult receptors. The fact that Merkel cells were always closely associated with nerve fibers throughout development seems to indicate that they derive from neural elements and therefore are probably of neural crest origin. Similar conclusions have been proposed for the lamellar cells in the Meissner corpuscle. These cells can be considered as specialized Schwann cells (IDÉ, 1977). Probably related to the developmental origin of Merkel cells is the question of the so-called "transitional" cells. These epidermal cells have features of both keratinocytes and Merkel cells and were described at first in mammals (see review by ENGLISH, 1977). Recently this author (SAXOD, 1978b) found "transitional" cells in the basal epidermal layer of

[3] It must be pointed out that there is a controversy about terminology. IDÉ and MUNGER (in press) have proposed the use of the general term "Grandry corpuscle" for all avian dermal sensory receptors consisting of secretory granule-containing cells and associated nerve fibers, whereas this author suggested (SAXOD, 1978b) to keep on using the eponym "Merkel corpuscle" for the present time to designate these receptors of chickens, until further investigations permit to determine their correct classification and nomenclature.

chickens, not far away from the dermal Merkel corpuscles. At the present time, although several hypothesis have been proposed, the role and significance of so-called "transitional" cells remain unclear, and the developmental origin of Merkel cells is still unknown.

References

Akoev, G.N., Chelyshev, Y.A., Elman, S.I.: Effects of the catecholamines on the excitability of single mechanoreceptors in Pacinian corpuscles. Neurophysiology **6**, 312—317 (1974).

Andersen, A.E., Nafstad, P.H.J.: An electron microscopic investigation of the sensory organs in the hard palate region of the hen *(Gallus domesticus)*. Z. Zellforsch. mikr. Anat. **91**, 391—401 (1968).

Andres, K.H.: Über die Feinstruktur der Rezeptoren an Sinushaaren. Z. Zellforsch. mikr. Anat. **75**, 339—365 (1966).

Andres, K.H.: Zur Ultrastruktur verschiedener Mechanorezeptoren von höheren Wirbeltieren. Anat. Anz. **124**, 551—565 (1969).

Andres, K.H.: Morphological criteria for the differentiation of mechanoreceptors in vertebrates. In: Mechanoreception. Proc. Rhein. Westf. Akad. Wissenschaften **53**, 135—152: Westdeutscher Verlag 1974.

Andres, K.H., Düring, M.V.: Morphology of cutaneous receptors. In: Handbook of Sensory Physiology, Vol. 2. Berlin-Heidelberg-New York: Springer 1973, pp. 1—28.

Bailey, S.E.R.: The responses of sensory receptors in the skin of green lizard *Lacerta viridis* to mechanical and thermal stimulation. Comp. Biochem. Physiol. **29**, 161—172 (1969).

Boeke, J.: Die Beziehungen der Nervenfasern zu den Bindegewebselementen und Taskzellen. Z. Zellforsch. mikr. Anat. **4**, 448—509 (1926).

Botezat, E.: Morphologie, Physiologie und phylogenetische Bedeutung der Geschmacksorgane der Vögel. Anat. Anz. **36**, 428—461 (1910).

Bourlond, A.: L'innervation cutanée. Paris: Masson 1968.

Breathnach, A.S., Robins, J.: Ultrastructural observations on Merkel cells in human foetal skin (abstract). J. Anat. **106**, 411 (1970).

Brown, A.G., Iggo, A.: The structure and function of cutaneous "touch corpuscles" after nerves crush. J. Physiol. **165**, 28—29 (1963).

Burgess, P.R., English, K.B., Horch, K.W., Stensaas, L.J.: Patterning in the regeneration of type I cutaneous receptors. J. Physiol. **236**, 57—82 (1974).

Burns, R.B., Wight, P.A.: The distribution of Herbst corpuscles in the foot of the domestic fowl *(Gallus domesticus)*. Res. vet. Sci. **11**, 585—587 (1970).

Catton, W.T.: Some properties of frog skin mechanoreceptors. J. Physiol. (Lond.) **141**, 305—322 (1958).

Cauna, N.: The distribution of cholinesterase in the cutaneous receptor organs, especially touch corpuscles of the human finger. J. Histochem. Cytochem. **8**, 367—375 (1960).

Cauna, N.: Cholinesterase activity in cutaneous receptors of man and some quadrupeds. Bibl. anat. (Basel) **2**, 86—96 (1961).

Cauna, N.: Functional significance of the submicroscopical, histochemical and microscopical organization of the cutaneous receptor organs. Anat. Anz. **111**, 181—197 (1962).

Cauna, N.: The effects of aging on the receptor organs of the human dermis. In: Advances in Biology of the Skin, Vol. 6. New York: Pergamon 1965, pp. 63—96.

Cauna, N., Mannan, G.: Structure of human digital Pacinian corpuscles. J. Anat. **92**, 1—20 (1958).

Chen, S., Gerson, S., Meyer, J.: The fusion of Merkel cell granules with a synapse-like structure. J. invest. Derm. **61**, 290—292 (1973).

Chouchkov, H.N.: Cytological distribution of the cholinesterase activity in the capsulated Vater-Pacini receptors. C.R. Acad. Bulg. Sci. **21**, 387—390 (1968).

Chouchkov, H.N.: Experimental histochemical investigation of the cholinesterases and of the primary catecholamines in the Pacinian corpuscles. C.R. Acad. Bulg. Sci. **23**, 863—866 (1970).

Chouchkov, H.N.: Ultrastructure of Pacinian corpuscle after the section of nerve fibres. Z. mikr. anat. Forsch. **83**, 33—46 (1971).

Chouchkov, H. N.: On the fine structure of free nerve endings in human digital skin, oral cavity and rectum. Z. mikr. anat. Forsch. **86**, 273—288 (1972).

Chouchkov, H. N.: On the ultrastructure of the Herbst corpuscles in the bill skin of the Pekin duck. C.R. Acad. Bulg. Sci. **26**, 1705—1708 (1973).

Chouchkov, H. N.: On the ultrastructure of the Grandry's corpuscles in the bill skin of the Pekin duck. C.R. Acad. Bulg. Sci. **27**, 129—132 (1974).

Chouchkov, H. N.: An electron microscopic study of the intraepidermal innervation of human glabrous skin. Acta anat. **88**, 84—92 (1974).

Ciges, M., Diaz Flores, L., Gonzalez, M., Rama, J.: Ultrastructural study of taste buds at rest and after stimulation, and comparative study between type III cell and Merkel cells. Acta Otolaryngol. **81**, 209—219 (1976).

Couteaux, R.: Principaux critères morphologiques et cytochimiques utilisables aujourd'hui pour définir les divers types de synapses. In: Actualités Neurophysiologiques, Vol. 3. Paris: Masson 1961, pp. 145—173.

Dhouailly, D.: Déterminisme de la différenciation spécifique des plumes néoptiles et téléoptiles chez le poulet et le canard. J. Embryol. exp. Morph. **24**, 73—94 (1970).

Dhouailly, D.: Formation of cutaneous appendages in dermo-epidermal recombinations between reptiles, birds and mammals. Wilhelm Roux Arch. Dev. Biol. **177**, 323—340 (1975).

Dhouailly, D., Saxod, R.: Les stades du développement de *Lacerta muralis* Laur. entre la ponte et l'éclosion. Bull. Soc. Zool. Fr. **99**, 489—494 (1974).

Diamond, J., Gray, J. A., Sato, M.: Initiation of impulses in Pacinian corpuscles. J. Physiol. (Lond.) **133**, 54—67 (1956).

Dijskstra, C.: De- und Regeneration der sensiblen Endkörperchen des Entenschnabels. Z. mikr. anat. Forsch. **34**, 74—158 (1933).

Dorward, P. K.: Response patterns of cutaneous mechanoreceptors in domestic duck. Comp. Biochem. Physiol. **35**, 729—735 (1970).

Duncan, C. J.: Excitatory mechanisms in dermo- and mechanoreceptors. J. theoret. Biol. **5**, 114—126 (1963).

Düring, M. V.: Zur Feinstruktur der intraepidermalen Nervenfasern von Nattern *(Natrix)*. Z. Anat. Entwickl.-Gesch. **141**, 339—350 (1973a).

Düring, M. V.: The ultrastructure of lamellated mechanoreceptors in the skin of reptiles. Z. Anat. Entwickl. Gesch. **143**, 81—94 (1973b).

Düring, M. V.: The ultrastructure of cutaneous receptors in the skin of *Caiman crocodilus*. In: Mechanoreception. Proc. Rhein. Westf. Akad. Wissenschaft **53**, 123—134 (1974a).

Düring, M. V.: The radiant heat receptor and other tissue receptors in the scales of the upper jaw of boa constrictor. Z. Anat. Entwickl. Gesch. **145**, 299—319 (1974b).

Düring, M. V., Andres, K. H.: The ultrastructure of taste and touch receptors of the frog's taste organ. Cell Tiss. Res. **165**, 185—198 (1976).

Düring, M. V., Seiler, W.: The fine structure of lamellated receptors in the skin of *Rana esculenta*. Z. Anat. Entwickl.-Gesch. **144**, 165—172 (1974).

English, K. B.: Cell types in cutaneous type I, mechanoreceptors (Haarscheiben) and their alterations with injury. Amer. J. Anat. **141**, 105—126 (1974).

English, K. B.: The ultrastructure of cutaneous type I mechanoreceptors (Haarscheiben) in cats following denervation. J. Comp. Neurol. **172**, 137—164 (1977).

Farquhar, M. G., Palade, G. E.: Junctional complexes in epithelia. J. Cell Biol. **17**, 375—412 (1963).

Fjällrant, N., Iggo, A.: The effect of histamine, 5-hydroxytryptamine and acetylcholine on cutaneous afferent fibres. J. Physiol. (Lond.) **578**, 156 (1961).

Goglia, C.: Ricerche elettron-microscopiche ed istochimiche sopra i corpuscoli di Herbst del becco dell'anitra prima e dopo stimulazione meccanica e prima e dopo resezione delle fibre sensitive. Acta Med. Rom. **7**, 33—58 (1969a).

Goglia, C.: Ricerche elettron-microscopiche ed istochimiche sopra i corpuscoli di Grandry del becco dell'anitra prima e dopo alcuni interventi sperimentali. Acta Med. Rom. **7**, 59—77 (1969b).

Goglia, C.: Ricerche ultrastrutturali ed istochimiche sui recettori nervosi sensitivi. In: Seminari Biologici della Facolta di Medicina e Chirurgia, 1—56, Milano: Vita e Pensiero 1973.

Goldman, D. E.: The transducer action of mechanoreceptor membranes. In: Sensory Receptors. Cold Spr. Harb. Symp. Quant. Biol. **30**, 59—68 (1965).

Gottschaldt, K. M.: The physiological basis of tactile sensibility in the beak of geese. J. comp. Physiol. **95**, 29—47 (1974).
Gottschaldt, K. M., Lausmann, S.: The peripheral morphological basis of tactile sensibility in the beak of geese. Cell Tiss. Res. **153**, 477—496 (1974).
Grandry, M.: Recherches sur les corpuscles de Pacini. J. Anat. Physiol. **6**, 390—395 (1869).
Gray, E. G.: Problems of interpreting the fine structure of vertebrate and invertebrate synapses. Int. Rev. Gen. exp. Zool. **2**, 139—170 (1966).
Gray, J. A. B., Malcolm, J. L.: The excitation of touch receptors in frog skin. J. Physiol. (Lond.) **115**, 1—15 (1951).
Gregory, J. E.: An electrophysiological investigation of the receptor apparatus in the duck's bill. J. Physiol. (Lond.) **229**, 151—164 (1973).
Guarino, M., Milazzo, F.: Attivita acetilcolinesterasica nei corpuscoli di Pacini. Minerva Med. **65**, 2438—2444 (1974).
Guth, L.: Taste buds on the cat's circumvallate papilla after reinnervation by glosso-pharyngeal, vagus, and hypoglossal nerves. Anat. Rec. **130**, 25—38 (1958).
Guth, L.: Histochemical changes following partial denervation of the circumvallate papilla of the rat. Exp. Neurol. **6**, 129—141 (1963).
Guth, L.: "Trophic" effects of vertebrate neurons. Neurosci. Res. Progr. Bull. **7**, 1—73 (1969).
Halata, Z.: Zu den Nervenendigungen (Merkelsche Endigungen) in der haarlosen Nasenhaut der Katze. Z. Zellforsch. **106**, 51—60 (1970).
Halata, Z.: Die Ultrastruktur der Lamellenkörperchen bei Wasservögeln (Herbstsche Endigungen). Acta anat. **80**, 362—371 (1971a).
Halata, Z.: Ultrastructure of Grandry nerve endings in the beak skin of some aquatic birds. Fol. morph. **3**, 225—232 (1971b).
Halata, Z.: Innervation der unbehaarten Nasenhaut des Maulwurfs (*Talpa europea*). I. Intraepidermale Nervenendigungen. Z. Zellforsch. **125**, 108—120 (1972).
Halata, Z.: Innervation der unbehaarten Nasenhaut des Maulwurfs (*Talpa europea*). II. Innervation der Dermis (einfache eingekapselte Körperchen). Z. Zellforsch. **125**, 121—131 (1972).
Halata, Z.: The mechanoreceptors of the mammalian skin. Ultrastructure and morphological classification. Berlin-Heidelberg-New York: Springer 1975.
Hashimoto, K.: The ultrastructure of the skin of human embryos. X. Merkel tactile cells in the finger and nail. J. Anat. **111**, 99—120 (1972).
Hashimoto, K.: Fine structure of perifollicular nerve endings in human hair. J. Invest. Dermatol. **59**, 432—440 (1973).
Hensel, H., Iggo, A., Witt, I.: A quantitative study of sensitive cutaneous thermoreceptors with afferent fibres. J. Physiol. **153**, 113—126 (1960).
Herbst, G.: Die Pacinischen Körperchen und ihre Bedeutung. Ein Beitrag zur Kenntnis der Nervenprimitivfasern. Göttingen: Vandenhoeck und Ruprecht 1848.
Heringa, G. C.: Le développement des corpuscules de Grandry et de Herbst. Arch. néerl. Sci. ex. nat. **3**, 235—315 (1918).
Holtfreter, J.: Mesenchyme and epithelia in inductive and morphogenetic processes. In: Epithelial mesenchymal interactions. 1—30, Baltimore, Md: Williams & Wilkins 1968.
Horch, K. W., Whitehorn, D., Burgess, P. R.: Impulse generation in type I cutaneous mechanoreceptors. J. Neurophysiol. **37**, 367—281 (1974).
Horridge, G. A.: Touch, pain and the chemosense of vertebrates. In: Interneurons. New York: Freeman 1968, pp. 195—218.
Hughes, A. F.: Aspects of neural ontogeny. London: Logos 1968.
Hunt, C. C.: The Pacinian corpuscle. In: The peripheral nervous system. New York-London: Plenum 1974, pp. 405—420.
Hurley, H. J.: Non-specific cholinesterase in specialize sensory nerve ending of human genital skin. Brit. J. Derm. **70**, 284—287 (1958).
Hurley, H. J., Koelle, G. B.: The effect of inhibition of non-specific cholinesterase on perception of tactile sensation in human volar skin. J. invest. Derm. **31**, 243—245 (1958).
Idé, C.: Development of Meissner corpuscle of mouse toe pad. Anat. Rec. **188**, 49—68 (1977).
Idé, C., Munger, B. L.: A cytologic study of Grandry corpuscle in chicken toe skin. J. Comp. Neurol. (in press).
Iggo, A.: New specific sensory structures in hairy skin. Acta Neurovegetativa. **24**, 175—180 (1963).

Iggo, A.: Cutaneous receptors with a high sensitivity to mechanical displacement. In: Touch, Heat, and Pain. London: Churchill 1966, pp. 237—256.

Iggo, A.: Cutaneous receptors. In: The Peripheral Nervous System. New York-London: Plenum 1974, pp. 347—404.

Iggo, A., Gottschaldt, K. M.: Cutaneous mechanoreceptors in simple and in complex sensory structures. In: Mechanoreception. Proc. Rhein. Westf. Akad. Wissenschaft. **53**, 153—176 (1974).

Iggo, A., Muir, A. R.: The structure and function of a slowly adapting touch corpuscle in hairy skin. J. Physiol. (Lond.) **200**, 736—796 (1969).

Ilyinsky, O. B., Chalisova, N. C., Kuznetsov, V. F.: Development of the new Pacinian corpuscles. Studies on the foreign innervation of mesentery. Experientia **29**, 1129—1131 (1973).

Jacobson, M.: Development Neurobiology. New York: Plenum 1978.

Jones, D. G.: The morphology of the contact region of vertebrate synaptosomes. Z. Zellforsch. **95**, 263—279 (1969).

Jones, D. P., Singer, M.: Neurotrophic dependence of the lateral-line sensory organs of the newt, *Triturus viridescens*. J. exp. Zool. **171**, 433—442 (1969).

Kadanoff, D.: Über das Auftreten von Unterschieden in der Innervation der Endflächenhaut der beiden Extremitäten bei den Säugetieren. Acta anat. **75**, 423—442 (1970).

Kadanoff, D.: Über intraepitheliale nervöse Endformationen in mehrschichtigem Plattenepithel. Z. Zellforsch. **121**, 171—180 (1971).

Kamrin, R.P., Singer, M.: Influence of sensory neurons isolated from the central nervous system on maintenance of taste buds and regeneration of barbels in the catfish *Ameiurus nebulosus*. Amer. J. Physiol. **174**, 146—148 (1953).

Karnovsky, M.J.: The localization of cholinesterase activity in rat cardiac muscle by electron microscopy. J. Cell Biol. **23**, 217—232 (1964).

Kelly, E.: Fine structure of cell contact and the synapse. Anesthesiology **28**, 6—30 (1967).

Kenton, B., Kruger, L., Wood, M.: Two classes of slowly adapting mechanoreceptors fibres in reptile cutaneous nerve. J. Physiol. **212**, 21—44 (1971).

Klein, M.: Sur la différenciation d'éléments tactiles dans le névrome d'amputation du bec de canard. Arch. Anat. Hist. Embryol. **14**, 267—300 (1932).

Kurosumi, K., Suzuki, H.: Fine structure of the dendritic cells and Merkel cells in the epidermis of various mammals. Jap. J. Dermatol. **81**, 343 (1971).

Lane, F. B., Whitear, M.: On the occurence of Merkel cells in the epidermis of teleost fishes. Cell Tiss. Res. **182**, 235—246 (1977).

Le Douarin, N.: Caractéristiques ultrastructurales du noyau interphasique chez la caille et le poulet et utilisation de cellules de caille comme "marqueurs biologiques" en embryologie expérimentale. Ann. Emb. Morph. **4**, 125—135 (1971).

Le Douarin, N.: A Feulgen-positive nucleolus. Exp. Cell Res. **77**, 459—468 (1973).

Le Douarin, N., Barq, G.: Sur l'utilisation des cellules de la caille japonaise comme "marqueurs biologiques" en embryologie expérimentale. C.R. Acad. Sci. Paris Sér. D **269**, 1543—1546 (1969).

Leitner, L. M., Roumy, M.: Thermosensitive units in the tongue and in the skin of the duck's bill. Pflügers Arch. **346**, 151—156 (1974a).

Leitner, L. M., Roumy, M.: Mechanosensitive units in the upper bill and in the tongue of the domestic duck. Pflügers Arch. **346**, 141—150 (1974b).

Leitner, L. M., Roumy, M., Saxod, R.: Activités afférentes ayant leur origine au niveau du bec inférieur du canard domestique. C.R. Acad. Sci. Paris Sér. D **277**, 1909—1911 (1973).

Loewenstein, W. R.: Excitation and changes in adaptation by stretch of mechanoreceptors. J. Physiol. **133**, 588—602 (1956).

Loewenstein, W. R.: Biological transducers. Sci. Amer. **203**, 98—108 (1960).

Loewenstein, W. R.: Facets of a transducer process. In: Sensory Receptors. Cold Spr. Harb. Symp. quant. Biol. **30**, 29—43 (1965).

Loewenstein, W. R.: Mechano-electric transduction in the Pacinian corpuscle. Initiation of sensory impulses in mechanoreceptors. In: Handbook of Sensory Physiology, Vol. 1. Principles of Receptor Physiology. Berlin-Heidelberg-New York: Springer 1971, pp. 523—556.

Loewenstein, W. R., Altamirano-Orrego, R.: Enhancement of activity in a Pacinian corpuscle by sympathomimetic agents. Nature (Lond.) **178**, 1292—1293 (1956).

Loewenstein, W. R., Molins, D.: Cholinesterase in a receptor. Science **128**, 1284 (1958).

Lyne, A. G., Hollis, D. E.: Merkel cells in sheep epidermis during foetal development. J. Ultrastruct. Res. **34**, 464—472 (1971).
Machita, H., Giacometti, L., Allegra, F.: The cholinesterase-positive papillary nerve end-organ of the skin of sub-human primates. J. invest. Derm. **48**, 337—341 (1967).
Mahrle, G., Orfanos, C. E.: Merkel cells as human cutaneous neuroreceptor cells. Arch. Derm. Forsch. **251**, 19—27 (1974).
Malamud, D.: Differentiation and the cell cycle. In: The Cell Cycle and Cancer. New York: Dekker 1971, pp. 132—141.
Malinovsky, L.: Die Nervenendkörperchen in der Haut von Vögeln und ihre Variabilität. Z. mikr. anat. Forsch. **77**, 279—303 (1967a).
Malinovsky, L.: Some problems connected with the evolution of skin receptors and their classification. Fol. morph. **15**, 18—25 (1967b).
Malinovsky, L.: Types of sensory corpuscles common to mammals and birds. Fol. morph. **16**, 67—73 (1968).
Malinovsky, L.: Ein Beitrag zur Entwicklung einfacher sensibler Körperchen bei der Hauskatze *(Felis silvestris f. catus* L.*)*. Acta anat. **76**, 220—235 (1970).
Malinovsky, L., Sommerova, J.: Die postnatale Entwicklung der Vater-Pacinischen Körperchen in den Fußballen der Hauskatze (*Felis silvestris f. catus* L.). Acta anat. **81**, 183—201 (1972a).
Malinovsky, L., Sommerova, J.: Sensory nerve endings in clitoris and penis in macaques. Scripta med. **43**, 67—81 (1972b).
Malinovsky, L., Sommerova, J.: Contribution to the quantitative evaluation of the sensory nerve endings in human labium minus pudendi. Scripta med. **46**, 291—304 (1973).
Malinovsky, L., Zemanek, R.: Sensory corpuscles in the beak skin of the domestic pigeon. Fol. morph. **17**, 241—250 (1969).
Malinovsky, L., Zemanek, R.: Sensory nerve endings in the joint capsules of the large limb joints in the domestic hen *(Gallus domesticus)* and the rook *(Corvus frugilegus)*. Fol. morph. **18**, 206—212 (1970).
Malinovsky, L., Zemanek, R.: Sensory innervation of the skin and mucosa of some parts of the head in the domestic fowl. Fol. morph. **19**, 18—23 (1971).
May, R. M.: Culture et greffe de cellules nerveuses chez les vertébrés supérieurs. Paris: Gauthier-Villars 1966.
Miller, M., Kashara, M.: Studies on the cutaneous innervation of lizards. Proc. Calif. Acad. Sci. **34**, 549—568 (1967).
Monneron, A., Bernhard, W.: Actions de certaines enzymes sur des tissus inclus en épon. J. Microsc. **5**, 697—714 (1966).
Montagna, W.: Cholinesterases in the cutaneous nerves of man. In: Advances in the Biology of Skin, Vol. 1. New York: Pergamon 1960, pp. 74—86.
Munger, B. L.: The intraepidermal innervation of the snout skin of the opossum. A light and electron microscopic study, with observations on the nature of Merkel's Tastzellen. J. Cell Biol. **26**, 79—96 (1965).
Munger, B. L.: Patterns of organization of peripheral sensory receptors. In: Handbook of Sensory Physiology, Vol. 1. Berlin-Heidelberg-New York: Springer 1971a, pp. 523—556.
Munger, B. L.: The comparative ultrastructure of slowly and rapidly adapting mechanoreceptors. In: Oral-Facial Sensory and Motor Mechanisms. New York: Appleton-Century-Crofts 1971b, pp. 83—103.
Munger, B. L.: Cytology and ultrastructure of sensory receptors in the adult and newborn primate tongue. In: Fourth Symposium on Oral Sensation and Perception: Development in the Fetus and Infant. Washington, D.C.: U.S. Government Printing Office 1973, pp. 75—95.
Munger, B. L.: Specificity in the development of sensory receptors in primate oral mucosa. In: Development of Upper Respiratory Anatomy and Function: Implications for Sudden Infant Death Syndrome, 96—120, Washington, D.C.: U.S. Government Printing Office 1976.
Munger, B. L., Pubols, L. M., Pubols, B. H.: The Merkel rete papilla, a new sensory receptor in mammalian glabrous skin. Brain Res. **29**, 47—61 (1971).
Mustakallio, K., Kiistala, V.: Electron microscopy of Merkel "Taskzellen". A potential monoamine storing cell of human epidermis. Acta derm. venereol. **47**, 323—326 (1967).
Nafstad, P. H. J.: Comparative ultrastructural study on Merkel cells and dermal basal cells in poultry *(Gallus domesticus)*. Z. Zellforsch. mikr. Anat. **116**, 342—348 (1971a).

Nafstad, P. H. J.: On the ultrastructure of neuro-epithelial interactions. The dermal innervation in the snout of the pig. Z. Zellforsch. mikr. Anat. **122**, 528—537 (1971b).
Nafstad, P. H. J.: On the dermal innervation. Z. Anat. Entwickl. Gesch. **135**, 337—349 (1972).
Nafstad, P. H. J., Andersen, A. E.: Ultrastructural investigation of the innervation of the Herbst corpuscle. Z. Zellforsch. mikr. Anat. **103**, 109—114 (1970).
Nafstad, P. H. J., Baker, R. E.: Comparative ultrastructural study of normal and grafted skin in the frog, *Rana pipiens*, with special reference to neuroepithelial connections. Z. Zellforsch. mikr. Anat. **139**, 451—462 (1973).
Nier, K.: Cutaneous sensitivity to touch and low-frequency vibration in selachians. J. comp. Physiol. **109**, 345—355 (1976).
Nilsson, B. Y.: Effects of sympathetic stimulation on mechanoreceptors of cat vibrissae. Acta physiol. scand. **85**, 390—397 (1972).
Pac, L.: Sensory nerve endings in the joint capsules of the green lizard. Scripta med. **45**, 115—120 (1972).
Palay, S. L.: Synapses in the central nervous system. J. Biophys. Biochem. Cytol. **2**, 193—202 (1956).
Palmer, P.: Ultrastructural alterations of the Merkel cells following denervation. Anat. Rec. **151**, 396—397 (1965).
Parker, G. H.: On the trophic influence, so-called, its rate and nature. Amer. Nat. **60**, 147—158 (1932).
Pearse, A. G. E.: Histochemistry. London: Churchill 1961.
Pease, D. C., Quilliam, T. A.: Electron microscopy of the Pacinian corpuscle. J. Biophys. Biochem. Cytol. **3**, 331—342 (1957).
Perl, E. R.: Étude sur la sensibilité tactile et la sensibilité vibratoire. In: Actualités neurophysiologiques, Vol. 5. Paris: Masson 1964, pp. 91—110.
Polacek, P.: Receptors of the joints. Acta Fac. Med. Univ. Brun. **23**, 1—107 (1966).
Polacek, P., Malinovsky, L.: Die Ultrastruktur der Genitalkörperchen in der Clitoris. Z. mikr. anat. Forsch. **84**, 293—310 (1971).
Polacek, P., Sklenska, A., Malinovsky, L.: Contribution to the problem of joint receptors in birds. Fol. morph. **14**, 33—42 (1966).
Poritsky, R. L., Singer, M.: The fate of taste birds in tongue transplants to the orbit in the urodele *Triturus*. J. exp. Zool. **153**, 211—218 (1963).
Quilliam, T. A.: Structure of receptor organs. In: Touch, Heat, and Pain. London: Churchill 1966a, pp. 86—116.
Quilliam, T. A.: Unit design and array patterns in receptor organs. In: Touch, Heat, and Pain. London: Churchill 1966b, pp. 1—16.
Quilliam, T. A., Armstrong, J.: Structural and denervation studies on the Herbst corpuscle. In: Cytology of the Nervous Tissue (Proc. Anat. Soc. Great Britain and Ireland). London: Taylor & Francis 1961, pp. 33—38.
Quilliam, T. A., Armstrong, J.: Mechanoreceptors. Endeavour **22**, 55—60 (1963).
Rhodin, J. A.: An atlas of ultrastructure. Philadelphia: Saunders 1963.
Ridley, A.: A biopsy study of the innervation of forearm skin grafted to the fingertip. Brain **93**, 547—554 (1970).
Roumy, M., Leitner, L. M.: Activités afférentes provenant du bec supérieur de la poule domestique. C.R. Acad. Sci. Paris sér. D **277**, 1791—1794 (1973).
Santini, M.: Noradrenergic fibers in Pacinian corpuscles. Anat. Rec. **160**, 494 (1968).
Santini, M., Yasuhiko, I., Pappas, G. D.: The fine structure of the sympathetic axons within the Pacinian corpuscle. Brain Res. **33**, 279—287 (1971).
Saxod, R.: Histogenèse des corpuscles sensoriels cutanés chez le poulet et le canard. Arch. Anat. micr. Morph. exp. **56**, 153—166 (1967).
Saxod, R.: Ultrastructure des corpuscules sensoriels cutanés de Herbst et de Grandry chez le canard. Arch. Anat. micr. Morph. exp. **57**, 379—400 (1968).
Saxod, R.: Etude au microscope électronique de l'histogenèse du corpuscule sensoriel cutané de Herbst chez le canard. J. ultrastruct. Res. **33**, 463—492 (1970a).
Saxod, R.: Etude au microscope électronique de l'histogenèse du corpuscule sensoriel cutané de Grandry chez le canard. J. ultrastruct. Res. **32**, 477—496 (1970b).
Saxod, R.: Rôle du nerf et du territoire cutané dans le développement des corpuscules de Herbst et de Grandry. J. Embryol. exp. Morph. **27**, 277—300 (1972a).
Saxod, R.: Interactions morphogènes au cours de l'histogenèse du corpuscule de Herbst, étudiées à l'aide de transplantations hétérochrones. J. Embryol. exp. Morph. **27**, 585—601 (1972b).

Saxod, R.: Activité cholinestérasique des corpuscules sensoriels cutanés de Herbst et de Grandry. Étude histochimique en microscopie photonique et électronique. Histochemie **34**, 43—63 (1973a).

Saxod, R.: Developmental origin of the Herbst cutaneous sensory corpuscle. Experimental analysis using cellular markers. Develop. Biol. **32**, 167—178 (1973b).

Saxod, R.: Les organites périnucléaires du corpuscule sensoriel cutané de Grandry. Organisation ultrastructurale et formation. Tissue and Cell **5**, 269—280 (1973c).

Saxod, R.: Etude ultrastructurale et analyse expérimentale de l'histogenèse des corpuscules sensoriels cutanés chez les oiseaux. Thesis, University Grenoble, 1973d.

Saxod, R.: Combination of cholinesterase staining of nerves and stereo viewing for three-dimensional study of skin innervation in whole mounts. J. Invest. Derm. **70** (1978a) in press.

Saxod, R.: Ultrastructure of Merkel Corpuscles and So-called "Transitional Cells" in the White Leghorn Chicken. Amer. J. Anat. (1978b) in press.

Saxod, R.: Ultrastructural study of Merkel corpuscles development in chicken beak (in preparation).

Schiff, J. D.: Role of the sympathetic innervation of the Pacinian corpuscle. J. gen. Physiol. **63**, 601—608 (1974).

Schwartzkopff, J.: Mechanoreception. In: Avian Biology, Vol. 3. New York-London: Academic 1973, pp. 417—477.

Sengel, P.: Déterminisme de la différenciation régionale des phanères chez l'embryon de poulet. Bull. Soc. Zool. Fr. **83**, 82—86 (1958).

Sengel, P.: The organogenesis and arrangement of cutaneous appendages in birds. In: Advances in Morphogenesis, Vol. 9. New York-London: Academic 1971, pp. 181—230.

Shanta, T. R., Bourne, G. H.: The "perineural epithelium", a metabolically active, continuous protoplasmic cell barrier surrounding peripheral nerve fasciculi. J. Anat. **96**, 527—537 (1962).

Shanta, T. R., Bourne, G. H.: New observations on the structure of the Pacinian corpuscule and its relation to the perineural epithelium of peripheral nerves. Amer. J. Anat. **112**, 97—109 (1963).

Shanta, T. R., Bourne, G. H.: The perineural epithelium, a new concept. In: The Structure and Function of Nervous Tissue, Vol. 1. New York-London: Academic 1968, pp. 379—459.

Siminoff, R., Kruger, L.: Properties of reptilian cutaneous mechanoreceptors. Exp. Neurol. **20**, 403—414 (1968).

Singer, M.: Nervous mechanisms in the regeneration of body parts in vertebrates. In: Developing Cell Systems and their Control. New York: Ronald 1960, pp. 115—133.

Singer, M.: The trophic quality of the neuron: some theorical considerations. In: Progress in Brain Research, Vol. 13. Amsterdam: Elsevier 1964, pp. 175—213.

Smith, K. R.: The structure and function of the Haarscheibe. J. comp. Neurol. **131**, 459—474 (1967).

Smith, K. R.: The ultrastructure of the human Haarscheibe and Merkel cell. J. Invest. Derm. **54**, 150—159 (1970).

Smith, K. R., Creech, B. J.: Effects of pharmacological agents on the physiological responses of hair discs. Exp. Neurol. **19**, 477—482 (1967).

Speidel, C. C.: Correlated studies on sense organs and nerves of the lateral-line in living frog tadpoles. II. The trophic influence as revealed by prolonged observations of denervated and reinnervated organs. Amer. J. Anat. **82**, 277—320 (1948).

Spencer, P. S., Schaumburg, H. H.: An ultrastructural study of the inner core of the Pacinian corpuscule. J. Neurocytol. **2**, 217—237 (1973).

Straile, E. W.: Encapsulated nerve end-organs in the rabbit, mouse, sheep and man. J. comp. Neurol. **136**, 317—336 (1969).

Szymonowicz, W.: Über den Bau und die Entwicklung der Nervenendigungen im Entenschnabel. Arch. mikr. Anat. **48**, 329—358 (1897).

Tello, J. F.: Die Entstehung der motorischen und sensiblen Nervendigungen. Z. Anat. **64**, 348—440 (1922).

Torrey, T. W.: The influence of nerve fibers upon taste buds during embryonic development. Proc. nat. Acad. Sci. (Wash.) **26**, 627—634 (1940).

Tsuji, T.: Free nerve endings of the epidermis in hairy and hairless mice. J. Invest. Derm. **57**, 247—255 (1971).

Tuttle, R., MacCleary, M.: Pharmacology of the Pacinian receptor. Arch. int. Pharmacodyn. **208**, 213—223 (1974).

Valenzuela, Y., Chacon, J.: Etude histochimique de l'activité acetylcholinestérasique dans les corpuscules de Wagner-Meissner. Ann. Histochim. **14**, 247—260 (1968).

Verley, R., Garma, L., Scherrrer, J.: Conceptions récentes sur le développement du système nerveux des mammifères. Ann. Psychol. **69**, 455—489 (1969).

Walter, P., Hebel, R.: Zur Morphologie sensibler Rezeptoren. Anat. Anz. **118**, 436—443 (1966).

Whitear, M.: The free nerve endings in fish epidermis. J. Zool. (Lond.) **163**, 231—236 (1971a).

Whitear, M.: Cell specialization and sensory function in fish epidermis. J. Zool. (Lond.) **163**, 237—264 (1971b).

Whitear, M.: The nerves in the frog skin. J. Zool. (Lond.) **172**, 503—529 (1974a).

Whitear, M.: The vesicle population in frog skin nerves. J. Neurocytol. **3**, 49—58 (1974b).

Wight, P. A. L., Siller, W. G., MacKenzie, G. M.: The distribution of Herbst corpuscles in the beak of domestic fowl. Brit. Poultry Sci. **11**, 165—170 (1970).

Will, J. G.: Vatersche Körperchen der Vögel, einige Bemerkungen über dieselben. Sitz. Ber. Akad. Wiss. **1**, 213—225 (1850).

Winkelmann, F. K.: The effect of sciatic nerve section on enzymatic reactions of sensory end-organs. J. Neuropathol. **21**, 655—657 (1962).

Winkelmann, F. K.: Innervation of the skin: notes on a comparison of primate and marsupial nerve endings. In: Biology of the Skin and Hair Growth. Sydney: Angus & Robertson 1965, pp. 171—182.

Winkelmann, F. K., Breathnach, M. S.: The Merkel cell. J. Invest. Derm. **60**, 2—15 (1973).

Winkelmann, F. K., Fortman, D. D. S.: The Merkel cell. Jap. J. Derm. **82**, 98—105 (1972).

Winkelmann, F. K., Myers, T. T.: The histochemistry and morphology of the cutaneous sensory end-organs of the chicken. J. comp. Neurol. **1117**, 27—35 (1961).

Wright, M. R.: The lateral line system of sense organs. Quart. Rev. Biol. **26**, 264—280 (1951).

Zalewski, A. A.: Role of nerve and epithelium in the regulation of alkaline phosphatase activity in gustatory papillae. Exp. Neurol. **23**, 18—28 (1969).

Zalewski, A. A.: Regeneration of taste buds after transplantation of tongue and ganglia grafts to the anterior chamber of the eye. Exp. Neurol. **35**, 519—528 (1972).

Zalewski, A. A.: Neuronal and tissue specifications involved in taste buds formation. Ann. N.Y. Acad. Sci. **228**, 344—349 (1974).

Zelena, J.: Development, degeneration and regeneration of receptor organs. In: Progress in Brain Research, Vol. 13. Amsterdam: Elsevier 1964, pp. 175—213.

Ziswiler, W., Trnka, V.: Tastkörperchen im Schlundbereich der Vögel. Suisse Zool. **79**, 307—318 (1972).

Chapter 9

Cell Death During Development of the Nervous System

JERRY SILVER, Cleveland, Ohio (USA)

With 5 Figures

Contents

A. Introduction

Normal embryogenesis of the vertebrate nervous system, like the development of numerous other organ systems, is often associated with the death and resorption of large numbers of cells (GLÜCKSMANN, 1951; SAUNDERS, 1966). In the developing brain two main periods of cell death can be distinguished (KÄLLÉN, 1955). The first occurs during the earliest morphogenetic stages of development (i.e., from the time of formation of the neural plate and tube through the neuromeric stage), when the proliferating neuroepithelium undergoes a series of cellular movements that mold the basic form of the definitive neural apparatus. The second period of necrosis occurs later, during the histogenetic phase of neural ontogeny, when the young nerve cells migrate from the ependymal zone and undergo their cytological differentiation. Thus, cell deaths consistently appear within the undifferentiated neuroepithelium during invaginations, evaginations, separations of rudiments or seam formations (classified as "morphogenetic" degenerations by GLÜCKSMANN, 1951), and neuronal cells also tend to degenerate during the period of axo-dendritic outgrowth (classified as "histogenetic" degenerations by GLÜCKSMANN, 1951). A third class of cell death results in the complete regression of a vestigial organ, such as the degeneration of the region in mouse and pig embryos that is homologous to that of the paraphysis of more primitive vertebrates (ERNST, 1926; FROBOESE, 1926). This "phylogenetic" type of necrosis (after GLÜCKSMANN, 1951) will not be considered in detail here.

Although the utility of some kinds of degeneration during development might be expected on general principles, such as in the complete elimination of larval organs or obsolete tissues, the function of cell death in other areas is not so obvious. In these other areas, experimental or genetic manipulation is necessary to bring about selected alterations in the normal complement of cellular resorptions before one can begin to assess whether a functional relationship exists between a particular developmental process and the cell degeneration associated with it. Although many discrete areas of the developing nervous system have been observed to undergo degenerative changes of various degree, only a few necrotic regions are readily amenable to experimentation. Thus, for the most part, the role of cell death in neurogenesis (and in other developing systems) remains largely a matter of speculation. The following consists of a description of the more prominent necrotic loci in the embryonic brain of vertebrates and a discussion of the possible consequences of this degeneration during normal and abnormal development. Some additional areas of degeneration outside of the CNS will also be discussed.

B. The Pattern of Cell Death During the Early Morphogenetic Stage of Neural Development

In many species of vertebrates numerous cells degenerate during formation of the elongate cells of the neural plate, during the transformation of the neural plate into the neural tube by the invagination, meeting, and fusion of the neural folds in the midline, and during the subsequent detachment of the neural tube from the overlying ectoderm (GLÜCKSMANN, 1930, 1951; KLIKA and JELINEK, 1969). GLÜCKSMANN (1930) observed that these degenerations appear near the ventral midline early in development. Later they are found primarily in the middle parts of the tube, and finally pycnotic cells are scattered in the dorsal region and in the connecting stalk between the neural tube and ectoderm. Necrotic remnants of the stalk are subsequently found between the closed tube and the superficial ectoderm. The number of cell deaths varies over the length of the tube and is particularly high in the region of the anterior neuropore.

In a quantitative histological study of the developing nervous system in the rabbit, KÄLLÉN (1955) observed degenerating cells in the most rostral portion of the neural tube before, during, and immediately after closure of the anterior neuropore. Following pore closure the rate of cell death in this area decreases rapidly. GLÜCKSMANN (1951) believed that portions of the embryo that are destined to detach or form seams often produce excess numbers of cells especially, as in the case of the neural tube and ectoderm, when a zone of connecting tissue develops prior to final separation or closure. The intervening cells later become superfluous and degenerate.

The roof of the telencephalon is the portion of the forebrain caudal to the anterior neuropore, which, after closure and detachment from the surface ectoderm, is invaginated with vasculature of the meninges to form the choroid plexus of the lateral ventricles (BAILEY, 1916). Throughout morphogenesis of the telencephalic roof plate, but before its invagination into the ventricles, cell necrosis is a consistent occurrence (GRAUMANN, 1950; KÄLLÉN, 1965; MARUYAMA and D'AGOSTINO, 1967; MATTANZA, 1973). In KÄLLÉN'S (1955) study of the embryonic rabbit brain, dying cells were found in this region (dorso-median part) of the archencephalon up to and including the 11-day stage (33 somites). The largest increase in number and extent of these degenerations, however, occurred during the 27–28 somite stage (9 days, 6 hrs), long after separation from the epidermis and closure of this part of the neural tube was complete. Thus, KÄLLÉN believed that these midline resorptions may be remnants of the separation process of the neural tube and epidermis, and also might be related to the formation of the choroid plexuses of the lateral, as well as the third and fourth ventricles, and also of the dorsal raphe in the mesencephalon. This interpretation was upheld by others (MARUYAMA and D'AGOSTINO, 1967; MATTANZA, 1973). It has also been suggested that degenerations in the roof of the neural tube may be involved in the reduction of the upper portion of the rhombencephalon and diencephalon to a thin membrane (KÄLLÉN, 1965).

In human fetuses, the transformation of the caudal part of the spinal cord into *conus medullaris* and *filum terminale* is accompanied by many cellular degenerations (GAMBLE, 1971). It has been proposed that the resorption of these cells may result in the decrease in circumference of this part of the cord.

During evagination of the optic vesicle in several species of vertebrates, a distinct area of cell death is found ventral and often rostral to the insertion of the optic stalk with the prosencephalon (KÄLLÉN, 1955, 1965). It was suggested that this intensely necrotic region (suboptic death center) may possibly be concerned with the approach of the pair of optic evaginations and thus in the formation of the chiasm system or, in part, with the ventral shifting of the optic stalk (KÄLLÉN, 1965).

In the lateral part of the rhombencephalon during the neuromeric stage, cell degenerations are found adjacent to the developing cranial ganglia (KÄLLÉN, 1955, 1965). The earliest of these dying cells to appear (23-somite stage in rabbit embryos) lie level with the trigeminal ganglion, dorsal and ventral to its attachment with the neural tube in neuromere VII. At later stages (27–28 somites) these necrotic cells are situated largely in the middle regions of the brain wall immediately opposite the ganglion. During the later stage similar areas of cell death are found in the neural tube adjacent to the facial-acoustic ganglion in neuromere IX and to the glossopharyngeal-vagus ganglion in neuromere XI. These necroses disappear by day 11 (33 somites). KÄLLÉN found it difficult to characterize these degenerations as either histogenetic of morphogenetic since (1) these centers of cell death are resorbed before histogenesis begins and, therefore, would not appear to be involved with differentiation, and (2) at least for the most rostral centers, necrosis begins to develop dorsally and ventrally in the neural tube, uneven with the point of attachment of the ganglion crest. Thus, correlations between cell death and ganglion development on purely mechanical grounds are equally obscure. The possibility that the resorption of cells may play a part in neuromere formation should also be considered.

GLÜCKSMANN (1930, 1951) and others (VON SZILY, 1912; ERNST, 1926; MANN, 1969; SILVER, 1972; KUWABARA and WEIDMAN, 1974; KUWABARA, 1975) observed that during normal vertebrate ocular morphogenesis large numbers of cells become necrotic and are resorbed within the eye rudiment prior to and during invagination of the optic cup. In addition, during invagination, formation of a vesicle and detachment from the ectoderm (or palate in the case of the nose), the lens, otocyst, and olfactory pit contain numerous degenerations (GLÜCKSMANN, 1951). These resorptions are largely analogous to those described during invagination of the neural tube and its separation from the surface ectoderm. Cell deaths also appear within the junctional epithelium during fusion of the palatine shelves and may be causally related to that process (ANDERSEN, 1967).

During the formation of Rathke's pocket (the rudiment of the anterior hypophysis) and its subsequent detachment from the oral epithelium, numerous cells degenerate, especially within the connecting stalk that forms transiently between the two structures prior to separation (STOECKEL et al., 1973). These degenerations are also similar to those found in other regions (lens, neural tube) where embryonic rudiments of ectodermal origin separate from the epithelial layer from which they were derived.

C. Cell Death During Ocular Morphogenesis

The formation of the vertebrate eye progresses through a complicated but orderly maze of linked morphogenetic and histogenetic events (for reviews see LOPASHOV and STROEVA, 1964; COULOMBRE, 1965). During its early development a series of invaginations transform the optic vesicle into the optic cup. Concurrently, the lens forms dorsally within the eye cup proper. Later on, the most ventral portion of the optic cup and eye stalk invaginates to form the retinal and optic fissures. This system of continuous channels subsequently allows for communication between the posterior pole of the retina (vitreal surface) and the optic stalk, providing a passage for the optic nerve and intraocular blood vessels (MANN, 1969). Invagination of the optic fissure is an important event for ensuing ocular development, since ganglion cell differentiation, optic nerve outgrowth, and growth of the ventral hemisphere of the eye do not begin until formation of the fissure is nearly complete (LOPASHOV and STROEVA, 1964).

In a detailed study of the pattern of degenerate cells in the developing mammalian eye rudiment, at least nine distinct necrotic loci were found (SILVER, 1972; SILVER and HUGHES, 1973). These degenerations are mainly concentrated in the ventral portions of the retina, optic stalk, and pigment epithelium (Figs. 1 and 2), and can be correlated with formation of the optic fissure, which subsequently invaginates amid these death centers (SILVER and HUGHES, 1973). The relationship between abnormal optic fissure development and the production of a group of

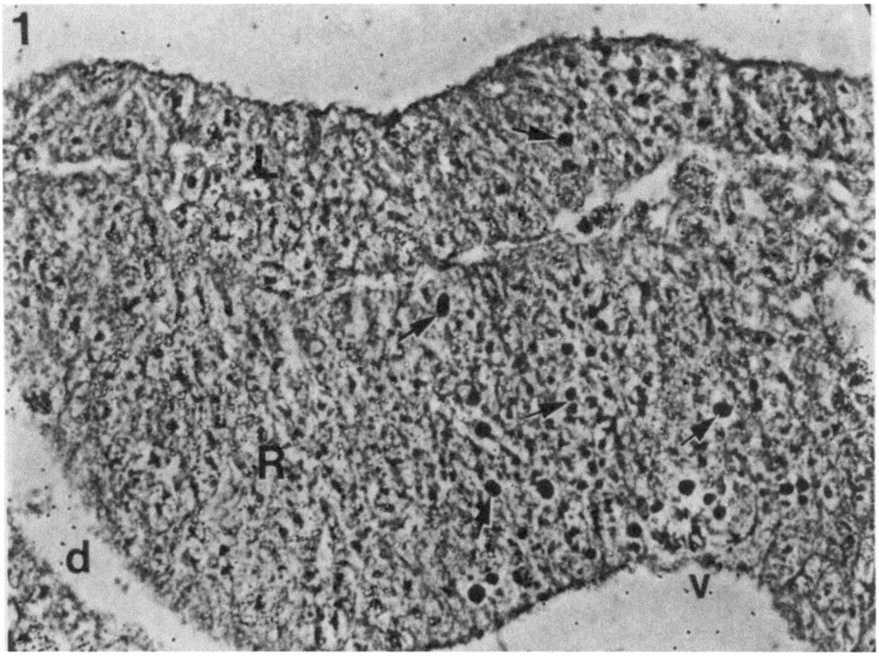

Fig. 1. Mid-sagittal section (phase-contrast micrograph) through optic vesicle of rat embryo (day 11), showing ventral location of pycnotic nuclei (arrows) in retina (R) and lens (L). Pycnotic nuclei unlabeled following 1-hr pulse injection of ^{3}H-thymidine. After subsequent incubation for 24–48 hrs, pycnotic nuclei become heavily labeled (SILVER, 1976). (D, dorsal; V, ventral). ×500

gross abnormalities of the eye (including *anophthalmia*, *microphthalmia*, orbital cysts, typical coloboma of the retina and iris, and congenital detachment of the retina) has been well established (VON HIPPEL, 1903; BROWMAN and RAMSEY, 1943; MANN, 1957; LOPASHOV and STROEVA, 1964; WARKANY, 1971).

D. Some Functional Considerations of Morphogenetic Cell Death

How cells are marshalled into a geometrically precise configuration during invaginations within the neural anlage has attracted a considerable amount of mechanistic speculation. GLÜCKSMANN (1930) proposed that strategically oriented cell degenerations and mitoses could produce the morphogenetic movements that shape the primordium of the vertebrate brain. He also believed that the yolk content of the egg was an important factor in the necessity for cells to degenerate during morphogenesis (GLÜCKSMANN, 1940, 1951). In the amphibian eye

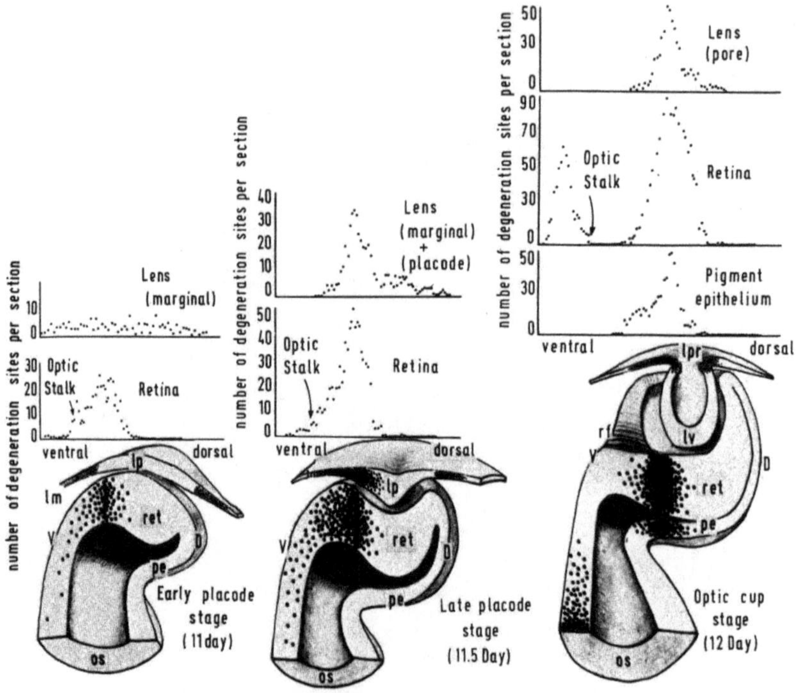

Fig. 2. Pattern of cell death in developing rat eye during morphogenetic stages. Drawings are mid-sagittal views of serially reconstructed eyes during optic vesicle stage (11 and 11.5 days' gestation) and optic cup stage (12 days' gestation). They show quantitative distribution of degeneration sites during invagination of eye rudiment. During latter part of day 12 and during day 13 optic fissure (see Fig. 3) forms disto-ventrally between necrotic regions of retina and optic stalk (indicated by bracket on 12-day eye) (lm, marginal lens ectoderm; 1p, lens placode; ret, retina; pe, pigment epithelium; os, optic stalk; lpr, lens pore; lv, lens vesicle; rf, retinal fissure; D, dorsal; V, ventral)

GLÜCKSMANN (1940) claimed that the abundance of yolk and the loose association of cells enable invaginations to occur without concomitant cell degenerations, whereas in the mammalian eye, where there is a relative absence of yolk and the cells are packed tightly, little room remains for lateral cell displacement during changes in shape of the eye rudiment, and the resorption of cells is a necessity. Therefore, according to GLÜCKSMANN (1951), one of the important functions of morphogenetic cell death, in addition to bringing about the removal of superfluous parts after the separation of one rudiment from another (see also WRENN and WESSELS, 1969), may be to facilitate the movement of cells within a structure.

While the role of cell necrosis in these situations may be a mechanical one, recent electron microscopic analyses of the developing neural tube and lens in several species of vertebrates suggest that cell movements may result largely from the action of cytoplasmic structures within individual cells (BYERS and PORTER, 1964; MCKEEHAN, 1951; ZWAAN and HENDRIX, 1973). Thus, a system of fine filaments appears within the apical portion of the lens and neural tube during (but not prior to or after) invagination (BALINSKY, 1961; BAKER and SCHROEDER, 1967;

WRENN and WESSELLLS, 1969; SCHROEDER, 1970; FREEMAN, 1972). It has been suggested that these filaments are contractile since (1) they resemble actin biochemically (ISHIKAWA et al., 1969) and (2) during neurulation the thickness of the filament bundles increases, while the circumference of the apical end of the neural epithelial cells diminishes (BURNSIDE, 1971).

ERNST (1926) has suggested that exogenous factors, such as vascular insufficiencies or pressure from neighboring cells during changes in the shape of embryonic organs, are responsible for cell death during morphogenetic processes. However, tissue culture experiments of Fell and co-workers on the avian sternum (FELL, 1939), knee joint (FELL and CANTI, 1934) and mandible (JACOBSON and FELL, 1941), which all produce large numbers of midline cell deaths during their development, have suggested that the normal pattern of cellular degeneration and morphogenesis may occur *in vitro* without the presence of neighboring body components. What would occur *in vivo* if death were prevented is unknown.

The suboptic death center, which was mentioned earlier, is a discrete zone of intense degeneration situated below the optic evagination in early avian and mammalian embryos (KÄLLÉN, 1955). To study the possible influence of the ventrally migrating eye anlage upon the production of necrosis in the suboptic center, KÄLLÉN and OLIVECRONA (1963; cited by KÄLLÉN, 1965) performed the following experiment on chick embryos. One eye rudiment was surgically ablated during the 14-somite stage and the embryos were permitted to survive until Hamburger-Hamilton stage 16–19 (HAMBURGER and HAMILTON, 1951), when the rate of cell death in the suboptic zone is normally quite marked. At that time, in most cases, no remnants of the eye could be found on the operated side. The lesion produced in the brain by the surgery had healed completely. However, the degree of degeneration on the operated side was of the same magnitude as on the control side. On the basis of this experiment, KÄLLÉN suggested that formative pressure from the eye rudiment was unlikely to cause the degeneration in the suboptic area. He also ruled out vascular insufficiency in these early stages, since resident blood vessels do not form until a later phase of development. Thus, it appeared to KÄLLÉN that cells within the suboptic locus were free of extraneous forces and might be genetically predestined to die.

In the developing chick embryo a distinct zone of mesodermal necrosis, the posterior necrotic zone (PNZ), occurs at the posterior junction of the wing bud and body wall at stage 24 (SAUNDERS et al., 1962; DAWD and HINCHLIFFE, 1971). Studies by SAUNDERS and co-workers (SAUNDERS, 1966; SAUNDERS and FALLON, 1967; SAUNDERS et al., 1962) and FALLON and SAUNDERS (1968) on the posterior necrotic zone of the chick limb bud have shown that cells may, in fact, be genetically programmed for death. Thus, when prospective PNZ cells of embryos at stage 17 through 23 were grafted to the somite region of the same embryo or to another embryo of an earlier or later developmental stage, or explanted in culture medium with or without somite tissue, death occurred on schedule (at stage-24, donor age). However, it was also suggested that environmental factors may also play a part in controlling the magnitude of cell death within this area, since PNZ cells excised prior to stage 22 and transplanted to the dorsal side of the wing bud or placed in tissue culture and associated with wing-bud mesoderm across a millipore filter, did not degenerate. Interestingly, normal development of the limb often occurred when the PNZ cells were allowed to remain *in situ* without dying as they did when a graft of dorsal wing tissue was implanted between the mesoderm of the PNZ and the ectoderm overlying it posteriorly (SAUNDERS, 1966). FALLON and SAUNDERS (1968) suggested that diffusible material from wing-bud mesoderm can regulate the intensity of cell death within the PNZ. Unfortunately, the substances that mediate the death sentence are unknown. It is noteworthy that certain chemicals (acetamide and cytosine) will modify the amount of cell necrosis in the Bar Eye mutant of *Drosophila* (FRISTROM, 1972). The actions of these substances on normal cell death in vertebrates is, as yet, untested.

Although the mechanical utility of morphogenetic cell death has been questioned (FELL, 1939; KÄLLÉN, 1965; SAUNDERS, 1966), analyses of abnormally developing systems, where there is an increase or decrease in the regular number of degeneration sites, suggest that cell death may indeed play a necessary role during normal morphogenetic events. In an investigation of this type one attempts to correlate (as cause A and effect B) the finding of (A) abnormal amounts of pycnosis in the embryo, if present early on and if limited to one of the normally occurring necrotic loci, with (B) the production of morphogenetic anomalies of the same area during subsequent stages. Thus, the talpid mutation of the fowl, as described by HINCHLIFFE and co-workers (HINCHLIFFE and EDE, 1967; HINCHLIFFE and THOROGOOD, 1974), results in abnormally shaped legs with a greatly increased number of unseparated digits. They observed that the normally occurring cell death in the posterior and anterior necrotic regions and in the interdigital zones was completely lacking in the mutant strain. Therefore, the anomalous shape of the limb in the neonate could be correlated with the absence of preliminary morphogenetic cell death in the limb bud of the embryo. In contrast to the situation in the talpid mutant is the wingless condition in chicks, which arises from cellular degeneration in excess of the usual amount. In most of such cases the precocious appearance of cell death in the anterior necrotic zone and its extension beyond the normal degenerate area results in complete resorption of the forelimb (HINCHLIFFE and EDE, 1973).

Necrosis of the interdigital tissue has been described in many vertebrates including duck, mouse, rat, mole, and man, and may be causally related to separation of the digits (MILAIRE, 1963; MENKES and DELEANU, 1964; MENKES et al., 1965; SAUNDERS, 1966; BALLARD and HOLT, 1968). This idea is substantiated by the observation that during development of the webbed foot of the duck, degenerating cells appear only in those interzones between digits where webbing is absent (DELEANU, 1965). Moreover, during development of limb chimera composed of chick mesoderm and duck ectoderm, interdigital necrosis is greatly reduced and all the toes usually remain connected by a narrow web (SAUNDERS, 1966). Furthermore, in syndactylous and polysyndactylous mutant mice (MILAIRE, 1965; JOHNSON, 1969) and in polysyndactylous chicks (HINCHLIFFE and THOROGOOD, 1974), it has been reported that interdigital resorptions are mostly absent.

In the normal development of the lower extremity, the future knee joint is delineated by a zone of necrotic cells. Administration of insulin to chick embryos is known to cause micromelia. ZWILLING (1959) has shown that the treatment results in an increase in the intensity and spread of the normal degenerating zone.

ZWILLING (1942) has also analyzed the rumpless mutation in the chick which results in missing or aberrant caudal vertebrae and associated tail structures. In normal chick embryos, one of the regularly forming zones of cell death is found posterior to the undifferentiated tail bud. Histologic examination of a series of mutant embryos revealed that this necrotic locus was more extensive than it was in normal animals and that the extent of the degenerating tissue corresponded to the degree of reduction of the caudal vertebrae. This condition can also be produced by the administration of insulin at critical periods (LANDAUER, 1945; MOSELEY, 1947). Abnormal concentrations of degenerate cells in the end and tail buds may also be related to dysrhaphic malformations of the spinal cord (JELINEK et al., 1971).

In the developing nervous system of the mouse, several lethal T-alleles involving the neural tube and associated skeletal elements are expressed by large excesses of degenerate cells. DUNN and BENNETT (1964) described the homozygous t^{w1} mutant, which is known to effect the ventral portion of the neural tube resulting in a syndrome of anomalies including microcephaly, microphthalmia, hydrocephalus and malformation of the brain and spinal cord. During day 8 or 9 of gestation, only a low level of pycnosis was observed in the ventral portion of the neural tube. Apparently, these early forming necrotic cells correspond to those seen in the young neural tube of normal animals (GLÜCKSMANN, 1951). However, subsequent to this initial stage, degeneration increases dramatically, beginning in the cervical regions and spreading in all directions. The distribution of cell deaths remains strictly confined to the ventral regions of the neural tube and leads in many instances to virtual destruction of the entire ventral portion of the brain and spinal cord.

E. The Anophthalmic Mutation of the Mouse

Some Consequences of the Absence of Morphogenetic Cell Death on Normal Development

One of the central questions concerning the role of cell death during ocular morphogenesis left unanswered by GLÜCKSMANN'S (1951) and SILVER and HUGHES' (1973) descriptive studies of normal embryos was whether or not the resorption of cells within the eye vesicle was a necessary prerequisite for the subsequent formation of the optic cup (i.e., whether invagination of the eye cup and ventral fissures would still occur if preliminary necrosis were abolished). In 1974 the present author was fortunate to receive[1] several mating pairs of a congenitally anophthalmic strain of mice in which the mutation of a single gene (or perhaps two genes) results in the production of total eyelessness in approximately 80–90%

[1] From Dr. H. B. CHASE and Dr. I. K. ABRAMOF.

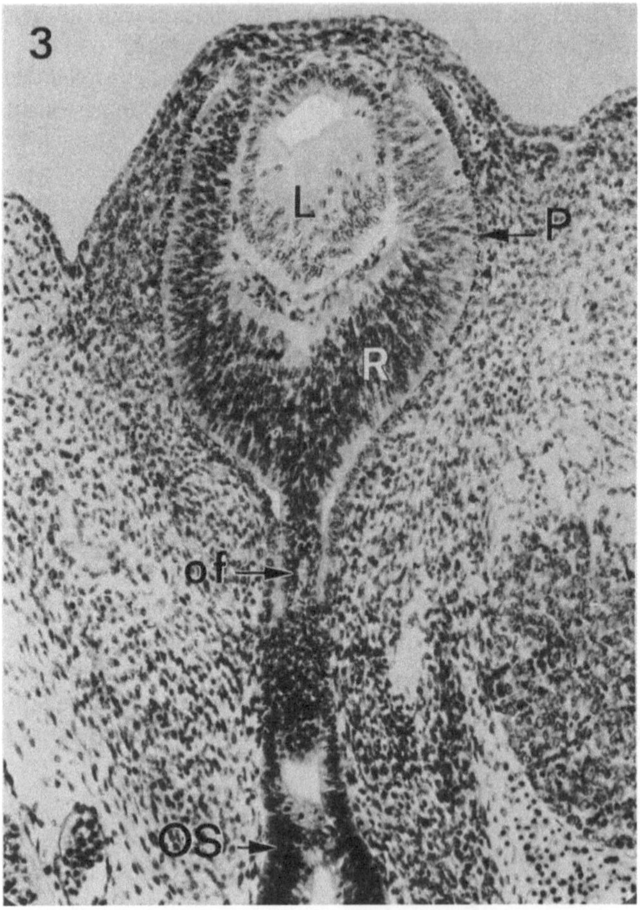

Fig. 3. Horizontal section through ventral aspect of optic cup of normal (C57BL/6J) developing mouse eye (12 days, 12 hrs) showing optic fissure (OF). (R, retina; P, pigment epithelium; L, lens; OS, optic stalk). ×300

of the neonates, while the remaining 10–20% have varying degrees of microphthalmia with or without orbital cysts (CHASE et al., 1941; CHASE, 1942, 1944, 1945). CHASE et al. (1941) maintained that in anophthalmic embryos the optic evagination forms normally until day 10, when there is an inhibition in the growth of the eye vesicle and cup, which prevents the optic rudiment from reaching the surface ectoderm and inducing a lens placode. Because of the lack of the lens the eye cup collapses and the optic fissure does not close. He proposed that inhibition of growth continues until day 13, after which whatever tissue remains usually grows at a normal rate but is too late to reorganize into a functional organ. However, CHASE also suggested that if the vesicle or cup is large enough to come in contact with the surface ectoderm, a lens will probably form, the optic fissure will close, and an eye of variable size will develop.

Using a more complete series of embryos than was studied by CHASE, the present author observed that during the earliest phases of ocular morphogenesis in

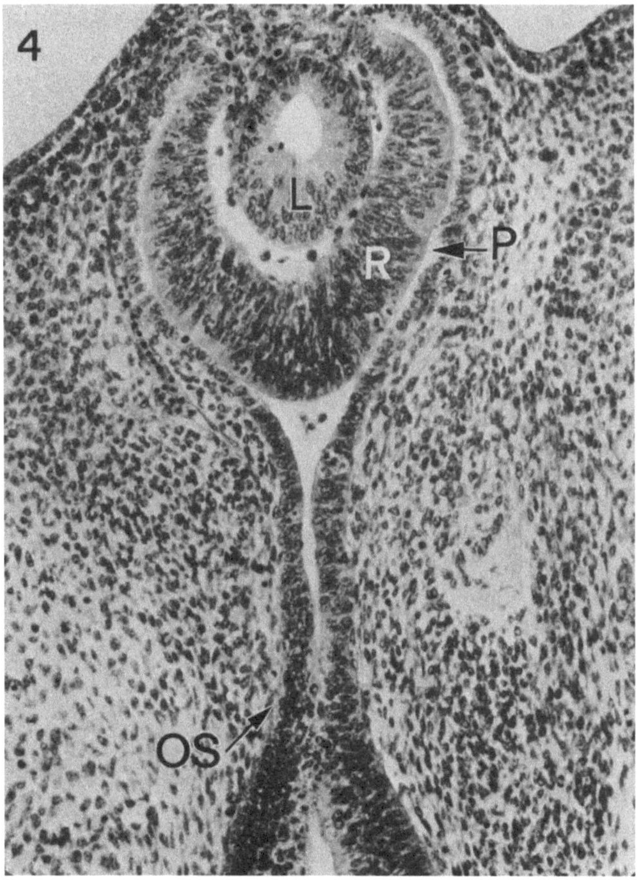

Fig. 4. Horizontal section through ventral optic cup of anophthalmic (ZRDCT-AN) mouse eye rudiment (12 days, 12 hrs), showing absence of formation of optic fissure. Normal cell death in eye is lacking during earlier stages in this mutant. (R, retina; L, lens; P, pigment epithelium; OS, optic stalk). × 400

the anophthalmic strain the regular production of necrotic loci in the optic vesicle and cup is entirely lacking (SILVER, 1974). In a subsequent investigation of the eyeless mutant, it was revealed in several animals in which a lens of normal proportions was present within the secondary eye chamber, that involution of the greater part of the optic cup can indeed occur independently of preliminary cell death. Thus, in the absence of degeneration, morphogenesis is regular (i.e., independent of cell death) up to and including formation of the retinal fissure; however, beyond this point normal development is severely retarded (i.e., dependent on cell death) and the optic fissure does not form (SILVER and HUGHES, 1974) (Figs. 3 and 4). Intraocular necrosis would, therefore, appear to be associated with one small portion of the invagination process, rather than with involution of the whole of the eye cup as was suggested by GLÜCKSMANN (1951).

Abnormal cell death may also be a causative factor in the reduction of the size of the lens and retina, which occurs in the large majority of anophthalmic embryos

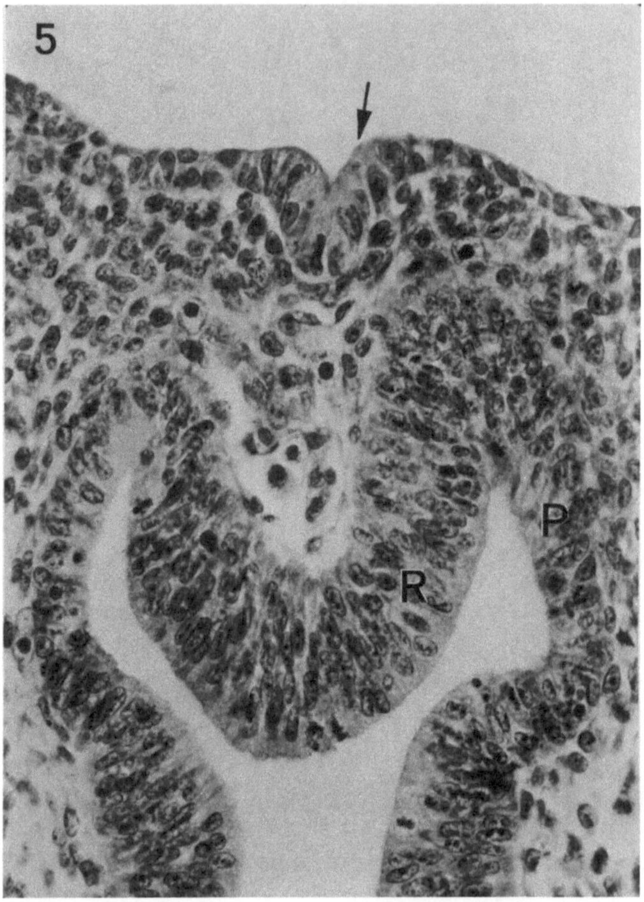

Fig. 5. Anophthalmic eye rudiment at 10 days and 12 hrs of gestation. Note explusion of lens from optic cup (arrow). (R, retina; P, pigment epithelium). × 500

(Silver and Hughes, 1974). An event of critical importance for normal eye development appears to be the degeneration and resorption of the residual mesenchymal cells, which become entrapped within the retina-lens interface following evagination of the optic vesicle. Bartlemez (1954) first observed this phenomenon in human embryos, although he ascribed no functional significance to it. In control mouse embryos (strain C57BL/6J) these resorptions occur on day 10, during the brief period following apposition of the optic vesicle and overlying ectoderm but prior to development of the retinal thickening and lens placode. In eyeless embryos during this stage, however, the surfaces of the prospective retina and lens epithelia, while closely apposed, are separated by variable numbers of mesenchymal cells, which remain alive and can undergo mitoses throughout development. Although a lens of varying size is induced in all anophthalmic embryos, in most cases it is dislocated from the eye cavity by proliferation of the persisting mesenchymal cell mass, which immediately prior to induction is displaced peripherally and comes to lie beneath the abortive portions of the presumptive lens and retinal rudiments (Fig. 5). This observation is in opposition to that of CHASE (1941), who believed it was improbable that a lens, once induced, would subsequently disappear.

Various experiments on induction, of the lens by MCKEEHAN (1958) and MUTHUKKARUPPAN (1965), of kidney tubules by GROBSTEIN (1956), and of the neural tube by SAXÉN (1961, 1976), have shown that interacting tissues can develop normally when separated by porous, synthetic barriers. However, it was proposed (SILVER and HUGHES, 1974) that failure of the normally transient mesenchyme (a viable barrier) to degenerate might diminish the normal area of induction between the presumptive lens ectoderm and optic vesicle, resulting in the eventual reduction of both lens and retina. In animals with dislocated lenses the entire eye rudiment is eventually resorbed and the anophthalmic condition results, while the remaining embryos that retain lenses will presumably develop varying degrees of microphthalmia, probably due to abnormal formation of the optic fissure.

These observations in the anophthalmic strain can be compared with experiments on amphibia, where early eye development is accomplished in the complete absence of morphogenetic cell death (GLÜCKSMANN, 1940). BECKWITH (1927) studied the effect of early lens extirpation on optic fissure development in the eye of *Amblystoma punctatum* at stages following contact between the presumptive retina and lens but prior to placode formation. It was found that in lensless eyes the eye cup and optic fissure formed but closure of the fissure was usually incomplete. In many embryos subjected to surgery lenses regenerated from the dorsal iris (for a review of lens regeneration see REYER, 1962) and in these cases, the optic fissure fused completely and a normal eye resulted immediately after and independently of the time lag before the onset of regeneration.

F. Histogenetic Cell Death

In several discrete portions of the normally developing nervous system, quantitative histological studies have shown that many more neurons are produced than are allowed to survive to maturity (for reviews see HUGHES, 1968; JACOBSON, 1970, 1978; COWAN, 1973). These excess cells are rendered superfluous and degenerate. For example, in the ciliary ganglion of the embryonic chick, approximately 50% of the cells undergo degeneration synchronously with the establishment of connections in the iris (LANDMESSER and PILAR, 1974). It was observed that prior to visible necrosis there was a failure of synaptic transmission again in half of the ganglionic cells and therefore it was suggested, as it has been for other systems (PIATT, 1946; ROGERS and COWAN, 1973; HAMBURGER, 1975), that the cells that die do so because they fail to form adequate connections in the periphery (LANDMESSER and PILAR, 1974).

Experimental studies have demonstrated that a reciprocal relationship may exist between local areas of the developing central nervous system and the size of the periphery that they innervate, in that decreases or increases in the normal area of the peripheral field often result in the hypoplasia or hyperplasia respectively of the associated centers within the CNS. That the effects seen centrally may, along with other factors, be mediated by modifications in the normal level of cellular degeneration has been considered for several of these systems. Thus, any modification in the area of the jaw musculature may alter the magnitude of cellular death and, in turn, the size of the mesencephalic nucleus (PIATT, 1946; ROGERS and COWAN, 1973; ALLEY, 1974). Grafting an extra limb (DETWILER, 1920, 1933) or a naturally occurring third hindlimb (BUEKER, 1945a, b) will result in hyperplasia of the associated spinal sensory ganglia and ventral motor cells, whereas amputation of the limb bud will result in hypoplasia of the spinal sensory and motor neurons (DETWILER, 1920, 1923, 1933; HAMBURGER, 1934; HUGHES and TSCHUMI, 1958; HUGHES, 1961). Excision of the optic vesicle (KOLLROS, 1953; COWAN et al., 1968) or of the anlage of the superior oblique muscle (DUNNEBACKE, 1953; COWAN and WENGER, 1967; COWAN, 1973) will result in hypoplasia by increased cell death in the optic tectum or trochlear nucleus. Conversely, implanting a large eye in place of a small one will result in hyperplasia of the optic tectum (TWITTY, 1932).

During the differentiation of the tadpole retina, cells degenerate at predictable times and locations (GLÜCKSMANN, 1940, 1951). Each of these zones of cell death precedes the differentiation of a new retinal layer and corresponds in its localization to that differentiation. Thus, the first wave of necrosis is followed by the appearance of ganglion cells, the second by that of the inner nuclear cells, and the third by that of the photoreceptor cells. GLÜCKSMANN believed that the coincidence of each wave of degeneration with the onset of a new type of differentiation was significant, and suggested that cells degenerated because their fate was already partially determined and they could not respond to an "histogenetic stimulus" in an alternative direction. There is little of this type of degeneration in the mammalian eye, however, and why the death of these cells is a necessary outcome in the frog retina while they are spared in other forms is not known. However, it is possible that, in mammals, the various subpopulations of retinal neurons may be diminished later on during the period of synaptogenesis.

From the foregoing discussion, it is evident that cellular death in controlled amounts may be a vital component of normal embryonic development, whereas abnormal concentrations of degenerate cells may result in morphogenetic or histogenetic anomalies leading to gross defects in the newborn. Why certain neurons remain alive during morphogenesis and are subsequently able to establish viable connections in the periphery or among themselves within the central nervous system, while others either die early on during morphogenetic movements or remain healthy for a time and only later are denied synaptic contact and condemned to death, remains one of the many unsolved problems in developmental neurobiology.

References

Alley, K. E.: Morphogenesis of the trigeminal mesencephalic nucleus in the hamster: Cytogenesis and neurone death. J. Embryol. exp. Morph. **31**, 99—121 (1974).

Andersen, H.: Histochemistry of early development of the human central face and nasal cavity, with special reference to movement and fusion of the palatine processes. Acta. anat. **68**, 473—508 (1967).

Bailey, P.: Morphology of the roof plate of the forebrain and the lateral choroidal plexus in the human embryo. J. comp. Neurol. **26**, 79—120 (1916).

Baker, P. C., Schroeder, T. E.: Cytoplasmic filaments and morphogenetic movement in the amphibian neural tube. Dev. Biol. **15**, 432—450 (1967).

Balinsky, B. I.: Ultrastructural mechanisms of gastrulation and neurulation (Symp. on Germ Cells and Development, Inst. Intern. Embryol. Fondaz). Pavia: Baselli 1961.

Ballard, K., Holt, S. J.: Cytological and cytochemical studies on cell death digestion in the foetal rat foot: the role of macrophages and hydrolytic enzymes. J. Cell Sci. **3**, 245—262 (1968).

Bartlemez, G. W.: The formation of neural crest from the primary optic vesicle in man. (Carnegie Inst. Wash. Publ. 603) Contrib. Embryol. **35**, 55—71 (1954).

Beckwith, C. J.: The effect of the extirpation of the lens rudiment on the development of the eye in *Amblystoma punctatum*, with especial reference to the choroid fissure. J. exp. Zool. **49**, 217—259 (1927).

Browman, L. G., Ramsey, F.: Embryology of microphthalmos in *Rattus Norvegicus*. Arch. Ophthal. **30**, 338—351 (1943).

Bueker, E. D.: The influence of a growing limb on the differentiation of somatic motor neurons in transplanted avian spinal cord segments. J. comp. Neurol. **82**, 335—361 (1945a).

Bueker, E. D.: Hyperplastic changes in the nervous system of a frog (*Rana*) as associated with multiple functional limbs. Anat. Rec. **93**, 323—331 (1945b).

Burnside, B.: Microtubules and microfilaments in newt neurulation. Dev. Biol. **26**, 416—441 (1971).

Byers, B., Porter, K. R.: Oriented microtubules in elongating cells of the developing lens rudiment after induction. Proc. nat. Acad. Sci. (Wash.) **52**, 1091—1099 (1964).

Chase, H. B.: Studies on an anophthalmic strain of mice. III. Results of crosses with other strains. Genetics **27**, 339—348 (1942).

Chase, H. B.: Studies on an anophthalmic strain of mice. IV. A second major gene for anophthalmia. Genetics **29**, 264—269 (1944).

Chase, H. B.: Studies on an anophthalmic strain of mice. V. Associated cranial nerves and brain centers. J. comp. Neurol. **83**, 121—136 (1945).

Chase, H. B., Chase, E. S.: Studies on an anopthalmic strain of mice. I. Embryology of the eye region. J. Morph. **68**, 279—301 (1941).

Coulombre, A. J.: The eye. In: Organogenesis. (R. L. DeHaan and H. Ursprung, Eds). New York: Holt, Rinehart and Winston 1965.

Cowan, W. M.: Neuronal death as a regulative mechanism in the control of cell number in the nervous system. In: Development and aging in the nervous system. (M. Rockstein, Ed.). New York: Academic 1973.

Cowan, W. M., Wenger, E.: Cell loss in the trochlear nucleus of the chick during normal development and after radical extirpation of the optic vesicle. J. exp. Zool. **164**, 267—280 (1967).

Cowan, W. M., Martin, A. H., Wenger, E.: Mitotic patterns in the optic tectum of the chick during normal development and after early removal of the optic vesicle. J. exp. Zool. **169**, 71—92 (1968).

Dawd, D. S., Hinchliffe, J. R.: Cell death in the "opaque patch" in the central mesenchyme of the developing chick limb: a cytological, cytochemical and electron microscope analysis. J. Embryol. exp. Morph. **26**, 401—424 (1971).

Deleanu, M.: Toxic action upon physiological necrosis and macrophage reaction in the chick embryo leg. Rev. Roumaine Embryol. Cytol. **2**, 45—56 (1965).

Detwiler, S. R.: On the hyperplasia of nerve centers resulting from excessive peripheral loading. Proc. nat. Acad. Sci. (Wash.) **6**, 96—101 (1920).

Detwiler, S. R.: Experiments on the transplantation of the spinal cord in *Amblystoma* and their bearing upon the stimuli involved in differentiation of nerve cells. J. exp. Zool. **37**, 339—393 (1923).

Detwiler, S. R.: Experimental studies upon the development of the amphibian nervous system. Biol. Rev. **8**, 269—310 (1933).

Dunn, L.C., Bennett, D.: Abnormalities associated with a chromosome region in the mouse. Science **144**, 260—267 (1964).

Dunnebacke, T.H.: The effects of the extirpation of the superior oblique muscle on the trochlear nucleus in the chick embryo. J. comp. Neurol. **98**, 155—177 (1953).

Ernst, M.: Über Untergang von Zellen während der normalen Entwicklung bei Wirbeltieren. Z. ges. Anat.: I. Z. Anat. Entw.-Gesch. **79**, 228—262 (1926).

Fallon, J.F., Saunders, J.W.: *In vitro* analysis of the control of cell death in a zone of prospective necrosis from the chick wing bud. Dev. Biol. **18**, 553—570 (1968).

Fell, H.B.: The origin and developmental mechanics of the avian sternum. Phil. Trans. roy. Soc. B. **229**, 407—463 (1939).

Fell, H.B., Canti, R.G.: Experiments on the development *in vitro* of the avian kneejoint. Proc. roy. Soc. B. **116**, 316—347 (1934).

Freeman, B.B.: Surface modifications of neural epithelial cells during formation of the neural tube in the rat embryo. J. Embryol. exp. Morph. **28**, 437—448 (1972).

Fristrom, D.: Chemical modification of cell death in the bar eye drosophila. Molec. gen. Genet. **115**, 10—18 (1972).

Froboese, C.: Über das Vorkommen von Fett in Embryonen. Z. mikr.-anat. Forsch. **7**, 527—641 (1926).

Gamble, J.H.: Electron microscopic observations upon the conus medullaris and filum terminale of human fetuses. J. Anat. **110**, 173—180 (1971).

Glücksmann, A.: Über die Bedeutung von Zellvorgängen für die Formbildung epithelialer Organe. Z. ges. Anat.: I. Z. Anat. Entw.-Gesch. **93**, 35—92 (1930).

Glücksmann, A.: Development and differentiation of the tadpole eye. Brit. J. Ophthal. **24**, 153—178 (1940).

Glücksmann, A.: Cell deaths in normal vertebrate ontogeny. Biol. Rev. **26**, 59—86 (1951).

Graumann, W.: Zelldegeneration in Telencephalon medium und Paraphysenentwicklung bei der weißen Maus. Z. ges. Anat.: I. Z. Anat. Entw. Gesch. **115**, 19—31 (1950).

Grobstein, C.: Trans-filter induction of tubules in mouse metanephrogenic mesenchyme. Exp. Cell Res. **10**, 424—440 (1956).

Hamburger, V.: The effects of wing bud extripation on the development of the central nervous system in chick embryos. J. exp. Zool. **68**, 449—494 (1934).

Hamburger, V.: Cell death in the development of the lateral motor column of the chick embryo. J. comp. Neurol. **160**, 536—546 (1975).

Hamburger, V., Hamilton, H.: A series of normal stages in the development of the chick embryo. J. Morph. **88**, 49—92 (1951).

Hinchliffe, J.R., Ede, D.A.: Limb development in the polydactylous talpid mutant of the fowl. J. Embryol. exp. Morph. **17**, 385—404 (1967).

Hinchliffe, J.R., Ede, D.A.: Cell death and the development of limb form and skeletal pattern in normal and wingless (*ws*) chick embryos. J. Embryol. exp. Morph. **30**, 753—772 (1973).

Hinchliffe, J.R., Thorogood, P.V.: Genetic inhibition of mesenchymal cell death and the development of form and skeletal pattern in the limbs of talpid^3 (ta^3) mutant chick embryos. J. Embryol. exp. Morph. **31**, 747—760 (1974).

Hughes, A.F.: Cell degeneration in the larval ventral horn of *Xenopus laevis* (daudin). J. Embryol. exp. Morph. **9**, 269—284 (1961).

Hughes, A.F.: Aspects of Neural ontogeny. London: Logos 1968.

Hughes, A.F., Tschumi, P.A.: The factors controlling the development of the dorsal root ganglia and ventral horn in *Xenopus laevis*. J. Anat. (Lond.) **92**, 498—527 (1958).

Ishikawa, H., Bischoff, R., Holtzer, H.: Formation of arrowhead complexes with heavy meromyosin in a variety of cell types. J. Cell Biol. **43**, 312—328 (1969).

Jacobson, M.: Developmental Neurobiology. New York: Holt, Rinehart, and Winston 1970.

Jacobson, M.: Developmental Neurobiology. New York: Plenum 1978.

Jacobson, W., Fell, H.B.: The developmental mechanics and potencies of the undifferentiated mesenchyme of the mandible. Quart. J. micr. Sci. **82**, 563—586 (1941).

Jelinek, R., Pychter, Z., Klika, E.: Syndrome of caudal regression and dysrhaphic malformations of the spinal cord. Fol. morph. **19**, 58—70 (1971).

Johnson, D.R.: Polysyndactyly, a new mutant gene in the mouse. J. Embryol. exp. Morph. **21**, 285—294 (1969).

Källén, B.: Cell degeneration during normal ontogenesis of the rabbit brain. J. Anat. **89**, 153—162 (1955).

Källén, B.: Degeneration and regeneration in the vertebrate central nervous system during embryogenesis. Prog. Brain Res. **14**, 77—96 (1965).

Klika, E., Jelinek, R.: The structure of the end and tail bud of the chick embryo. Fol. morph. **17**, 29—40 (1969).

Kollros, J.J.: The development of the optic lobes in the frog. I. The effects of unilateral enucleation in embryonic stages. J. exp. Zool. **123**, 153—187 (1953).

Kuwabara, T.: Development of the optic nerve of the rat. Invest. Ophthalmol. **14**, 732—745 (1975).

Kuwabara, T., Weidman, T.A.: Development of the prenatal rat retina. Invest. Ophthalmol. **13**, 725—739 (1974).

Landauer, W.: Rumplessness of chicken embryos produced by the injection of insulin and other chemicals. J. exp. Zool. **98**, 65—77 (1945).

Landmesser, L., Pilar, G.: Synaptic transmission and cell death during normal ganglionic development. J. Physiol. **241**, 737—749 (1974).

Lopashov, G.V., Stroeva, O.G.: Development of the eye: experimental studies. New York: Dániel Davey 1964.

McKeehan, M.S.: Cytological aspects of embryonic lens induction in the chick. J. exp. Zool. **117**, 31—64 (1951).

McKeehan, M.S.: Induction of portions of the chick lens without contact with the optic cup. Anat. Rec. **132**, 297—305 (1958).

Mann, I.: Developmental Abnormalities of the Eye. Philadelphia, Pa.: Lippincott 1957.

Mann, I.: The Development of the Human Eye. New York: Grune and Stratton 1969.

Maruyama, S., D'Agostino, A.N.: Cell necrosis in the central nervous system of normal rat fetuses. An electron microscopic study. Neurology **17**, 550—558 (1967).

Mattanza, G.G.: Significance of embryonic cell necrosis in the forebrain. II. Histochemical studies on the mouse. Acta anat. **85**, 206—215 (1973).

Menkes, B., Deleanu, M.: Leg differentiation and experimental syndactyly in chick embryos. Rev. Roumaine Embryol. Cytol. **1**, 69—77 (1964).

Menkes, B., Deleanu, M., Ilies, A.: Comparative study of some areas of physiological necrosis at the embryo of man, some laboratory mammalians and fowl. Rev. Roumaine Embryol. Cytol. **2**, 161—171 (1965).

Milaire, J.: Étude morphologique et cytochimique du développement des membres chez la souris et chez la taupe. Arch. Biol. **74**, 131—137 (1963).

Milaire, J.: Étude morphogénétique de trois malformations congénitales de l'autopède chez la souris (syndactylisme-brachypodisme-hémimélie dominante) par des méthodes cytochimiques. Mem. Acad. roy. belg. Classe Sci. **16**, 3: 119 (1965).

Moseley, H.R.: Insulin-induced rumplessness of chicks. J. exp. Zool. **105**, 279—316 (1947).

Muthukkaruppan, V.: Inductive tissue interaction in the development of the mouse lens *in vitro*. J. exp. Zool. **159**, 269—288 (1965).

Piatt, J.: The influence of the peripheral field on the development of the mesencephalic V nucleus in *Amblystoma*. J. exp. Zool. **102**, 109—141 (1946).

Reyer, R.W.: Regeneration in the amphibian eye. In: Regeneration (D. Rudnick, Ed.). New York: Roland 1962.

Rogers, L., Cowan, W.M.: The development of the mesencephalic nucleus of the trigeminal nerve in the chick. J. comp. Neurol. **147**, 291—320 (1973).

Saunders, J.W.: Death in embryonic systems. Science **154**, 604—612 (1966).

Saunders, J.W., Fallon, J.F.: Cell death in morphogenesis. In: Major problems in developmental biology. New York: Academic 1967.

Saunders, J.W., Gasseling, M.T., Saunders, L.C.: Cellular death in morphogenesis of the avian wing. Dev. Biol. **5**, 147—178 (1962).

Saxén, L.: Transfilter neural induction of amphibian ectoderm. Dev. Biol. **3**, 140—152 (1961).

Saxén, L.: Are morphogenetic tissue interactions mediated by transmissible signal substances or through cell contacts? Nature **259**, 662—663 (1976).

Schroeder, T.E.: Neurulation in Xenopus laevis. An analysis and model based on light and electron microscopy. J. Embryol. exp. Morph. **23**, 427—462 (1970).

Silver, J.: The role of cell degeneration in the developing rat eye. Anat. Rec. **172**, 406 (1972).

Silver, J.: The relationship between morphogenetic cell death and congenital eyelessness. Anat. Rec. **178**, 463 (1974).
Silver, J.: A study of ocular morphogenesis in the rat using ^{3}H-thymidine autoradiography: evidence for thymidine recycling in the developing retina. Dev. Biol. **49**, 487—495 (1976).
Silver, J., Hughes, A. F.: The role of cell death during morphogenesis of the mammalian eye. J. Morph. **140**, 159—170 (1973).
Silver, J., Hughes, A. F.: The relationship between morphogenetic cell death and the development of congenital anophthalmia. J. comp. Neurol. **157**, 281—302 (1974).
Stoeckel, M. E., Porte, A., Hindeland-Gertner, C., Dillman, H. D.: A light and electron microscopic study of the pre- and postnatal development and secretory differentiation of the pars tuberalis of the rat hypophysis. Z. Zellforsch. **142**, 347—365 (1973).
Twitty, V. C.: Influence of the eye on the growth of its associated structures, studied by means of heteroplastic transplantation. J. exp. Zool. **61**, 333—374 (1932).
Von Hippel, E.: Embryologische Untersuchungen über die Entstehungsweise der typischen angeborenen Spätbildungen (colobome) des Augapfels. Graefes Arch. Ophthal. **55**, 507—548 (1903).
Von Szily, A.: Über die einleitenden Vorgänge bei der ersten Entstehung der Nervenfasern im N. opticus. Graefes Arch. Ophthal. **81**, 67—86 (1912).
Warkany, J.: Congenital Malformations. Chicago: Year Book Med. Publ. 1971.
Wrenn, J. T., Wessels, N. K.: An ultrastructural study of lens invagination in the mouse. J. exp. Zool. **171**, 359—368 (1969).
Zwaan, J., Hendrix, R. W.: Changes in cell and organ shape during early development of the ocular lens. Amer. Zool. **13**, 1039—1049 (1973).
Zwilling, E.: The development of dominant rumplessness in chick embryos. Genetics **27**, 641—656 (1942).
Zwilling, E.: Micromelia as a direct effect of insulin. Evidence from *in vitro* and *in vivo* experiments. J. Morph. **104**, 159—179 (1959).

Author Index

Page numbers in *italics* refer to the bibliography

Subject Index

MIX
Papier aus verantwortungsvollen Quellen
Paper from responsible sources
FSC® C105338
FSC
www.fsc.org

If you have any concerns about our products,
you can contact us on
ProductSafety@springernature.com

In case Publisher is established outside the EU,
the EU authorized representative is:
Springer Nature Customer Service Center GmbH
Europaplatz 3, 69115 Heidelberg, Germany

Printed by Libri Plureos GmbH
in Hamburg, Germany